Soil Biology

Volume 30

Series Editor
Ajit Varma, Amity Institute of Microbial Technology,
Amity University Uttar Pradesh, Noida, UP, India

For further volumes:
http://www.springer.com/series/5138

Irena Sherameti • Ajit Varma
Editors

Detoxification of Heavy Metals

Springer

Editors
Dr. Irena Sherameti
Friedrich-Schiller Universität Jena
Institut für Allgemeine Botanik und
Pflanzenphysiologie
Dornburger Str. 159
07743 Jena
Germany
irenasherameti@yahoo.de

Prof. Dr. Ajit Varma
Amity University Uttar Pradesh
Amity Institute of Microbial Sciences
Noida Uttar Pradesh
Block A, Ground Floor, Sector 125
India
ajitvarma@aihmr.amity.edu

ISSN 1613-3382
ISBN 978-3-642-21407-3 e-ISBN 978-3-642-21408-0
DOI 10.1007/978-3-642-21408-0
Springer Heidelberg Dordrecht London New York

Library of Congress Control Number: 2011934686

Printed on acid-free paper

Springer is part of Springer Science+Business Media (www.springer.com)

Foreword

Many heavy metals are essential for all organisms, e.g. as active centres of enzymes. At higher than optimal concentrations, however, they become toxic. For non-essential metals, toxicity is observed above a range of tolerance. Because of the relevance of these phenomena for damage to nature in general and to humans in particular, heavy metal toxicity and mechanisms counteracting it in various ways are a subject of intensive research since many years (for a comprehensive recent review see, e.g. Küpper and Kroneck 2005).

Although I myself do not directly work on soil biology but on plant biology, I am writing this foreword because terrestrial plants certainly and heavily depend on the soil they are growing in. Thus, they suffer when this soil contains toxic compounds, such as excess levels of heavy metals. Such toxic heavy metal concentrations can have natural reasons; naturally heavy metal-rich soils are found in various locations around the world where metal ores come to the surface and decay due to weathering. A few examples of such locations are the Katangan copper belt in Kongo and Zaire (Duvigneaud 1958; Malaisse et al. 1999), nickel-rich serpentine soils in Cuba (Reeves et al. 1996), North America (Rajakaruna et al. 2009) as well as Sulawesi and New Caledonia (Proctor 2003) and some zinc and cadmium sites in Europe (Reeves and Brooks 1983). While these locations are usually not regarded as agriculturally relevant and usually no attempts are made to detoxify them (as it usually would be futile), the plants growing on them still have to detoxify the stream of nutrients they take up from such soils. This theme is dealt with in the Chaps. 3, 7, 8, 9, 10, 11, 15, 17 and 19 of this book and is also a theme of my own research for many years.

In terms of soil detoxification, sites that originally had low heavy metal levels but were contaminated by human activity are the main targets for soil detoxification as the main theme of this book, in particular when they are otherwise attractive for agriculture. Such contaminations can have various reasons and can be found in many countries of the world although a common misconception is that this would be mainly a problem for poor countries. The most obvious reason for anthropogenic heavy metal contamination of soils is the presence of ore-mining or -refining industry nearby, where emissions of dust particles and leakage of contaminated water (e.g. from dumps and storages) are the main causes of environmental pollution. Famous examples are Severonikel nickel–copper smelter at Monchegorsk in Russia (Barcan 2002), the Sudbury Area in Canada (Mandal et al. 2002) and the Cevennes region in France (Lombi et al. 2000). Heavy metal pollution of soils can also be a widespread and severe problem for countries that do not have a major metal-mining industry (any more). Again, industry may play a role, as metals are used in various processes, e.g. as catalysts, a famous case is the mercury poisoning in Minamata Bay, Japan (review by Harada 1995). But in many countries, like my home country Germany, the main source of heavy metal pollution of soils is the excessive use of metals in agriculture, as still many copper- and zinc-based pesticides are allowed, and especially copper is highly toxic for plants (much more than for most animals including humans). Copper concentrations in vineyard soil exceed legislative limits in the vast majority of studied vineyards (Komarek et al. 2010), and agricultural field runoff may reach micromolar levels (Gallagher et al. 2001), which is lethal for many sensitive plant species within days to weeks of exposure. In the USA, the metalloid arsenic became a problem in a similar way; arsenic compounds were used as insecticides in cotton industry (Osburn 1926) and caused severe contaminations of soils, surface water and groundwater in regions of intense cotton farming (Carbonell et al. 1998). Another way of heavy metal contamination by agriculture is the application of mineral fertilisers, as these often contain heavy metals and the metalloid arsenic as contaminants (McLaughlin et al. 1996; He et al. 2005). Sewage sludge is usually not a good fertiliser for the same reason (McBride et al. 1997). Another source of metal pollution in heavily industrialised countries like Germany is car traffic. The well-known case is lead that was banned from fuel many years ago and that was more toxic to animals than to plants (plants hardly take it up). Less known, but more toxic, is the release of cadmium from car tyres, which leads to significantly enhanced cadmium levels along busy roads (Lagerwerff and Specht 1970).

In all these cases of anthropogenic soil contamination with heavy metals, the highest heavy metal concentrations are found rather close to the surface, although not directly in the uppermost few millimetres to centimetres as these are leached by rain like in natural heavy metal sites (McBride et al. 2005; Mitani and Ogawa 1998). For this reason, decontamination of such areas is, in principle, possible in several ways. The classic way would be the removal of the topsoil and leaching of it in a chemical or microbial way in special facilities. Although this method is costly, this is the only realistic option for very small (and at the same time economically or socially very important) spots. For larger areas, decontamination by plants seems to be the most attractive option, as on fertile ground (which would be a most attractive kind of site for decontamination as it could be agriculturally valuable) plants will grow well without too much human effort. But it is hotly debated what kind of plants should be used for this task. In principle, three main strategies exist: (1) the use of naturally occurring metal hyperaccumulator plants, probably combined with classical breeding, (2) the use of high-biomass non-accumulator plants, (3) the transfer of genes from hyperaccumulator plants to turn originally non-accumulating high-biomass plants into high-biomass metal hyperaccumulators. While some of this is dealt with in more detail in chapters of this book, I would like to summarise work on these strategies from the perspective of my own work on metal metabolism in plants.

Many natural hyperaccumulators, i.e., plants that actively accumulate several percent of heavy metals in the dry mass of their above-ground parts, have a good potential to be used for phytoremediation, i.e., to extract and remove heavy metals from anthropogenically contaminated soils, which was first proposed by Chaney (1983). Some of them even allow for commercially profitable phytomining, i.e., the extraction of metals from naturally heavy metal-rich soils (that are not directly usable as metal ores) with subsequent burning of the plants, the ash of which can be used as a metal ore (first proposed by Baker and Brooks 1988). These applications of metal phytoextraction have been a subject to extensive research as reviewed, e.g. by Baker and Brooks (1989), Baker et al. (2000), McGrath et al. (1993), McGrath and Zhao (2003), Salt et al. (1995, 1998), Chaney (1983), Chaney et al. (2005, 2007) and Küpper and Kroneck (2005, 2007). For the metalloid arsenic, the fast-growing, high-biomass, As hyperaccumulating fern *Pteris vittata* and related species are very promising candidates for phytoremediating As-contaminated areas (Ma et al. 2001; Zhao et al. 2002; Meharg 2003). For cadmium, the Cd/Zn hyperaccumulator *T. caerulescens* seems to be the best known candidate for phytoremediation. Although it has a rather small biomass of 2–5 t ha^{-1} (Robinson et al. 1998; McGrath and Zhao 2003), the extreme bioaccumulation coefficient of its southern French ecotypes (Lombi et al. 2000; Zhao et al. 2003) yields Cd extraction rates high enough for cleaning up moderately Cd-contaminated soils within a few years as tested in the field by Robinson et al. (1998), Hammer and Keller (2003) and McGrath et al. (2006). The high copper sensitivity of *T. caerulescens*, however, may limit its use; copper concentrations that occur in multi-contaminated soils were found to strongly inhibit its growth (Walker and Bernal 2004). This might be alleviated by selection of copper-resistant individuals that occur in natural populations of this species (Mijovilovich et al. 2009). Nickel was the first metal for which the economic feasibility of phytomining was shown, and some nickel hyperaccumulators hyperaccumulate the even more valuable cobalt as well (Brooks and Robinson 1998; Robinson et al. 1999). Nicks and Chambers (1995) yielded a crop of nickel of equal value compared with an average crop of wheat by planting *Streptanthus polygaloides* on a metal-rich soil in California (USA). They furthermore showed that by burning these plants it is possible to yield, with low input of energy, a bio-ore (the plant ash) containing about 15% nickel. *Berkheya coddii* has been known as a high-biomass Ni hyperaccumulator since the work of Anderson et al. (1996). Robinson et al. (1999) carried out comprehensive studies of metal uptake and showed that fertilisation with sulfur and nitrogen greatly increased Ni and Co hyperaccumulation. Thus, their work has demonstrated that this species is a very promising candidate for both phytoremediation and phytomining. This has been confirmed by field trials in a recent study, which demonstrated that this species easily yields 110 kg of nickel per hectare and year (Brooks et al. 2001), and even 170 kg should be possible (Brooks et al. 1998). Similarly, *Alyssum bertolonii* has been shown to produce high enough nickel yields per hectare for phytomining (Robinson et al. 1997; Brooks and Robinson 1998; Brooks et al. 1998), which now has already been put into commercial operation (McGrath and Zhao 2003). For zinc, the Chinese plant *Sedum alfredii* may be the most promising candidate for phytoremediation and possibly even for commercial phytomining because of its correlation of high zinc accumulation with relatively high biomass (Long et al. 2002; Ye et al. 2003). In contrast, *Thlaspi caerulescens* has a rather low biomass and at high soil zinc concentrations also a low bioaccumulation coefficient (Robinson et al. 1998; Zhao et al. 2003), so that its use in zinc phytoremediation is generally limited to moderate levels of contamination. Indeed, while field trials on moderately contaminated soil by Baker et al. (1994) were successful, those on more heavily Zn-contaminated soil failed (Hammer and Keller 2003).

In addition to true hyperaccumulator plants, various other plants have been proposed for use in soil phytoremediation. One idea is to use high-biomass plants for absorbing the metals; it is argued that the much higher biomass will yield higher metal extraction per area of land compared with hyperaccumulators, despite the much lower metal content of non-accumulator plants (e.g. Salt et al. 1995, 1998; Pulford and Watson 2003). Those who argue for such an approach, however, mostly ignore that such a strategy would dilute the extracted metal in a much larger amount of toxic biomass compared with hyperaccumulator plants; this biomass would be too toxic for use as compost and would not contain enough metal to make a recycling of the phytoextracted metal feasible (discussed, e.g. by Chaney et al. 1997; Williams 2002). In addition, the bioaccumulation factor of metals in non-accumulator plants is usually so low that hundreds of crops would be required for phytoremediation of even a moderately contaminated soil (Baker et al. 1994; Chaney et al. 1997; McGrath and Zhao 2003). Those who argue for this approach because of the low biomass of many (not all, see above!) hyperaccumulators should also keep in mind the following facts.

(a) The biomass yield of non-accumulator plants on contaminated soils is reduced by phytotoxicity of the contaminating metal (Ebbs et al. 1997; Chaney et al. 1997).
(b) The biomass of hyperaccumulators can be rather easily improved by selecting suitable ecotypes and individuals within the natural population (Li et al. 2003; Schwartz et al. 2003), breeding (Brewer et al. 1999) and fertilisation (two to three times increase; Bennett et al. 1998; McGrath et al. 2000; Brooks et al. 2001; Li et al. 2003; Schwartz et al. 2003).
(c) The metal accumulation of hyperaccumulators can further be optimised by selection. Many recent studies pointed out more than 20-fold variation of bioaccumulation factors for the same metal between ecotypes/populations (e.g. Meerts and Van Isacker 1997; Bert et al. 2000, 2002; Escarré et al. 2000; Lombi et al. 2000; Macnair 2002; Roosens et al. 2003; Zhao et al. 2003). Furthermore, the accumulation efficiency is not directly correlated with the metal content of the habitat (Bert et al. 2002), and strong variation of metal bioaccumulation factors as well as metal resistance exists even within one population (Macnair 2002; Mijovilovich et al. 2009). Finally, accumulation is higher on the average moist agricultural land compared with their dry natural habitats (Angle et al. 2003), and fertilisation increases it further (Schwartz et al. 2003). In summary, presently it is not the phytoremediation

by hyperaccumulators that is a "hype," but the use of non-accumulating plants for this task. The only way that a non-hyperaccumulating plant species may become a better alternative would be by creating (by genetic engineering or traditional breeding) metal-accumulating cultivars.

It is often argued that instead of using natural hyperaccumulators for phytoremediation and phytomining, genetically engineered plants should be used. Looking at the results of classical selection breeding of hyperaccumulators vs. attempts to create transgenic hyperaccumulators, the former approach appears much more promising, for the following reasons. Research on the mechanisms of hyperaccumulation has revealed that this process involves many different steps in diverse parts of the plant, starting from enhanced uptake into the roots (e.g. Lasat et al. 1996) and continuing via enhanced xylem loading (e.g. Papoyan and Kochian 2004), translocation to the shoots possibly by transport ligands (e.g. Trampczynska et al. 2010), unloading from the veins and finally sequestration into vacuoles of usually epidermal storage cells (Küpper et al. 1999, 2001; Frey et al. 2000; Leitenmaier and Küpper 2011) – as reviewed e.g. by Küpper and Kroneck (2005, 2007) and Chaps. 3, 7, 8, 11 and 19 of this book. Furthermore, individual members of metal transport protein families display vastly different tissue-, age-, and metal nutrition-dependent regulation in the same plant (Küpper and Kochian 2010). Therefore, to re-create a hyperaccumulator by genetic engineering, one would have to modify the expression of many genes in a tissue-specific way and probably at particular stages of plant and leaf ontogenesis. This has not been achieved, not even in an approximation, in any study so far (review, e.g. by Chaney et al. 2007). Therefore, it is not surprising that in all attempts of creating hyperaccumulators by genetic engineering at best a few times enhancement of metal accumulation compared with the original non-accumulator wildtype was achieved, while true (natural) hyperaccumulators usually have hundreds of times higher metal bioaccumulation coefficients than those non-accumulators (Küpper and Kroneck 2005, 2007; Chaney et al. 2007). And such transgenics are not useful to apply, for the same reasons as explained for wildtype non-accumulators. Unless someone finds a general "switch gene" that leads to the changed expression pattern of all the other genes involved in hyperaccumulation, transgenic plants that really accumulate as much metal as hyperaccumulators will remain a science fiction.

In contrast, field trials have shown that the biomass of natural hyperaccumulators can be dramatically increased by addition of fertiliser, natural selection and classical breeding to reach levels that are economically attractive (reviewed by Chaney et al. 2005). As a source for selecting species that are suitable for a specific phytoextraction tasks, conservation of metallophyte biodiversity is of prime importance (Whiting et al. 2004).

Universität Konstanz
Fachbereich Biologie
78457 Konstanz
Germany

Hendrik Küpper

References

Anderson TR, Howes AW, Slater K, Dutton MF (1996) Studies on the nickel hyperaccumulator, *Berkheya coddii*. In: Jaffre T, Reeves RD and Becquer T (eds) The ecology of ultramafic and metalliferous areas. Proceedings of the second international Conference on Serpentine Ecology, Noumea, New Caledonia, July 31–Aug. 5, 1995. OSTROM, New Caledonia, pp 261–266

Angle JS, Baker AJM, Whiting SN, Chaney RL (2003) Soil moisture effects on uptake of metals by Thlaspi, Alyssum and Berkheya. Plant Soil 256:325–332

Baker AJM, Brooks RR (1989) Terrestrial higher plants which hyperaccumulate metallic elements – a review of their distribution, ecology and phytochemistry. Biorecovery 1:81–126

Baker AJM, Brooks RR, Reeves R (1988) Growing for gold... and copper... and zinc. New Sci 117:44–48

Baker AJM, McGrath SP, Sidoli CMD, Reeves RD (1994) The possibility of in situ heavy metal decontamination of polluted soils using crops of metal-accumulating plants. Resour Conservat Recycl 11:41–49

Baker AJM, McGrath SP, Reeves RD, Smith JAC (2000) Metal hyperaccumulator plants: a review of the ecology and physiology of a biochemical resource for phytoremediation of metal-polluted soils. In: Terry N, Bañuelos G, Vangronsveld J (eds) Phytoremediation of contaminated soil and water. Lewis Publishers, Boca Raton, FL, USA, pp 85–107

Barcan V (2002) Nature and origin of multicomponent aerial emissions of the copper-nickel smelter complex. Environ Int 28:451–456

Bennett FA, Tyler EK, Brooks RR, Gregg PEH, Stewart RB (1998) Fertilisation of hyperaccumulators to enhance their potential for phytoremediation and phytomining. In: Brooks RR (ed) Plants that hyperaccumulate heavy metals. CAB International, Wallingford, UK, pp 249–259

Bert V, Macnair MR, de Laguérie P, Saumitou-Laprade P, Petit D (2000) Zinc tolerance and accumulation in metallicolous and nonmetallicolous populations of *Arabidopsis halleri* (Brassicaceae). New Phytol 146:225–233

Bert V, Bonnin I, Saumitou-Laprade P, de Laguérie P, Petit D (2002) Do *Arabidopsis halleri* from nonmetallicolous populations accumulate zinc and cadmium more effectively than those from metallicolous populations? New Phytol 155:47–57

Brewer EP, Saunders JA, Angle JS, Chaney RL, McIntosh MS (1999) Somatic hybridization between the zinc accumulator *Thlaspi caerulescens* and *Brassica napus*. Theor Appl Genet 99:761–771

Brooks RR, Robinson BH (1998) The potential use of hyperaccumulators and other plants for phytomining. In: Brooks RR (ed) Plants that hyperaccumulate heavy metals. CAB International, Wallingford, UK, pp 203–226

Brooks RR, Chambers MF, Nicks LJ, Robinson BH (1998) Phytomining. Trends Plant Sci 3:359–362

Brooks RR, Robinson BH, Howes AW, Chiarucci A (2001) An evaluation of *Berkheya coddii* Roessler and *Alyssum bertolonii* Desv for phytoremediation and phytomining of nickel. South Afr J Sci 97:558–560

Carbonell AA, Aarabi MA, DeLaune RD, Gambrell RP, Patrick WH (1998) Arsenic in wetland vegetation: availability, phytotoxicity, uptake and effects on plant growth and nutrition. Sci Total Environ 217:189–199

Chaney RL (1983) Plant uptake of inorganic waste. In: Parr JE, Marsh PB, Kla JM (eds) Land treatment of hazardous wastes. Noyes Data Corp., Park Ridge, IL, pp 50–76

Chaney RL, Malik M, Li YM, Brown SL, Brewer EP, Angle JS, Baker AJM (1997) Phytoremediation of soil metals. Curr Opin Biotechnol 8:279–284

Chaney RL, Angle JS, McIntosh MS, Reeves RD, Li YM, Brewer EP, Chen KY, Roseberg RJ, Perner H, Synkowski EC, Broadhurst CL, Wang S, Baker AJM (2005) Using hyperaccumulator plants to phytoextract soil Ni and Cd. Zeitschrift für Naturforschung 60c:190–198

Chaney RL, Angle JS, Broadhurst CL, Peters CA, Tappero RV, Sparks DL (2007) Improved understanding of hyperaccumulation yields commercial phytoextraction and phytomining technologies. J Environ Qual 36:1429–1443

Duvigneaud P (1958) La végétation du Katanga et de ses sols métallifères. Bull Soc Roy Bot Belg 90:127–286

Ebbs S, Lau I, Ahner B, Kochian LV (2002) Phytochelatin synthesis is not responsible for Cd tolerance in the Zn/Cd hyperaccumulator *Thlaspi* (J.&C. Presl). Planta 214:635–640

Escarré J, Lefèbvre C, Gruber W, Leblanc M, Lepart J, Rivière Y, Delay B (2000) Zinc and cadmium hyperaccumulation by *Thlaspi caerulescens* from metalliferous and nonmetalliferous sites in the Mediterranean area: implications for phytoremediation. New Phytol 145:429–437

Frey B, Keller C, Zierold K, Schulin R (2000) Distribution of Zn in functionally different epidermal cells of the hyperaccumulator *Thlaspi caerulescens*. Plant Cell Environ 23:675–687

Gallagher DL, Johnston KM, Dietrich AM (2001) Fate and transport of copper-based crop protectants in plasticulture runoff and the impact of sedimentation as a best management practise. Water Res 35:2984–2994

Hammer D, Keller C (2003) Phytoextraction of Cd and Zn with *Thlaspi caerulescens* in field trials. Soil Use Manage 19:144–149

Harada M (1995) Minamata disease – methylmercury poisoning in Japan caused by environmental-pollution. Crit Rev Toxicol 25:1–24

He ZLL, Yang XE, Stoffella PJ (2005) Trace elements in agroecosystems and impacts on the environment. J Trace Elem Med Biol 19:125–140

Komarek M, Cadkova E, Chrastny V, Bordas F, Bollinger JC (2010) Contamination of vineyard soils with fungicides: a review of environmental and toxicological aspects. Environ Int 36:138–151

Küpper H, Kochian LV (2010) Transcriptional regulation of metal transport genes and mineral nutrition during acclimation to Cadmium and Zinc in the

Cd/Zn hyperaccumulator, *Thlaspi caerulescens* (Ganges population). New Phytol 185:114–129

Küpper H, Kroneck PMH (2005) Heavy metal uptake by plants and cyanobacteria. In: Sigel A, Sigel H, Sigel RKO (eds) Metal ions in biological systems, Band 44, Kapitel 5. Marcel Dekker, Inc., New York, pp 97–142

Küpper H, Kroneck PMH (2007) Nickel in the environment and its role in the metabolism of plants and cyanobacteria. In: Sigel A, Sigel H, Sigel RKO (Hrsg) Metal ions in life sciences, vol 2, Chap 1. Wiley, Chichester, pp 31–62

Küpper H, Zhao F, McGrath SP (1999) Cellular compartmentation of zinc in leaves of the hyperaccumulator *Thlaspi caerulescens*. Plant Phys 119:305–311

Küpper H, Lombi E, Zhao FJ, Wieshammer G, McGrath SP (2001) Cellular compartmentation of nickel in the hyperaccumulators *Alyssum lesbiacum*, *Alyssum bertolonii* and *Thlaspi goesingense*. J Exp Bot 52:2991–2300

Lagerwerff JV, Specht AW (1970) Contamination of roadside soil and vegetation with cadmium, nickel, lead, and zinc. Environ Sci Technol 4:583–586

Lasat MM, Baker AJM, Kochian LV (1996) Physiological characterization of root Zn2+ absorption and translocation to shoots in Zn hyperaccumulator and nonaccumulator species of Thlaspi. Plant Physiol 112:1715–1722

Leitenmaier B, Küpper H (2011) Cadmium uptake and sequestration kinetics in individual leaf cell protoplasts of the Cd/Zn hyperaccumulator *Thlaspi caerulescens*. Plant Cell Environ 34:208–219

Li YM, Chaney R, Brewer E, Roseberg R, Angle JS, Baker A, Reeves R, Nelkin J (2003) Development of a technology for commercial phytoextraction of nickel: economic and technical considerations. Plant Soil 249:107–115

Lombi E, Zhao FJ, Dunham SJ, McGrath SP (2000) Cadmium accumulation in populations of *Thlaspi caerulescens* and *Thlaspi goesingense*. New Phytol 145:11–20

Long XX, Yang XE, Ye ZQ, Ni WZ, Shi WY (2002) Differences of uptake and accumulation of zinc in four species of Sedum. Acta Bot Sin 44:152–157

Ma LQ, Komar KM, Tu C, Zhang WH, Cai Y, Kennelley ED (2001) A fern that hyperaccumulates arsenic – a hardy, versatile, fast-growing plant helps to remove arsenic from contaminated soils. Nature 409:579

Macnair MR (2002) Within an between population genetic variation for zinc accumulation in *Arabidopsis halleri*. New Phytol 155:59–66

Malaisse F, Baker AJM, Ruelle S (1999) Diversity of plant communities and leaf heavy metal content at Luiswishi copper/cobalt mineralization, Upper Katanga, Dem. Rep. Congo. Biotechnol Agron Soc Environ 3:104–114

Mandal R, Hassan NM, Murimboh J, Chakrabarti CL, Back MH (2002) Chemical speciation and toxicity of nickel species in natural waters from the Sudbury Area (Canada). Environ Sci Technol 36:1477–1484

McBride MB, Richards BK, Steenhuis T, Russo JJ, Sauve S (1997) Mobility and solubility of toxic metals and nutrients in soil fifteen years after sludge application. Soil Sci 162:487–500

McBride MB, Barrett KA, Martinez CE (2005) Zinc and cadmium distribution and leaching in a metalliferous peat. Water Air Soil Pollut 171:67–80

McGrath SP, Zhao FJ (2003) Phytoextraction of metals and metalloids from contaminated soils. Curr Opin Biotechnol 14:277–282

McGrath SP, Sidoli CMD, Baker AJM, Reeves RD (1993) The potential for the use of metal-accumulating plants for the in situ decontamination of metal-polluted soils. In: Eijsackers HJP, Hamers T (eds) Integrated soil and sediment research: a basis for proper protection. Kluwer Academic Publishers, Dordrecht, pp 673–677

McGrath SP, Dunham SJ, Correll RL (2000) Potential for phytoextraction of zinc and cadmium from soils using hyperaccumulator plants. In: Terry N, Bañuelos G (eds) Phytoremediation of contaminated soil and water. Lewis Publishers, Boca Raton, FL, pp 109–128

McGrath SP, Lombi E, Gray CW, Caille N, Dunham SJ, Zhao FJ (2006) Field evaluation of Cd and Zn phytoextraction potential by the hyperaccumulators *Thlaspi caerulescens* and *Arabidopsis halleri*. Environ Pollut 141:115–125

McLaughlin MJ, Tiller KG, Naidu R, Stevens DP (1996) The behaviour and environmental impact of contaminants in fertilizers. Aust J Soil Res 34:1–54

Meerts P, Van Isacker N (1997) Heavy metal tolerance and accumulation in metallicolous and non-metallicolous populations of *Thlaspi caerulescens* from continental Europe. Plant Ecol 133:221–231

Meharg AA (2003) Variation in arsenic accumulation – hyperaccumulation in ferns and their allies. New Phytol 157:25–31

Mijovilovich A, Leitenmaier B, Meyer-Klaucke W, Kroneck PMH, Götz B, Küpper H (2009) Complexation and toxicity of copper in higher plants (II): different mechanisms for Cu vs. Cd detoxification in the Cu-sensitive Cd/Zn hyperaccumulator *Thlaspi caerulescens* (Ganges ecotype). Plant Physiol 151:715–731

Mitani T, Ogawa M (1998) Cadmium leaching by acid rain from cadmium-enriched activated sludge applied in soil. J Environ Sci Health A33:1569–1581

Nicks L, Chambers MF (1995) Farming for metals. Mining Environ Manage 3(3):15–18

Osburn MR (1926) Comparative tests with sodium fluosilicate and calcium arsenate for the control of the cotton boil weevil (*Anthonomus grandis*). J Econ Entomol 19:643–644

Papoyan A, Kochian LV (2004) Identification of *Thlaspi caerulescens* genes that may be involved in heavy metal hyperaccumulation and tolerance. Characterization of a novel heavy metal transporting ATPase. Plant Physiol 136:3814–3823

Proctor J (2003) Vegetation and soil and plant chemistry on ultramafic rocks in the tropical Far East. Perspect Plant Ecol Evol Syst 6:105–124

Pulford ID, Watson C (2003) Phytoremediation of heavy metal-contaminated land by trees – a review. Environ Int 29:529–540

Rajakaruna N, Harris TB, Alexander EB (2009) Serpentine geoecology of eastern North America: a review. Rhodora 111:21–108

Reeves RD, Brooks RR (1983) European species of *Thlaspi* L. Cruciferae) as indicators of nickel and zinc. J Geochem Explor 18:275–283

Reeves RD, Baker AJM, Borhidi A, Berazain R (1996) Nickel-accumulating plants from the ancient serpentine soils of Cuba. New Phytol 133:217–224

Robinson BH, Chiarucci A, Brooks RR, Petit D, Kirkman JH, Gregg PEH, De Dominicis V (1997) The nickel hyperaccumulator plant *Alyssum bertolonii* as a potential agent for phytoremediation and phytomining of nickel. J Geochem Explor 59:75–86

Robinson BH, Leblanc M, Petit D, Brooks RR, Kirkman JH, Gregg PEH (1998) The potential of *Thlaspi caerulescens* for phytoremediation of contaminated soils. Plant Soil 203:47–56

Robinson BH, Brooks RR, Clothier BE (1999) Soil amendments affecting nickel and cobalt uptake by *Berkheya coddii* – potential use for phytomining and phytoremediation. Ann Bot 84:689–694

Roosens N, Verbruggen N, Meerts P, Ximénez-Embun P, Smith JAC (2003) Natural variation in cadmium tolerance and its relationship to metal hyperaccumulation for seven populations of *Thlaspi caerulescens* from western Europe. Plant Cell Environ 26:1657–1672

Salt DE, Blaylock M, Nanda Kumar PBA, Dushenkov V, Ensley BD, Chet I, Raskin I (1995) Phytoremediation: a novel strategy for removal of toxic metals from the environment using plants. Biotechnology 13:468–474

Salt DE, Smith RD, Raskin I (1998) Phytoremediation. Annu Rev Plant Physiol Plant Mol Biol 49:643–668

Schwartz C, Echevarria G, Morel JL (2003) Phytoextraction of cadmium with *Thlaspi caerulescens*. Plant Soil 249:27–35

Trampczynska A, Küpper H, Meyer-Klaucke W, Schmidt H, Clemens S (2010) Nicotianamine forms complexes with Zn(II) in vivo. Metallomics 2:57–66

Walker DJ, Bernal MP (2004) The effects of copper and lead on growth and zinc accumulation of *Thlaspi caerulescens* J. and C. Presl: implications for phytoremediation of contaminated soils. Water Air Soil Pollut 151:361–372

Whiting SN, Reeves RD, Richards D, Johnson MS, Cooke JA, Malaisse F, Paton A, Smith JAC, Angle JS, Chaney RL, Ginocchio R, Jaffre T, Johns R, McIntyre T, Purvis OW, Salt DE, Schat H, Zhao FJ, Baker AJM (2004) Research priorities for conservation of metallophyte biodiversity and their potential for restoration and site remediation. Restor Ecol 12:106–116

Williams JB (2002) Phytoremediation in wetland ecosystems: progress, problems and potential. Crit Rev Plant Sci 21:607–635

Ye HB, Yang XE, He B, Long XX, Shi WY (2003) Growth response and metal accumulation of *Sedum alfredii* to Cd/Zn complex-polluted ion levels. Acta Bot Sin 45:1030–1036

Zhao FJ, Dunham SJ, McGrath SP (2002) Arsenic hyperaccumulation by different fern species. New Phytol 156:27–31

Zhao FJ, Lombi E, McGrath SP (2003) Assessing the potential for zinc and cadmium phytoremediation with the hyperaccumulator *Thlaspi caerulescens*. Plant Soil 249:37–43

Preface

The volume Soil Heavy Metals duly edited by Irena Sherameti and Ajit Varma, published in 2010, was a success story. This was nicely celebrated in typical German style in the house of Professor Dr. Ralf Oelmüller, Institute of General Botany and Plant Physiology, University of Jena. Over a glass of wine I proposed to Irena to edit a volume on detoxification of heavy metals in soil. After a short discussion, we agreed to work together on this volume.

This volume summarises the ongoing scientific activities in the field of detoxification of heavy metals in soils, plants and microorganisms. The chapters are arranged in such a way that first we get an introduction about the art of detoxification of heavy metals and the heavy metal plants. The second group of chapters deals with the phytoremediation in general and phytoremediation of special ions. The next section describes several aspects of plant responses to heavy metals and the responses of special organisms/groups to heavy metals. At last different methodologies for detoxification of heavy metals in soils and plants are discussed.

Soil, one of the most important natural resources, is becoming degraded due to anthropogenic activities such as mining, agricultural activities, sewage sludge, fossil fuel combustion, metallurgical and chemical industries and electronics. As described in Chap. 1 written by Jyoti Agrawal, Irena Sherameti and Ajit Varma each source of contamination has its own damaging effects to plants, animals and humans, but the pollution from heavy metals is of serious concern and a big potential threat to the environment and human health. This chapter gives a general overview of some of the sources of heavy metal contaminants in soil, soil–plant relationships regarding heavy metals and heavy metal tolerance mechanism(s) in plants. In Chap. 2, Hermann Bothe directs us to the heavy metal soils and heavy metal plants (Metallophytes) of Central Europe showing that the adaptations of these metallophytes to the adverse conditions of heavy metal soils differ from one plant species to the next. Further we get introduced to some strategies employed by the metallophytes to cope with high concentrations of heavy metals at the whole plant level and gene expressions upon heavy metal stress in plants. Functional significance of metal ligands in hyperaccumulating plants is analysed by Marjana Regvar and Katarina Vogel-Mikuš in Chap. 3. This chapter focuses on ligands (organic acids, histidine, metallothioneins, low-molecular-weight thiols, etc.) that have roles in the immobilisation, transport and/or storage of accumulated metals in plant organs, tissues and cells.

Chapter 4, written by Shao Hongbo, Chu Liye, Xu Gang, Yan Kun, Zhang Lihua and Sun Junna is a progress in phytoremediating heavy metal-contaminated soils, that introduces the latest development in the field of phytoremediation as one of the main methods for removing hazardous heavy metal from contaminated soils. Using plants and microbes is preferred because of its cost-effectiveness, environmental friendliness and fewer side effects. So far all plant species recognised as useful for phytoremediation belong to angiosperm phylogeny group that is classified into 63 orders and 413 families. The authors of Chap. 5, Stanislaw W. Gawronski, Maria Greger and Helena Gawronska, show that among all only species from 8 orders and 18 families are identified as well-tolerating pollutants and useful for phytoremediation having advantages and limitations in their usefulness as phytoremediants. The authors of Chap. 6, Dora M. Carmona, Raúl Zornoza, Ángel Faz, Silvia Martínez-Martínez and Jose A. Acosta, describe the environmental impacts of mining activities in Southeast Spain. A field trial was established and experimental plots were designed, using marble wastes, pig manure and sewage sludge as amendments to reclaim the mine soils. The authors monitored the dynamics of heavy metals, soil properties and vegetation along 5 years after reclamation.

Zinc is an essential micronutrient with various cellular functions, but excess Zn in plants is toxic and causes chlorosis and growth disorders. To ensure Zn homeostasis the transport machinery is responsible for uptake and export of Zn that includes members of the metal tolerance protein (MTP), ZRT1/IRT1-like protein (ZIP) and heavy metal ATPase (HMA) families. Their roles in the acquisition, distribution, homeostasis and signalling of Zn are described in Chap. 7 by Miki Kawachi, Yoshihiro Kobae, Rie Tomioka and Masayoshi Maeshima. Copper, trace amounts of which are required to sustain plant life (so-called essential elements), in high concentrations causes plant death. Discussing current methods and approaches used for quantification of apoplastic and symplastic copper pools has a significant place in Chap. 8 written by Valentina P. Kholodova, Elena M. Ivanova and Vladimir V. Kuznetsov. The role of arbuscular mycorrhizal fungi producing an extraradical mycelium in metal ion immobilisation is also considered in this chapter. Arsenic is a ubiquitously distributed and an extremely toxic metalloid affecting the health of many people in more than 23 countries. On land arsenic is relatively immovable through binding of soil particle; however, most arsenic can readily dissolve in water and in soluble form may leach into surface and ground waters. Chapter 9, written by Dharmendra K. Gupta, Sudhakar Srivastava, H.G. Huang, Maria C. Romero-Puertas and Luisamaria M. Sandalio, focuses on arsenic contamination, accumulation, tolerance and detoxification mechanisms in plants. Chapter 10 of Kavita Shah presents an overview of the research information on sources and effects of cadmium metal on plants in particular. The knowledge of metal hyperaccumulation physiology and the molecular and genetic basis of Cd tolerance and detoxification in plants forms a major part of this chapter. The prospects and the future applications of hyperaccumulators in phytoremediation of Cd metal are also discussed. Dieter Rehder deals in Chap. 11 with the transport, accumulation and physiological effects of vanadium. Industrial and volcanic exhalation of vanadium oxides can cause locally a vanadium overload in soil surface areas. Soil bacteria such as *Geobacter metallireducens* and *Shewanella oneidensis* reduce vanadate to insoluble and comparatively harmless vanadium (IV) hydroxide. The remobilization of vanadium (IV) can occur by strong chelators excreted by other bacteria such as Azotobacter. Tapan Jyoti Purakayastha deals in Chap. 12 with the remediation of arsenic-contaminated soil. The use of engineered microbes as selective biosorbents is an attractive green cure technology for the low-cost and efficient removal of arsenic from soil. Fate of cadmium in calcareous soils under salinity conditions is discussed in Chap. 13 by Ali Khanmirzaei. The chemistry of calcareous and saline soils, the application of fractionation and speciation analysis for investigating the mobility and environmental ecotoxicity of this element in calcareous soils and some examples on Cd detoxification in carbonate rich soils are outlined in this chapter.

The current status of organellar proteomics as a high-throughput approach for obtaining a better understanding of heavy metal accumulation and detoxification in plants is analysed in detail in Chap. 14 by Nagib Ahsan, Byung-Hyun Lee and Setsuko Komatsu. To identify the proteins involved in organ-specific heavy metal response pathways is a fundamental step in the process of understanding the molecular mechanisms leading to accumulation and detoxification of toxic heavy metals in plant cells. Chapter 15, written by Laura A. Hardulak, Mary L. Preuss and Joseph M. Jez, provides an overview of sulfur metabolism in plants, how it plays a critical role in heavy metal tolerance and how efforts to engineer these pathways may improve bioremediation efforts. Metabolically, sulfur metabolism is a core pathway for the synthesis of molecules required for heavy metal tolerance in plants. Etsuro Yoshimura in Chap. 16 discusses Cd(II)-activated synthesis of phytochelatins. Phytochelatins are implicated in heavy metal tolerance in higher plants, algae, and a fungal species. Synthesis of the peptides is mediated by an enzyme designated as PC synthase (PCS) from the tripeptide glutathione (GSH).

Elsholtzia splendens has been proven to be a Cu-tolerant plant and can remarkably influence the behaviour of Cu in root–soil interface by root exudates, rhizosphere bacteria and arbuscular mycorrhizal fungi. *E. splendens* has evolved a series of defensive strategies against Cu stress such as Cu compartmentation and speciation transformation, which are discussed in detail in Chap. 17 by Yingxu Chen, Mingge Yu and Dechao Duan.

The role of aquatic macrophytes in biogeochemical cycling of heavy metals, the relevance to soil-sediment continuum detoxification and ecosystem health is presented in Chap. 18 by Przemysław Malec, Beata Mysliwa-Kurdziel, M.N.V. Prasad, Andrzej Waloszek and Kazimierz Strzałka. The wetland sediments and soils of flood plains play an important role in the biogeocycling of heavy metals. The role of both photosynthetic activity and competitive/synergistic effects of the elements available to aquatic macrophytes in the circulation and deposition of metals are discussed in terms of the functioning of wetland ecosystems and phytoremediation. To stimulate phytoremediation, fast growing plants with high metal uptake and high biomass are required. Alternatively, soil microorganisms such as fungi and bacteria are used in heavy metal detoxification. The recent advances in effect and significance of fungi and rhizobacteria in heavy metal detoxification is reviewed in Chap. 19 by Sema Camci Cetin, Ayten Karaca, Ridvan Kizilkaya and Oguz Can Turgay. The same group of authors contributed Chap. 21 in which the detoxification of heavy metals using earthworms is discussed. Earthworms can effect either available or total metal concentrations in soil because of their capability for accumulating heavy metals in their tissues and hence reduce their involvement in soil food chain. D.V. Yadav, Radha Jain and R.K. Rai, authors of Chap. 20, deal with the phytoremediation/ detoxification of heavy metals from soils through sugar crops, especially sugar cane, sugar beet and sweet sorghum. The potential of these sugar crops is presented. At Chap. 22, Roberto Terzano and Matteo Spagnuolo discuss the stabilisation of heavy metals by promoting zeolite synthesis in soil which can be easily done at low temperatures by adding Si- and Al-containing materials in alkaline conditions. This methodology is a promising one and in combination with other physico-chemical or biological remediation processes can effectively stabilise heavy metals in polluted sites.

This volume promises to be useful for researchers, students and other academicians involved in understanding the basics of detoxification of heavy metals in soils.

We are very thankful to all authors for contributing to this volume and we hope that their contribution will stimulate further high-quality teaching and research. It has been a pleasure to edit this book, primarily due to the stimulating cooperation of the contributors.

We wish to thank Hanna G. Hensler-Fritton, Editorial Director Life Sciences/ Biomedicine Europe II, Jutta Lindenborn and Dieter Czeschlik (former Life science Head, Springer Heidelberg) for generous assistance and patience in finalising the volume. A special thanks goes to our families.

Finally, we would like to thank Dr Sebastian Steiner from the Institute of General Botany and Plant Physiology, Friedrich-Schiller University of Jena, for his kind support on computer assistance.

Jena, Germany
New Delhi, India
June 2011

Irena Sherameti
Ajit Varma

Contents

Contributors

Acosta, Jose A. Sustainable Use, Management, and Reclamation of Soil and Water Research Group. Department of Agrarian Science and Technology, Technical University of Cartagena, Paseo Alfonso XIII, 52, 30203 Cartagena, Murcia, Spain; Institute for Biodiversity and Ecosystem Dynamics, University of Amsterdam, Nieuwe Achtergracht 166, 1018 WV Amsterdam, The Netherlands, j.a.acostaa viles@uva.nl

Agrawal, Jyoti Amity Institute of Microbial Technology, Amity University Uttar Pradesh, Block 'E 3' Fourth Floor, Sector 125, Noida, Uttar Pradesh, India, jagrawal@amity.edu

Ahsan, Nagib National Institute of Crop Science, 2-1-18 Kannondai, Tsukuba 305-8518, Japan, nagib_ahsan@yahoo.com

Bothe, Hermann Botanical Institute, The University of Cologne, Zülpicher Str. 47b, 50923 Cologne, Germany, Hermann.Bothe@uni-koeln.de

Carmona, Dora M. Sustainable Use, Management, and Reclamation of Soil and Water Research Group. Department of Agrarian Science and Technology, Technical University of Cartagena, Paseo Alfonso XIII, 52, 30203 Cartagena Murcia, Spain; Universidad Pontificia Bolivariana, Environmental Research Group, Circular 1ª #70-01, Medellín, Colombia

Cetin, Sema Camci Faculty of Foresty, Cankiri Karatekin University, Cankiri, Turkey, semacamcicetin@hotmail.com

Chen, Yingxu Institute of Environmental Science and Technology, Zhejiang University, Hangzhou 310029, China, yxchen@zju.edu.cn

Duan, Dechao Ministry of Agriculture, Key Laboratory of Non-point Source Pollution Control, Hangzhou 310029, China; Department of Environmental

Engineering, Huajiachi Campus, Zhejiang University, 268 Kaixuan Road, Hangzhou Zhejiang, 310029, China, 23292422@qq.com

Faz, Ángel Sustainable Use, Management, and Reclamation of Soil and Water Research Group. Department of Agrarian Science and Technology, Technical University of Cartagena, Paseo Alfonso XIII, 52, 30203 Cartagena, Murcia, Spain, angel.fazcano@upct.es

Gang, Xu Yantai Institute of Costal Zone Research, Chinese Academy of Sciences, Yantai 264003, China, gxu@yic.ac.cn

Gawronska, Helena Laboratory for Basic Research in Horticulture, Faculty of Horticulture and Landscape Architecture, Warsaw University of Life Sciences-SGGW, Warsaw, Poland, helena_gawronska@sggw.pl

Gawronski, Stanislaw W. Laboratory for Basic Research in Horticulture, Faculty of Horticulture and Landscape Architecture, Warsaw University of Life Sciences-SGGW, Warsaw, Poland, stanislaw_gawronski@sggw.pl

Greger, Maria Department of Botany, Stockholm University, Stockholm Sweden, Maria.Greger@botan.su.se

Gupta, Dharmendra Kumar Departamento de Bioquimica, Biologia Cellular y Molicular de Plantas, Estacion Experimental Del Zaidin, CSIC, Apartado 419, Granada 18008, Spain, guptadk1971@gmail.com

Hardulak, Laura A. Department of Biological Sciences, Webster University, 470 East Lockwood Avenue, Webster Groves, MO 63119, USA

Hongbo, Shao Yantai Institute of Costal Zone Research, Chinese Academy of Sciences, Yantai 264003, China; Institute for Life Sciences, Qingdao University of Science & Technology (QUST), Qingdao 266042, China, shaohongbochu@126.com

Huang, H.G. Ministry of Education Key Laboratory of Environmental Remediation and Ecosystem Health, Zhejiang University, Hangzhou 310029, China, huanghg82@gmail.com

Ivanova, Elena M. Laboratory of Physiological and Molecular Mechanisms of Adaptation, Timiryazev Institute of Plant Physiology RAS, Botanicheskaya ul., 35, Moscow 127276, Russia, Ilma25@mail.ru

Jain, Radha Indian Institute of Sugarcane Research, Lucknow 226002, Uttar Pradesh, India, radha_arihant@yahoo.com

Jez, Joseph M. Department of Biology, Washington University, One Brookings Drive, Campus Box 1137, St. Louis, MO 63130, USA, jjez@biology2.wustl.edu

Junna, Sun Yantai Institute of Costal Zone Research, Chinese Academy of Sciences, Yantai 264003, China; The Graduate University of Chinese Academy of Sciences, Beijing 100049, China, jnsun@yic.ac.cn

Karaca, Ayten Faculty of Agriculture, Department of Soil Science and Plant Nutrition, Ankara University, Diskapi, Ankara, Turkey, akaraca@agri.ankara.edu.tr

Kawachi, Miki Department of Biology, Graduate School of Science, Kobe University, Rokkodai 1-1, Nada-ku, Kobe 657-8501, Japan, kawachi.miki@gmail.com

Khanmirzaei, Ali Department of Soil Science, Karaj branch, Islamic Azad Universirty, Karaj, Iran, alikhanmirzai@yahoo.com

Kholodova, Valentina P. Laboratory of Physiological and Molecular Mechanisms of Adaptation, Timiryazev Institute of Plant Physiology RAS, Botanicheskaya ul., 35, Moscow 127276, Russia, v-p-kholodova@ippras.ru

Kizilkaya, Ridvan Faculty of Agriculture, Department of Soil Science and Plant Nutrition, Ondokuz Mayis University, Samsun Turkey, ridvank@omu.edu.tr

Kobae, Yoshihiro Graduate School of Bioagricultural Sciences, Nagoya University, Nagoya 464-8601, Japan, kobae@agr.nagoya-u.ac.jp

Komatsu, Setsuko National Institute of Crop Science, 2-1-18 Kannondai, Tsukuba 305-8518, Japan, skomatsu@affrc.go.jp

Kun, Yan Yantai Institute of Costal Zone Research, Chinese Academy of Sciences, Yantai 264003, China, kyan@yic.ac.cn

Kuznetsov, Vladimir V. Laboratory of Physiological and Molecular Mechanisms of Adaptation, Timiryazev Institute of Plant Physiology RAS, Botanicheskaya ul., 35, Moscow 127276, Russia, vlkuzn@ippras.ru

Lee, Byung-Hyun Division of Applied Life Sciences (BK21 Program), IALS, PMBBRC, College of Agriculture, Gyeongsang National University, Jinju 660-701, Republic of Korea, hyun@gnu.ac.kr

Lihua, Zhang Yantai Institute of Costal Zone Research, Chinese Academy of Sciences, Yantai 264003, China, lhzhang@yic.ac.cn

Liye, Chu Institute for Life Sciences, Qingdao University of Science & Technology (QUST), Qingdao 266042, China, chuliye1965@126.com

Maeshima, Masayoshi Laboratory of Cell Dynamics, Graduate School of Bioagricultural Sciences, Nagoya University, Nagoya 464-8601, Japan, maeshima@agr.nagoya-u.ac.jp

Malec, Przemyslaw Faculty of Biochemistry, Department of Plant Physiology and Biochemistry, Biophysics and Biotechnology, Jagiellonian University, ul. Gronostajowa 7, 30-387, Kraków Poland, malecpk@gmail.com

Martínez-Martínez, Silvia Sustainable Use, Management, and Reclamation of Soil and Water Research Group. Department of Agrarian Science and Technology, Technical University of Cartagena, Paseo Alfonso XIII, 52, 30203 Cartagena, Murcia, Spain, silvia.martinez@upct.es

Mysliwa-Kurdziel, Beata Faculty of Biochemistry, Department of Plant Physiology and Biochemistry, Biophysics and Biotechnology, Jagiellonian University, ul. Gronostajowa 7, 30-387, Kraków, Poland, b.mysliwa@wp.pl

Prasad, M.N.V. Department of Plant Sciences, University of Hyderabad, Prof. C.R. Rao Road, Gachhibowli, Central University P.O., Hyderabad 500 046, India, prasad_mnv@yahoo.com

Preuss, Mary L. Department of Biological Sciences, Webster University, 470 East Lockwood Avenue, Webster Groves MO 63119, USA, marypreuss34@webster.edu

Purakayastha, Tapan Jyoti Division of Soil Science and Agricultural Chemistry, Indian Agricultural Research Institute, New Delhi 110012, India, tpurakayastha@gmail.com

Rai, R.K. Indian Institute of Sugarcane Research, Lucknow 226002, Uttar Pradesh, India, ramakantrai@rediffmail.com

Regvar, Marjana Department of Biology, Biotechnical Faculty, University of Ljubljana, Večna pot 111, SI-1000, Ljubljana Slovenia, marjana.regvar@bf.uni-lj.si

Rehder, Dieter Institut für Anorganische und Angewandte Chemie, Martin-Luther-King-Platz 6, 20146 Hamburg Germany, rehder@chemie.uni-hamburg.de

Romero-Puertas, Maria C. Departamento de Bioquimica, Biologia Cellular y Molicular de Plantas, Estacion Experimental Del Zaidin, CSIC, Apartado 419, Granada 18008, Spain, maria.romero@eez.csic.es

Sandalio, Luisamaria M. Departamento de Bioquimica, Biologia Cellular y Molicular de Plantas, Estacion Experimental Del Zaidin, CSIC, Apartado 419, Granada 18008, Spain, luisamaria.sandalio@eez.csic.es

Shah, Kavita Environmental Science, Institute of Environment and Sustainable Development, Banaras Hindu University, Varanasi 221005, U.P, India, kavita-shah@bhu.ac.in

Sherameti, Irena Institut für Allgemeine Botanik und Pflanzenphysiologie, Friedrich-Schiller-Universität Jena, Dornburger Str. 159, 07743 Jena, Germany, irenasherameti@yahoo.de

Spagnuolo, Matteo Dipartimento di Biologia e Chimica Agroforestale e Ambientale, Università degli Studi di Bari, Via Amendola 165/A, 70126, Bari Italy

Strzałka, Kazimierz Faculty of Biochemistry, Department of Plant Physiology and Biochemistry, Biophysics and Biotechnology, Jagiellonian University, ul. Gronostajowa 7, 30-387, Kraków, Poland, kazimierzstrzalka@gmail.com

Srivastava, Sudhakar Nuclear Agriculture and Biotechnology Division, Bhabha Atomic Research Centre, Mumbai 400085, India, sudhakar.srivastava@gmail.com

Terzano, Roberto Department of Agroforestal and Environmental Biology and Chemistry, University of Bari "Aldo Moro", Via Amendola 165/A, 70126, Bari Italy, r.terzano@agr.uniba.it

Turgay, Oguz Can Faculty of Agriculture, Department of Soil Science and Plant Nutrition, Ankara University, Diskapi, Ankara, Turkey, turgay@agri.ankara.edu.tr

Tomioka, Rie Graduate School of Bioagricultural Sciences, Nagoya University, Nagoya 464-8601, Japan, tomiokar@agr.nagoya-u.ac.jp

Varma, Ajit Amity Institute of Microbial Technology, Amity University Uttar Pradesh, Block 'E 3' Fourth Floor, Sector 125, Noida, Uttar Pradesh, India, ajitvarma@aihmr.amity.edu

Vogel-Mikuš, Katarina Department of Biology, Biotechnical Faculty, University of Ljubljana, Večna pot 111, SI-1000, Ljubljana Slovenia, katarina.vogel@bf.uni-lj.si

Waloszek, Andrzej Faculty of Biochemistry, Department of Plant Physiology and Biochemistry, Biophysics and Biotechnology, Jagiellonian University, ul. Gronostajowa 7, 30-387, Kraków, Poland, andrzej.waloszek@uj.edu.pl

Yadav, D.V. Indian Institute of Sugarcane Research, Lucknow 226002, Uttar Pradesh, India, dvyadav_2008@rediffmail.com

Yoshimura, Etsuro Department of Applied Biological Chemistry, School of Agriculture and Life Sciences, The University of Tokyo, 1-1-1 Yayoi, Bunkyo Tokyo 113-8657, Japan, ayoshim@mail.ecc.u-tokyo.ac.jp

Yu, Mingge Institute of Environmental Science and Technology, Zhejiang University, Hangzhou 310029, China, mgyu_369@zju.edu.cn

Zornoza, Raúl Sustainable Use, Management, and Reclamation of Soil and Water Research Group. Department of Agrarian Science and Technology, Technical University of Cartagena, Paseo Alfonso XIII, 52, 30203 Cartagena, Murcia, Spain, raul.zornoza@upct.es

Chapter 1
Detoxification of Heavy Metals: State of Art

Jyoti Agrawal, Irena Sherameti, and Ajit Varma

1.1 Introduction

Land and water are precious natural resources on which rely the sustainability of agriculture and the civilization of mankind. Unfortunately, they have been subjected to maximum exploitation and are severely degraded or polluted due to anthropogenic activities. The pollution includes point sources such as emission, effluents, and solid discharge from industries, vehicle exhaustion, and metals from smelting and mining, and nonpoint sources such as soluble salts (natural and artificial), use of insecticides/pesticides, disposal of industrial and municipal wastes in agriculture, and excessive use of fertilizers (McGrath et al. 2001; Nriagu and Pacyna 1988; Schalscha and Ahumada 1998). Each source of contamination has its own damaging effects to plants, animals, and ultimately to human health, but those that add heavy metals to soil and water are of serious concern due to their persistence in the environment and carcinogenicity to human beings. They cannot be destroyed biologically but are only transformed from one oxidation state or organic complex to another (Garbisu and Alkorta 2001; Gisbert et al. 2003). Therefore, heavy metal pollution poses a great potential threat to the environment and human health.

In order to maintain good quality of soil and water and keep them free from contamination, continuous efforts have been made to develop technologies that are easy to use, sustainable, and economically feasible. Physicochemical approaches

J. Agrawal • A. Varma (✉)
Amity Institute of Microbial Technology, Amity University Uttar Pradesh, Block 'E 3' Fourth Floor, Sector 125, Noida, Uttar Pradesh, India
e-mail: ajitvarma@aihmr.amity.edu

I. Sherameti
Institut für Allgemeine Botanik und Pflanzenphysiologie, Friedrich-Schiller-Universität Jena, Dornburger Str. 159, 07743 Jena, Germany
e-mail: irenasherameti@yahoo.de

I. Sherameti and A. Varma (eds.), *Detoxification of Heavy Metals*, Soil Biology 30,
DOI 10.1007/978-3-642-21408-0_1, © Springer-Verlag Berlin Heidelberg 2011

have been widely used for remedying polluted soil and water, especially at a small scale. However, they experience more difficulties for a large scale of remediation because of high costs and side effects. Some microorganism-based remediation techniques, such as bioremediation, show potential for their ability to degrade and detoxify certain contaminants. Although these biological systems are less amenable to environmental extremes than other traditional methods, they have the perceived advantage of being more cost effective (Cunningham et al. 1997). The use of plant species for cleaning polluted soil and water named as phytoremediation has gained increasing attention since last decade, as an emerging cheaper technology. Many studies have been conducted in this field in the last two decades. Numerous plant species have been identified and tested for their traits in the uptake and accumulation of different heavy metals. Mechanisms of metal uptake at whole plant and cellular levels have been investigated. Progresses have been made in the mechanistic and practical application aspects of bioremediation and phytoremediation. They are reviewed in this chapter.

1.2 The Origin of Heavy Metals in Soil

1.2.1 Geochemical Origins of Heavy Metals

Ten 'major elements', O, Si, Al, Fe, Ca, Na, K, Mg, Ti, and P, constitute over 99% of the total element content of the earth's crust. The remaining elements of the periodic table are called 'trace elements' and their individual concentrations do not normally exceed 1,000 mg/kg (0.1%); in fact, most have average concentrations of less than 100 mg/kg (Mitchell 1964).

Trace elements occur as trace constituents of primary minerals in igneous rocks (which crystallize from molten magma). They become incorporated into these minerals by isomorphously substituting in the crystal lattice for ions of one of the major elements at the time of crystallization.

Sedimentary rocks comprise approximately 75% of the rocks outcropping at the earth's surface and are therefore more important than igneous as soil parent materials. They are formed by the lithification of sediments comprising rock fragments or resistant primary minerals, secondary minerals such as clays, or chemical precipitates such as $CaCO_3$. In general, clays and shales tend to have relatively high concentrations of many elements due to their ability to adsorb metal ions. Black (or bituminous) shales contain high concentrations of several metals and metalloids, including Ag, As, Cd, Cu, Pb, Mo, U, V, and Zn. The sediments from which they are formed act both as an adsorbent for heavy metals and as a substrate for microorganisms. The latter catalyzes the development of reducing conditions, which lead to further heavy metal accumulation through the precipitation of metal sulfides.

1.2.2 Sources of Heavy Metals Contaminants in Soils

Although heavy metals are ubiquitous in soil parent materials, the major anthropogenic sources of metals to soils and the environment are given in the following sections.

1.2.2.1 Metaliferous Mining and Materials

Metals utilized in manufacturing are obtained from either the mining of ore bodies in the earth's crust, or the recycling of scrap metal. Ores are naturally occurring concentration of minerals with a sufficiently high concentration of metals to render them economically worthwhile exploiting. With increase in demand for metals and improvements in mineral extraction technology, a higher proportion of ore bodies are being mined and consequently much greater quantities of waste are produced, especially tailings (finely milled fragments of rock and residual particles of ore not removed by the separation process). These tailing particles, which can be transported by either wind or water, constitute a significant source of metal contamination in soils surrounding the mines and in alluvial soils downstream.

Accidental flooding resulting from the failure of dams in tailing lagoons has been responsible for many severe pollution events in several countries. Once in the soil, the ore mineral fragments undergo oxidation and other weathering reactions and as a consequence the metal ions become distributed within the soil system and hence potentially bioavailable. Most ore minerals contain several other metals as minor inclusions as shown in Table 1.1.

1.2.2.2 Agricultural and Horticultural Materials

Agricultural practices constitute very important nonpoint sources of metals which make significant contributions to their total concentrations in soils in many parts of the world, especially in regions of intensive farming. According to the data obtained by the Lowestoft Laboratory of the Ministry of Agriculture, Fisheries and Food (MAFF 1986), the main sources are as follows:

Impurities in fertilizers: Cd, Cr, Mo, Pb, U, V, and Zn
Sewage sludge: especially Cd, Ni, Cu, Pb, and Zn (and many other elements)
Manures from intensive animal production, especially pigs and poultry: Cu, As, and Zn
Pesticides: Cu, As, Hg, Pb, Mn, and Zn
Refuse derived composts (not widely used in agriculture): Cd, Cu, Ni, Pb, and Zn
Desiccants: As
Wood preservatives: As, Cu, and Cr
Corrosion of metal objects (galvanized metal roofs and wire fences): Zn and Cd

Table 1.1 Common ore minerals

Metal	Ore minerals	Associated heavy metals
Ag	Ag_2S, PbS	Au, Cu, Sb, Zn, Pb, Se, Te
As	FeAsS, AsS, Cu Ores	Au, Ag, Sb, Hg, U, Bi, Mo, Sn, Cu
Au	Au[a], $AuTe_2$, (Au, Ag)Te_2	Te, Ag, As, Sb, Hg, Se
Ba	$BaSO_4$	Pb, Zn
Cd	ZnS	Pb, Zn, Cu
Cr	$FeCr_2O_4$	Ni, Co
Cu	$CuFeS_2$, Cu_5FeS_4, Cu_2S, Cu_3AsS_4, CuS, Cu[a]	Zn, Cd, Pb, As, Se, Sb, Ni, Pt, Mo, Au, Te
Hg	HgS, Hg[a], Zn Ores	Sb, Se, Te, Ag, Zn, Pb
Mn	MnO_2	Fe, Co, Ni, Zn, Pb
Mo	MoS_2	Cu, Re, W, Sn
Ni	$(Ni, Fe)_9S_8$, NiAs, $(Co,Ni)_3S_4$	Co, Cr, As, Pt, Se, Te
Pb	PbS	Ag, Zn, Cu, Cd, Sb, Ti, Se, Te
Sb	Sb_2S_3, Ag_3SbS_3	Ag, Au, Hg, As
Se	Cu Ores	As, Sb, Cu, Ag, Au
Sn	SnO_2, $Cu_2(Fe, Zn)SnS_4$	Nb, Ta, W, Rb
U	U_2O_3	V, As, Mo, Se, Pb, Cu, Co, Ag
V	C_2O_5, VS_4	U
W	WO_3, $CaWO_4$	Mo, Sn, Nb
Zn	ZnS	Cd, Cu, Pb, As, Se, Sb, Ag, Au, In

[a]Native metal deposits (adapted from Peters 1978)

It should be noted that not all of these sources relate to current practices and materials.

Most agricultural and horticultural soils in technologically advanced countries are regularly amended with fertilizers and many also receive organic manures (usually based on livestock feces and urine) and lime. Typical ranges of heavy metal concentrations found in these materials are given in Table 1.2 (Kabata-Pendias and Pendias 1992; Webber et al. 1984).

1.2.2.3 Sewage Sludge

Sewage sludge is the residue produced from the treatment of domestic and industrial wastewaters and large amounts are produced worldwide.

Sewage sludge are a significant source of plant nutrients and organic matter and some specially treated sludge, such as those containing lime or cement kiln dust, have useful liming properties as well. However, the beneficial properties of sludge are limited by their contents of potentially harmful substances such as heavy metals and organic micropollutants (PAHs, PCBs, and pesticides). Although all the sludge contain a wide range of metals and other contaminants in varying concentrations, those from industrial catchments generally have higher metal contents than those from mainly suburban domestic areas. The heavy metals most likely to cause problems for crop production on sludge-amended soils are Cd, Cu, Ni, and Zn (Davis and Calton-Smith 1980; McGrath et al. 1994).

Table 1.2 Typical ranges of heavy metal concentration in fertilizers, farmyard manure, lime, and compost (mg/kg)

Metal	Phosphate fertilizers	Nitrate fertilizers	Farmyard manure	Lime	Composted refuse
Ag	–	–	–	–	–
As	2–1,200	2.2–120	3–25	0.1–25	2–52
B	5–115	–	0.3–0.6	10	–
Cd	0.1–170	0.05–8.5	0.1–0.8	0.04– 0.1	0.01–100
Co	1–12	5.4–12	0.3–24	0.4–3	–
Cr	66–245	3.2–19	1.1–55	10–15	1.8–410
Cu	1–300	–	2–172	2–125	13–3,580
Hg	0.01–1.2	0.3–2.9	0.01–0.36	0.05	0.09–21
Mn	40–2,000	–	30–969	40–1,200	–
Mo	0.1–60	1–7	0.05–3	0.1–15	–
Ni	7–38	7–34	2.1–30	10–20	0.9–279
Pb	7–225	2–27	1.1–27	20–1,250	1.3–2,240
Sb	<100	–	–	–	–
Se	0.5	–	2.4	0.08–0.1	–
U	30–300	–	–	–	–
V	2–1,600	–	–	20	–
Zn	50–1,450	1–42	15–566	10–450	82–5,894

Adapted from Kabata-Pendias and Pendias (1992) and Webber et al. (1984)

Due to the relatively high concentrations of heavy metals, the sludges can be major sources of heavy metals in the soil to which they are applied. Several authors have shown that the transfer of heavy metals from sewage sludge-amended soils to crops is significantly lower than those from inorganic sources, such as metal salts or mining wastes (Alloway and Jackson 1991).

1.2.2.4 Fossil Fuel Combustion

In general, fossil fuel combustion results in the dispersion of a wide range of heavy metals, which can include: Pb, Cd, Cr, Zn, As, Sb, Se, Ba, Cu, Mn, U, and V, over a very large area, although not all these elements are present in significant concentrations in all types of coal and petroleum. The metals accumulate in the coal and petroleum deposits as they formed and are either emitted into the environment as airborne particles during combustion or accumulated in the ash which may itself be transported and contaminate soils or waters or may be leached in situ. The combustion of petrol (gasoline) containing Pb additives has been the largest source of this metal in this environment and has affected soils over a high proportion of the earth's terrestrial surface. Pb is emitted in the exhaust of vehicles running on Pb containing petrol as aerosol particles 0.01–0.1 μm in diameter, but these primary particles can cluster to form large particles (0.3–1 μm). These particles comprise mainly PbBrCl but can react with other air pollutants to form more complex compounds such as a-2PbBrCl·NH_4Cl.

1.2.2.5 Metallurgical Industries

Metallurgical industries can contribute to soil pollution in several ways: (a) by the emission of aerosols and dusts which are transported in air and eventually deposited on soils or vegetation; (b) by liquid effluents which may pollute soils at times of flooding; (c) by the creation of waste dumps in which metals become corroded and leached into the underlying soils. Many heavy metals are used in specialist alloys and steels – V, Mn, Pb, W, Mo, Cr, Ni, Cu, Zn, Sn, Si, Ti, Te, Ir, Ge, Tl, Sb, In, Cd, Be, Bi, Li, As, Ag, Sb, Pr, Os, Nb, Nd, and Gd (MAFF 1986). Hence, the manufacture of these materials, their disposal, or recycling in scrap metal can lead to environmental pollution by a wide range of metals.

1.2.2.6 Electronics

A large number of heavy metals are used in the manufacture of semiconductors, cables, contacts, and other electrical components. These include: Cu, Zn, Au, Ag, Pb, Sn, Y, W, Cr, Se, Sm, Ir, In, Ga, Ge, Re, Sn, Tb, Co, Mo, Hg, Sb, As, and Gd (MAFF 1986). Environmental pollution can occur from the manufacture of the components, their accidental contact with soils, and their disposal in waste. In addition to metals, old electronic equipment may also often include capacitors and transformers containing polychlorinated biphenyls (PCBs), which are persistent organic pollutants in soils.

1.2.2.7 Chemical and Other Industrial Sources

Other significant sources of heavy metal pollution of soils and the environment can be the manufacture and/or use and disposal of the following (MAFF 1986):

Chloride manufacture, Hg
Batteries, Pb, Sb, Zn, Cd, Ni, and Hg
Pigments and paints, Pb, Cr, As, Sb, Se, Mo, Cd, Ba, Zn, and Co
Catalysts, Pt, Sm, Sb, Ru, Co, Rh, Re, Pd, Os, Ni, and Mo
Polymer stabilizers, Cd, Zn, Sn, and Pb
Printing and graphics, Se, Pb, Cd, Zn, Cr, and Ba
Medical Uses, Ag, Sn, Ba, Cu, Hg, Sb, Se, Pt, and Zn
Additives in fuels and lubricants, Se, Te, Pb, Mo, and Li

1.2.2.8 Waste Disposable

The disposal of household, municipal, and industrial wastes can lead to soil pollution with heavy metals in various ways. The landfilling of municipal solid waste can lead to several metals including Cd, Cu, Pb, Sn, and Zn being dispersed into soil, groundwater, and surface water in leachates if the landfill is not managed properly. Temporary waste stockpiles can cause significant soil contamination which may not be discovered until analysis at a later time when the land is no longer used for this purpose. This has been the cause of metal contamination in several urban allotment gardens in the UK.

1.3 Soil–Plant Relationships of Heavy Metals

1.3.1 Soil–Plant System

The major interrelationships affecting the dynamics of heavy metals between the soil and the plant are shown in Fig. 1.1. The soil–plant system is an open system subject to inputs, such as contaminants, fertilizers, and pesticides, and to losses, such as the removal of metals in harvested plant material, leaching, erosion, and vitalization.

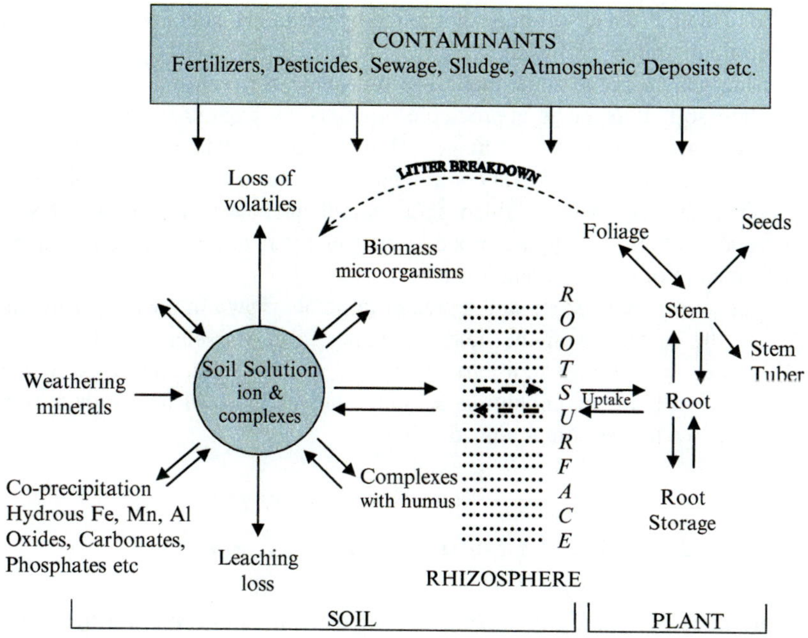

Fig. 1.1 The soil–plant system showing the key components concerned with the dynamics of heavy metals (modified from Peterson and Alloway 1979)

1.3.2 Plant Uptake of Metals

The factors affecting the amounts of metal absorbed by a plant are those controlling: (a) the concentration and speciation of the metal in the soil solution, (b) the movement of the metal from the bulk soil to the root surface, (c) the transport of the metal from the root surface in the root, and (d) its translocation from the root to the shoot (Peterson and Alloway 1979). Plant uptake of mobile ions present in the soil solution is largely determined by the total quantity of this ion in the soil, but, in the case of strongly absorbed ions, absorption is more dependent on the amount of root produced (Krauskopf 1967). Mycorrhizae are symbiotic fungi which effectively increase the absorptive area of the root and can assist in the uptake of nutrient ions, such as orthophosphates and micronutrients.

Absorption of metals by plant roots can be by both passive and active processes. Passive uptake involves diffusion of ions in the soil solution into the root endodermis. On the other hand, active uptake takes place against a concentration gradient but requires metabolic energy and can therefore be inhibited by toxins. The mechanisms appear to differ between metals: for instance, Pb uptake is generally considered to be passive, while that of Cu, Mo, and Zn is thought to be either active metabolic uptake, or a combination of both active and passive uptake (Kabata-Pendias and Pendias 1992).

Absorption mechanism can vary for different metal ions, but ions that are absorbed in the roots by the same mechanism are likely to compete with each other. For example, Zn absorption is inhibited by Cu and H^+, but not by Fe and Mn; Cu absorption is inhibited by Zn, NH_4^+, Ca, and K (Barber 1984).

The rhizosphere is the zone about 1–2 mm wide between plant roots and the surrounding soil. It receives appreciable amounts of organic material from the roots, including exudates, mucilage, sloughed-off cells, and their lysates (Marschner 1986). These organic compounds give rise to intense microbiological and biochemical activity in the rhizosphere, which enables roots to mobilize some of the metals that are strongly absorbed in the soil, by acidification, redox changes, or the formation of organic complexes.

The uptake of metals from soil is greater in plants grown in pots in greenhouse than from the same soil in the field (Graham 1981; Marschner 1986). This is probably due to differences in microclimate and soil moisture, and to the roots of container-grown plants in contaminated soil, whereas those of field-grown plants may reach down to less contaminated soil.

1.3.3 The Biological Essentiality of Trace Elements

There are three criteria to say whether a trace element is essential for the normal growth of plants or not:

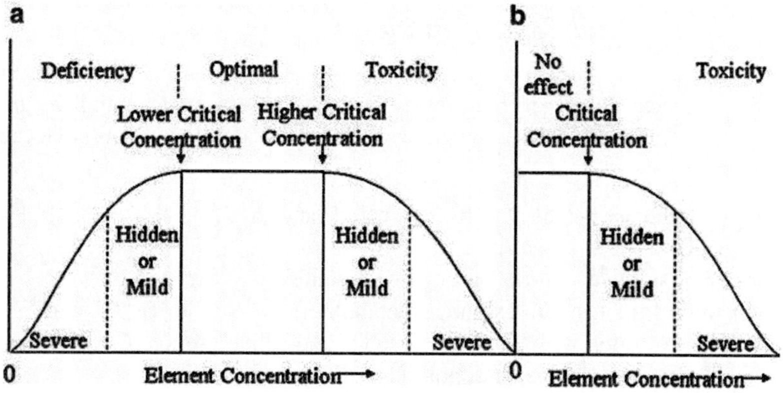

Fig. 1.2 Typical dose–response curves for (**a**) essential trace elements (micronutrients) and (**b**) nonessential trace elements (modified from Alloway 1995)

(a) The plant can neither grow nor complete its life cycle without an adequate supply of the element.
(b) The element cannot be wholly replaced by any other element.
(c) The element has a direct influence on the plant and is involved in its metabolism (Bowen 1979).

Apart from C, H, O, N, P, K, and S, the elements which have been shown to be essential for plants are Al, B, Br, Ca, Cl, Co, Cu, F, Fe, I, K, Mg, Mn, Mo, Na, Ni, Rb, Si, Ti, V, and Zn (Kabata-Pendias and Pendias 1992).

Essential trace elements are frequently referred to as micronutrients. If a plant supply of a micronutrient is inadequate, its growth is adversely affected. At the other extreme, an excessive supply of a micronutrient will cause toxicity. Typical dose–response curves for micronutrients and for nonessential trace elements (Fig. 1.2) show that when the supply of a micronutrient to a plant is inadequate, growth and yield are severely reduced and symptoms of deficiency are manifested. With an increasing supply of micronutrient, the yield reduction becomes progressively lower and the symptoms are less marked. As the supply of the micronutrient increases beyond the lower critical concentration, there is a zone of luxury consumption with no effect on yield. The upper critical concentration heralds the commencement of yield reductions due to toxicity, which becomes more severe until the lethal dose is reached.

The curve of nonessential elements in Fig. 1.2 shows that there is no deficiency effect with low concentration of the element; yield is not affected until the upper concentration limit is reached, after which toxicity occurs in the same way as with an excess of a micronutrient.

1.3.4 Heavy Metal Toxicity in Plants

Excessive concentrations of both essential and nonessential metals result in phytotoxicity. Kabata-Pendias and Pendias (1992) list the following possible causal mechanisms:

1. Changes in the permeability of the cell membrane: Ag, Au, Br, Cd, Cu, F, Hg, I, Pb, and UO_2
2. Reaction of sulphydryl (–SH) groups with cations: Ag, Hg, and Pb
3. Competition for sites with essential metabolites: As, Sb, Se, Te, W, and F
4. Affinity for reacting with phosphate groups and active groups of ADP or ATP: Al, Be, Y, Zr, lanthanides, and, possibly, all heavy metals
5. Replacement of essential ions (mainly major cations): Cs, Li, Rb, Se, and Sr
6. Occupation of sites for essential groups such as phosphate and nitrate: arsenate, fluorate, borate, bromate, selenate, tellurate, and tungstate.

In excessive amounts, the relative toxicity of different metals to plants can vary with plant genotype and experimental conditions. The most toxic metals to higher plants and microorganisms are Hg, Cu, Ni, Pb, Co, Cd, and possibly also Ag, Be, and Sn (Mench and Martin 1991). Food plants which tolerate relatively high concentrations of these potentially hazardous metals are likely to create a greater health risk than those which are more sensitive and show definite symptoms of toxicity.

1.3.5 Effects of Heavy Metals on the Soil Microbial Mass

Several authors have shown that high concentrations of various heavy metals in soils had inhibitory effects on microbial activity. Tyler et al. (1989) reported that the normal decomposition of conifer litter and recycling of plant nutrients were inhibited in the forest surrounding a brass foundry which had emitted large amounts of Cu, Zn, and other metals as aerosols over many years. The reason for the inhibition of microbial activity was that the growth of trees in the area was retarded due to deficiencies in plant macronutrients. However, other authors, such as Olson and Thornton (1982), have reported that soils from severely metal contaminated sites, such as Shipham in Somerset, contained bacteria that showed tolerance to Cd relative to bacteria in uncontaminated soils. Doelman and Haanstra (1979) showed that Pb inhibited both microbial respiration and dehydrogenase activity in polluted soils. Although tolerant populations of microorganisms were found in polluted soils, there was a change in the balance of the different types of the microorganisms present which could have an impact on soil fertility.

It was found that the metals from sewage sludge had a marked inhibitory effect on symbiotic nitrogen fixation in the roots of white clover due to toxicity affecting *Rhizobium leguminosum bv trifolii*. Experiments conducted in vitro showed a decreasing order of toxicity as being: Cu > Cd > Ni > Zn (Chaudri et al. 1992). Other workers studying rhizobia in more recent field experiments elsewhere in UK concluded that the low number of these bacteria present was probably due to inhibitory effect of Cd (Obbard and Jones 1993). An independent study in Japan on the effects of Cd, Cr, Cu, Ni, and Pb on organic decomposition in gley and adosol soils showed that all the metals inhibited the evolution of CO_2. In this study, Cd and Cu showed the highest inhibitory effect, while Pb showed the smallest one (Hattori 1992).

1.4 Heavy Metal Detoxification of Soil

Lone et al. (2008) classified different approaches used to reclaim metal polluted soils into physicochemical and biological ones.

1.4.1 Physiochemical Methods of Remediating Metal Polluted Soil

The physiochemical approaches involved in soil remediation include as follows.

1.4.1.1 Excavation Method

This involves the excavation and reburial of polluted soils in special landfills (Conder et al. 2001; Jing et al. 2007; Lombi et al. 2002; Neilson et al. 2003; Bennett et al. 2003). Although the excavation method is the most used approach to reclaim contaminated soils (Lombi et al. 2001), it does not remediate the soil (Neilson et al. 2003).

1.4.1.2 Capping of the Polluted Soil

This involves top soiling of the polluted soils with uncontaminated soils from offsite to a depth that would minimize uptake of heavy metals by vegetation (Neilson et al. 2003; Okoronkwo et al. 2005). Still, this does not give a permanent solution to the problem since the metal can still be leached into the underground water.

1.4.1.3 Fixation and Inactivation (Stabilization) of the Polluted Soil

This involves the conversion of the heavy metals into those forms that are less mobile and available for plants and microflora (Lone et al. 2008; Conder et al. 2001). Usually, the essence of stabilization is to reduce the amount of phytoavailable metal and thus reduce their toxicities to plants, animals, and soil organisms. Some commonly used chemical immobilization agents include zeolite, gravel sludge, beringite, alkaline materials, organic materials (sewage sludge and compost), phosphate (Conder et al. 2001), and lime stabilized municipal biosolids (Stuczynski et al. 2007; Conder et al. 2001).

1.4.1.4 Soil Washing

This technique involves the use of acids (HCl and HNO_3), chelators (EDTA, Nitriloacetic acid, DTPA, etc.), and other anionic surfactant (biosurfactant) (Neilson et al. 2003) to solubilize the polluting metals. It may take the form of in situ treatment, which involves soil flushing with pumps (Neilson et al. 2003), or ex situ treatment, which involves washing an excavated portion of the contaminated site with these agents followed by the return of clean soil residue to the site (Lone et al. 2008). This method is generally expensive and it is fraught with lots of side effects (Lone et al. 2008).

Other physicochemical methods include: thermal treatment (Jing et al. 2007), precipitation, or flocculation followed by sedimentation, ion exchanges, reverse osmosis, and microfiltration (Lone et al. 2008). These physicochemical approaches are not suitable for practical purposes because of their high cost, low efficiency, destruction of soil structure, and fertility (Lone et al. 2008; Jing et al. 2007).

1.4.2 Biological Approaches of Remediating Metal Polluted Soil

The biological approaches involved in soil remediation include:

- Use of microorganisms to detoxify metal by valence transformation, extracellular chemical precipitation, or volatilization (some microorganisms can enzymatically reduce a variety of metals in metabolic processes that are not related to metal assimilation) (Lone et al. 2008).
- Use of special type of plants to decontaminate soil or water by inactivating metals in the rhizosphere or translocating them in their aerial parts. This approach is called phytoremediation.

These new techniques are cheaper, efficient, and more environmental friendly (Lone et al. 2008; Jing et al. 2007).

1.4.2.1 Microorganism-Based Remediation

Selective pressures from a metal-containing environment have led to the development of resistance systems in microorganisms to virtually all toxic metals (Rouch et al. 1995). These systems are mostly plasmid-mediated and very specific and have been found in virtually all eubacterial groups studied (Silver and Misra 1984; Ji and Silver 1995). These reports included mostly aerobic microorganisms, with prominent examples being resistance in *Staphylococcus* sp., *Escherichia coli*, *Pseudomonas aeruginosa*, and *Bacillus* sp. (Nakahara et al. 1977; Marques et al. 1979; Harnett and Gyles 1984; Schwarz and Hobel 1989; Belliveau et al. 1991; Wang and Shen 1995). Resistance has been reported for mercury [Hg(II)] and organomercurials in obligate anaerobes like Bacteroides and Clostridium species.

There are differences between chromosomal and plasmid-based metal resistance systems. Essential metal resistance systems are usually chromosome-based and more complex than plasmid systems. Plasmid-encoded systems, on the other hand, are usually toxic-ion efflux mechanism.

A cell may develop metal resistance systems in an attempt to protect sensitive cellular components. Limiting metal access or altering cellular components decreases their sensitivity to metals. Several factors determine the extent of resistance in a microorganism: the type and number of mechanisms for metal uptake, the role each metal plays in normal metabolism, and the presence of genes located on plasmids, chromosomes, or transposons that control metal resistance. Six mechanisms are postulated to be involved in resistance to metals (Silver 1992; Rouch et al. 1995).

Metal Exclusion by Permeability Barrier

Alterations in the cell wall, membrane, or envelope of a microorganism are examples of metal exclusion by permeability barrier. This mechanism is an attempt by the organism to protect metal-sensitive, essential cellular components. A prominent example is the exclusion of Cu(II) resulting from altered production of the membrane channel protein porin by *E. coli* B (Rouch et al. 1995). This is usually a single gene mutation, which decreases the permeability of the membrane to metal ions (Ji and Silver 1995). Another example is nonspecific binding of metals by the outer membrane or envelope. This offers limited metal protection due to saturation of binding sites (Beveridge and Murray 1976; Hoyle and Beveridge 1983).

Bacteria that naturally form an extracellular polysaccharide coating demonstrate the ability to bioabsorb metal ions and prevent them from interacting with vital cellular components (Scott et al. 1988; Scott and Palmer 1990). The exopolysaccharide coating of these bacteria may provide sites for the attachment of metal cations (Scott and Palmer 1988). Exopolysaccaride by itself is not as efficient in binding Cd(II) as an organism with intact extracellular coating

(Scott and Palmer 1988). This protective layer appears to prevent uptake, keeping metal ions away from sensitive cellular components.

Periplasmic binding of Cu(II) is found in *Pseudomonas* sp. where resistance is coded for by an operon of four genes: *copA*, *copB*, *copC*, and *copD*; *copA* and *copB* confer partial resistance with the addition of *copC* and *copD*, providing for full Cu(II) resistance (Silver and Walderhaug 1992; Silver and Ji 1994). In some species of *Staphylococcus aureus*, penicillinase plasmids can mediate resistance by changing cell membrane permeability to Cd(II) as well as to other metals. In this case in the membrane appear conformational changes that prevent metal ions from entering.

Active Transport of the Metal Away from the Microorganism

Active transport or efflux systems represent the largest category of metal resistance systems. Microorganisms use active transport mechanisms to export toxic metals from their cytoplasm. These mechanisms can be chromosomal or plasmid-encoded. These efflux systems can be non-ATPase or ATPase-linked and highly specific for the cation or anion they export (Silver et al. 1989; Nies and Silver 1995). Evidence for ATP-dependence comes from the use of energy uncouplers and ionophore antibiotics (Silver and Misra 1984; Rensing et al. 1998). Uncouplers prevent the formation of ATP, which results in a decline in the efflux of cations or anions. Prominent examples are inducible plasmid-encoded resistance for As(V), arsenite [As(III)], and antimonite mediated by the *ars* operon in *E. coli* and *S. aureus*; Cd(II) resistance encoded by the *cad* operon in *S. aureus*, *Bacillus* sp., and *Listeria* sp. or the *czc* operon found in *Alcaligenes eutrophus*; and Pb(II) resistance mediated by *ZntA* in *E. coli* and *CadA* in *S. aureus* (Silver and Walderhaug 1992; Ji and Silver 1995; Rensing et al. 1998).

Intracellular Sequestration of Metals by Protein Binding

Intracellular sequestration is the accumulation of metals within the cytoplasm to prevent exposure to essential cellular components. Metals commonly sequestered are Cd(II), Cu(II), and Zn(II). Two examples exist for this form of metal resistance: metallothionein production in *Synechococcus* sp. and cysteine-rich proteins in *Pseudomonas* sp. (Rouch et al. 1995; Silver and Phung 1996). The metal resistance system in *Synechococcus* sp. consists of two genes: *smtA* and *smtB*. *smtA* encodes a metallothionein that binds to Cd(II) and Zn(II) (Silver et al. 1989; Silver 1992). Cysteine residues in SmtA metallothionein may act as a sink for excess toxic cations.

Extracellular Sequestration

Metal resistance based on extracellular sequestration has been hypothesized only in bacteria, but it is found in several species of yeast and fungi (Joho et al. 1995). One of the forms of Ni(II) resistance in yeast may be based on this mechanism. *Saccharomyces cerevisiae* may reduce absorption of Ni(II) by excreting large amounts of glutathione (Murata et al. 1985). Glutathione binds with great affinity to heavy metals. Yeast carrying the methylglyoxal resistance gene demonstrates the ability to form extracellular metal glutathione complexes in metal-rich media (Murata et al. 1985). Resistance results when the toxic metal is bound in a complex and cannot enter the cell membrane. A similar mechanism exists in Cu(II)-resistant fungi (Murphy and Levy 1983). These fungi secrete oxalate to form a metal–oxalate complex.

Enzymatic Detoxification of a Metal to a Less Toxic Form

Mercury resistance is a model example of an enzymatic detoxification system in microorganisms. Both Gram-positive (*S. aureus*, *Bacillus* sp.) and Gram-negative bacteria (*E. coli*, *P. aeruginosa*, *Serratia marcescens*, and *Thiobacillus ferrooxidans*) demonstrate resistance to Hg(II) (Misra 1992). Mercury is toxic because it binds to and inactivates essential thiols that are part of enzymes and proteins. Some bacteria contain a set of genes that form a Hg(II) (*mer*) resistance operon. This operon not only detoxifies Hg(II) but also transports and self-regulates resistance (Misra 1992; O'Halloran 1993; Ji and Silver 1995). The same set of genes also encodes the production of a periplasmic binding protein and membrane-associated transport proteins. The periplasmic binding protein collects Hg(II) in the surrounding environment and transport proteins take it to the cytoplasm for detoxification.

The Hg(II) resistance or *mer* operon codes for the production of two enzymes. The first organomercurial lyase catalyzes the following reaction (Weiss et al. 1977; Misra 1992):

$$RHgX + H^+ + X^- \Leftrightarrow RH + HgX_2$$

Mercuric ion reductase catalyzes the following reaction (Weiss et al. 1977; Misra 1992):

$$Hg(SR)_2 + NADPH + H^+ \Leftrightarrow Hg(0) + NADP^+ + 2RSH$$

Another enzymatic detoxification system includes plasmid-mediated As(V) resistance in *B. subtilis*, *S. aureus*, and *E. coli* species (Rouch et al. 1995; Tsutomu and Kobayashi 1998). The *arsC* gene of the *ars* operon codes for arsenate reductase, which reduces intracellular As(V) to the more toxic As(III) form (Nies and Silver 1995). As(III) is then removed from the organism by an efflux pump encoded by other genes in the *ars* operon (Wang and Shen 1995). This enzyme does not function on its own; instead it requires a coupling protein to reduce arsenate. These coupling proteins can vary among microorganisms; *S. aureus*, for example, uses thioredoxin, whereas *E. coli* uses glutaredoxin (Silver and Walderhaug 1992).

Reduction in Metal Sensitivity of Cellular Targets

Some microorganisms adapt to the presence of toxic metals by altering the sensitivity of essential cellular components; this provides a degree of natural protection (Rouch et al. 1995). Protection is achieved by mutations that decrease sensitivity but do not alter basic function or by increasing production of a particular cellular component to keep ahead of metal inactivation. DNA repair mechanisms also provide limited protection to plasmid and genomic DNA. The microorganism may also protect itself by producing metal-resistant components or alternate pathways in an effort to bypass sensitive components. Adaptation has been found in *E. coli*. Upon exposure to Cd(II), unadapted *E. coli* demonstrate considerable DNA damage; however, after subculture the same organisms show resistance (McEntee et al. 1986; Mergeay 1991). Natural resistance can result from normal cellular functions that give the organism a basic level of tolerance (Rouch et al. 1995). An example is glutathione, which may offer protection to metal ions such as Ag(I), Cu(I,II), Cd(II), and Hg(II) (Ni'bhriain et al. 1983) by suppressing free radical formation (Rouch et al. 1995).

1.4.2.2 Phytoremediation of Heavy Metal Polluted Soil

Phytoremediation is an integrated multidisciplinary approach to the cleanup of contaminated soils, which combines the disciplines of plant physiology, soil chemistry, and soil microbiology (Cunningham and Ow 1996). These contaminants include heavy metals, radionuclides, chlorinated solvents, petroleum hydrocarbons, PCBs, PAHs, organophosphate insecticides, explosives, and surfactants (Khan et al. 2004).

This technique involves the use of green plants to decontaminate soil, water, and air. Its application spans through the remediation of both organic and inorganic pollutants (Lone et al. 2008).

Certain species of higher plants can accumulate very high concentration of metals in their tissues without showing toxicity (Klassen et al. 2000; Bennett et al. 2003). Such plants can be used successfully to clean up heavy metal polluted soils. For having a feasible cleanup method, the plants must (1) extract large concentrations of heavy metals into their roots, (2) translocate the heavy metal into the surface biomass, and (3) produce a large quantity of plant biomass. In addition, remediative plants must have mechanism(s) to detoxify and/or tolerate high metal concentrations accumulated in their shoots. In the natural setting, certain plants that have the potential of uptaking heavy metals have been identified. Indian mustard (*B. juncea*) is a high biomass, rapidly growing plant that has an ability to accumulate Ni and Cd in its shoots. It is a promising plant for phytoremediation (Terry et al. 1992).

In the following section, we mention different categories of phytoremediation such as phytoextraction, phytostabilization, rhizofiltration, and phytovolatilization (Lone et al. 2008).

Phytoextraction

This technology involves the extraction of metals by plant roots and the translocation thereof to shoots. The roots and shoots are subsequently harvested to remove the contaminants from the soil. With successive cropping and harvesting, the levels of contaminants in the soil can be reduced (Vandenhove et al. 2001). Usually, the shoot biomass is harvested for proper disposal in special site or is burnt to recover the metal (Bennett et al. 2003; Islam et al. 2007; Peciulytė et al. 2006). Researchers at the University of Florida have discovered the ability of the Chinese brake fern, *P. vittata*, to hyperaccumulate arsenic (Ma et al. 2001a). Sunflower, *Halianthus annus*, has proved effective in the remediation of radionuclides and certain other heavy metals (Schnoor 1997).

Phytostabilization

Phytostabilization, also referred to as in-place inactivation, is primarily used for the remediation of soil, sediment, and sludges (United States Protection Agency 2000). It is the use of plant roots to limit contaminant mobility and bioavailability in the soil. The plants' primary purposes are to (1) decrease the amount of water percolating through the soil matrix, which may result in the formation of a hazardous leachate, (2) act as a barrier to prevent direct contact with the contaminated soil, and (3) prevent soil erosion and the distribution of the toxic metal to other areas (Raskin and Ensley 2000). Phytostabilization can occur through the sorption, precipitation, complexation, or metal valence reduction. It is useful for the treatment of lead (Pb) as well as arsenic (As), cadmium (Cd), chromium (Cr), copper (Cu), and zinc (Zn). Some of the advantages associated with this technology are that the disposal of hazardous material/biomass is not required (United States Protection Agency 2000) and it is very effective when rapid immobilization is needed to preserve groundwater and surface water.

Rhizofiltration

Rhizofiltration is primarily used to remediate extracted groundwater, surface water, and wastewater with low contaminant concentrations (Ensley 2000). It is defined as the use of plants, both terrestrial and aquatic, to absorb, concentrate, and precipitate contaminants from polluted aqueous sources in their roots. Rhizofiltration can be used for Pb, Cd, Cu, Ni, Zn, and Cr, which are primarily retained within the roots (United States Protection Agency 2000; Jing et al. 2007). The advantages associated with rhizofiltration are the ability to use both terrestrial and aquatic plants for either in situ or ex situ applications. Another advantage is that contaminants do not have to be translocated to the shoots. Thus, species other than hyperaccumulators may be used. Terrestrial plants are preferred because they have a fibrous and much longer root system, increasing the amount of root area (Raskin and Ensley 2000). Dushenkov et al. (1995) observed that roots of many hydroponically grown terrestrial plants such as Indian mustard (*B. juncea* L.) and sunflower (*H. annuus* L.) effectively removed the potentially toxic metals, Cu, Cd, Cr, Ni, Pb, and Zn, from aqueous solutions. An experiment on rhizofiltration by Karkhanis et al. (2005) was conducted in a greenhouse, using *Pistia*, duckweed, and water hyacinth (*Eichornia crassipes*), to remediate aquatic environment contaminated by coal ash containing heavy metals.

Phytovolatilization

Phytovolatilization is the uptake and release into the atmosphere of volatile material such as mercury or arsenic containing compound (Jing et al. 2007; Lone et al. 2008). It involves the use of plants to take up contaminants from the soil, transforming them into volatile forms and transpiring them into the atmosphere (United States Protection Agency 2000). This process primarily has been used to remediate Hg^{2+} contaminated soil. The advantage of this method is that the contaminant, mercuric ion, may be transformed into a less toxic substance (i.e., elemental Hg). The disadvantage to this is that the mercury released into the atmosphere is likely to be recycled by precipitation and then redeposited back into lakes and oceans, repeating the production of methyl-mercury by anaerobic bacteria. Indian mustard and canola (*Brassica napus*) may be effective for phytovolatilization of selenium, and, in addition, accumulate the selenium (Bañuelos et al. 1997).

Lombi et al. (2001) suggested two approaches for the phytoextraction of heavy metals. The first is the continuous or natural phytoextraction. This involves the use of natural hyperaccumulator plants with exceptional metal accumulating capacity to remediate the soil.

The second approach suggested (Lombi et al. 2001) is the chemically enhanced phytoextraction. This involves the use of high biomass crops that are induced to take up large amount of metals when their mobility in soil is enhanced by chemical treatment. The chemicals employed are mostly chelating agents such as EDTA, NTA, and citric acid (Lombi et al. 2001).

1.5 Heavy Metal Tolerance Mechanism(s) in Plants

Heavy metals such as Cu and Zn are essential for normal plant growth, although elevated concentrations of both essential and nonessential metals can result in growth inhibition and toxicity symptoms. Some plant species, however, have evolved tolerant races that can survive and thrive on such metalliferous soils, presumably by adapting mechanisms that may also be involved in the general homeostasis of, and constitutive tolerance to, essential metal ions as found in all plants. Plants have a range of potential mechanisms at the cellular level that might be involved in the detoxification and thus tolerance to heavy metal stress. The strategies for avoiding heavy metal buildup are diverse. Extracellularly they include roles for mycorrhizas and for cell wall and extracellular exudates. Tolerance could also involve the plasma membrane, either by reducing the uptake of heavy metals or by stimulating the efflux pumping of metals that have entered the cytosol. Within the protoplast a variety of potential mechanisms exist, for example, for the repair of stress-damaged proteins involving heat shock proteins or metallothioneins, and for the chelation of metals by organic acids, amino acids, or peptides, or their compartmentation away from metabolic processes by transport into the vacuole. This range of mechanisms is summarized in Fig. 1.3.

1.5.1 Extracellular Avoidance of Metal Buildup

1.5.1.1 Mycorrhizas

Although mycorrhizas are not always considered as plant metal tolerance mechanisms, they, and particularly ectomycorrhizas that are characteristic of trees and shrubs, can be effective in ameliorating the effects of metal toxicity on the host plant (Marschner 1995; Hüttermann et al. 1999; Jentschke and Godbold 2000). However, the mechanisms involved in conferring this increase in tolerance are difficult to understand as they may be quite diverse. Since large differences in response to metals have been observed, both between fungal species and to different metals within a species, these mechanisms show considerable species and metal specificity (Hartley et al. 1997; Hüttermann et al. 1999). For example, Colpaert and van Assche (1992) showed that the ectomycorrhizal fungus *Paxillus involutus* retained Zn and that this reduced the Zn content of *Pinus sylvestris*, whereas another species *Thelephora terrestris* retained little Zn and even increased the Zn content of the host (Colpaert and van Assche 1992). The mechanisms employed by the fungi at the cellular level to tolerate heavy metals are probably similar to some of the

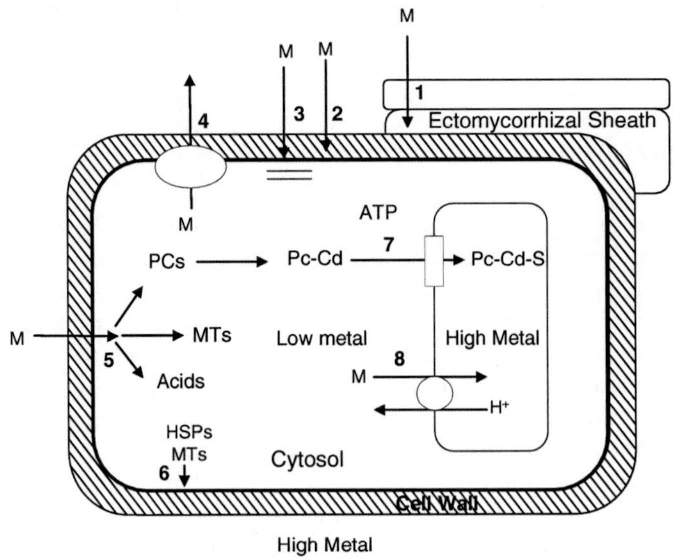

Fig. 1.3 Summary of potential cellular mechanisms available for metal detoxification and tolerance in higher plants. 1. Restriction of metal movement to roots by mycorrhizas. 2. Binding to cell wall and root exudates. 3. Reduced influx across plasma membrane. 4. Active efflux into apoplast. 5. Chelation in cytosol by various ligands. 6. Repair and protection of plasma membrane under stress condition. 7. Transport of PC–Cd complex into the vacuole. 8. Transport and accumulation of metals in vacuole (modified from Marschner 1995)

strategies employed by higher plants, namely binding to extracellular materials or sequestration in the vacuolar compartment. Thus in the fungus *Pisolithus tinctorius*, tolerance to Cu and Zn was achieved by binding to extrahyphal slime (Tam 1995), whereas detoxification of Cd in *P. involutus* involved binding of Cd to the cell walls and accumulation of Cd in the vacuole (Blaudez et al. 2000).

In relation to the role of ectomycorrhizas in metal tolerance by the host plant, most mechanisms that have been proposed involve various exclusion processes that restrict metal movement to the host roots. These have been extensively reviewed and assessed (Jentschke and Godbold 2000) and include absorption of metals by the hyphal sheath, reduced access to the apoplast due to the hydrophobicity of the fungal sheath, chelation by fungal exudates, and adsorption onto the external mycelium. Clearly, from the variation between species described above, these different exclusion mechanisms are likely to vary in significance between different plant and fungal interactions.

There are fewer reports on the role played by arbuscular mycorrhizas in metal tolerance. Weissenhorn et al. (1995) showed that the effects of maize root colonization by arbuscular mycorrhiza could either reduce the heavy metal content of the plants or increase metal absorption from polluted soils, depending on growth conditions, the fungus, and the metal. However, a *Glomus* isolate (Br1) obtained from zinc violets (*Viola calaminaria*) growing on a heavy metal soil was shown to support the growth of maize and alfalfa on heavy metal soils more effectively than a commonly used *Glomus* isolate (Hildebrandt et al. 1999).

1.5.1.2 The Cell Wall and Root Exudates

The binding properties of the cell wall and the role on metal tolerance have been controversial. Although the root cell wall is directly in contact with metals in the soil solution, adsorption onto the cell wall must be of limited capacity and thus have a limited effect on metal activity at the surface of the plasma membrane. It is also difficult to explain metal-specific tolerance by such a mechanism (Ernst et al. 1992). However, Bringezu et al. (1999) reported that the heavy metal-tolerant *Silene vulgaris* ssp. *humilis* accumulated a range of metals in the epidermal cell walls, either bound to a protein or as silicates.

One related process concerns the role of root exudates in metal tolerance. Root exudates have a variety of roles (Marschner 1995) including that of metal chelators that may enhance the uptake of certain metals. In an investigation into the role of Ni-chelating exudates in Ni-hyperaccumulating plants, it was observed that the Ni-chelating histidine and citrate accumulated in the root exudates of nonhyper-accumulating plants and thus could help to reduce Ni uptake and so play a role in a Ni-detoxification strategy (Salt et al. 2000). Since the range of compounds exuded is wide, other exudates could play a role in tolerance to other metals. The clearest example of a role for root secretions in tolerance is in relation to organic acids and the detoxification of the light metal Al (Ma et al. 2001b). Buckwheat, for example, secretes oxalic acid from the roots in response to Al stress and accumulates nontoxic Al-oxalate in the leaves (Ma et al. 1997); thus, detoxification occurs both externally and internally.

1.5.1.3 Plasma Membrane

The plant plasma membrane may be regarded as the first 'living' structure that is a target for heavy metal toxicity. Plasma membrane function may be rapidly affected by heavy metals as seen by an increased leakage from cells in the presence of high concentrations of metals, particularly of Cu. For example, it was shown that Cu, but not Zn, caused increased K^+ efflux from excised roots of *Agrostis capillaris* (Wainwright and Woolhouse 1977). Similarly, others concluded that damage to the cell membrane, monitored by ion leakage, was the primary cause of Cu toxicity in roots of *Silene vulgaris*, *Mimulus guttatus*, and wheat, respectively (De Vos et al. 1991; Strange and Macnair 1991; Quartacci et al. 2001). Certainly direct effects of Cu and Cd treatments on the lipid composition of membranes have been reported (Ros et al. 1990; Fodor et al. 1995; Hernandez and Cooke 1997; Quartacci et al. 2001), which may have a direct effect on membrane permeability. In addition, Cd treatments have been shown to reduce the ATPase activity of the plasma membrane fraction of wheat and sunflower roots (Fodor et al. 1995), while, in *Nitella*, Cu-induced changes in cell permeability were attributed to nonselective conductance increases and inhibition of the light-stimulated H^+-ATPase pump (Demidchik et al. 1997).

Thus, tolerance may involve the protection of plasma membrane integrity against heavy metal damage that would produce increased leakage of solutes from cells (De Vos et al. 1991; Strange and Macnair 1991; Meharg 1993). However, there is little evidence to show how this might be achieved. For example, metal-tolerant plants do not appear to possess enhanced tolerance to free radicals or reactive oxygen species, but rather rely on improved mechanisms for metal homeostasis (Dietz et al. 1999). Again these effects on membranes are metal-specific since, in contrast to Cu, Zn protects membranes against oxidation and generally does not cause membrane leakage (Ernst et al. 1992; Cakmak 2000). Another factor that may be involved in the maintenance of plasma membrane integrity in the presence of heavy metals could be enhanced membrane repair after damage (Salt et al. 1998). This could involve heat shock proteins or metallothioneins, and evidence for this is discussed in the following sections.

Apart from tolerance involving a more resistant plasma membrane or improved repair mechanisms, the cell membrane may play an important role in metal homeostasis, either by preventing or reducing entry into the cell or through efflux mechanisms. Many of these cations, of course, are essential and so complete exclusion is not possible; selective efflux may be more realistic. In bacteria, most resistance systems are based on the energy-dependent efflux of toxic ions (Silver 1996). It appears that the metabolic penalty for having more specific uptake mechanisms, and thus restricting the entry of toxic ions, is greater than that of having inducible efflux systems (Silver 1996).

The number of examples of exclusion or reduced uptake mechanisms in higher plants is quite limited. The clearest example of reduced uptake as an adapted tolerance mechanism is in relation to arsenic toxicity (Meharg and Macnair 1990, 1992). In *Holcus lanatus* roots, phosphate and arsenate appear to be taken up by the same systems. However, an arsenate-tolerant genotype showed a much lower rate of uptake for both anions than the nontolerant genotype, and also showed an absence of the high-affinity uptake system. The altered phosphate and arsenate uptake system was genetically correlated to arsenate tolerance (Meharg and Macnair 1992). More recently, a plasma membrane transporter in tobacco that confers Ni tolerance and Pb hypersensitivity has been described (Arazi et al. 1999). The transporter, designated Nt CBP4, is a calmodulin-binding protein that is structurally similar to certain K^+ and nonselective cation channels. Transgenic plants that overexpressed this transporter showed improved Ni tolerance and hypersensitivity to Pb, which were associated with reduced Ni accumulation and enhanced Pb accumulation.

An alternative strategy for controlling intracellular metal levels at the plasma membrane involves the active efflux of metal ions; although there is no direct evidence for a role for plasma membrane efflux transporters in heavy metal tolerance in plants, recent research has revealed that plants possess several classes of metal transporters that must be involved in metal uptake and homeostasis in general and thus could play a key role in tolerance. These include the heavy metal CPx-ATPases, the Nramps, and the CDF (cation diffusion facilitator) family

(Williams et al. 2000), and the ZIP family (Guerinot 2000). Recently, a role for the Nramps in Fe and Cd uptake has been reported (Thomine et al. 2000); interestingly, disruption of an *AtNramp 3* gene slightly increased Cd resistance, whereas overexpression resulted in Cd hypersensitivity in Arabidopsis.

1.5.2 Intracellular Detoxification Pathways

1.5.2.1 Heat Shock Proteins

Heat shock proteins (HSPs) characteristically show increased expression in response to the growth of a variety of organisms at temperatures above their optimal growth temperature. They are found in all groups of living organisms, can be classified according to molecular size, and are now known to be expressed in response to a variety of stress conditions including heavy metals (Vierling 1991; Lewis et al. 1999); they act as molecular chaperones in normal protein folding and assembly, but may also function in the protection and repair of proteins under stress conditions.

Several authors have reported an increase of HSP expression in plants in response to heavy metal stress. Tseng et al. (1993) showed that, in rice, both heat stress and heavy metal stress increased the levels of mRNAs for low molecular mass HSPs (16–20 kDa), while Neumann et al. (1995) indicated that HSP17 is expressed in roots of *Armeria maritima* plants grown on Cu-rich soils. Small heat shock proteins (e.g., HSP17) were also shown to increase in cell cultures of *Silene vulgaris* and *Lycopersicon peruvianum* in response to a range of heavy metal treatments (Wollgiehn and Neumann 1999); however, no or very low amounts of HSPs were found in plants growing on metalliferous soils, suggesting that HSPs are not responsible for the heritable metal tolerance of *Silene*.

1.5.2.2 Phytochelatins

Chelation of metals in the cytosol by high-affinity ligands is potentially a very important mechanism of heavy metal detoxification and tolerance. Potential ligands include amino acids and organic acids, and two classes of peptides, the phytochelatins and the metallothioneins (Rauser 1999; Clemens 2001). The phytochelatins have been the most widely studied in plants, particularly in relation to Cd tolerance (Cobbett 2000; Goldsbrough 2000).

The phytochelatins (PCs) are a family of metal complexing peptides that have a general structure (γ Glu Cys)$_n$ Gly, where $n = 2$–11, and are rapidly induced in plants by heavy metal treatments (Rauser 1995; Zenk 1996; Cobbett 2000; Goldsbrough 2000). PCs are synthesized nontranslationally using glutathione as a substrate by PC synthase (Grill et al. 1989; Rauser 1995), an enzyme that is activated in the presence of metal ions (Cobbett 2000). The genes for PC synthase have now been identified in Arabidopsis and yeast (Clemens et al. 1999; Ha et al. 1999; Vatamaniuk et al. 1999).

It is shown that in *Brassica juncea* Cd accumulation is accompanied by a rapid induction of PC biosynthesis and that the PC content is theoretically sufficient to chelate all Cd taken up; this protects photosynthesis but did not prevent a decline in transpiration rate (Haag-Kerwer et al. 1999). In Arabidopsis, Xiang and Oliver (1998) showed that treatment with Cd and Cu resulted in increased transcription of the genes for glutathione synthesis. Zhu et al. (1999) overexpressed the γ-glutamylcysteine synthetase gene from *E. coli* in *Brassica juncea* resulting in increased biosynthesis of glutathione and PCs and an increased tolerance to Cd.

The final step in Cd detoxification, certainly in the fission yeast and probably in higher plants, involves the accumulation of Cd and PCs in the vacuole (Salt et al. 1998; Schat et al. 2000). This accumulation appears to be mediated by Cd/H^+ antiporter and an ATP-dependent ABC transporter, located at the tonoplast (Salt and Wagner 1993; Salt and Rauser 1995; Rea et al. 1998); the stabilization of the Cd–PC complex in the vacuole involves the incorporation of acid-labile sulfide.

1.5.2.3 Metallothioneins

Higher plants contain two major types of cysteine-rich metal binding peptides: the metallothioneins (MTs) and the phytochelatins. MTs are gene-encoded polypeptides that are usually classified into two groups. Class 1 MTs possess cysteine residues that align with a mammalian (equine) renal MT; Class 2 MTs also possess similar cysteine clusters, but these cannot be easily aligned with Class 1 MTs (de Miranda et al. 1989; Robinson et al. 1993; Prasad 1999). In plants, there is a lack of information concerning the metals likely to be bound by MTs, although Cu, Zn, and Cd have been the most widely studied (Tomsett and Thurman 1988; Robinson et al. 1993; Goldsbrough 2000).

Although MTs can be induced by Cu treatments and there is evidence for a role in heavy metal tolerance in fungi and animals (Hamer 1986), the role of MTs in heavy metal detoxification in plants remains to be established (Zhou and Goldsbrough 1994; Zenk 1996; Giritch et al. 1998; Schat et al. 2000). However, it has been reported that MT2 mRNA was strongly induced in Arabidopsis seedlings by Cu, but only slightly by Cd and Zn (Zhou and Goldsbrough 1994); when genes for MT1 and MT2 from Arabidopsis were expressed in an MT-deficient yeast mutant, both genes complemented the mutation and provided a high level of resistance to Cu. Van Vliet et al. (1995) showed that MT genes can be induced by Cu and that the expression of MT2 RNA is increased in a Cu-sensitive mutant of Arabidopsis that accumulates high concentrations of Cu. In contrast, in a study of the effects of Cd exposure on *Brassica juncea*, it was reported that MT2 expression was delayed relative to PC synthesis (Haag-Kerwer et al. 1999) and they argued against a role for MT2 in Cd detoxification. MTs could clearly play a role in metal metabolism, but their precise function is not clear; they may have distinct functions for different metals (Hamer 1986).

1.5.2.4 Organic Acids and Amino Acids

Carboxylic acids and amino acids such as citric, malic, and histidine are potential ligands for heavy metals and so could play a role in tolerance and detoxification (for reviews, see Rauser 1999; Clemens 2001); however, strong evidence for a function in tolerance, such as a clear correlation between amounts of acid produced and exposure to a metal, is not yet existing. For example, a 36-fold increase was reported in the histidine content of the xylem sap on exposure to Ni in the Ni-hyperaccumulating plant *Alyssum lesbiacum* (Krämer et al. 1996). In addition, supplying histidine to a nonaccumulating species greatly increased both its Ni tolerance and the capacity for Ni transport to the shoot. However, the histidine response may not be a widespread mechanism of Ni tolerance since it was not observed in another Ni hyperaccumulator, *Thlaspi goesingense* (Persans et al. 1999). A possible role of the histidine in the root exudates as a Ni detoxifying agent has been discussed earlier (see Sect. 1.5.1.2).

1.5.2.5 Antioxidative Defense Mechanism

Buildup of toxic concentration of heavy metals within the plant tissues results, at some stage of stress exposure, in an increased formation of reactive oxygen species (ROS) (Shah et al. 2001; Verma and Dubey 2003). In general, ROS (O_2^-, OH, and H_2O_2) are products of normal cellular metabolism, production of which is under tight control due to cellular antioxidative defense system. Presence of ROS causes oxidative damage to biomolecules such as lipids, proteins, nucleic acids, etc. (Shah et al. 2001; Blokhina et al. 2003). Induction in the activities of antioxidative enzymes and increase in the level of nonenzymic antioxidants are strategies that plants have adopted to scavenge and to reduce oxidative damage caused due to ROS under heavy metal stress (Shah et al. 2001; Fecht-Christoffers et al. 2003; Verma and Dubey 2003).

Recent reports suggest that though antioxidative defense system is not directly involved in heavy metal detoxification, yet ROS play important role as intermediate signaling molecules to regulate the expression of genes for plant's defense system (Orozco-Cardenas et al. 2001; Vranova et al. 2002).

1.5.2.6 Heavy Metal Sequestration

In Vacuoles

Various metabolites and ions are stored inside the vacuoles. Vacuolar sequestration of a number of heavy metals such as Cd, Ni, As, and Zn is known, which diverts metal ions from metabolically active compartments (cytosol, chloroplasts, and mitochondria) and minimizes the harmful effects of metal ions to vital cellular processes. Transporters are present in internal membranes to allow regulation of stored metals in organelles. Active accumulation of most of the metal ions is driven by the electrochemical potential by electrogenic proton influxes via the vacuolar H^+-ATPase (Kakinuma et al. 1993). Cd is transported across the tonoplast by a Cd^2/H^+ antiport mechanism (Carrier et al. 2003). Both vacuolar (Bidwell et al. 2004) and extravacuolar localization of Ni ion occurs via a pH-gradient-dependent manner in yeast (Nishimura et al. 1998), whereas in Ni-hyperaccumulator plant *Thlaspi goesingense* vacuolar metal transport proteins termed as metal-tolerance proteins (TgMTPs) are involved in the compartmentalization of Ni in vacuoles (Lombi et al. 2002). Transport of Zn to the vacuoles is mediated by 'Zn-malate shuttle', malate being liberated in exchange for oxalate or citrate and is shuttled back to cytoplasm (Ernst et al. 1992).

In Trichome and Hydropotes

Apart from vacuolar sequestration, plants possess additional morphological features that are also involved in heavy metal sequestration and detoxification. Several reports have confirmed the involvement of glandular trichomes and epidermal structures (hydropotes) in the chelation, sequestration, and detoxification of the metals.

Trichomes are epidermal hairs present at the surface of plant leaves and have diversified roles in exudation of various molecules, protection against wind and sunlight, storage of metals, etc. Retardation in growth and about twofold increases in the number of trichomes were observed in Cd-exposed tobacco seedlings (Choi et al. 2001). A significant proportion of Ni has been found in trichomes of *Alyssum lesbiacum* plants (Krämer et al. 1996). At the bases of *Arabidopsis halleri* trichomes, elevated concentrations of Zn have been found (Sarret et al. 2002). Specific overexpression of a gene coding for a metallothionein (MT2) has been reported in trichomes (Garcia–Hernandez et al. 1998), which suggests that trichomes constitute important sites for accumulation as well as detoxification of toxic metal ions.

In the semiaquatic and aquatic plants of the families Menyanthaceae and Nymphaceae, hydropotes located on the abaxial epidermis of the leaf laminae accumulate Cd (Lavid et al. 2001a). It is suggested that usual polymerization of polyphenols by peroxidase in hydropotes gets enhanced after uptake of heavy metals and thereby detoxification of metals occurs by their binding with polyphenols in these glands (Lavid et al. 2001b, c).

1.6 Conclusion

The pollution of soil and water with heavy metals is an environmental concern today. Metals and other inorganic contaminants are among the most prevalent forms of contamination found at waste sites, and their remediation in soils and sediments is among the most technically difficult. The high cost of existing cleanup technologies led to the search for new cleanup strategies that have the potential to be low cost, low impact, visually benign, and environmentally sound.

Phytoremediation is a potential remediation strategy that can be used to decontaminate soils contaminated with inorganic pollutants. But phytoremediation technology is still in its early development stages and full-scale applications are still limited. For widespread future use of this technique, it is important that public awareness about this technology is considered and clear and precise information is made available to the general public to enhance its acceptability as a global sustainable technology to be widely used.

The recent advances in plant biotechnology have created a new hope for the development of hyperaccumulating species. However, research work is needed in this respect such as metal uptake studies at cellular level including efflux and influx of different metal ions by different cell organelles and membranes. Rhizosphere studies under the control and field conditions are also needed to examine the antagonistic and synergistic effects of different metal ions in soil solution and the polluted waters. In-depth soil microbial studies are required to identify the microorganisms highly associated with metal solubility and/or precipitations. To date, the available methods for the recovery of heavy metals from plant biomass of hyperaccumulators are limited. Traditional disposal approaches such as burning and ashing are not applicable to volatile metals; therefore, investigations are needed to develop new methods and strategies for effective recovery of metals from the hyperaccumulator plant biomass.

New approaches of genetic manipulation techniques increase the possibility of creating super-accumulator bacteria or plants that can help decontaminate polluted soils, water, and sewage.

References

Alloway BJ (1995) Heavy metals in soils. Blackie Academic and Professional, London

Alloway BJ, Jackson AP (1991) The behaviour of heavy metals in sewage sludge-amended soil. Sci Total Environ 100:151–176

Arazi T, Sunkar R, Kaplan B, Fromm H (1999) A tobacco plasma membrane calmodulin-binding transporter confers Ni^{2+} tolerance and Pb^{2+} hypersensitivity in transgenic plants. Plant J 20:171–182

Bañuelos GS, Ajwa HA, Mackey B, Wu LL, Cook C, Akohoue S, Zambrzuski S (1997) Evaluation of different plant species used for phytoremediation of high soil selenium. J Environ Qual 26(3):639–646

Barber SA (1984) Soil nutrient bioavailability. A mechanistic approach. Wiley, New York, pp 188–238

Belliveau BH, Staradub ME, Trevor JT (1991) Occurrence of antibiotic and metal resistance and plasmids in *Bacillus* strains isolated from marine sediment. Can J Microbiol 37(5):513–520

Bennett LE, Burkhead JL, Hale KL, Terry N, Pilon M, Pilon-Smits EAH (2003) Bioremediation and biodegradation: analysis of transgenic Indian mustard plant for phytoremediation of heavy metal-contaminated mine tailings. J Environ Qual 32:432–440

Beveridge TJ, Murray GE (1976) Uptake and retention of metals by cell walls of *Bacillus subtilis*. J Bacteriol 127:1502–1518

Bidwell SD, Crawford SA, Woodrow IE, Sommer-Knudsen J, Marshall AT (2004) Subcellular localization of Ni in the hyperaccumulator, *Hybanthus floribundus* (Lindely) F. Muell. Plant Cell Environ 27:705–716

Blaudez D, Botton B, Chalot M (2000) Cadmium uptake and subcellular compartmentation in the ectomycorrhizal fungus *Paxillus involutus*. Microbiology 146:1109–1117

Blokhina O, Virolainen E, Fegerstedt KV (2003) Antioxidants, oxidative damage and oxygen deprivation stress: a review. Ann Bot 91:179–194

Bowen HJM (1979) Environmental chemistry of the elements, vol 147. Academic, London, pp 6–10

Bringezu K, Lichtenberger O, Leopold I, Neumann D (1999) Heavy metal tolerance of *Silene vulgaris*. J Plant Physiol 154:536–546

Cakmak I (2000) Possible roles of zinc in protecting plant cells from damage by reactive oxygen species. New Phytol 146:185–205

Carrier P, Baryla A, Havaux M (2003) Cadmium distribution and microlocalization in oilseed rape (*Brassica napus*) after long-term growth on cadmium-contaminated soil. Planta 216:939–950

Chaudri AM, McGrath SP, Giller KE (1992) Survival of the indigenous population of *Rhizobium leguminosarum* biovar trifolii in soil spiked with Cd, Zn, Cu and Ni salts. Soil Biol Biochem 24:625–632

Choi YE, Harada E, Wada M, Tsuboi H, Morita Y, Kusano T, Sano H (2001) Detoxification of cadmium in tobacco plants: formation and active excretion of crystals containing cadmium and calcium through trichomes. Planta 213:45–50

Clemens S (2001) Molecular mechanisms of plant metal tolerance and homeostasis. Planta 212:475–486

Clemens S, Kim EJ, Neumann D, Schroeder JI (1999) Tolerance to toxic metals by a gene family of phytochelatin synthases from plants and yeast. EMBO J 18:3325–3333

Cobbett CS (2000) Phytochelatin biosynthesis and function in heavy-metal detoxification. Curr Opin Plant Biol 3:211–216

Colpaert J, van Assche J (1992) Zinc toxicity in ectomycorrhizal *Pinus sylvestris*. Plant Soil 143:201–211

Conder JM, Lanno RP, Basta NT (2001) Assessment of metal availability in smelter soil using earthworms and chemical extractions. J Environ Qual 30:1231–1237

Cunningham SD, Ow DW (1996) Promises and prospects of root zone of crops phytoremediation. Plant Physiol 110:715–719

Cunningham SD, Shann JR, Crowley DE, Anderson TA (1997) Phytoremediation of contaminated water and soil. In: Kruger EL, Anderson TA, Coats JR (eds) Phytoremediation of soil and water contaminants. ACS symposium series 664. American Chemical Society, Washington, DC, pp 2–19

Davis RD, Calton-Smith C (1980) Crop as indicators of the significance of contamination of soil by heavy metals; Technical Report TR140. Water Research Centre, Stevenage, UK

de Miranda JR, Thomas MA, Thurman DA, Tomsett AB (1989) Metalliothionein genes from the flowering plant *Mimulus guttatus*. FEBS Lett 260:277–280

De Vos CHR, Schat H, De Waal MAM, Vooijs R, Ernst WHO (1991) Increased resistance to copper-induced damage of the root cell plasmalemma in copper tolerant *Silene cucubalus*. Physiol Plant 82:523–528

Demidchik V, Sokolik A, Yurin V (1997) The effect of Cu^{2+} on ion transport systems of the plant cell plasmalemma. Plant Physiol 114:1313–1325

Dietz K-J, Baier M, Krämer U (1999) Free radicals and reactive oxygen species as mediators of heavy metal toxicity in plants. In: Prasad MNV, Hagemeyer J (eds) Heavy metal stress in plants: from molecules to ecosystems. Springer, Berlin, pp 73–97

Doelman P, Haanstra L (1979) Effect of lead on soil respiration and dehydrogenase activity. Soil Biol Biochem 11:475–479

Dushenkov V, Nanda Kumar PBA, Motto H, Raskin I (1995) Rhizofiltration: the use of plants to remove heavy metals from aqueous streams. Environ Sci Technol 29:1239–1245

Ensley BD (2000) Rationale for the use of phytoremediation. In: Raskin I, Ensley B (eds) Phytoremediation of toxic metals: using plants to clean-up the environment. Wiley, New York, pp 3–11

Ernst WHO, Verkleij JAC, Schat H (1992) Metal tolerance in plants. Acta Bot Neerl 41:229–248

Fecht-Christoffers MM, Maier P, Horst WJ (2003) Apoplastic peroxidases and ascorbate are involved in manganese toxicity and tolerance of *Vigna unguiculata*. Physiol Plant 117:237–244

Fodor E, Szabó-Nagy A, Erdei L (1995) The effects of cadmium on the fluidity and H^+-ATPase activity of plasma membrane from sunflower and wheat roots. J Plant Physiol 147:87–92

Garbisu C, Alkorta I (2001) Phytoextraction: a cost effective plant-based technology for the removal of metals from the environment. Bioresour Technol 77(3):229–236

Garcia–Hernandez M, Murphy A, Taiz L (1998) Metallothioneins 1 and 2 have distinct but overlapping expression patterns in Arabidopsis. Plant Physiol 118:387–397

Giritch A, Ganal M, Stephan UW, Baumlein H (1998) Structure, expression and chromosomal localization of the metallothionein-like gene family of tomato. Plant Mol Biol 37:701–714

Gisbert C, Ros R, de Haro A, Walker DJ, Pilar Bernal M, Serrano R, Avino JN (2003) A plant genetically modified that accumulates Pb is especially promising for phytoremediation. Biochem Biophys Res Commun 303(2):440–445

Goldsbrough P (2000) Metal tolerance in plants: the role of phytochelatins and metallothioneins. In: Terry N, Banuelos G (eds) Phytoremediation of contaminated soil and water. CRC, Boca Raton, FL, pp 221–233

Graham RD (1981) Absorption of copper by plant roots. In: Loneragan JF, Robson AD, Graham RD (eds) Copper in soils and plants. Academic, Sydney, pp 141–163

Grill E, Loffler S, Winnacker E-L, Zenk MH (1989) Phytochelatins, the heavy-metal-binding peptides of plants are synthesized from glutathione by a specific γ-glutamylcysteine dipeptidyl transpeptidase (phytochelatin synthase). Proc Natl Acad Sci USA 86:6838–6842

Guerinot ML (2000) The ZIP family of metal transporters. Biochim Biophys Acta 1465:190–198

Ha SB, Smith AP, Howden R, Dietrich WM, Bugg S, O'Connell MJ, Goldsbrough PB, Cobbett CS (1999) Phytochelatin synthase genes from Arabidopsis and the yeast *Schizosaccharomyces pombe*. Plant Cell 11:1153–1163

Haag-Kerwer A, Schäfer HJ, Heiss S, Walter C, Rausch T (1999) Cadmium exposure in *Brassica juncea* causes a decline in transpiration rate and leaf expansion without effect on photosynthesis. J Exp Bot 50:1827–1835

Hamer DH (1986) Metallothionein. Annu Rev Biochem 55:913–951

Harnett NM, Gyles CL (1984) Resistance to drugs and heavy metals, colicin production, and biochemical characteristics of selected bovine and porcine *Escherichia coli* strains. Appl Environ Microbiol 48:930–945

Hartley J, Cairney JWG, Meharg AA (1997) Do ectomycorrhizal fungi exhibit adaptive tolerance to potentially toxic metals in the environment? Plant Soil 189:303–319

Hattori H (1992) Influence of heavy metals on soil microbial activities. Soil Sci Plant Nutr 38 (1):93–100

Hernandez LE, Cooke DT (1997) Modification of the root plasma membrane lipid composition of cadmium-treated *Pisum sativum*. J Exp Bot 48:1375–1381

Hildebrandt U, Kaldorf M, Bothe H (1999) The zinc violet and its colonization by arbuscular mycorrhizal fungi. J Plant Physiol 154:709–711

Hoyle B, Beveridge TS (1983) Binding of metallic ions to the outer membrane of *Escherichia coli*. Appl Environ Microbiol 46:749–752

Hüttermann A, Arduini I, Godbold DL (1999) Metal pollution and forest decline. In: Prasad NMV, Hagemeyer J (eds) Heavy metal stress in plants: from molecules to ecosystems. Springer, Berlin, pp 253–272

Islam E, Yang X, Li T, Liu D, Jin X, Meng F (2007) Effect of Pb toxicity on root morphology, physiology and ultrastructure in the two ecotypes of *Elsholtzia argyi*. J Hazard Mater 147:806–816

Jentschke G, Godbold DL (2000) Metal toxicity and ectomycorrhizas. Physiol Plant 109:107–116
Ji G, Silver S (1995) Bacterial resistance mechanism for heavy metals of environmental concern. J Ind Microbiol 14:61–75
Jing Y, He Z, Yang X (2007) Role of soil rhizobacteria in phytoremediation of heavy metal contaminated soils. J Zhejiang Univ Sci B 8:192–207
Joho M, Inouhe M, Tohoyama H, Murayama T (1995) Nickel resistance in yeast and other fungi. J Ind Microbiol 14:64–168
Kabata-Pendias A, Pendias H (1992) Trace elements in soils and plants, 2nd edn. CRC, Boca Raton, FL
Kakinuma Y, Masuda N, Igarashi K (1993) Proton potential-dependent polyamine transport system in vacuolar membrane vesicles of *Saccharomyces cerevisiae*. Biochim Biophys Acta 1107:126–130
Karkhanis M, Jadia CD, Fulekar MH (2005) Rhizofilteration of metals from coal ash leachate. Asian J Water Environ Pollut 3(1):91–94
Khan FI, Hussain T, Hejazi R (2004) An overview and analysis of site remediation technologies. J Environ Manage 71:95–122
Klassen SP, McLean JE, Grossel PR, Sims RC (2000) Fate and behavior of lead in soils planted with metal-resistant species (River birch and smallwing sedge). J Environ Qual 29:1826–1834
Krämer U, Cotter-Howells JD, Charnock JM, Baker AJM, Smith JAC (1996) Free histidine as a metal chelator in plants that accumulate nickel. Nature 379:635–638
Krauskopf KB (1967) Introduction to geochemistry. McGraw Hill, New York
Lavid N, Schwartz A, Lewinsohn E, Tel-Or E (2001a) Phenol and phenol oxidases are involved in cadmium accumulation in the water plants *Nymphoides peltata* (Menyanthaceae) and *Nymphaeae* (Nymphaeaeceae). Planta 214:189–195
Lavid N, Barkay Z, Tel-Or E (2001b) Accumulation of heavy metals in epidermal glands of the waterlily (Nymphaeaeceae). Planta 212:313–322
Lavid N, Schwartz A, Yarden O, Tel-Or E (2001c) The involvement of polyphenols and peroxidase activities in heavy-metal accumulation by epidermal glands of the waterlily (Nymphaeaeceae). Planta 212:323–331
Lewis S, Handy RD, Cordi B, Billinghurst Z, Depledge MH (1999) Stress proteins (HSPs): methods of detection and their use as an environmental biomarker. Ecotoxicology 8:351–368
Lombi E, Zhao FJ, Dunham SJ, McGrath SP (2001) Phytoremediation of heavy metal-contaminated soils: natural hyperaccumulator versus chemically enhanced phytoextraction. J Environ Qual 30:1919–1926
Lombi E, Zhao FJ, Fuhrman M, Ma L, McGrath SP (2002) Arsenic distribution and speciation in the fronds of the hyperaccumulator *Pteris vittata*. New Phytol 156:195–203
Lone MI, He Z, Stoffella PJ, Yang X (2008) Phytoremediation of heavy metal polluted soils and water: progress and perspectives. J Zhejiang Univ Sci B 9:210–220
Ma JF, Zheng SJ, Matsumoto H (1997) Detoxifying aluminium with buckwheat. Nature 390:569–570
Ma LQ, Komar KM, Tu C, Zhang W, Cai Y, Kennelley ED (2001a) A fern that hyperaccumulates arsenic. Nature 409:579
Ma JF, Ryan PR, Delhaize E (2001b) Aluminium tolerance in plants and the complexing role of organic acids. Trends Plant Sci 6:273–278
MAFF (1986) Advice on avoiding pollution from manures and other slurry wastes, Maff Booklet 2200. MAFF Publication, London
Marques AM, Congregado F, Simon-Pajol DM (1979) Antibiotic and heavy metal resistance of *Pseudomonas aeruginosa* isolated form soils. J Appl Bacteriol 47:347–350
Marschner H (1986) Mineral nutrition in higher plants. Academic, London
Marschner H (1995) Mineral nutrition in higher plants, 2nd edn. Academic, London
McEntee JD, Woodrow JR, Quirk AV (1986) Investigation of cadmium resistance in *Alcaligenes* sp. Appl Environ Microbiol 51:515–520

McGrath SP, Chang AC, Page AL, Witter A (1994) Land application of sewage sludge: scientific perspectives of heavy metal loading limits in Europe and the United States. Environ Rev 2:108–118

McGrath SP, Zhao FJ, Lombi E (2001) Plant and rhizosphere process involved in phytoremediation of metal-contaminated soils. Plant Soil 232(1/2):207–214

Meharg AA (1993) The role of the plasmalemma in metal tolerance in angiosperms. Physiol Plant 88:191–198

Meharg AA, Macnair MR (1990) An altered phosphate uptake system in arsenate-tolerant *Holcus lanatus*. New Phytol 116:29–35

Meharg AA, Macnair MR (1992) Genetic correlation between arsenate tolerance and the rate of influx of arsenate and phosphate in *Holcus lanatus*. Heredity 69:336–341

Mench M, Martin E (1991) Mobilization of cadmium and other metals from two soils by root exudates of *Zea mays* L., *Nicotiana tobaccum* L., and *Nicotiana rustica*L. Plant Soil 132:187–196

Mergeay M (1991) Towards an understanding of the genetics of bacterial metal resistance. Trends Biotechnol 9:17–24

Misra TK (1992) Bacterial resistance to inorganic mercury salts and organomercurials. Plasmid 27:4–16

Mitchell RL (1964) Trace elements in soil. In: Bear FE (ed) Chemistry of the soil, 2nd edn. Reinhold Publishing, New York, pp 320–368

Murata K, Fukuda Y, Shimosaka M, Wantanabe K, Saikusa T, Kimura A (1985) Phenotypic character of the methylglycoxal resistance gene in *Saccharomyces cerevisae*: expression in *Escherichia coli* and application to breeding wild-type yeast strains. Appl Environ Microbiol 50:1200–1207

Murphy RJ, Levy JF (1983) Production of copper oxalate by some copper tolerant fungi. Trans Br Mycol Soc 81:165–168

Nakahara H, Ishikawa T, Yasunaga S, Kondo I, Kozukue H, Silver S (1977) Linkage of mercury, cadmium, and arsenate and drug resistance in clinical isolates of *Pseudomonas aeruginosa*. Appl Environ Microbiol 33:975–976

Neilson JW, Artiola JF, Maier RM (2003) Characterization of lead removal from contaminated soils by non toxic washing agents. J Environ Qual 32:899–908

Neumann D, Nieden UZ, Lichtenberger O, Leopold I (1995) How does *Armeria maritima* tolerate high heavy metal concentrations? J Plant Physiol 146:704–717

Ni'bhriain NN, Silver S, Foster TJ (1983) Tn5 insertion mutation in the mercuric ion resistance genes derived from plasmid R100-1. J Bacteriol 155:690–703

Nies DH, Silver S (1995) Ion efflux systems involved in bacterial metal resistances. J Ind Microbiol 14:189–199

Nishimura K, Igarashib K, Kakinuma Y (1998) Proton gradient driven nickel uptake by vacuolar membrane vesicles of *Saccharomyces cerevisiae*. J Bacteriol 180:1962–1964

Nriagu JO, Pacyna JM (1988) Quantitative assessment of worldwide contamination of air water and soils by trace metals. Nature 333(6169):134–139

O'Halloran T (1993) Transition metals in control of gene expression. Science 261:715–725

Obbard JP, Jones KC (1993) The effect of heavy metals on dinitrogen fixation by *Rhizobium* – white clover in a range of long-term sewage sludge amended and metal-contaminated soils. Environ Pollut 79(2):105–112

Okoronkwo NE, Igwe JC, Onwuchekwa EC (2005) Risk and health implication of polluted soils for crop production. Afr J Biotechnol 4:1521–1524

Olson BH, Thornton I (1982) The resistance patterns to metals of bacterial populations in contaminated land. J Soil Sci 33:271–277

Orozco-Cardenas ML, Narvaez-Vasquez J, Ryan CA (2001) Hydrogen peroxide act as a second messenger for the induction of defence genes in tomato plants in response to wounding, systemin and methyl jasmonate. Plant Cell 13:179–191

Peciulytė D, Repečkienė J, Levinskaitė L, Lugauskas A, Motuzas A, Prosyčevas I (2006) Growth and metal accumulation ability of plants in soil polluted with Cu, Zn and Pb. Ekologija 1:48–52
Persans MW, Yan X, Patnoe J-MML, Krämer U, Salt DE (1999) Molecular dissection of the role of histidine in nickel hyperaccumulation in *Thlaspi goesingense* (Hálácsy). Plant Physiol 121:1117–1126
Peters WC (1978) Exploration and mining geology. Wiley, New York
Peterson PJ, Alloway BJ (1979) Cadmium in soils and vegetation. In: Webb M (ed) The chemistry, biochemistry and biology of cadmium. Elsevier/North-Holland Biomedical Press, Amsterdam, NY, pp 45–92
Prasad MNV (1999) Metallothioneins and metal binding complexes in plants. In: Prasad MNV, Hagemeyer J (eds) Heavy metal stress in plants: from molecules to ecosystems. Springer, Berlin, pp 51–72
Quartacci MF, Cosi E, Navari-Izzo F (2001) Lipids and NADPH-dependent superoxide production in plasma membrane vesicles from roots of wheat grown under copper deficiency or excess. J Exp Bot 52:77–84
Raskin I, Ensley BD (2000) Rationale for use of phytoremediation. In: Raskin I, Ensley BD (eds) Phytoremediation of toxic metals: using plants to clean up the environment. Wiley, New York, pp 53–70
Rauser WE (1995) Phytochelatins and related peptides. Structure, biosynthesis and function. Plant Physiol 109:1141–1149
Rauser WE (1999) Structure and function of metal chelators produced by plants – the case for organic acids, amino acids, phytin and metallothioneins. Cell Biochem Biophys 31:19–48
Rea PA, Li Z-S, Lu Y-P, Drozdowicz YM (1998) From vacuolar GS-X pumps to multispecific ABC transporters. Annu Rev Plant Physiol Plant Mol Biol 49:727–760
Rensing C, Sun Y, Mitra B, Rosen BP (1998) Pb(II) translocating *p*-type ATPases. J Biol Chem 273:32614–32617
Robinson NJ, Tommey AM, Kuske C, Jackson PJ (1993) Plant metallothioneins. Biochem J 295:1–10
Ros ROC, Cooke DT, Burden RS, James CS (1990) Effects of the herbicide MCPA and the heavy metals, cadmium and nickel on the lipid composition, Mg^{2+-}-ATPase activity and fluidity of plasma membranes from rice, *Oryza sativa* (cv. Bahia) shoots. J Exp Bot 41:457–462
Rouch DA, Lee BTD, Morby AP (1995) Understanding cellular responses to toxic agents: a model for mechanism choice in bacterial metal resistance. J Ind Microbiol 14:132–141
Salt DE, Rauser WE (1995) MgATP-dependent transport of phytochelatins across the tonoplast of oat roots. Plant Physiol 107:1293–1301
Salt DE, Wagner GJ (1993) Cadmium transport across tonoplast of vesicles from oat roots. Evidence for a Cd^{2+}/H^{+} antiport activity. J Biol Chem 268:12297–12302
Salt DE, Smith RD, Raskin I (1998) Phytoremediation. Annu Rev Plant Physiol Plant Mol Biol 49:643–668
Salt DE, Kato N, Krämer U, Smith RD, Raskin I (2000) The role of root exudates in nickel hyperaccumulation and tolerance in accumulator and nonaccumulator species of Thlaspi. In: Terry N, Banuelos G (eds) Phytoremediation of contaminated soil and water. CRC, Boca Raton, FL, pp 189–200
Sarret G, Saumitou-Laprade P, Bert V, Proux O, Hazemann JL, Traverse A, Marcus MA, Manceau A (2002) Forms of zinc accumulated in the hyperaccumulator *Arabidopsis halleri*. Plant Physiol 130:1815–1826
Schalscha BE, Ahumada TI (1998) Heavy metals in rivers and soils of central Chile. Water Sci Technol 37(8):251–255
Schat H, Llugany M, Bernhard R (2000) Metal–specific patterns of tolerance, uptake and transport of heavy metals in hyperaccumulating and nonhyperaccumulating metallophytes. In: Terry N, Banuelos G (eds) Phytoremediation of contaminated soil and water. CRC, Boca Raton, FL, pp 171–188

Schnoor JL (1997) Phytoremediation. Ground-water remediation technology analysis center, Technology evaluation report, TE-98-01. University of Lowa, Department of Civil and Environmental Engineering and Center for Global and Regional Environmental Research Iowa City, Iowa

Schwarz ST, Hobel H (1989) Plasmid and resistance to antimicrobial agents and heavy metals in *Staphylococcus hyicus* from pigs and cattle. J Vet Med 36:669–673

Scott JA, Palmer SJ (1988) Cadmium bio-sorption by bacterial exopolysaccharide. Biotechnol Lett 10:21–24

Scott JA, Palmer SJ (1990) Sites of cadmium uptake in bacteria used for biosorption. Appl Microbiol Biotechnol 33:221–225

Scott JA, Sage GK, Palmer SJ (1988) Metal immobilization by microbial capsular coatings. Biorecovery 1:51–58

Shah K, Kumar RG, Verma S, Dubey RS (2001) Effect of cadmium on lipid peroxidation, superoxide anion generation and activities of antioxidant enzymes in growing rice seedlings. Plant Sci 61:1135–1144

Silver S (1992) Plasmid-determined metal resistance mechanisms: range and overview. Plasmid 27:1–3

Silver S (1996) Bacterial resistance to toxic metal ions. A review. Gene 179:9–19

Silver S, Ji G (1994) Newer systems for bacterial resistance's to toxic heavy metals. Environ Health Perspect 102:107–113

Silver S, Misra TK (1984) Bacterial transformation of and resistances to heavy metals. Basic Life Sci 28:23–46

Silver S, Phung LT (1996) Bacterial heavy metal resistance: new surprises. Annu Rev Microbiol 50:753–789

Silver S, Walderhaug M (1992) Gene regulation and chromosome determined inorganic ion transport in bacteria. Microbiol Rev 56:195–228

Silver S, Nucifors G, Chu L, Misra TK (1989) Bacterial resistance ATPases: primary pumps for exporting toxic cations and anions. Trends Biochem Sci 14:76–80

Strange J, Macnair MR (1991) Evidence for a role for the cell membrane in copper tolerance of Mimulus guttatus Fischer ex DC. New Phytol 119:383–388

Stuczynski TI, Siebielec G, Daniels WL, McCarty GC, Chaney RL (2007) Biological aspects of metal waste reclamation with sewage sludge. J Environ Qual 36:1154–1162

Tam PCF (1995) Heavy metal tolerance by ectomycorrhizal fungi and metal amelioration by *Pisolithus tinctorius*. Mycorrhiza 5:181–187

Terry N, Carlson C, Raab TK, Zayed A (1992) Rates of Se volatilization among crop species. J Environ Qual 21:341–344

Thomine S, Wang R, Ward JM, Crawford NM, Schroeder JI (2000) Cadmium and iron transport by members of a plant metal transporter family in Arabidopsis with homology to Nramp genes. Proc Natl Acad Sci USA 97:4991–4996

Tomsett AB, Thurman DA (1988) Molecular biology of metal tolerances of plants. Plant Cell Environ 11:383–394

Tseng TS, Tzeng SS, Yeh CH, Chang FC, Chen YM, Lin CY (1993) The heat-shock response in rice seedlings-isolation and expression of cDNAs that encode class-I low-molecular-weight heat-shock proteins. Plant Cell Physiol 34:165–168

Tsutomu S, Kobayashi Y (1998) The *ars* operon in the skin element of *Bacillus subtilis* confers resistance to arsenate and arsenite. J Bacteriol 180:1655–1661

Tyler G, Balsberg Pahlsson A-M, Bengtsson G, Baath E, Tranvik L (1989) Heavy-metal ecology of terrestrial plants, microorganisms and invertebrates. Water Air Soil Pollut 47:189–215

United States Protection Agency (USPA) (2000) Introduction to phytoremediation. EPA 600/R-99/107. US Environmental Protection Agency, Office of Research and Development, Cincinnati, OH

van Vliet C, Anderson CR, Cobbett CS (1995) Copper-sensitive mutant of *Arabidopsis thaliana*. Plant Physiol 109:871–878

Vandenhove H, van Hees M, van Winkel S (2001) Feasibility of phytoextraction to clean up low-level uranium-contaminated soil. Int J Phytoremediation 3:301–320

Vatamaniuk OK, Mari S, Lu Y-P, Rea PA (1999) AtPCS1, a phytochelatin synthase from Arabidopsis: isolation and in vitro reconstitution. Proc Natl Acad Sci USA 96:7110–7115

Verma S, Dubey RS (2003) Lead toxicity induces lipid peroxidation and alerts the activities of antioxidant enzymes in growing rice plants. Plant Sci 164:645–655

Vierling E (1991) The roles of heat shock proteins in plants. Annu Rev Plant Physiol Plant Mol Biol 42:579–620

Vranova E, Inze D, Van Breusegem F (2002) Signal transduction during oxidative stress. J Exp Bot 53:1227–1236

Wainwright SJ, Woolhouse HW (1977) Some physiological aspects of copper and zinc tolerance in *Agrostis tenuis* Sibth.: cell elongation and membrane damage. J Exp Bot 28:1029–1036

Wang YT, Shen H (1995) Bacterial reduction of hexavalent chromium. J Ind Microbiol 14:159–163

Webber MD, Kloke A, Tjell JC (1984) A review of current sludge use guidelines for the control of heavy metal contamination in soil. In: L'Hermite P, Ott H (eds) Processing and use of sewage sludge. Reidel Publishing Company, Dordrecht, pp 371–386

Weiss A, Murphy S, Silver S (1977) Mercury and organomercurial resistance determined by plasmids in *Staphylococcus aureus*. J Bacteriol 132:197–208

Weissenhorn I, Leyval C, Belgy G, Berthelin J (1995) Arbuscular mycorrhizal contribution to heavy-metal uptake by maize (Zea mays L.) in pot culture with contaminated soil. Mycorrhiza 5:245–251

Williams LE, Pittman JK, Hall JL (2000) Emerging mechanisms for heavy metal transport in plants. Biochim Biophys Acta 77803:1–23

Wollgiehn R, Neumann D (1999) Metal stress response and tolerance of cultured cells from *Silene vulgaris* and *Lycopersicon peruvianum*: role of heat stress proteins. J Plant Physiol 154:547–553

Xiang C, Oliver DJ (1998) Glutathione metabolic genes coordinately respond to heavy metals and jasmonic acid in Arabidopsis. Plant Cell 10:1539–1550

Zenk MH (1996) Heavy metal detoxification in higher plants – a review. Gene 179:21–30

Zhou J, Goldsbrough PB (1994) Functional homologs of fungal metallothionein genes from Arabidopsis. Plant Cell 6:875–884

Zhu YL, Pilon-Smits EAH, Tarun AS, Weber SU, Jouanin L, Terry N (1999) Cadmium tolerance and accumulation in Indian mustard is enhanced by overexpressing gamma-glutamylcysteine synthetase. Plant Physiol 121:1169–1177

Chapter 2
Plants in Heavy Metal Soils

Hermann Bothe

2.1 Heavy Metal Soils

Heavy metals have a molecular mass >5.0 g cm^{-3} which is distinctly higher than the average particle density of soils (2.65 g cm^{-3}). Several heavy metals such as iron (Fe), manganese (Mn), zinc (Zn), copper (Cu), cobalt (Co), or molybdenum (Mo) are essential for the growth of organisms. Others have a single function and only in some organisms such as vanadium (V) in some peroxidases and in V-nitrogenases or nickel (Ni) in hydrogenases. The remainder of the heavy metals is always toxic to organisms: cadmium (Cd), lead (Pb), uranium (U), thallium (Tl), chromium (Cr), silver (Ag), and mercury (Hg). Arsenic (As) and selenium (Se) are nonheavy metals. However, since they partly share toxicity features with heavy metals, they are often referred to as "metalloids" in publications.

All soils contain heavy metals. In nonheavy metal soils, the concentrations of Zn, Cu, Pb, Ni, Cd and Cr range between 0.0001 and 0.065%, whereas Mn and Fe can reach 0.002% and 10.0%, respectively (Ernst 1974). With the exception of iron, all heavy metals above a concentration of 0.1% in the soil become toxic to plants and therefore change the community structure of plants in a polluted habitat. However, each plant species has a specific threshold value for each heavy metal where it exerts toxicity (Ernst 1982). Plants specifically adapted to life on heavy metal-rich soils ("heavy metal soils") are termed metallophytes.

Zinc-rich soils (0.1–10.0% Zn) often contain a high content of Pb, but virtually no Cd, whereas soils with a high copper content (0.1–3.2% Cu) have higher concentrations of other heavy metals such a Zn, Pb, Co, Ni including Cd (Ernst 1974).

Soils that carry metallophytes can be classified by the content of their main heavy metal: serpentine soils are rich in Ni, seleniferous soils carry Se, calamine

H. Bothe (✉)
Botanical Institute, The University of Cologne, Zülpicher Str. 47b, 50923 Cologne, Germany
e-mail: Hermann.Bothe@uni-koeln.de

I. Sherameti and A. Varma (eds.), *Detoxification of Heavy Metals*, Soil Biology 30,
DOI 10.1007/978-3-642-21408-0_2, © Springer-Verlag Berlin Heidelberg 2011

soils have Zn as major pollutant and soils of the African copper belt are rich in Cu, Co, Cr, Ni, and Zn (Alford et al. 2010).

Special cases appear to be the serpentine soils (ultramafic soils) that are rock-forming hydrous magnesium phyllosilicate $(Mg,Fe)_3Si_2O_5(OH)_4$ minerals worldwide. They are also characterized by low levels of potassium and phosphorus and a low ratio of calcium/magnesium (Kinzel and Weber 1982). Because they carry high concentrations of Ni, Cr and Co, plants growing there have to cope with similar adverse conditions as those on heavy metal soils.

Most heavy metal soils worldwide have been exploited by human activities from early times. However, due to the insufficient technologies of medieval time and earlier periods, the remnants (heavy metal heaps) are still rich in heavy metals and carry a typical metallophyte vegetation. The concentration of heavy metals can be particularly high around smelters. When close to rivers, the heavy metals are gradually washed off leading to the diminution of the heavy metal vegetation with special concern about the risk of losing endangered plants (Becker and Dierschke 2008; Lucassen et al. 2010).

A special case appears to be soils contaminated by mercury (Hg). Plants growing on Hg-polluted soils show a reduced vitality, as observed at sites in Slovenia (Idrija) by the author. Specific Hg-adapted metallophytes do not seem to exist. However, 13 plant species accumulated Hg out of 87 examined from a highly contaminated waste area originating from a chemical plant in Northern Italy (Massa et al. 2010). In this study, *Polygonum aviculare* was the best accumulator and contained higher levels of Hg in roots than in shoots. The low productivity of this plant prevents its use for phytoextraction of Hg.

2.2 The Heavy Metal Plants (Metallophytes)

A botanist can immediately identify a habitat as heavy metal soil polluted by high concentrations of Zn, Pb or other heavy metals by the composition of the plants. In Central Europe, most heavy metal soils carry four to six typical metallophytes. Highest Zn-concentrations are endured by spring sandwort (*Minuartia* (*Alsine*) *verna*, Fig. 2.1h) of the Caryophyllaceae which otherwise occurs only, although abundantly, on chalk meadows and rocks in alpine regions of Europe above the timberline. The thrifts or sea pinks (*Armeria maritima*) of the Plumbaginaceae form a somewhat difficult plant complex. The subspecies *A. maritima* ssp. *halleri* (Fig. 2.1f) grows only on heavy metal heaps in European plains, and can be differentiated morphologically only with difficulty from the subspecies *maritima* of the coastal salt marshes or the alpine ssp. *alpina*. The genus *Armeria* does not appear to be finally resolved taxonomically which is also true for its occurrence in different plant associations (Becker and Dierschke 2008). The same applies to pennycresses of the Brassicaceae (*Thlaspi alpestre* agg.) on heavy metal soils (Koch et al. 1998). *Thlaspi caerulescens* (Fig. 2.1k) and *T. praecox* form ecotypes on heavy metal heaps. The plants on some heavy metal locations are often termed

Fig. 2.1 (Continued)

T. caerulescens ssp. *calaminare*, but the differentiation to *T. montanum* and the "genuine" *T. alpestre* is not always obvious. A revision has recently been made on the genus *Thlaspi*; *T. caerulescens* is now supposed to be *Noccaea caerulescens*

Fig. 2.1 A selection of European heavy metal plants. (**a**) A meadow on a Zn-rich heavy metal heap at Schlangenberg/Stolberg close to Aachen, Germany. The yellow zinc violet (*Viola lutea* ssp. *calaminaria*) dominates and also *Armeria maritima* ssp. *halleri* is detectable as red flowering spots. Photo from June 1998. (**b**) The blue-zinc violet (*Viola lutea* ssp. *westfalica*) occurs only in the Pb-ditch and its surrounding heaps at Blankenrode close to Paderborn, Germany. The photo also shows *Cardaminopsis* (*Arabidopsis*) *halleri* (white flowers) at this location. May 1987. (**c**) The blue-zinc violet (*Viola lutea* ssp. *westfalica* also grows happily on nonpolluted soils. Photo from the author's own allotment at D-Erftstadt. June 2009. (**d**) The alpine *Viola lutea* is the ancestor of the zinc violets (Hildebrandt et al. 2006). Zweisimmen/Bern, Switzerland. July 1997. (**e**) *Viola tricolor* forms specific ecotypes on heavy metal heaps. Siebertal, Harz mountains, Germany. June 1992. (**f**) *Armeria maritima* ssp. *halleri* from Schlangenberg/Stolberg close to Aachen. June 2002. (**g**) *Silene vulgaris* var. *humilis* which shows the bent shoots typical of the heavy metal ecotype of this plant species. Schlangenberg/Stolberg close to Aachen. June 2002. (**h**) *Minuartia* (*Alsine*) *verna* on heavy metal polluted gravel in the Gailitz river bed close to Arnoldstein, Southern Austria. (**i**) *Thlaspi cepaeifolium* on a heap close to a Zn-smelter at Cave del Predil (Raibl) in Northern Italy, close to the Slovenian border. This is an almost extinct species, but it is not related to *Thlaspi caerulescens*. May 2001. (**j**) *Alyssum wulfenianum*, an extremely rare heavy metal plant in the Gailitz river close to Arnoldstein, Southern Austria. May 2001. (**k**) *Thlaspi caerulescens* ssp. *calaminare* (=*Thlaspi calaminare*). Schlangenberg/Stolberg close to Aachen. April 2002. (**l**) *Thlaspi goesingense* from a serpentine soil at Pernegg/Mur in Steiermark, Austria. June 2002

(Meyer 2006). However, the taxonomy of the genus *Thlaspi* remains controversial (Koch et al. 1998) and it presumably is polyphyletic (Broadley et al. 2007). A separate tribe is the Ni-hyperaccumulator *Thlaspi goesingense* (Fig. 2.1l) which thrives on serpentine soils in South-Eastern European countries. This latter is fairly productive and could be a good candidate for phytoremediation purposes on heavy metal heaps.

The bladder campion (*Silene vulgaris*, Fig. 2.1g) of the Caryophyllaceae forms a characteristic ecotype on heavy metal heaps. Whereas the shoots of this plant are straight on nonpolluted soils, they are bent or curved in specimens found on heavy metal heaps (forma *humilis*). I express from my own experience of diverse heavy metal heaps in Germany, Poland and alpine areas in Austria, Slovenia and Italy that this is an indicative feature of the heavy metal ecotype. Any taxonomic relationship of the *humilis* ecotype to the alpine *S. vulgaris* ssp. *prostrata* does not appear to be resolved at present. *S. vulgaris* has frequently been used for physiological and biochemical studies on heavy metal tolerance (e.g. Schat and Tenbookum 1992; Schat and Vooijs 1997; Kovacik et al. 2010).

Haller's rockcress (*Cardaminopsis halleri*, Fig. 2.1b) is even more in the center of interest of experimental studies nowadays. It is closely related to the model plant of modern molecular biology, *Arabidopsis thaliana*, and therefore termed *Arabidopsis halleri* in recent publications (Koch and Matchinger 2007; Macnair et al. 1999; Meyer et al. 2009, 2010). The gene compositions of both *Arabidopsis thaliana* and *Cardaminopsis halleri* are similar, and molecular techniques to identify and manipulate the genes can be applied to both species (Pauwels et al. 2005; Willems et al. 2007; Roosens et al. 2008). In the field, *Cardaminopsis halleri* occurs on heavy metal heaps, but not when the content of heavy metals, particularly Cu, is high in soils (Becker and Dierschke 2008). In contrast to other metallophytes already mentioned, it escapes from heavy metal heaps to neighboring nonpolluted sites and is therefore sometimes called "pseudometallophyte" (e.g. Pauwels et al. 2005, 2006) and its abundance is positively affected by soil depth and moisture (Becker and Dierschke 2008).

My personal comment on the nomenclature *Arabidopsis–Cardaminopsis* may be added here. The former genus *Cardaminopsis* is now abandoned to *Arabidopsis* (Koch and Matchinger 2007). All members of the former *Cardaminopsis* possess eight chromosomes in the haploid state whereas *A. thaliana* has only five. Molecular information is based only on ITS and cpDNA sequencing but not on nuclear DNA properties. The evolutionary split between the x-5 *A. thaliana* and the x-8 *Cardaminopsis* occurred around 5 Ma (Bechsgaard et al. 2006), a long time ago. However, in all these considerations, morphological criteria are completely ignored. From knowing natural populations of *A. thaliana* and *Cardaminopsis* species (*C. halleri*, *petraea*, *arenosa*), I would immediately believe that *Arabidopsis* and *Cardaminopsis* belong to different genera, although, as is known, no absolute criteria exist to define a species or a genus. On the other hand, the juxtaposition of members of *Cardaminopsis* to the model plant *A. thaliana* places them more into the focus of current interest. The sequence identity between *A. thaliana* and *C. halleri* is around 94% (Becher et al. 2004). For comparison, chimpanzee and human share a sequence identity of 98.8% (Cyranowski 2001; Mikkelsen et al. 2005), so a common genus for both species might not be accepted by many people. Another relative of *A. thaliana*, the salt cress *Thellungiella halophila*, is 95% identical, at the cDNA level, but both genera are considered to be separate (Verbruggen et al. 2009).

Other plant ecotypes exist in Central Europe that are particularly adapted to a life on heavy metal soils. For example, a subspecies thriving on heavy metal heaps and termed *Festuca aquisgranensis* has been separated from the *Festuca ovina* aggregate (Patzke and Brown 1990). However, the *Festuca ovina* agg. is a fairly difficult taxon which is not yet finally resolved by morphological criteria. Other plants, particularly grasses (Bradshaw 1952; Wu et al. 1975; Humphreys and Nicholls 1984) may also have evolved heavy metal-adapted ecotypes (subspecies) which can, however, only be identified as separate species by current molecular techniques.

The zinc violets are special beauties (Fig. 2.1a) that occur only on heavy metal sites of very restricted distribution in Western Central Europe (Schwickerath 1944). The yellow zinc violet (*V. lutea* ssp. *calaminaria*) lives on Zn-rich soils in the area between Aachen, Germany and Liège, Belgium, and the blue form (*V. lutea* ssp. *westfalica*, Fig. 2.1b, c) thrives in a ditch and the surrounding heaps in an area of some 1 km × 0.5 km at Blankenrode, Eastern Westphalia, Germany. Both taxa belong to the most endangered plants in Central Europe and are probably the only unambiguous endemic taxa in Central Europe outside the alpine regions. Lack of competitiveness forces them to survive on heavy metal heaps. Molecular analyses showed that they are closely related to the alpine *V. lutea* (Fig. 2.1d) which occurs in Vosges or in the Sudeten mountains and more rarely in the Alps (Hildebrandt et al. 2006). The zinc violets might be relicts of the glacial period where they and their parents *V. lutea* seemingly had a wider distribution. Their patchy occurrence in Central Europe and also in Great Britain is an indication for such a history. Isolation on heavy metal heaps of the zinc violets may have resulted in the separation from their parents into the two separate entities.

In more Eastern and Southern European countries the zinc violets seem to be replaced by wild pansy (heartsease), *Viola tricolor* (Fig. 2.1e), which has developed ecotypes that can cope with fairly high concentrations of heavy metals (Slomka et al. 2011). This plant is, however, not restricted to heavy metal soils but occurs on gravel or sandy soils throughout the areas just mentioned.

When leaving Central Europe, the vegetation on heavy metal sites changes. Heavy metal soils in the area between Carinthia, Austria, Friaul in Italy and Northern Slovenia carry two very remarkable and almost extinct metallophytes: *Alyssum wulfenianum* (Fig. 2.1j) and *Thlaspi cepaeifolium* ssp. *cepaeifolium* (Fig. 2.1i) which is unrelated to the *Thlaspi alpestre* agg, mentioned above but has close taxonomic affinities to *Thlaspi rotundifolium*. The need to preserve these extremely rare metallophytes is obvious. A detailed account of the metallophytes and their role in plant associations particularly in Northern and Southern Europe is given by Ernst (1974), and a list of heavy metal plants of the world can be found in Brooks (1998) or Prasad and Hagemeyer (1999). It is estimated that approximately 500 angiosperms representing about 0.2% of all higher plants are metallophytes (Baker and Brooks 1989; cited in Krämer 2010).

A differentiation is often made between absolute (strict or eu-) metallophytes and facultative (pseudo-) metallophytes according to their occurrence either only on polluted sites or on both contaminated and noncontaminated habitats (Lambinon and Augier 1964; Willems et al. 2007). However, as just said, all metallophytes of Central Europe occur outside heavy metal sites. On their natural stands in the European plains, they cannot compete with faster growing plants on nonpolluted soils. Their patchy distribution in Europe may be the result of glacial époques and the postglacial warm period. Such a differentiation between strict and pseudo-metallophytes is possibly justified at the subspecies level, for example for *Viola lutea* ssp. *calaminaria* and *V. l.* ssp. *westfalica* or *Armeria maritima* ssp. *halleri* on one side and *V. lutea* ssp. *lutea* or *Armeria maritima* ssp. *maritima* on the other. Such a differentiation merges into the discussion what is a subspecies in botany.

2.3 Accumulating and Hyperaccumulating Metallophytes

Heavy metal plants differ largely in their heavy metal contents as indicated by analysis of the plants that grew on a Zn-rich heavy metal soil close to Aachen, Germany (Table 2.1). Leaves of *Thlaspi caerulescens* (*T. alpestre* ssp. *calaminare*) and *Minuartia verna* contained the highest amount of Zn. The ratio between Cd and Zn is low in *Minuartia verna* and *Silene cucubalus* var. *humilis* in contrast to the situation in *T. caerulescens*. *Armeria maritima* in particular accumulates Pb. Thus

Table 2.1 Heavy metal content in mmol × kg^{-1} in leaves of plants living on the Zn-polluted soil at Breinigerberg close to Aachen, Germany

Plant species	Zn	Pb	Cd
Thlaspi alpestre ssp. *calaminare* (=*T. caerulescens*)	159.0	8.21	**4.83**
Minuartia verna	151.3	6.52	0.65
Armeria maritima ssp. *calaminaria*	112.8	**11.60**	1.10
Silene cucubalus var. *humilis*	40.8	0.29	0.02
Plantago lanceolata	39.2	2.35	0.19
Lotus corniculatus	30.4	0.05	0.02
Anthyllis vulneraria	28.8	0.15	0.03
Festuca ovina	28.3	0.97	0.13
Campanula rotundifolia	24.8	4.70	0.98
Thymus serpyllum agg.	22.9	3.96	0.33
Cladonia rangifera (podetium) (lichen)	21.4	8.08	0.40
Rumex acetosa	21.4	2.12	0.16
Agrostis tenuis	17.4	0.88	0.10
Achillea millefolium	14.8	1.38	0.02
Euphrasia stricta	14.3	0.94	0.10
Viola lutea ssp. *calaminaria*	**8.9**	**0.19**	**0.02**
Pimpinella saxifraga	8.2	0.26	0.03

The data are taken from Ernst (1982). Extreme values are given in bold there is no tolerance to heavy metal in general, but each metallophyte has evolved a strategy to cope with an individual heavy metal.

The amount of heavy metals taken up by a plant is dependent on the concentration of heavy metals in the polluted soil. In most plants, heavy metals are predominantly accumulated in roots. The shoot/root ratio is generally below unity in most plants but not in metallophytes. In some heavy metal plants, the concentration of heavy metals in shoots and leaves can be particularly high and the partitioning of heavy metals between shoots and roots differs from one metallophyte to the next and with each individual heavy metal (Krämer 2010).

The zinc violet *V. lutea* ssp. *calaminaria* has very low amounts of Zn, Pb, and Cd in its leaves (Table 2.1). Although conflicting data have been published on the levels of heavy metals in zinc violets (Jedrzejczyk et al. 2002; Noret et al. 2007), these violets have distinctly less heavy metals in their organs and thus show a different pattern of adaptation to heavy metal stress in soils than *Thlaspi caerulescens*, *Minuartia verna*, or *Armeria maritima*. The two basic strategies to respond to heavy metal toxicity emerge from the comparison of the data of Table 2.1. Metallophytes can be either excluders of heavy metals or they can be accumulators (Ramirez-Rodriguez et al. 2005). The zinc violet is a good example of an excluder. The other three species mentioned (*Thlaspi*, *Minuartia*, and *Armeria*) are accumulators. Some plants are hyperaccumulators, however, only for a single or few specific heavy metals. *Thlaspi caerulescens*, *Cardaminopsis halleri*, and the Crassulaceae *Sedum alfredii* (Sun et al. 2007) hyperaccumulate Zn and Cd, but not Pb (Broadley et al. 2007). The Katanga species *Haumaniastrum katangense* is a hyperaccumulator for copper (Chipeng et al. 2010). The Ni-hyperaccumulating species *Alyssum murale* (Ernst 2005), *Alyssum bertolonii* (Boominathan et al. 2004), *Berkheya coddii* (Robinson et al. 2003), or several endemic species of the serpentine flora of Zimbabwe (Brooks and Yang 1984) were suggested as potential enrichers of this element for human exploitation (leaching purposes). The fern *Pteris vittata* accumulates Se from seleniferous soils (Liao et al. 2004) and the violet *Viola baoshanensis* is a Cd-hyperaccumulator (Wu et al. 2010). Hyperaccumulators of Pb do not seem to exist (or are rare) due to the fact that Pb is extremely immobile and is not easily accumulated by plants (Bert et al. 2002). An extreme example for hyperaccumulation is *Sebertia acuminata* which can store up to 26% (w/w) Ni in its latex (Jaffré et al. 1976; Verbruggen et al. 2009).

The degree of tolerance of plants to heavy metals was divided into the three categories: hypotolerance, basal tolerance, and hypertolerance in a more recent review (Ernst et al. 2008). Although some plant species are clear hyperaccumulators for Cu, Zn or Cd, no species can store excess of *all* heavy metals. The borderline between basal tolerance and hyperaccumulation is blurred. A hyperaccumulator may be defined by the threshold value which is approximately a 10 times higher concentration of a heavy metal in the aerial parts compared to the content in a nonhyperaccumulator growing on the same polluted habitat (Bert et al. 2002). Approximately 400 taxa worldwide are hyperaccumulators (Bert et al. 2002). To give more precise values a Zn hyperaccumulator is defined as a plant that contains >10,000 μg g^{-1} dry weight (1%, w/w), and a Cd hyperaccumulator should have >100 μg g^{-1} dry weight (0.01%, w/w) (Baker 1981; Bert et al. 2002).

Halophytes can be divided into salt-resistant and salt-tolerant species. *Salicornia europaea* and *Suaeda maritima* need a high soil concentration of NaCl for germination and growth and are therefore salt-resistant species. All other halophytes grow better in non-NaCl enriched soils (gardens) than in saline habitats and are therefore salt-tolerant (Hildebrandt et al. 2007). All metallophytes of Central Europe are metal-tolerant (Hildebrandt et al. 2007). Lack of competitiveness forces them to live on heavy metal soils where they are not overgrown by more productive glycophytes. This is also true of the blue-zinc violet (*V. lutea* ssp. *westfalica*) where its germination and growth was claimed to require polluted soils with excess of heavy metals (Nauenburg 1986). The blue-zinc violet was shown to grow with no impaired vitality in garden soils at several locations in Germany in the last years. Its germination requires darkness, but is even more effective in nonpolluted garden earth than in heavy metals soils (Slomka et al. 2011). I am not aware of any clear-cut heavy-metal-resistant plant species worldwide that would be strictly dependent on the presence of excess of heavy metals in its growth substratum.

2.4 Strategies Employed by the Metallophytes to Cope with High Concentrations of Heavy Metals at the Whole Plant Level

Heavy metal tolerance has been developed by plants of totally unrelated taxonomic affinities. It is frequent in Brassicaceae, and also seen in Caryophyllaceae, Plumbaginaceae, Violaceae, Asteraceae, Poaceae, and others. Over 34 different plant families have developed heavy metal tolerant species (Verbruggen et al. 2009). Heavy metal tolerance is thus one of the clear-cut examples of convergence in biology. Therefore, it is not surprising that the strategies to cope with the excess of heavy metals differ from one plant taxon to the next. Similarities in the mechanisms exist between metallophytes and halophytes.

Armeria maritima can thrive both in coastal salt marshes with a high NaCl content (ssp. *maritima*) and on heavy metal heaps (ssp. *halleri*). This plant has special glands, developed from stomata, which apparently serve to excrete excess NaCl (Lüttge 1975; Rozema et al. 1981). Similar glands might also be used to transfer the surplus of heavy metals to the outside by the metallophyte *A. maritima* ssp. *halleri*. However, this has not yet been verified to my knowledge. Halophytes of the genus *Atriplex* secrete NaCl into hairs of the leaves and stems and discard them when overloaded (Lüttge 1975). Such a strategy seems also to be employed by metallophytes, as recently reassessed for an ecotype of *Viola tricolor* from the Bukowno heavy metal heap in Southern Poland (Slomka et al. 2011). *Minuartia verna* has the vegetation point in the center of a bundle of small, relatively long leaves. The vegetation point continuously forms new maiden leaves essentially free of heavy metals. These leaves are used to deposit heavy metals during the vegetation period and then fall off when overloaded with heavy metals. Most *Thlaspi* species such as *T. caerulescens* are fast growing annuals that rapidly die after flower and seed formation with an overload of heavy metals. Their seeds are essentially free of heavy metals which, therefore, cannot affect them during germination next spring.

The zinc violet as a heavy metal excluder utilizes special arbuscular mycorrhizal fungi (AMF) as recently reviewed in a chapter of another book of this series (Bothe et al. 2010) and elsewhere (Hildebrandt et al. 2007). Roots of zinc violet samples collected from the natural sites are yellow due to the deposition of the yellow pigment, termed mycorradicin (Klingner et al. 1995a). This is a visible indication of mycorrhizal colonization in violets and other plants (Klingner et al. 1995b), and this pigment is deposited into the plant cells due to the action of the fungi. Counting of the degree of mycorrhizal colonization indicated that roots of the zinc violets, either the yellow morph from the Aachen-Liège area or the blue form from Blankenrode, are heavily colonized by AMF. A fungus, *Glomus intraradices* Br1, had been isolated that consistently enabled diverse plants to grow on diverse heavy metal soils, provided fertilization was optimized. Biophysical determinations of the concentrations of heavy metals in roots of maize and also of the blue-zinc violet indicated that plants colonized by AMF contained considerably less heavy metals in their roots than nonfungal control plants, in line with the data of Table 2.1. Uptake of the major part of the heavy metals into the living compartments of the two symbiotic partners might be prevented due to the action of the AMF. It is suggested that the majority of the heavy metals are transferred to the cell walls and the vacuoles of the fungi. The heavy metals that reached the inside of the plant cells are deposited mainly into the inner root cortical cells where the fungal structures reside. The techniques employed did not allow us to discriminate whether the heavy metals in the inner cortical cells were deposited into plant or fungal structures.

Another example of a metallophyte colonized by AMF is the Asteraceae *Berkheya coddii* of South Africa (Turnau and Mesjasz-Przybylowicz 2003). The Brassicaceae *Biscutella laevigata* on heavy metal heaps at Olkucz, Southern Poland (Orłowska et al. 2002) or *Thlaspi* sp. (Regvar et al. 2003; Vogel-Mikuš et al. 2006) are occasionally colonized, particularly at the flowering state. However, it is doubtful whether the degree of root colonization by AMF is high enough for a significant contribution to heavy metal tolerance by these plants.

In most cases, heavy metal heaps are not covered by trees. Since heavy metals accumulate gradually in the cells, only some short-living herbs can cope with the adverse soil conditions. In Central Europe some crippled fir (*Pinus sylvestris*) or birch (*Betula pendula*) trees can be found on older heavy metal soils. Such trees may utilize ectomycorrhizal fungi for their survival. Evidence was presented that a copper-tolerant ectomycorrhizal fungus *Suillus luteus* was able to protect pine trees against Cu toxicity (Adriaensen et al. 2005).

2.5 Toxicity of the Heavy Metals in Cells and Responses of the Plant Cells

Heavy metals can have multiple effects in the plant cytoplasm:

1. They can bind to functionally essential SH-groups in enzymes and thereby inactivate them.
2. They can substitute functional elements in prosthetic groups of enzymes resulting in an inactive catalysis. This is particularly the case for Cd substituting Zn in proteins.
3. They may enhance the generation of ROS (reactive oxygen species) such as $O_2^{\bullet -}$, $OH^{\bullet}$, H_2O_2, and $^1O_2$. These ROS are generated in cells in either the Fenton reaction:

$$H_2O_2 + Fe^{2+} \rightarrow OH^- + OH^{\bullet} + Fe^{3+} \quad \text{or any other heavy divalent metal} \qquad (2.1)$$

or the Haber–Weiss reaction:

$$O_2^{\bullet -} + H_2O_2 + H^+ \rightarrow OH^{\bullet} + H_2O + O_2. \qquad (2.2)$$

The reaction velocity of the Haber–Weiss reaction is enhanced by any heavy metal cation which can change the oxidation state (Elstner 1990; Prasad and Hagemeyer 1999).

As indicated in Fig. 2.2, the plant can cope with these adverse affects by multiple responses (Salt 2001; Hall 2002):

(a) Formation of *Siderophores*: These are large, complex molecules that are excreted from the plant cells into the soil. These chelators bind heavy metals to form a large complex which is not taken up by the cells. Thus the surplus of heavy metals would not reach the cell cytoplasm. Such a role of siderophores is not accepted by all investigators working in the field.

(b) Synthesis of small molecules such as the *carboxylic acids*: malate, citrate, or oxaloacetate that bind heavy metals with their acid groups. These carboxylic acids may bind heavy metals outside the roots or in the root apoplasm which may prevent uptake of heavy metals. The root symplasm of plants excretes up to 20% of the organic carbon containing such organic acids. Root exudation may change the pH-value in soils and thereby may alter the accessibility and uptake rate of heavy metals into plant roots (Alford et al. 2010). Alternatively, such simple organic acids may be used to selectively deposit heavy metals. In particular, Zn may be bound to malate in the cytoplasm and transferred as a complex across the tonoplast to the vacuole. There Zn-malate may dissociate, then Zn^{2+} may bind to stronger chelators such as citrate and oxoaloacetate and finally malate may be retranslocated to the cytoplasm. This long-ago-postulated Zn-malate shuttle hypothesis (Ernst et al. 1992) is still quoted (Broadley et al. 2007) but is not in the focus of current interest.

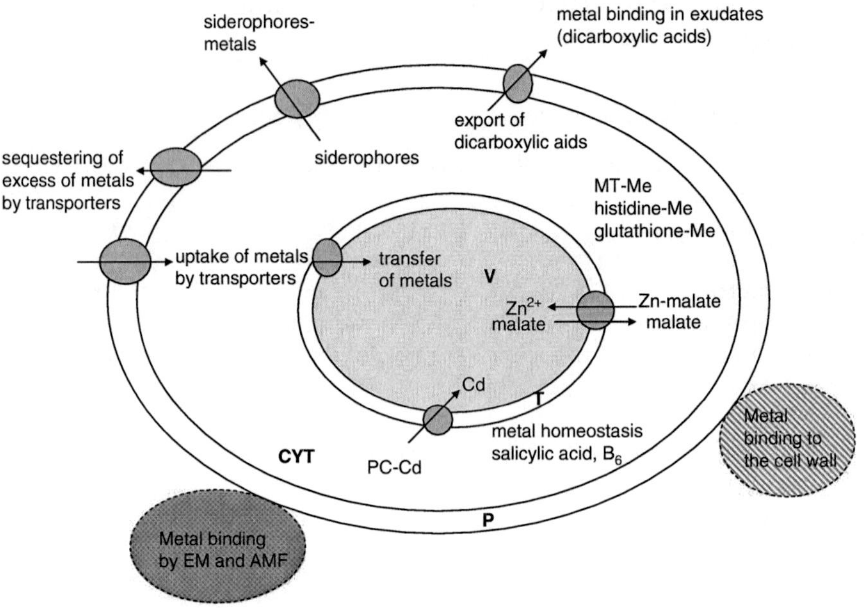

Fig. 2.2 Schematic summary of the factors involved in heavy metal tolerance of plants. Abbreviations: *P* plasma membrane (cytoplasmic membrane); *CYT* cytoplasm, cytosol; *T* tonoplast; *V* vacuole; *Me* metal; *PC* phytochelatin; *MT* metallothionein. Adapted from Hall (2002), however, significantly modified

(c) Other small molecules that may bind heavy metals at their SH-groups are *free histidines or glutathione*. The role of the free amino acid histidine in detoxifying heavy metals, particularly nickel, has recently been stressed (Ingle et al. 2005; Krämer et al. 1996; Krämer 2010). Excess of heavy metals, particularly Fe, may also be bound by nicotianamine (Krämer 2010). Other candidates for metal binding in the cytoplasm are polyamines (Cicatelli et al. 2010). Small molecules may be used to maintain heavy metal homeostasis in the plant cytoplasm.

(d) *Metallothioneins*. These are small cysteine-rich proteins. The SH-groups of the cysteines may bind heavy metals in the cytoplasm and may thereby prevent them from exerting toxic effects.

(e) *Phytochelatins*. These compounds contain cysteine in a more regular way than metallothioneins. The compositions of the phytochelatins are multiples of glutamate-cysteine ($n = 2$–11) which terminate with one glycine. The SH-groups of the cysteines in the phytochelatins may also serve to bind metals. Phytochelatins are synthesized nontranslationally by phytochelatine synthase with glutathione as substrate (Hall 2002).

(f) *Heavy metal transporters*. The plasma membrane surrounding the root parenchyma cells acts as a barrier against the uptake of heavy metals from the cytoplasm. This membrane and the tonoplast surrounding the vacuole possess several classes of transport proteins such as CPx-ATPases for Cu and Cd, ABC transporters for Cd transport into the vacuole, ZIP transporters (ZRT-, IRT-related proteins) for Fe and Zn, N*ramp* transporters with a broad-range affinity for heavy metals and the CDF (cation diffusion facilitator) family (Hall 2002). In general, a plant possesses several genes coding for each class of transporters which may individually be expressed depending on the soil toxicity (heavy metal content), the plant organ and organelle, the plant development state and the annual season. Therefore, the study of the role of an individual transporter amongst the realm of all others is complex. Investigators tend to put their favorite transporter and their encoding gene into the focus. Those heavy metals that are essential for the growth of organisms (Fe, Mn, Cu, Co, Mo) may be taken up in excess by their transporters in metallophytes. Always toxic elements such as Cd, Pb, Cr, U, Ag, Hg, Ni (in most organisms), As and Se might be taken up erroneously; often transporters cannot strictly discriminate between the element wanted and a toxic heavy metal.

(g) *ROS detoxifying enzymes*. Upon stress caused by high loads of heavy metals or salts, by drought or soil acidity, plants express stress responsive genes. Here glutathione S-transferase is of paramount importance, but, in addition, superoxide dismutases, cytochrome P450, thioredoxin are detected in SSH libraries where the gene expression from stressed and control plants are compared (see Hildebrandt et al. 2007). Such proteins might serve to detoxify ROS generated by excess of heavy metals [(2.1) and (2.2)]. In addition, heat shock proteins (HSPs) are expressed not only under high temperatures but also under a variety of stress conditions, among which is exposure to excess of heavy metals. HSPs not only function as chaperones in protein folding and assembly but may also serve in protecting and repairing proteins from oxidative stress (Hall 2002).

(h) *Alterations in the root morphology*. The Zn- and Cd-hyperaccumulator *Thlaspi caerulescens* has recently been shown to possess a peri-endodermal layer of cells with irregularly thickened tangential cell walls impregnated by lignin (Broadley et al. 2007; Zelko et al. 2008). The secondary compact cylinder commences close to the root tip and encircles the endodermis and is not seen in the nonmetallophyte *Thlaspi arvense*. Any specific role of this cylinder in the detoxification of heavy metals in *T. caerulescens* is not clear at present. To my knowledge, alterations in the root anatomy of other metallophytes have not been reported but are not unlikely. Alterations in the root structures occur in the halophyte *Aster tripolium* upon NaCl load and mycorrhizal colonization (Scheloske et al. 2004). Increases in root biomasses as well as root hair production and root lengths have been reported for high Zn-containing soil patches compared to Zn-deficient adjacent areas (see Broadley et al. 2007). However, this may not be a general phenomenon for all metallophytes and all heavy metal soils. The different metallophytes have all sorts of roots, from fine hairy root bundles to thick, long primary roots with few, marginally branched side roots at their natural habitats (my own observations).

2.6 Genes and Their Expressions Upon Heavy Metal Stress in Plants

It is often stated that metal tolerances are strictly metal-specific; however, the situation is often not so simple as this (see Schat and Vooijs 1997). Any combined tolerance to different heavy metals in a metallophyte could result from the sum of different metal-specific mechanisms or from less-specific mechanisms that pleiotropically confer tolerances to different metals. Work with *Silene vulgaris* (Schat and Vooijs 1997) showed that tolerance to Cu, Zn and Cd is under nonpleiotropic, independent genetic control. On the other hand, tolerance to Ni and Co seems to be a pleiotropic by-product of Zn tolerance. Hyperaccumulation of Zn and Cd is a constitutive trait in *Cardaminopsis* (*Arabidopsis*) *halleri* (Bert et al. 2002). Zn tolerance in *C. halleri* populations from heavy metal soils is, on average, higher than in those from nonpolluted habitats (Pauwels et al. 2006; Meyer et al. 2010). On the other hand, a population from a noncontaminated site accumulated Zn in its roots and shoots more rapidly and more effectively than plants from a polluted habitat (Bert et al. 2000).

Another issue concerns the debate of whether heavy metal tolerance and hyperaccumulation share a common genetic basis or are under control of different genes. Whereas a strict separation was claimed in earlier publications (Macnair et al. 1999), a simultaneous evolution of both traits on heavy metal soils and the phenotypic expression of shared genes was favored by others at that time (Krämer et al. 1997) which is corroborated in more recent communications (Frérot et al. 2010). The ecological role of hyperaccumulation of heavy metals is not understood. It is stated that the accumulation of heavy metals much in excess of the concentration of polluted soils may make metallophytes resistant against attack by carnivores or pathogens. However, experimental evidence for this selective advantage is missing (Noret et al. 2005).

In *Cardaminopsis halleri*, the tolerance to Zn and Cd is constitutive, meaning that all populations from either polluted or noncontaminated soils share the properties to be tolerant and to be able to accumulate these two heavy metals (Pauwels et al. 2006). In contrast, Cd and Ni tolerance may not be constitutive in *Thlaspi caerulescens* (Verbruggen et al. 2009), but this issue does not seem to be finally resolved. Studies on the genetic basis of Zn and Cd tolerances of *C. halleri* are favored by the fact that it form fertile crosses with the closely related *Arabidopsis lyrata* ssp. *petraea* = *Cardaminopsis petraea* = *C. hispida* {Macnair et al. 1999). This species did not develop tolerances to Zn and Cd and also does not occur on heavy metal soils. However, it also occurs in extreme habitats in nature. It is a typical element of dolomite soils in the Alps and has a disjunctive occurrence in the central European plains on gypsum soils where only few higher plant species can live. Thus, like *Arabidopsis halleri*, *Cardaminopsis* (*Arabidopsis*) *petraea* is also adapted to stress conditions such as drought and nutrient limitations in its natural habitats.

Earlier it was believed that tolerance to a range of heavy metals including Zn, Cu and As is controlled by a small number (one or two) of major genes, with additional modifiers for the degree of tolerance (Schat et al. 1993; Smith and Macnair 1998). This view has recently been changed by quantitative trait loci (QTL, Perez-Figueroa et al. 2010) mapping. The gene regions of complex adaptive traits such as heavy metal tolerance can be identified on the chromosomes by analyzing the crosses between *Cardaminopsis halleri* and *C. petraea* (Macnair et al. 1999; Willems et al. 2007). Three regions on different chromosomes of *C. halleri* were mapped which are responsible for Zn tolerance, Zn hyperaccumulation, or both (Willems et al. 2007). Individual genes cannot be localized by such an approach, since all three QTL regions were several cM long, and 1 cM corresponds to about 250 kb or about 40 genes (Mauricio 2001; Willems et al. 2007). In a more recent study, even five QTL regions were identified by analyzing crosses between *C. halleri* and *C. petraea* grown in a soil polluted by low Zn-concentrations, and three QTL were mapped both in plants from either low or high Zn-polluted soil (Frérot et al. 2010). Similar data with essentially the same results were published by Filatov et al. (2007). At least 23 genes are known to be involved in metal homeostasis of *A. thaliana* (Roosens et al. 2008), and Zn tolerance and hyperaccumulation make the molecular biology in *C. halleri* even more complex. For Cd-hyperaccumulation by *C. halleri*, only one QTL was identified, and Cd hyperaccumulation and tolerance are not independent (Willems et al. 2010). As said before, the latter has now been described as likely to be the case, at least partially, also for Zn (Frérot et al. 2010).

Transcription studies indicated that more than 30 different genes associated with metal hyperaccumulation showed an enhanced expression level in *C. halleri* compared to the situation in *A. thaliana* (Hanikenne et al. 2008; Verbruggen et al. 2009). One QTL of *C. halleri* contains the *HMA4* gene encoding a plasma membrane heavy metal ATPase of the P-type superfamily. This ATPase is involved in loading transition metal ions (particularly Zn and Cd) into the xylem from the surrounding vascular tissues (Verbruggen et al. 2009) and is thus an important control element in long distance transport. RNA silencing (RNAi interferences) of this gene in *C. halleri* showed that the gene product is required for Cd and also – to a lesser extent – for Zn tolerance in *C. halleri* (Hanikenne et al. 2008; Krämer 2010). *HMA4* had much higher transcript abundance in *C. halleri* than in *A. thaliana*. These enhanced transcript levels are due to a combination of the expansion to three almost identical copies of the gene present in tandem in the genome of *C. halleri*, and promoter mutations enhancing promoter strength in all the three copies, together with altered cis-regulations of the expressions of these genes (Hanikenne et al. 2008; Krämer 2010). For comparison, *A. thaliana* contains only one ortholog of this gene. The *HMA4* gene seems to be a major gene with its product serving in the translocation of Zn and Cd from the root symplasm to the xylem vessels, although it is not sufficient for Zn tolerance and hyperaccumulation in *C. halleri* (Hanikenne et al. 2008). The function of different parts of this gene, in particular the essential role of the C-terminal domain in translocation heavy metals, has recently been assigned (Mills et al. 2010).

Similar duplications occurred with the gene *MTP1* of the cation diffusion facilitator family with its products controlling Zn transport into the vacuole (Desbrosses-Fonrouge et al. 2005). Whereas *A. thaliana* contains only one copy of *MTP1*,

C. halleri possesses five paralogues of it which have apparently evolved recently (Shahzad et al. 2010). These five different *MTP1* genes underwent different evolutionary fates such as neo-functionalization, nonfunctionalization, and subfunctionalization (Shahzad et al. 2010). Also in *Thlaspi caerulescens* and in the Ni-hyperaccumulator *T. goesingense*, transcript levels of *MTP1* as well as of *HMA4* are higher than in related nonhyperaccumulating members of the Brassicaceae (Krämer 2010).

Members of the ZIP family (zinc-regulated transporters) mediate the uptake of Zn from the root apoplasm into the parenchyma cells at the cytoplasmic membrane. Of these, *ZNT1* is highly expressed in roots of *T. caerulenscens* but not in *Thlaspi arvense* (Pence et al. 2000). As reviewed by Verbruggen et al. (2009) or Krämer (2010), *ZNT1* mediates high-affinity Zn-transport and low-affinity Cd-uptake and is also found to be highly expressed by microarray analyses of large parts of the genome of *T. caerulescens*. Its ortholog in *A. thaliana*, termed *ZIP4*, is expressed only under Zn-deficiency, whereas this gene is transcribed independently of the Zn-supply and at a high level in hyperaccumulators. The individual role of the other members of the ZIP-transporter family in hyperaccumulation has not been assessed as yet. This is the case for the orthologues of *ZIP3*, *ZIP6*, *ZIP9*, *ZIP10* and *IRT3* of *A. thaliana* which were shown to be overexpressed in *C. halleri* and *T. caerulescens* by microassay analysis (Verbruggen et al. 2009). In addition, the roles of transporters such as members of the N_{RAMP} family, of the ABC group, of the Ca^{2+}/H^{+} antiporters (CaCA superfamily) and others (Verbruggen et al. 2009) remain to be assigned in future investigations. To illustrate the complexity of the subject, more than 2,000 genes were more than five times upregulated in *C. halleri* compared with *A. thaliana*, and 1,147 of these had an unknown function (Van de Mortel et al. 2008).

Expression studies of different metallothioneins in *T. caerulescens* indicated that their forms contribute in variable ways to heavy metal tolerance but not so much to hyperaccumulation by this plant (Hassinen et al. 2009).

Suppression subtractive hybridization (SSH) approaches are a tool to reveal the genes expressed in plants grown under heavy metal stress but not in nonaffected control plants. In such trials, genes coding for enzymes detoxifying reactive oxygen radicals are in the priority list of expressed candidates (Ouziad et al. 2005; Dai et al. 2010). Glutathione-S-transferase is of paramount importance in detoxifying such radicals generated by exposure of the cells to excess heavy metals or salts (Ouziad et al. 2005) and is distinctly overexpressed under such conditions. The roles of such detoxifying enzymes and the differential expression of their encoding genes do not seem to be in the focus of the current research.

Fine control of heavy metal uptake may be modulated by signal molecules. There is some evidence that salicylic acid stimulates Cd-uptake in soybean (Drazic and Mihaillovic 2009). Vitamin B6 has been implicated in defense against cellular oxidative stress caused by excess of heavy metals, and one gene of the vitamin B6 biosynthetic pathway has recently been identified in the arbuscular mycorrhizal fungus *Glomus intraradices* (Benabdellah et al. 2009). The field "role of signal molecules in heavy metal resistance of plants" is just emerging but might be extensively researched in the near future.

2.7 Aspects of the Use of Metallophytes to Remediate Soils Polluted by Heavy Metals

Phytoremediation has been reviewed (Salt et al. 1998; Pilon-Smits 2005); Maerques et al. 2009; Muthukumar and Bagyaraj 2010). It comprises stabilization of soils devoid of plants due to heavy metal toxicity and extraction of heavy metals by plants. Volatilization and degradations have a role in the removal of other pollutants but not in the case of heavy metals. However, Hg and the toxic metalloids As and Se may be chemically converted to more readily extractable or volatile compounds.

It is sometimes stated that metallophytes are often slow growing and are not productive enough to stabilize heavy metal soils against erosion, leaching, or run-off (Pilon-Smits 2005). This is, indeed, the case for most metallophytes. However, *Thlaspi goesingense* (Fig. 2.1) of South-Eastern European serpentine soils is fairly productive and offers good promises to achieve this goal. Edaphic factors of the soils to be stabilized play an important role in any attempt. It is in most cases not sufficient to sow or plant metallophytes in contaminated, almost vegetation-free soils. Such habitats are not only polluted by excess heavy metals, but are often also nutrient-limited (Becker and Dierschke 2008; Turnau et al. 2010; Bothe unpublished). Thus an extensive soil analysis is necessary before any stabilization projects can be started. Heavy rainfalls during a thunderstorm may destroy the laboriously planted and just growing metallophytes in a vegetation-free soil, as this author had to encounter in an aborted attempt some years ago. Likewise, sea-gulls from a neighboring garbage dump (or grazing animals) suddenly noticing a green spot of metallophytes within a larger vegetation-free area can also attack the site and ruin the best-planned experiment within minutes (also my own experience).

Phytoextraction of heavy metals by hyperaccumulating plants offers a perspective to enrich heavy metals in soils poor in these elements. As said before, hyperaccumulators enrich heavy metals such as Ni in their above-ground parts which can easily be harvested, and then the heavy metals could be concentrated by ashing. The Ni-hyperaccumulators *Alyssum bertolonii* and *Streptanthus polygaloides* have been suggested as best candidates for Ni-phytoextraction in the field (Li et al. 2003).

Outside of heavy metals, metallophytes or other plants, either alone or in combination with mycorrhizal fungi, could be used to enrich radioactive compounds such as ^{137}Cs, ^{90}Sr, or ^{238}U (Westhoff 1999). Plants take up Cs instead of K which is highly mobile and is mainly transferred to the shoot. In contrast, Sr can serve as a substitute for Ca which is deposited mainly in cell walls and is concentrated in roots. Whole plants must therefore be harvested to enrich radionuclides from soils and to collect the plants in heaps for ashing and concentrating the radioactive material. Such an idea seems to be attractive for soils around the Chernobyl reactor but it would be necessary to treat large areas and to convince local authorities of the need.

Plant life in soils is dependent on interactions with microorganisms which may better exploit water and nutrients than roots and may provide growth-promoting substances to the plants. The literature is filled with positive effects of rhizobacteria on plant growth, and a review is required to summarize the subject. In general, bacterial inocula work fine in laboratory experiments, but they are frequently outcompeted by indigenous bacteria in the field. Arbuscular mycorrhizal fungi (AMF) exist that are better adapted to the harsh conditions in heavy metal soils than fungal isolates from nonpolluted soils (Weissenhorn and Leyval 1993). In our experiments, a mycorrhizal fungus, *Glomus intraradices* Br1, was isolated from the roots of the yellow zinc violet grown in a calamine soils near Aachen, Germany. This isolate was consistently more effective than a conventional AMF isolate in promoting growth of diverse plants in several heavy metal soils, provided fertilization was optimized (Hildebrandt et al. 1999; Kaldorf et al. 1999). Other specifically adapted AMF exist in heavy metal soils that could be exploited for phytoremediation purposes (Tonin et al. 2001). As said, metallophytes are concentrated within the Brassicaceae. Unfortunately, members of this family are generally AMF-negative or are at best poorly colonized (De Mars and Boerner 1996) which might exclude any application of these plants in combination with AMF. The zinc violets are strongly AMF-positive (Hildebrandt et al. 1999) and are fairly productive perennials. Due to the beauty of these plants, any phytoremediation attempt is faced with the risk of human rapture. Thus the best metallophyte/fungal combination has still to be found.

Phytoremediation appears to be an attractive field which is in infancy due to its complexity. Up to the present, remediation of soils polluted by heavy metals is governed mainly by chemists, and biologically based applications await future directives.

2.8 Conclusions

Plants of diverse families can grow in heavy metal soils, and their different biology offers fascinating research perspectives. Molecular studies of heavy metal tolerance and hyperaccumulation are currently focused on the two metallophytes *Cardaminopsis* (*Arabidopsis*) *halleri* and *Thlaspi caerulescens*. This is conceivable because of their close relatedness to the model plant *Arabidopsis thaliana*. The complete sequence of a metallophyte genome has not been published as yet. When such knowledge becomes available it will largely facilitate future research of heavy metal tolerance and hyperaccumulation. Publications in the last 3 years seem to indicate that any difference between heavy metal tolerance and hyperaccumulation is not so obvious as previously claimed. Another feature to emerge is that heavy metal tolerance might have arisen by gene duplications and the altered regulations of their expressions and not by the modifications of sequences of genes which originally had different functions in nonmetallophytes. The field of molecular biology of metallophytes is developing rapidly and exciting results are to be expected in the very near future.

The use of metallophytes, either alone or in combination with microorganisms, to stabilize heavy metal soils against erosion or to phytoextract heavy metals is an attractive idea. For various reasons, the field is in its infancy despite the many thoughts about it and experimental attempts in the past. Broad-scale applications are conceivable, but the cost–benefit outcome has not yet been properly assessed.

References

Adriaensen K, Vralstad T, Noben JP, Vangronsveld J, Colpaert JV (2005) Copper-adapted *Suillus luteus*, a symbiotic solution for pines colonizing Cu mine spoils. Appl Environ Microbiol 71:7279–7284

Alford EA, Pilon-Smits EAH, Paschke MW (2010) Metallophytes – a view from the rhizosphere. Plant Soil 337:33–50

Baker AJM (1981) Accumulators and excluders-strategies in the responses of plants to heavy metals. J Plant Nutr 3:643–654

Baker AJM, Brooks RR (1989) Terrestrial higher plants which hyperaccumulate metallic elements: a review of their distribution, ecology and phytochemistry. Biorecovery 1:81–126

Becher M, Talke IN, Krall L, Krämer U (2004) Cross-species microarray transcript profiling reveals high constitutive expression of metal homeostasis genes in shoots of the zinc hyperaccumulator *Arabidopsis halleri*. Plant J 37:251–268

Bechsgaard JS, Castric V, Charlesworth D, Vekemans X, Schierup MH (2006) The transition to self-compatibility in *Arabidopsis thaliana* and evolution within S-haplotypes over 10 Myrs. Mol Biol Evol 23:1741–1750

Becker T, Dierschke T (2008) Vegetation response to high concentrations of heavy metals in the Harz Mountains. Phytocoenologia 38:255–265

Benabdellah K, Azcon-Aguilar G, Valderas A, Speziga D, Fitzpatrick TB, Ferrol N (2009) *GintPDX1* encodes a protein involved in vitamin B6 biosynthesis that is up-regulated by oxidative stress in the arbuscular mycorrhizal fungus *Glomus intraradices*. New Phytol 184:682–693

Bert V, Macnair MR, DeLaguérie P, Saumitou-Laprade P, Petit D (2000) Zinc tolerance and accumulation in metallicolous and nonmetallicolous populations in *Arabidopsis halleri* (Brassicaceae). New Phytol 146:225–233

Bert V, Bonnin I, Saumitou-Laprade P, De Laguérie P, Petit D (2002) Do *Arabidopsis halleri* from nonmetallicolous populations accumulate zinc and cadmium more effectively than those from metallicolous populations? New Phytol 155:47–57

Boominathan R, Saha-Chaudhury NM, Sahajwall V, Doran PM (2004) Production of nickel bio-ore from hyperaccumulator plant biomass: applications in phytomining. Biotechnol Bioeng 86:243–250

Bothe H, Regvar M, Turnau K (2010) Arbuscular mycorrhiza, heavy metal and salt tolerance. In: Sherameti I, Varma A (eds) Soil heavy metals. Springer, Heidelberg, pp 87–111

Bradshaw AD (1952) Populations of *Agrostis tenuis* resistant to lead and zinc poisoning. Nature 169:1098

Broadley MR, White P, Hammond JP, Zelko I, Lux A (2007) Zinc in plants. New Phytol 173:677–702

Brooks RR (1998) Plants that hyperaccumulate heavy metals. CABI Publishing, Wallingford

Brooks RR, Yang X-H (1984) Elemental levels and relationships in the endemic serpentine flora of the Great Dyke, Zimbabwe, and their significance as controlling factors for the flora. Taxon 33:392–399

Chipeng FK et al (2010) Copper tolerance in the cuprophyte *Haumaniastrum katagense* (S. Moore) PA Duvign Plancke. Plant Soil 328:235–244

Cicatelli A, Lingua G, Todeschini V, Biondi S, Torrigiani P, Castiglione S (2010) Arbuscular mycorrhizal fungi restore normal growth in a white poplar clone grown on heavy metal-

contaminated soil, and this is upregulated of foliar methallothionein and polyamine biosynthetic gene expression. Ann Bot 106:791–802
Cyranowski D (2001) Almost human. Nature 418:910–912
Dai QL, Huang BF, Yang ZY, Yuan JG, Yang JZ (2010) Identification of cadmium-induced genes in maize seedlings by suppression subtractive hybridization. Front Environ Sci Eng China 4:449–458
De Mars BG, Boerner REJ (1996) Vesicular arbuscular mycorrhizal development in the Brassicaceae in relation to plant life span. Flora 191:179–189
Desbrosses-Fonrouge AG, Voigt K, Schröder A, Arrivault S, Thomine S, Krämer U (2005) *Arabidopsis thaliana MTP1* is a Zn transporter in the vacuolar membrane which mediates Zn detoxification and drives leaf accumulation. FEBS Lett 570:4165–4174
Drazic G, Mihaillovic N (2009) Salicylic acid modulates accumulation of Cd in seedlings of Cd-tolerant and Cd-susceptible soybean genotypes. Arch Biol Sci 61:431–439
Elstner EF (1990) Der Sauerstoff, Biochemie, Biologie. Medizin. BI-Wissenschaftsverlag, Mannheim, Wien, Zürich
Ernst WHO (1974) Schwermetallvegetation der Erde. Gustav Fischer, Stuttgart
Ernst WHO (1982) Schwermetallpflanzen. In: Kinzel H (ed) Pflanzenökologie und Mineral-Stoffwechsel. Ulmer, Stuttgart, pp 472–506
Ernst WHO (2005) Phytoextraction of mine wastes – options and impossibilities. Chem Erde – Geochem 65:29–42
Ernst WHO, Verkleij JAC, Schat H (1992) Metal tolerance in plants. Acta Bot Neerl 41:229–248
Ernst WHO, Krauss GJ, Verkleij JAC, Wesenberg D (2008) Interaction of heavy metals with sulphur metabolism in angiosperms from an ecological point of view. Plant Cell Environ 31:123–143
Filatov V, Dowdle J, Smirnoff N, Ford-Lloyd B, Newbury HJ, Macnair MR (2007) A quantitative trait loci analysis of Zn hyperaccumulation in *Arabidopsis halleri*. New Phytol 174:580–590
Frérot H et al (2010) Genetic architecture of zinc hyperaccumulation in *Arabidopsis halleri*: the essential role of QTL × environment interactions. New Phytol 187:355–367
Hall JL (2002) Cellular mechanisms for heavy metal detoxification and tolerance. J Exp Bot 53:1–11
Hanikenne M et al (2008) Evolution of metal hyperaccumulation required *cis*-regulatory changes and triplication of *HMA4*. Nature 453:391–396
Hassinen VH, Tuomainen M, Peraniemi S, Schat H, Karenlamp SO, Tervahauta AL (2009) Metallothioneins 2 and 3 contribute to the metal-adapted phenotype but are not directly linked to Zn accumulation in the metal hyperaccumulator *Thlaspi caerulescens*. J Exp Bot 60:187–196
Hildebrandt U, Kaldorf M, Bothe H (1999) The zinc violet and its colonisation by arbuscular mycorrhizal fungi. J Plant Physiol 154:709–717
Hildebrandt U et al (2006) The rare endemic zinc violets of Central Europe originate from *Viola lutea* Huds. Plant Syst Evol 257:205–222
Hildebrandt U, Regvar M, Bothe H (2007) Arbuscular mycorrhiza and heavy metal tolerance. Phytochemistry 68:139–146
Humphreys MO, Nicholls MK (1984) Relationship between tolerance to heavy metals in *Agrostis-capillaris* L. (*Agrostis-tenuis* Sibth). New Phytol 98:177–190
Ingle RA, Mugford ST, Rees JD, Campbell MM, Smith JAC (2005) Constitutively high expression of the histidine biosynthetic pathway contributes to nickel tolerance in hyperaccumulator plants. Plant Cell 17:2089–2106
Jaffré T, Brooks RR, Lee J, Reeves RD (1976) *Sibertia acuminata*: a hyperaccumulator of nickel from New Caledonia. Science 193:579–580
Jedrzejczyk M, Rostanski A, Malkowski E (2002) Accumulation of zinc and lead in selected taxa from genus *Viola* L. Acta Biologica Cracoviensia Series Botanica 44:49–55
Kaldorf MO, Kuhn AJ, Schröder WH, Hildebrandt U, Bothe H (1999) Selective element deposits in maize colonized by a heavy metal tolerance conferring arbuscular mycorrhizal fungus. J Plant Physiol 154:718–728

Kinzel H, Weber M (1982) Serpentin-Pflanzen. In: Kinzel H (ed) Pflanzenökologie und Mineral-Stoffwechsel. Ulmer, Stuttgart, pp 381–410

Klingner A, Bothe H, Wray V, Marner FJ (1995a) Identification of a yellow pigment formed in maize roots upon mycorrhizal colonization. Phytochemistry 38:53–55

Klingner A, Hundeshagen B, Kernebeck H, Bothe H (1995b) Localization of the yellow pigment formed in roots of gramineous plants colonized by arbuscular fungi. Protoplasma 185:50–57

Koch MA, Matchinger M (2007) Evolution and genetic differentiation among relatives of *Arabidopsis thaliana*. Proc Nat Acad Sci USA 104:6272–6277

Koch M, Mummenhoff K, Hurka H (1998) Systematics and evolutionary history of heavy metal tolerant *Thlaspi caerulescens* in Western Europe: evidence from genetic studies based on isoenzyme analysis. Biochem Syst Ecol 26:823–838

Kovacik J, Klejdus B, Hedbavny J, Backor M (2010) Tolerance of *Silene vulgaris* to copper: population related comparison of selected physiological parameters. Environ Toxicol 25: 581–592

Krämer U (2010) Metal hyperaccumulation in plants. Annu Rev Plant Biol 61:517–534

Krämer U, Coster-Howells JD, Charnock JM, Baker AJM, Smith JAC (1996) Free histidine as metal chelator in plants that accumulate nickel. Nature 379:635–638

Krämer U, Smith RD, Wenzel WW, Raskin I, Salt RD (1997) The role of metal transport and tolerance in nickel hyperaccumulation by *Thlaspi goesingense*. Plant Physiol 115:1641–1650

Lambinon J, Augier P (1964) La flore et la végétation des terrains calaminaires de la Wallonie septentrionale et de la Rhénanie axoise. Types chorologiques et groupes écologiques. Natura Monsana 16:113–130

Li Y-M et al (2003) Development of a technology for commercial phytoextraction of nickel: economic and technical considerations. Plant Soil 249:107–115

Liao XY, Chen TB, Lei M, Huang ZC, Xiao XY, An ZZ (2004) Root distributions and elemental accumulations of Chinese brake (*Pteris vittata* L.) from As-contaminated soils. Plant Soil 261:109–116

Lucassen E, Van Kempen MML, Roelofs JGM, Van der Velde G (2010) Decline in metallophytes in tertiary polluted floodplain grassland in the Netherlands, experimental evidence for metal and nutritional changes in soil as driver factors. Chem Ecol 26:273–287

Lüttge U (1975) Salt glands. In: Baker DA, Hall JL (eds) Ion transport in plant cells and tissues. North-Holland Publishing Comp, Amsterdam, Oxford, pp 335–376

Macnair MR, Bert V, Huitson SB, Saumitou-Laprade P, Petit D (1999) Zinc tolerance and hyperaccumulation are genetically independent characters. Proc R Soc Lond B Biol Sci 266:2175–2179

Maerques APGC, Rangel AOSS, Castro PML (2009) Remediation of heavy metal contaminated soils: phytoremediation as a potentially promising clean-up technology. Crit Rev Environ Sci Technol 30:622–654

Massa N, Andreucci F, Poli M, Aceto M, Barbato R, Berta G (2010) Screening for heavy metal accumulators amongst autochthonous plants in a polluted site in Italy. Ecotoxicol Environ Saf 73:1988–1997

Mauricio R (2001) Mapping quantitative trait loci in plants: uses and caveats for evolutionary biology. Nat Rev Genet 2:370–380

Meyer KF (2006) Kritische Revision der *Thlaspi*-Arten Europas, Afrikas und Vorderasiens. Spezieller Teil IX-Noccaea Moench. Haussknechtia Beiheft, Jena, p 343

Meyer CL, Vitalis R, Saumitou-Laprade P, Castric V (2009) Genomic pattern of adaptive divergence in *Arabidopsis halleri*, a model species for tolerance to heavy metal. Mol Ecol 18:2050–2062

Meyer CL et al (2010) Variability of zinc tolerance among and within populations of the pseudometallophyte species *Arabidopsis halleri* and possible role of directional selection. New Phytol 185:130–142

Mikkelsen TS et al (2005) Initial sequences of the chimpanzee genome and comparison with the human genome. Nature 437:69–87

Mills RF et al (2010) Functional significance of *AtHMA4* C-terminal domain in planta. PLoS One 5(10):e13388

Muthukumar T, Bagyaraj DJ (2010) Use of arbuscular mycorrhizal fungi in phytoremediation of heavy metal contaminated soils. Proc Natl Acad Sci India Sect B Biol Sci 80:103–121

Nauenburg JD (1986) Untersuchungen zur Variabilität, Ökologie und Systematik der Viola tricolor-Gruppe in Mitteleuropa, Thesis, The University of Göttingen, Germany, pp 126

Noret N, Meerts P, Tolrà RP, Poschenrieder C, Barceló D, Escarré J (2005) Palatability of *Thlaspi caerulescens* for snails: influence of Zn and glucosinolates. New Phytol 165:763–773

Noret N, Josens G, Escarré J, Lefébvre C, Panichelli S, Meerts P (2007) Development of *Issoria lathonia* (Lepidoptera: Nymphalidae) on zinc-accumulating and nonaccumulating *Viola* species (Violaceae). Environ Toxicol Chem 26:565–571

Orłowska E, Zubek S, Jurkiewicz A, Szarek-Lukaszewska G, Turnau K (2002) Influence of restoration on arbuscular mycorrhiza of *Biscutella laevigata* L. (Brassicaceae) and *Plantago lanceolata* L. (Plantaginaceae) from calamine spoil mounds. Mycorrhiza 12:153–160

Ouziad F, Hildebrandt U, Schmelzer E, Bothe H (2005) Differential gene expressions in arbuscular mycorrhizal-colonized tomato grown under heavy metal stress. J Plant Physiol 162:634–649

Patzke W, Brown G (1990) *Festua aquisgranensis* sp. *nova* ein neuer Vertreter der Kollektivart *Festuca ovina* L. Decheniana 143:194–195

Pauwels M, Saumitou-Laprade P, Holl AC, Petit D, Bonnin I (2005) Multiple origin of metallicolous populations of the pseudometallophyte *Arabidopsis halleri* (Brassicaceae) in central Europe: the cpDNA testimony. Mol Ecol 14:4403–4414

Pauwels M, Frérot H, Bonnin I, Saumitou-Laprade P (2006) A broad-scale analysis of population differentiation for Zn tolerance in an emerging model species for tolerance study: *Arabidopsis halleri* (Brassicaceae). J Evol Biol 19:1838–1850

Pence NS et al (2000) The molecular physiology of heavy metal transport in the Zn/Cd hyperaccumulator *Thlaspi caerulescens*. Proc Nat Acad Sci USA 97:4956–4960

Perez-Figueroa A, Garcia-Pereira MJ, Saura M, Rolan-Alvarez E, Caballero A (2010) Comparing three different methods detect selective loci using dominant markers. J Evol Biol 23:2267–2276

Pilon-Smits E (2005) Phytoremediation. Annu Rev Plant Biol 56:15–39

Prasad MNV, Hagemeyer J (eds) (1999) Heavy metal stress in plants – from molecules to ecosystems. Springer, Heidelberg

Ramirez-Rodriguez V, Lopez-Bucio J, Herrera-Estrella L (2005) Adaptive responses in plants to nonoptimal soil pH. In: Jenks MA, Hasegawa PM (eds) Plant abiotic stress. Blackwell, Oxford, pp 145–170

Regvar M et al (2003) Colonization of pennycresses *Thlaspi* sp. of the Brassicaceae by arbuscular mycorrhizal fungi. J Plant Physiol 160:615–626

Robinson RH, Lombi E, Zhao FJ, McGrath SP (2003) Uptake and distribution of nickel and other metals in the hyperaccumulator *Berkheya coddii*. New Phytol 158:279–285

Roosens N, Willems G, Gode C, Courseaux A, Saumitou-Laprade P (2008) The use of comparative genome analysis and systemic relationships allows extrapolating the position of Zn tolerance QTL regions from *Arabidopsis halleri* into *Arabidopsis thaliana*. Plant Soil 306:105–116

Rozema J, Gude H, Pollak G (1981) An ecophysiological study of the salt excretion of four halophytes. New Phytol 89:201–207

Salt D (2001) Responses and adaptations of plants to metal stress. In: Hawkesford MJ, Buchner P (eds) Molecular analysis of plant adaptation to the environment. Kluwer Academic Publishers, Dordrecht

Salt DE, Smith RD, Raskin RD (1998) Phytoremediation. Annu Rev Plant Physiol Plant Mol Biol 49:643–668

Schat H, Tenbookum VM (1992) Genetic control of copper tolerance in *Silene vulgaris*. Heredity 68:219–229

Schat H, Vooijs R (1997) Multiple tolerance and co-tolerance to heavy metals in *Silene vulgaris*: co-segregation analysis. New Phytol 136:489–496

Schat H, Kuiper E, TenBookum WM, Vooijs R (1993) A general model for the genetic control of copper tolerance in *Silene vulgaris*: evidence from crosses between plants from different tolerant populations. Heredity 70:142–147

Scheloske S, Maetz M, Schneider T, Hildebrandt U, Bothe H, Povh H (2004) Element distribution in mycorrhizal and nonmycorrhizal roots of the halophyte *Aster tripolium* determined by proton induced X-ray emission. Protoplasma 223:183–189

Schwickerath M (1944) Das Hohe Venn und seine Randgebiete. Gustav Fischer, Jena

Shahzad Z et al (2010) The five *AhMTP1* Zinc transporters undergo different evolutionary fates towards adaptive evolution to zinc tolerance in *Arabidopsis halleri*. PLoS Genet 6(4): e1000911

Slomka A et al (2011) Violets of the section Melanium, their colonization by arbuscular mycorrhizal fungi and their occurrence on heavy metal heaps. J Plant Physiol 168:1191–1199

Smith SE, Macnair MR (1998) Hypostatic modifiers cause variations in degree of copper tolerance in *Mimulus guttatus*. Heredity 80:760–768

Sun Q, Ye ZH, Wang X-R, Wong MH (2007) Cadmium hyperaccumulation leads to an increase of glutathione rather than phytochelatins in the cadmium hyperaccumulator *Sedum alfredii*. J Plant Physiol 164:1489–1498

Tonin C, Vandenkoornhuyse P, Joner EJ, Strczek J, Leyval C (2001) Assessment of arbuscular mycorrhizal fungi diversity in the rhizosphere of *Viola calaminaria* and effect of these fungi on heavy metal uptake by clover. Mycorrhiza 10:161–168

Turnau K, Mesjasz-Przybylowicz J (2003) Arbuscular mycorrhiza of *Berkheya coddii* and other Ni-hyperaccumulating members of Asteraceae from ultramafic soils in South Africa. Mycorrhiza 13:185–190

Turnau K, Ostachowicz B, Wojtczak G, Anielska T, Sobczyk L (2010) Metal uptake by xerothermic plants introduced into Zn-Pb industrial wastes. Plant Soil 337:299–311

Van de Mortel JE et al (2008) Expression differences for genes involved in lignin, glutathione and sulphate metabolism in response to cadmium in *Arabidopsis thaliana* and the related *Thlaspi caerulescens*. Plant Cell Environ 31:301–324

Verbruggen N, Hermans C, Schat H (2009) Molecular mechanisms of metal hyperaccumulation in plants. New Phytol 181:759–776

Vogel-Mikuš K, Pongrac P, Kump P, Necemer M, Regvar M (2006) Colonization of a Zn, Cd and Pb hyperaccumulator *Thlaspi praecox* Wulfen with indigenous arbuscular mycorrhizal mixture induces changes in heavy metal and nutrient uptake. Environ Pollut 139:362–371

Weissenhorn I, Leyval C (1993) Cd-tolerant arbuscular mycorrhizal (AM) fungi from heavy-metal polluted soils. Plant Soil 157:247–256

Westhoff A (1999) Mycorrhizal plants for phytoremediation of soils contaminated with radionuclides. Restor Reclam Rev (Student On-line Journal) 5(4)

Willems G, Drager DB, Courbot M, Gode C, Verbruggen N, Saumitou-Laprade P (2007) The genetic basis of zinc tolerance in the metallophyte *Arabidopsis halleri* ssp. *halleri* (Brassicaceae) An analysis of quantitative trait loci. Genetics 176:659–674

Willems G, Frérot H, Gennen J, Salis P, Saumitou-Laprade P (2010) Quantitative trait loci analysis of mineral element concentration in an *Arabidopsis halleri* × *Arabidopsis lyrata petraea* F_2-progeny grown on cadmium-contaminated soil. New Phytol 187:368–379

Wu L, Bradshaw AD, Thurman DA (1975) Potential for evolution of heavy-metal tolerance in plants. 3. Rapid evolution of the copper tolerance in *Agrostis stolonifera*. Heredity 34:165–172

Wu CA, Liao B, Wang SL, Zhang J, Li JT (2010) Pb and Zn accumulation in a Cd-hyperaccumulator (*Viola baoshanensis*). Int J Phytoremediation 12:574–585

Zelko I, Lux A, Czibula K (2008) Difference in the root structure of hyperaccumulator *Thlaspi caerulescens* and non-hyperaccumulator *Thlaspi arvense*. Int J Environ Pollut 33:123–132

Chapter 3
Functional Significance of Metal Ligands in Hyperaccumulating Plants: What Do We Know?

Marjana Regvar and Katarina Vogel-Mikuš

3.1 Introduction

Due to their physical and chemical properties, the transition metals (e.g., Mn, Fe, Ni, Cu, Zn) can act as specific cofactors for numerous metalloproteins that are involved in different metabolic processes in the cell. Such processes include respiration, nitrogen fixation, and photosynthesis, and thus these transition metals are involved in the maintenance of the functional and structural integrity of plant cells. On the other hand, when they are present in excess, the transition metals can interfere with many metabolic processes, mainly through inhibition of correct protein/enzyme functioning and by causing oxidative damage. At metal-polluted sites, plants also have to deal with the presence of nonessential metals and metalloids (e.g., Cd, Pb, Hg, As, Se), therefore the regulation of their metal homeostasis can be even more challenging.

Metal-hyperaccumulating plants represent a relatively minor group of plant species, and they can accumulate and tolerate metals in their shoots to levels that are several orders of magnitude higher than in other plants. The phenomenon of metal hyperaccumulation has puzzled scientists since its first description by Baumann in 1885 (Brooks 2000). The challenge of holding metal levels within vital limits in the cell requires metal homeostasis mechanisms that can balance the activities of the transporters that mediate import of these metals into the cell, their distribution to the organelles, and their export from the cell. This homeostasis needs metal sensors that must correctly distinguish between the particular inorganic elements (Waldron et al. 2009).

Transcriptional control is especially important in the regulation of this essential transition-metal homeostasis in plants (Pilon et al. 2009). Recent studies have

M. Regvar (✉) • K. Vogel-Mikuš
Department of Biology, Biotechnical Faculty, University of Ljubljana, Večna pot 111, SI-1000 Ljubljana, Slovenia
e-mail: marjana.regvar@bf.uni-lj.si

I. Sherameti and A. Varma (eds.), *Detoxification of Heavy Metals*, Soil Biology 30,
DOI 10.1007/978-3-642-21408-0_3, © Springer-Verlag Berlin Heidelberg 2011

indicated that the copy number and expression of specific metal-transporter genes can impact metal hyperaccumulation by triggering root metal deficiency, and consequentially enhancing metal uptake (Hanikenne et al. 2008). In addition, under such circumstances, preventing the association of the inappropriate metals with proteins might be even more challenging than the acquisition of the correct metal by the proteins (Waldron et al. 2009). Consequently, there is still considerable ambiguity around the basic questions of why and how metal hyperaccumulation actually evolved (Rascio and Navari-Izzo 2010).

3.2 Problems in Maintaining Metal Homeostasis

When in contact with organic material, metals tend to bind to specific functional groups, or ligands, of the organic molecules. The four classes of functional polypeptide-derived or *endogenous* ligands recognized in plants are: (a) amino groups at the N terminus; (b) carboxylate groups at the C terminus; (c) carbonyl and amide groups of the main amino acid chain; and (d) side groups of the main amino acid chain (e.g., amide, amino, carboxyl, hydroxyl, imidazole, phenol, selenol, sulfide, thiol groups). Ligands that do not derive from polypeptides are known as *exogenous*, and these range from simple inorganic entities (e.g., oxide, hydroxide, sulfide, water, and other solvent-derived molecules, and physiological ligands, such as dioxygen and nitric oxide) to polydentate organic compounds (e.g., porphyrins and corrins) (Degtyarenko 2000).

The natural order of stability of complexes with transient divalent metals is known as the Irving–Williams series. For high-spin complexes of divalent ions of the first-row transition metals, the stability constants for the formation of complexes follows the order: Mn(II) $<$ Fe(II) $<$ Co(II) $<$ Ni(II) $<$ Cu(II) $>$ Zn(II), and this order has been shown to hold for a wide variety of ligands. When compared with other divalent cations, such as Ca and Mg, the stabilities of these transition-metal complexes are much greater (Irving and Williams 1948). Biology must therefore overcome this trend for some proteins to bind uncompetitive metals and for others to bind competitive ones.

To achieve metal homeostasis, the amount of each metal within a cell has to be sensed, to allow adjustments of the actions of the transporters at the plasma membrane or on internal compartments. Metal importers, metal exporters, and metal stores maintain a limited supply of competitive metals. At low metal concentrations in plants, proteins compete with other molecules for these metals, rather than the metals competing with other metals for the proteins. Numerous proteins have therefore developed dependencies on metal species that cannot be readily replaced. In extreme cases only, metals with similar properties can be replaced with metals that might have once been plentiful, but are now scarce (Waldron et al. 2009). On metal-polluted sites, however, the metals that are in abundance can compete with the original metals for the active binding sites on proteins causing metal imbalances and misfunctioning. Obviously, metal-hyperaccumulating plants have found solutions to overcome these situations.

One way to ensure that the correct metal is acquired by a metalloenzyme is to exploit the delivery proteins, i.e., the metallochaperones. Metals are passed from metallochaperones to cognate apoproteins by means of ligand-exchange reactions that determine which proteins gain access to the metals supplied (Waldron et al. 2009). A complete inventory of *Arabidopsis thaliana* metallochaperone-like proteins reveals a large family of 67 proteins. These include proteins containing heavy-metal-associated (HMA) domains and heavy metal ATPases (Tehseen et al. 2010). Metallochaperones in plants provide the plant needs during the life cycle and contribute to metal redistribution to storage organs during plant senescence (Mira et al. 2001). The presumed roles of these metallochaperones at the cellular level is illustrated in Fig. 3.1.

When metals are present at toxic concentrations, the metallochaperones can bind the excess metals, and hence they contribute to metal detoxification. Homologs of all three groups of metallochaperones described in yeast have also been described in *Arabidopsis*. By analogy to their functions in other organisms, the ATX1-like Cu chaperones interact with the P-type ATPase HMA5, with a presumable role in Cu compartmentalization and detoxification (Andrés-Colás et al. 2006). This might

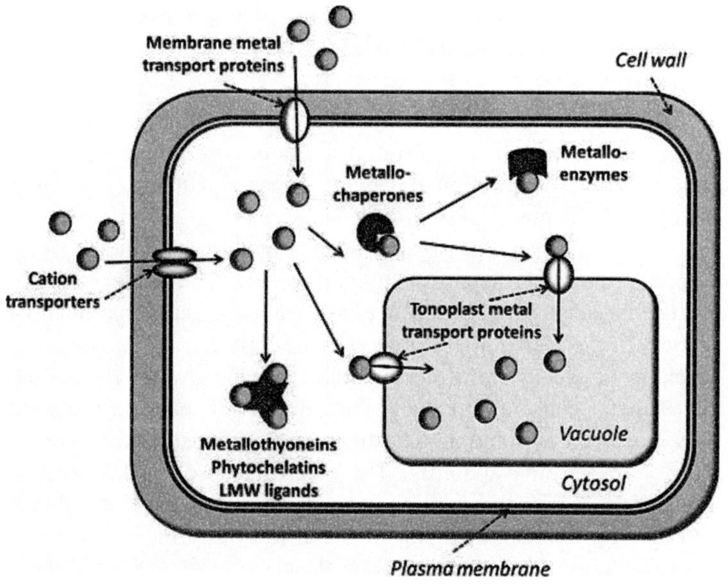

Fig. 3.1 A proposed role of metalchaperones in plant cell is trafficking metals to metalloenzymes. When metals are present in access metallochaperones are proposed to interact with tonoplast metal transporters to facilitate metal compartmentation and detoxification. In addition, low cytosolic free metal concentrations can also be maintained by scavenging systems including phytochelatins, metallothioneins, and low molecular weight (LMW) ligands such as amino acids, glutathione, and LMW thiols

also apply to metal-hyperaccumulating plant species. In the metal-hyperaccumulating *Thlaspi caerulescens*, the overexpression of the P-type ATPase HMA4 confers increased Cd tolerance through its functioning as a Cd-efflux transporter. This contribution to enhanced xylem loading makes HMA4 a key player in the metal-hyperaccumulation character of *T. caerulescens* (Papoyan and Kochian 2004), probably with the engagement of one of the metallochaperones. The involvement of HMA4 in metal hyperaccumulation has also been indicated by new insights into the metal hyperaccumulation and hypertolerance of *Arabidopsis halleri*. After a reduction in the expression of *AhHMA4* using RNA interference, Hanikenne et al. (2008) demonstrated that both, an increase in the copy number, and an elevated expression of the individual *AhHMA4* gene copies contribute to hyperaccumulation and hypertolerance of Zn by enhancing the xylem loading and transport of Zn to shoots, and acting as a physiological master switch in the up-regulation of Zn-deficiency-response gene expression in roots. Clearly, more studies are needed before we can determine whether metal hyperaccumulation per se developed as a metal homeostasis response to metal imbalances that are triggered by metal deficiency in plants, and decipher the role(s) of the metallochaperones in the metal homeostasis of metal-hyperaccumulating plants.

3.3 Soil-Metal Partitioning and Toxicity

Metal speciation and the related solid–liquid partitioning greatly affect the mobility and bioavailability of metals in the soil. The major determinants that affect soil metal speciation are: (a) adsorption of metals to organic matter, oxyhydroxides, and clay minerals; and (b) precipitation of metals as pure or mixed solids, and their complexation with inorganic or organic ligands or mineral colloids. Among the properties of the soil, pH is the most important factor affecting the retention of free metal ions (Degryse et al. 2009). Indeed, total metal concentrations are known to be poor predictors of metal toxicity (Smolders et al. 2009).

The leaching of metals from the soil into the groundwater is also related to the dissolved metal concentrations and the amounts of metal in the solid phase that buffers the metal in solution. The free metal is considered to be the major determinant of metal bioavailability. The part of the solid-phase metal that is rapidly exchangeable with the solution phase is known as the "labile" phase, and this should also be a better predictor of metal-ion buffering in solute transport models than the total metal concentrations (Degryse et al. 2009). Therefore, the uptake of trace metals from solution by plants is commonly believed to depend on the free metal ion activity, rather than on the total metal concentration, and this has been explained by the free-ion activity model (FIAM) of the toxicities of trace metals towards higher plants. Many strong synthetic chelators provide further evidence for this view, whereas the presence of weak soluble ligands (e.g., organic acids and amino acids) alters the aqueous speciation of the trace metals and their availability to higher plants (Parker et al. 2001). Consequently, the uptake of metals such as Cd and Zn from such solutions might actually be greater than the uptake from ligand-free solutions with the same Cd and Zn ion activities. This can be partially explained by intact ligand–metal uptake complexes, and partially by breaks in the endodermal barrier of root apices and of the lateral root initiation sites, where metal–ligand complexes are transported with the water flow via the apoplastic pathway. Unfortunately, reported exceptions to the FIAM do not appear to occur in a predictable and systematic way, with regard to metal or organic ligand strengths, or plant species (McLaughlin et al. 1997; Parker et al. 2001).

In bulk soils, the situation is even more complex because of limitations in metal transport to the root surface, complexation of metals with organic ligands in the soil, and the tendency of roots to alter the chemistry of the rhizosphere (Degryse et al. 2006). Metal-hyperaccumulating plants are known to form "hot spots," meaning that in the rhizosphere of metal-hyperaccumulating plants, the metal concentrations are greatly increased, compared to those of non-metal-accumulating plants. As metals are known to stabilize the soil organic metal content (Lützow et al. 2006), which is partially due to decreased activities of the soil microorganisms that are involved in organic-matter decomposition, it is of no particular surprise that more organic matter is found in the *Thlaspi goesingense* rhizosphere than in that of non-metal-accumulating plants from the same sampling site (Wenzel et al. 2003). An increase in the organic matter content might therefore enhance metal uptake further through the uptake of intact metal–organic complexes. Indeed, field-collected data reinforce the hypothesis relating to ligand-induced Ni solubility in the rhizosphere of *T. goesingense*, which was attributed to the formation of Ni-organic acid complexes (Wenzel et al. 2003). In addition, increased mobilization of Cd and Zn in the rhizosphere of the metal-hyperaccumulating *Sedum alfredii* can at least partially be attributed to the synthesis of short-chain organic acids (e.g., oxalic, tartaric, formic, acetic acids) by the rhizosphere bacteria associated with the roots, although the presence of metals in the substrate might have a large effect on this property (Li et al. 2010). Despite their importance, metal-tolerance mechanisms of arbuscular mycorrhizal fungi have been poorly documented, which is primarily due to the difficulties in their cultivation. Indeed, there are some discrepancies in the current literature regarding the ability of arbuscular mycorrhizal fungi to exude organic acids. It appears likely, however, that there are substantial differences between fungal isolates in this regard, which might also be related to metal-toxicity alleviation, as demonstrated in the case of aluminum (Klugh and Cumming 2007). Although there are now few hundreds of known metal-hyperaccumulating plants, it is estimated that to date only 10% of these have had their rhizospheres examined (Alford et al. 2010). Therefore, further research is needed to resolve the diversity of rhizosphere microorganisms of metal-hyperaccumulating plants and their interactions that are involved in metal uptake.

3.4 Metal Immobilization and Transport in the Root

From the large body of evidence that has accumulated over the last decade, it has become obvious that not one, but rather several mechanisms provide metal immobilization and promote tolerance to elevated metal concentrations in metal-hyperaccumulating plants. This system begins immediately in the poorly explored apoplasts, where metal ions are first adsorbed by the carbohydrate network of the root/cell wall. The dependence of metal uptake on pH indicates that acidic carboxyl (R–COOH) and hydroxyl (R–OH) groups of cell wall constituents, together with amines, phosphates, thiols, and others, are the most likely sites for ion exchange (Kratochvil and Volesky 1998). In addition to physical and chemical properties of a metal, metal–cell wall interactions also largely depend on the number of particular functional groups with which the metals can interact. When comparing monocotyledonous and dicotyledonous plant species, the later have higher pectin content and higher numbers of available free carboxyl and hydroxyl groups in their cell walls, and so these can adsorb more metals on the root surface because of this greater cation-exchange capacity (Mari and Lebrun 2005). The capacity of the root cell walls probably governs further uptake of metals by symplastic pathways up to the shoots, and it is well established that dicotyledonous plants are more efficient metal accumulators than monocotyledonous plants.

In the root cortex, metals and metal complexes have to enter the symplast before they can pass through the endodermis (Wang et al. 2010). In the cytoplasm, metals can either bind to free amino acids, the nonproteinogenic nicotianamine, protein ligands that are rich in Cys residues, such as metallothioneins, metallochaperones, phytochelatins, and low molecular weight thiols (Callahan et al. 2006; Krämer and Clemens 2005; Hernández-Allica et al. 2006; Trampczynska et al. 2010).

The chelation of metals with specific ligands can influence metal partitioning and transport. Root Zn in *T. caerulescens* was shown to be coordinated to histidine, which is believed to have an important role in Zn homeostasis in the root (Salt et al. 1999). Histidine also has a relatively high association constant for Ni. In addition, exposure of the hyperaccumulating *Allysum lesbiacum* to Ni elicitates high levels of free histidine synthesis and enhanced xylem loading. Histidine has therefore been proposed to be an important determinant of Ni accumulation in this species. However, the same property was not confirmed for *T. goesingense* (Kerkeb and Krämer 2003; Persans et al. 1999).

Root phytochelatin synthesis is accompanied by higher levels of Cd retention in the roots of the nonhyperaccumulating *Thlaspi arvense* compared to the metal-hyperaccumulating *T. caerulescens*, which indicates the importance of phytochelatin synthesis in accelerated metal sequestration in roots, in the vacuoles, of nonaccumulating plant species, when compared to metal hyperaccumulators (Ebbs et al. 2002; Wong and Cobbett 2008). It is generally believed that extensive phytochelatin synthesis that would significantly contribute to hypertolerance in metal-hyperaccumulating plants would be too costly. Instead, the elevated levels of glutathione and low molecular weight thiols that are found in the root tips of *T. goesingense* and *T. caerulescens* (Hernández-Allica et al. 2006) are believed to contribute to reduced production of reactive oxygen species, and the withstanding of relatively high cytosolic metal concentrations (Freeman et al. 2004). Thus, these elevated levels of glutathione probably also have an important role in the enhancement of xylem loading.

The metallothioneins are metal-binding proteins that show differential expression patterns in differing plant organs. It was proposed that Cys arrangements in the metallothioneins confer differential metal-binding properties to the different molecules. Although their exact functions remain a matter of debate, they appear to function in metal tolerance and/or homeostasis. They can bind six to eight Cu, Zn, Cd, Hg, and Pb ions as M_3 or M_4 clusters. In addition to their functions as safe depositories of metal ions, they might also allow metal delivery to transport proteins (Callahan et al. 2006). Expression of metallothionein genes in metal-hyperaccumulating plants can also exceed the expression found in nonaccumulating plant species, which supports their role in metal hyperaccumulation. In addition, the expression patterns of the metallothionein genes can differ across hyperaccumulators and nonaccumulators. However, it has been suggested that these are not directly involved in metal accumulation, but rather contribute to metal tolerance, presumably by contributing to the elemental homeostasis under excess metal conditions (Rosens et al. 2004).

In the xylem sap of *T. caerulescens*, Zn can be found as its free ion or it can be chelated to organic acids that are responsible for root-to-shoot transport. Apparently, at low pH values of the xylem sap (pH ~5.5), the imidazole nitrogen of histidine becomes protonated, which results in a decrease in the stability of the Zn–histidine complex, which favors complexation of Zn with citrate (Salt et al. 1999). About half of the cellular Zn has also been shown to be bound to nicotianamine, which has been shown to be a predominant chelator for Zn shoot transport rather than for Zn storage (Mari et al. 2006; Trampczynska et al. 2010). In addition to forming stable complexes with Zn, nicotianamine can also bind Fe, Ni, Cu, Mn, and Co. Transporters from the yellow stripe-like (YSL) transporter family are believed to be involved in long-distance transport of metal–nicotianamine complexes in plants, and in roots they are believed to mediate metal uptake into cells that are directly involved in xylem loading (Curie et al. 2009). On the other hand, root xylem loading of Cd by HMA2 and HMA4 was recently described as the decisive mechanism for Cd translocation to the shoot in *A. thaliana* (Wong and Cobbett 2008). Although small differences can be seen even between different populations, Cd in the *Thlaspi* species is mostly transported coordinated to thiols and organic acids (Küpper et al. 2004; Ebbs et al. 2008; Vogel-Mikuš et al. 2010).

3.5 Metal Immobilization and Compartmentalization in the Shoot

Metal-localization studies have shown that, in general, mechanisms of metal accumulation at the plant organ, tissue, and cellular levels lead to metal immobilization within tissues that are less metabolically active, thus maintaining plant functions across a wide range of metal concentrations (Küpper et al. 1999; Küpper et al. 2001; Vogel-Mikuš et al. 2008). Epidermal cells of leaves are recognized as efficient sinks for Ni, Zn, and Cd, and they represent important tolerance mechanism in the *Thlaspi* species (Küpper et al. 1999; Küpper et al. 2001; Krämer et al. 2000; Vogel-Mikuš et al. 2008). However, as the mesophyll is composed of considerably larger volume percentage of the leaf, the total mesophyll Cd and Zn concentrations might be even more important as metal sinks, despite the lower mesophyll concentrations when compared to epidermal tissues, with the vacuoles of mesophyll cells as the predominant stores (Ma et al. 2005; Vogel-Mikuš et al. 2008). In plants with trichomes, trichome cells might accommodate the majority of Cd and Zn, as demonstrated for *A. halleri* (Küpper et al. 2000). At the cellular level, the most pronounced mechanism of metal accumulation is sequestration into the vacuole, although in aluminum-accumulating tea plants, for example, the predominant mechanism of metal immobilization is through binding of the aluminum to the cell walls of epidermal cells (Tolrà et al. 2011).

Immobilization and compartmentalization of metals in metabolically less active parts of tissues and cells indicate that the processes of metal homeostasis are highly regulated and involve several metal-chelating molecules that can be found in the cytoplasm, the vacuole, and even in the cell wall. Ligands that bind metals more tightly (strong ligands) are considered to be mainly in the cytosol, especially in young plant tissues, where the vacuoles are not yet fully developed (Küpper et al. 2004; Vogel-Mikuš et al. 2010). In the vacuoles, however, organic acids that are considered to be weak metal ligands usually provide the majority of the binding sites for metal ions.

Early studies of the coordination of metals in metal-hyperaccumulating plants showed that histidine coordinates Ni, but it was not recognized as the primary determinant of Ni hyperaccumulation in *T. goesingense* (Persans et al. 1999). Rather, the Ni–histidine complexes appeared to be present in the cytoplasm, whereas most of the intracellular Ni found in the vacuoles was coordinated to citrate (Krämer et al. 2000), thus supporting the above-mentioned hypothesis. In addition, complexation of Ni with nicotianamine can facilitate vacuolar Ni transport (Trampczynska et al. 2010). The ability of *T. goesingense* to hyperaccumulate Ni seems to be governed by enhanced accumulation of Ni within the leaf vacuoles. The transporter responsible for this vacuolar accumulation, TgMTP1, has also been recognized to confer tolerance to Ni, Cd, Co, and Zn (Persans et al. 2001). Enhanced cellular protection of *T. goesingense* from lipid peroxidation is additionally provided by elevated glutathione levels (Freeman et al. 2004). Similarly, most of the Zn in *T. caerulescens* is coordinated to citrate, whereas chelation with histidine, the cell wall, and oxalate contribute little to Zn shoot immobilization (Salt et al. 1999). High organic acid concentrations are a constitutive property of *T. caerulescens*, and these might be responsible for the high Zn requirement in this species due to extensive vacuolar sequestration (Tolrà et al. 1996). Instead, shoot Cd appears to be coordinated mainly with oxygen or sulfur ligands depending on the developmental stage of the plant (Küpper et al. 2004; Vogel-Mikuš et al. 2010). As the storage of metals coordinated with oxygen ligands in vacuoles of leaves is presumably less costly than the synthesis of strong metal-specific ligands, vacuolar Cd storage is the predominant tolerance mechanism of mature and senescent leaves. In young tissues, however, sulfur ligands predominate, which might also reflect smaller proportions of metals being sequestered in vacuoles. For phloem transport, Cd is bound to thiol (–SH) ligands, most likely glutathione, although it might also be bound to nicotianamine, since in *T. praecox*, around one third of the ligands present in the phloem belong to O/N groups. In seeds, as well as the sulfur ligands that represented two-thirds of all of the Cd ligands, phytate has been suggested as an additional Cd-coordination molecule that is not found in vegetative tissues. Before vacuoles become fully functional, coordination of Cd to strong ligands in embryonic tissues seems to be a reasonable solution related to ontogenesis. In addition, larger portions of Zn and Cd coordination to histidine in old and senescent leaves reinforce the hypothesis on ontogenesis as an important determinant for metal coordination, which is presumably related to metal remobilization during senescence and the related deterioration of vacuoles (Küpper et al. 2004; Vogel-Mikuš et al. 2010).

Metal coordination has also been reported to change with metal concentration. By increasing the shoot Cd concentrations in *T. caerulescens*, the proportion of oxygen ligands in mature leaves decreases relative to the proportions of histidine and sulfur ligands (Küpper et al. 2004). This indicates that increasing Cd concentrations in tissues lead to increases in the proportion of cytosolic Cd. However, more studies will be needed to reveal the primary ligands of the different metals over the wider range of metal concentrations.

3.6 In Situ Analyses of Metal Ligands in Plants

Due to the reactivities of these metals towards biological components, metal immobilization is considered to be the key step in metal hypertolerance. The mechanisms of metal immobilization have, however, received significantly less attention than the mechanisms of metal uptake and transport, primarily because of the need for the development and adoption of noninvasive and nondestructive techniques that can be aimed at precisely identifying the localization of metal ions within plant tissues and the ligands involved in their chelation (Mari and Lebrun 2005).

Even today, with all of the advantages of the modern analytical equipment that is available to us, the determination of metal ligands in plant tissues still represents a challenge. This is due to the relatively low metal concentrations in the tissues and the ease with which metal–ligand complexes can be lost or modified during sample preparation. The speciation and chemical form of metals in plants can be determined by specialized techniques, such as HPLC coupled to element-specific detection by inductively coupled plasma mass spectrometry (ICP-MS) or electron-spray ionization mass spectrometry (ESI-MS). The most popular technique, HPLC-ICP-MS, uses conventional HPLC with either reverse-phase or ion-exchange columns to separate the metal–ligand compounds, which are then detected by an ICP-MS tuned to specific metal isotopes. However, to be able to propose a reliable structure of the metal–ligand complexes detected, in addition to this mass spectrometry data, chemical and physiochemical data are often required.

From the plant physiology point of view, the main disadvantage of these techniques is that the metal–ligand compounds have to be removed from the matrix (extracted). During the extraction procedure, the metal–ligand environment can be altered by chemical pretreatment of the samples, as the release of the contents from different cell compartments can result in the formation of complexes that are not present in vivo (Callahan et al. 2006). To avoid the studying of artifacts due to sample preparation procedures, the use of complementary techniques, such as nuclear magnetic resonance (NMR), and direct speciation methods using either X-ray absorption or diffraction techniques (e.g., extended X-ray absorption fine structure [EXAFS], X-ray absorption near-edge structure [XANES], X-ray diffraction [XRD], and microprobe analysis) that give access to structural information of solid samples is of great importance in these types of studies. These techniques mentioned have become increasingly popular because of the minimal sample preparation needed and the resulting decreased risk of species transformation (Caruso et al. 2003). In metal-hyperaccumulating plants, the majority of studies that have dealt with identification of metal ligands have used the EXAFS and XANES X-ray absorption techniques, e.g., Persans et al. (1999) for determination of Ni coordination in *T. goesingense*; Küpper et al. (2004) for determination of Cd and Zn ligands in leaves of *T. caerulescens*; and Vogel-Mikuš et al. (2010) for determination of Cd ligands in vegetative tissues and seeds of *T. praecox*. With the development of synchrotron radiation stations with beam lines that are dedicated to (sub)micro-X-ray fluorescence mapping, it now even becomes possible to localize the metals in tissues at the cellular and subcellular levels. XANES and/or EXAFS spectra can also be recorded to provide information concerning the chemical oxidation stages and coordination geometries of elements in complexes (XANES), and information on the local environment of the atoms investigated in the sample (i.e., the number and species of neighboring atoms, their distance from the selected atom, and the thermal or structural disorder of their positions) (Gardea-Torresday et al. 2005 and ref. therein, Fakuda et al. 2008). Typical examples of these types of studies can be found in Isaure et al. (2006) and Fakuda et al. (2008), where micro-XRF was used for the localization of Cd, and micro-XANES for the determination of the Cd ligands in *A. thaliana* and *A. halleri*. In addition, Kachenko et al. (2010) used the same techniques for localization and arsenic speciation in the leaves of the fern *Pityrogramma calomelanos* var. *austroamericana*.

3.7 Conclusions

There are thus several levels of metal immobilization in plants. Some appear to have developed from the need to deliver metals to metalloproteins, whereas others appear to result from more specific responses of metal-hyperaccumulating plants to high metal concentrations. There are clear differences in the regulation of metal homeostasis between non-metal-accumulating and metal-hyperaccumulating plants. To date, little is known about the plant metal perception and metal sensing, although the greater metal requirements of metal-hyperaccumulating plants might arise from reduced metal sequestration in the root and more intensive root-to-shoot metal transport. Recent discoveries lead us to believe that metal hyperaccumulation is connected to the self-imposed metal deficiency imbalance responses that are regulated by mechanisms of metal homeostasis and that are just beginning to emerge. Intense metal sequestration in the shoot, on the other hand, provides mechanisms for intense metal tolerance of hyperaccumulating plants. Taken together, it appears likely that metal-hyperaccumulating plants have what it takes for successful remediation of metal-enriched soils. The question remains, do we know how to successfully explore these plant traits to achieve our goals?

References

Alford ÉR, Pilon-Smits EAH, Paschke MW (2010) Metallophytes – a view from the rhizosphere. Plant Soil 337:33–50

Andrés-Colás N, Sancenón V, Rodríguez-Navarro S, Mayo S, Thiele DJ, Ecker JR, Puig S, Peñarrubia L (2006) The Arabidopsis heavy metal P-type ATPase HMA5 interacts with metallochaperones and functions in copper detoxification of roots. Plant J 45:225–236

Brooks RR (2000) Plants that hyperaccumulate heavy metals. CAB International, Wallingford

Callahan DL, Baker AJ, Kolev SD, Wedd AG (2006) Metal ion ligands in hyperaccumulating plants. J Biol Inorg Chem 11:2–12

Caruso JA, Klaue B, Michalke B, Rocke DM (2003) Group assessment: elemental speciation. Ecotoxicol Environ Saf 56:32–44

Curie C, Cassin G, Couch D, Divol F, Higuchi K, Jean ML, Misson J, Schikora A, Czernic P, Mari S (2009) Metal movement within the plant: contribution of nicotianamine and yellow stripe 1-like transporters. Ann Bot 103:1–11

Degryse F, Smolders E, Parker DR (2006) Metal complexes increase uptake of Zn and Cu by plants: implications for uptake and deficiency studies in chelator-buffered solutions. Plant Soil 289:171–185

Degryse F, Smolders E, Parker DR (2009) Partitioning of metals (Cd, Co, Cu, Ni, Pb, Zn) in soils: concepts, methodologies, prediction and applications – a review. Eur J Soil Sci 60:590–612

Degtyarenko K (2000) Bioinorganic motifs: towards functional classification of metalloproteins. Bioinformatics 6:851–864

Ebbs S, Lau I, Ahner B, Kochian L (2002) Phytochelatin synthesis is not responsible for Cd tolerance in the Zn/Cd hyperaccumulator *Thlaspi caerulescens* (J. & C. Presl). Planta 214:635–640

Ebbs SD, Zambrano MC, Spiller SM, Newiville M (2008) Cadmium sorption, influx, and efflux at the mesophyll layer of leaves from ecotypes of the Zn/Cd hyperaccumulator *Thlaspi caerulescens*. New Phytol 181:626–636

Fakuda N, Hokure A, Kitajima N, Terada Y, Saito H, Abe T, Nakai I (2008) Micro X-ray fluorescence imaging and micro X-ray absorption spectroscopy of cadmium hyper-accumulating plant, *Arabidopsis halleri* ssp. gemmifera, using high-energy synchrotron radiation. J Anal Atom Spectrom 23:1068–1075

Freeman JL, Persans MW, Nieman K, Albrecht C, Peer W, Pickering IJ, Salt DE (2004) Increased glutathione biosynthesis plays a role in nickel tolerance in Thlaspi nickel hyperaccumulators. Plant Cell 16:2176–2191

Gardea-Torresday JL, Peralta-Videa JR, de la Rosa G, Parsons JG (2005) Phytoremediation of heavy metals and study of the metal coordination by X-ray absorption spectroscopy. Coord Chem Rev 249:1797–1810

Hanikenne M, Talke IN, Haydon MJ, Lanz C, Nolte A, Motte P, Kroymann J, Weigel D, Krämer U (2008) Evolution of metal hyperaccumulation required *cis*-regulatory changes and triplication of *HMA4*. Nature 453:391–396

Hernández-Allica J, Garbisu C, Becerril JM, Barrutia O, Garcia-Plazola JI, Zhao FJ, McGrath SP (2006) Synthesis of low molecular weight thiols in response to Cd exposure in *Thlaspi caerulescens*. Plant Cell Environ 29:1422–1429

Irving H, Williams RJP (1948) Order of stability of metal complexes. Nature 162:746–747

Isaure MP, Fayard B, Sarret G, Pairis S, Bourguignon J (2006) Localization and chemical forms of cadmium in plant samples by combining analytical electron microscopy and X ray spectromicroscopy. Spectrochim Acta 61:1242–1252

Kachenko AG, Gräfe M, Singh B, Heald SM (2010) Arsenic speciation in tissues of the hyperaccumulator *P. calomelanos var. austroamericana* using X-ray absorption spectroscopy. Environ Sci Technol 44:4735–4740

Kerkeb L, Krämer U (2003) The role of free histidine in xylem loading of nickel in *Alyssum lesbiacum* and *Brassica juncea*. Plant Physiol 131:716–724

Klugh KR, Cumming JR (2007) Variations in organic acid exudation and aluminum resistance among arbuscular mycorrhizal species colonizing *Liriodendron tulipifera*. Tree Physiol 27:1103–1112

Krämer U, Clemens S (2005) Functions and homeostasis of zinc, copper, and nickel in plants. In: Tamás MJ, Martinoia E (eds) Molecular Biology of Metal Homeostasis and Detoxification. Springer-Verlag, Berlin

Krämer U, Pickering IJ, Prince RC, Raskin I, Salt DE (2000) Subcellular localization and speciation of nickel in hyperaccumulator and non-accumulator *Thlaspi* species. Plant Physiol 122:1343–1353

Kratochvil D, Volesky B (1998) Advances in the biosorption of heavy metals. Tibtech 16:291–300

Küpper H, Zhao F-J, McGrath SP (1999) Cellular compartmentation of zinc in leaves of the hyperaccumulator *Thlaspi caerulescens*. Plant Physiol 119:305–311

Küpper H, Lombi E, Zhao F-J, McGrath SP (2000) Cellular compartmentation of cadmium and zinc in relation to other elements in the hyperaccumulator *Arabidopsis halleri*. Planta 212:75–84

Küpper H, Lombi E, Zhao F-J, Wieshammer G, McGrath SP (2001) Cellular compartmentation of nickel in the hyperaccumulator *Alysum lesbiacum*, *Alyssum berolonii* and *Thlaspi goesingense*. J Exp Bot 52:2291–2300

Küpper H, Mijovilovich A, Meyer-Klaucke W, Kroneck PMH (2004) Tissue- and age-dependent differences in the complexation of cadmium and zinc in the cadmium/zinc hyperaccumulator *Thlaspi caerulescens* (Ganges ecotype) revealed by X-ray absorption spectroscopy. Plant Physiol 134:748–757

Li WC, Ye ZH, Wong MH (2010) Metal mobilization and production of short-chain organic acids by rhizosphere bacteria associated with a Cd/Zn hyperaccumulating plant, *Sedum alfredii*. Plant Soil 326:453–467

Lützow MV, Kögel-Knabner I, Ekschmitt K, Matzner E, Guggenberger G, Marcschner B, Flessa H (2006) Stabilization of organic matter in temperate soils: mechanisms and their relevance under different soil conditions – a review. Eur J Soil Sci 57:426–445

Ma JF, Ueno D, Zhao F-J, McGrath SP (2005) Subcellular localization of Cd and Zn in the leaves of a Cd-hyperaccumulating ecotype of *Thlaspi caerulescens*. Planta 220:731–736

Mari S, Lebrun M (2005) Metal immobilization: where and how? In: Tamás MJ, Martinoia E (eds) Molecular biology of metal homeostasis and detoxification, vol 14. Springer- Verlag, Berlin

Mari S, Gendre D, Pianelli K, Ouerdane L, Lobinski R, Bryat J-F, Lebrun M, Czernic P (2006) Root-to-shoot long-distance circulation of nicotianamine and nicotianamine-nickel chelates in the metal hyperaccumulator *Thlaspi caerulescens*. J Exp Bot 57:4111–4122

McLaughlin MJ, Smolders E, Merckx R, Maes A (1997) Plant uptake of Cd and Zn in chelator-buffered nutrient solution depends on ligand type. In: Ando T et al (eds) Plant nutrition for sustainable food production and environment. Kluwer, Dordrecht

Mira H, Martínez-Garcíja F, Peñarrubia L (2001) Evidence for the plant-specific intercellular transport of the *Arabidopsis* copper chaperone CCH. Plant J 25:521–528

Papoyan A, Kochian LV (2004) Identification of *Thlaspi caerulescens* genes that may be involved in heavy metal hyperaccumulation and tolerance. Characterization of a novel heavy metal transporting ATPase. Plant Physiol 136:3814–3823

Parker DR, Pedler J, Ahnstrom ZAS, Resketo M (2001) Reevaluating the free-ion activity model of trace metal toxicity toward higher plants: experimental evidence with copper and zinc. Environ Toxicol Chem 20:899–906

Persans MW, Yan X, Patnoe J-MML, Krämer U, Salt D (1999) Molecular dissection of the role of histidine in nickel hyperaccumulation in *Thlaspi goesingense* (Hálácsy). Plant Physiol 121:1117–1126

Persans MW, Nieman K, Salt DE (2001) Functional activity and role of cation-efflux family members in Ni hyperaccumulation in *Thlaspi goesingense*. Proc Natl Acad Sci USA 98: 9995–10000

Pilon M, Cohu CM, Ravet K, Abddel-Ghany SE, Gaymard F (2009) Essential transition metal homeostasis in plants. Curr Opin Plant Biol 12:347–357

Rascio N, Navari-Izzo F (2010) Heavy metal hyperaccumulating plants: how and why they do it? And what makes them so interesting? Plant Sci 180:169–181

Rosens NH, Leplae R, Bernard C, Verbruggen N (2004) Evidence for copper homeostasis function of metallothionein (MT3) in the hyperaccumulator *Thlaspi caerulescens*. FEBS Lett 577:9–16

Salt DE, Prince RC, Baker AJM, Raskin I, Pickering IJ (1999) Zinc ligand in the metal hyperaccumulator *Thlaspi caerulescens* as determined using X-ray absorption spectroscopy. Environ Sci Technol 33:713–717

Smolders E, Oorts K, Van Sprang P, Schoeters I, Jansen CR, McGrath SP, McLaughlin MJ (2009) Toxicity of trace metals in soil as affected by soil type and aging after contamination: using calibrated bioavailability models to set ecological soil standards. Environ Toxicol Chem 28:1633–1642

Tehseen M, Carins N, Sherson S, Cobbett C (2010) Metallochaperone-like genes in *Arabidopsis thaliana*. Metallomics 2:556–564

Tolrà RP, Poschenrieder C, Barceló J (1996) Zinc hyperaccumulation in *Thlaspi caerulescens*. II. Influence on organic acids. J Plant Nutr 19:1541–1550

Tolrà RP, Vogel-Mikus K, Hajiboland R, Kump P, Pongrac P, Kaulich B, Gianoncelli A, Babin V, Barceló J, Regvar M, Poschenrieder C (2011) Localization of aluminium in tea (*Camelia sinensis*) leaves using low energy X-ray fluorescence spectro-microscopy. J Plant Res 124 (1):165–172

Trampczynska A, Küpper H, Meyer-Klaucke W, Schmidt H, Clemens S (2010) Nicotianamine forms complexes with Zn(II) *in vivo*. Metallomics 2:57–66

Vogel-Mikuš K, Regvar M, Mesjasz-Pzybyłowicz W, Simčič J, Pelicon P, Budnar M (2008) Spatial distribution of Cd in leaves of metal hyperaccumulating *Thlaspi praecox* using micro-PIXE. New Phytol 179:712–721

Vogel-Mikuš K, Arčon I, Kodre A (2010) Complexation of cadmium in seeds and vegetative tissues of the cadmium hyperaccumulator *Thlaspi praecox* as studied by X-ray absorption spectroscopy. Plant Soil 331:439–451

Waldron KJ, Rutherford JC, Ford D, Robinson NJ (2009) Metalloproteins and metal sensing. Nature 460:823–830

Wang P, Zhou D-M, Li L-Z, Luo X-S (2010) Evaluating the biotic ligand model for toxicity and the alleviation of toxicity in terms of cell membrane surface potential. Environ Toxicol Chem 29:1503–1511

Wenzel WW, Bunkowski M, Pushenreiter M, Horak O (2003) Rhizosphere characteristics of indigenously growing nickel hyperaccumulator and excluder plants on serpentine soil. Environ Pollut 123:131–138

Wong CKE, Cobbett CS (2008) HMA P-type ATPase are the major mechanism for root-to-shoot Cd translocation in *Arabidopsis thaliana*. New Phytol 181:71–78

Chapter 4
Progress in Phytoremediating Heavy-Metal Contaminated Soils

Shao Hongbo, Chu Liye, Xu Gang, Yan Kun, Zhang Lihua, and Sun Junna

4.1 Introduction

Since the industrial revolution, anthropogenic impacts have caused more and more hazardous heavy metals' releasing to environment. Soils, being the basic and most essential part of the ecological system, are heavily contaminated, too. Until now over 20,000,000 acres of farmland in China have been contaminated by heavy metals such as Sn, Cr, Pb, and Zn, which account for almost one-fifth of the total arable farmland area. Every year, China suffers a 10,000,000 tons' loss of crop output due to deteriorating heavy metal pollution (Wang and Ma 2008; Shao et al. 2010; Bi et al. 2010; Wu et al. 2010; Shi et al. 2009b). Different from other organic pollutants, hazardous heavy metals are indestructible, as they cannot be chemically or biologically degraded. Even worse, some heavy metals can concentrate along the food chain and eventually accumulate in the human body, because we are at the top of the food chain (Baker et al. 2000; Crowley et al. 1991; Bizily et al. 1999; Gisbert et al. 2003; Shi et al. 2009a, b; Yang et al. 2009; Li et al. 2009;

S. Hongbo (✉)
Yantai Institute of Costal Zone Research, Chinese Academy of Sciences, Yantai 264003, China

Institute for Life Sciences, Qingdao University of Science & Technology (QUST), Qingdao 266042, China
e-mail: shaohongbochu@126.com

C. Liye
Institute for Life Sciences, Qingdao University of Science & Technology (QUST), Qingdao 266042, China

X. Gang • Y. Kun • Z. Lihua
Yantai Institute of Costal Zone Research, Chinese Academy of Sciences, Yantai 264003, China

S. Junna
Yantai Institute of Costal Zone Research, Chinese Academy of Sciences, Yantai 264003, China

The Graduate University of Chinese Academy of Sciences, Beijing 100049, China

I. Sherameti and A. Varma (eds.), *Detoxification of Heavy Metals*, Soil Biology 30,
DOI 10.1007/978-3-642-21408-0_4, © Springer-Verlag Berlin Heidelberg 2011

Shao et al. 2008, 2009, 2010; Wang and Ma 2008; Bi et al. 2010; Wu et al. 2010). Therefore, increasing attention has been paid in recent years to the remediation of polluted soils, among which the use of plants and microbes to remove hazardous metal ions is particularly emphasized (Pence et al. 2000; Mehra et al. 1991; Winge et al. 1985).

Phytoremediation is the use of plants to remove pollutants from environment, whereas microremediation refers to the use of microbes. Usually if a plant can accumulate more than 1,000 mg/kg (or 1,000 ppm) of Cu, Co, Cr, Ni, or Pb, or more than 10,000 mg/kg (or 10,000 ppm) of Mn or Zn, it is defined as a hyperaccumulator (Kagi 1991; Loeffler et al. 1989). Microbes have a larger specific surface area and are more efficient to activate and remove heavy metals. Plants, however, have harvestable stem and leaves aboveground, which is convenient for subsequent post processing. Using a transgenic technology to combine the two methods, namely symbiotic system, would be an optimum way to remove and collect hazardous metals efficiently (Kramer et al. 1996; Salt et al. 1995a, b).

However, transgenic methods sometimes cause gene flow (the transfer of alleles of genes from one individual to another), which is latent but dangerous for a nonnative gene segment by spreading (Kramer et al. 2000; Lasat et al. 1996). Several suggestions of minimizing the resulting potential gene pollution are available in this article. The authors think no country would risk their crop output to plant noncrop hyperaccumulators. So the "crop accumulators" stand for the developmental tendency of hyperaccumulator plants. In the end, a novel and tentative plan is given by the authors to evaluate the feasibility and demonstrate what procedures are needed before large-scale commercialization.

4.2 Mechanisms of Four Metal-Removing-Methods

To date, main four methods were proposed by researchers: chemical or physical remediation, animal remediation such as with earthworms, phytoremediation, and microremediation. Because of the obvious disadvantages and deficiency in feasibility, wide application of the former two methods is restricted. Summarized aspects of disadvantages of these two methods are given in the Sect. 4.3. In this part, mechanism of each four methods for removing hazardous heavy metal is explained and compared, and we concerns more about the latter two-phytoremediation and microremediation (Bioremediation).

4.2.1 Mechanism of Chemical or Physical Remediation

Chemical or physical method is early used and even endemically commercialized in America. Physical methods (e.g., soil leaching method and absorbent fixation) and chemical methods (e.g., bioreduction and chelate extraction) are used in practice. In these methods, the use of chelators cannot be avoided. By adding synthetic

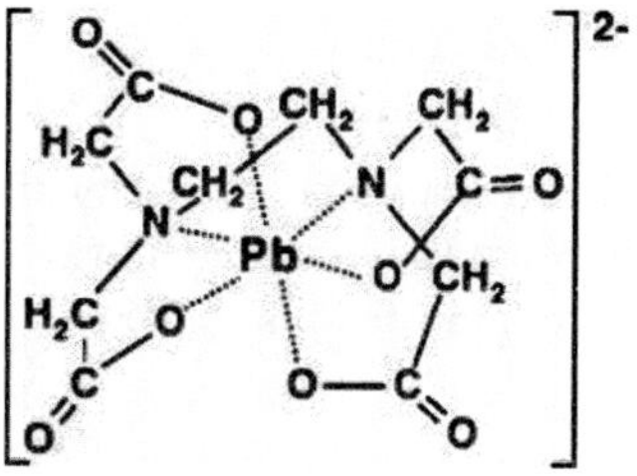

Fig. 4.1 EDTA-Pb complexes. Dotted bonds to Pb are coordinate

chelators such as ethylenediamine-tetraacetic acid (EDTA), both the solubility and bioavailability of heavy metals are improved. A chelating reagent's molecule can form several coordinative bonds to a certain metal atom, increasing its concentration in soil aqueous phase and mobility (Baker et al. 2000) (Fig. 4.1). Considering some metal ions strongly bonds to the soil phase and are less bioavailable, powerful chelating reagents are employed such as Na salt of EDTA. However, such approach needs not only expensive chemical reagent and machines but also many technicians. Worse, excessively usage of chemical chelates has been proven to pollute the ground water and negatively affect soil quality, for many necessary ions are also chelated unselectively. For example, elements Fe and Ca are usually lost after the spray of EDTA, because their concentration in the soil is much higher than those target heavy metals such as Pb and thus have more access and possibility to chelation. Wenzel et al. (2003) conducted an experiment using canola (*Brassica napus L*) and reported that leaching losses of Cu, Pb, and Zn(polluting ground water) far exceeded the amounts of metal taken up by plants after EDTA was applied, which indicated that under some certain circumstances the disadvantage of chelating reagent far outweighs its advantage. Therefore, taking reagent toxicity, unselectivity, and inefficacy into account, a careful consideration concerning ecology, economy, and human health is imperative before chelators are being put into practice (Bizily et al. 2000).

4.2.2 Mechanism of Animal Remediation

Animal here mainly refers to earthworm, because it is one of the most important soil organisms and plays an indispensable role in improving soil quality (Sriprang et al. 2002, 2003; Kashiwa et al. 2001; Fox et al. 1982). By their feeding, burrowing, excreting, and metabolic redox material, both soil texture and nutrition content are improved. Chemical groups such as –COOH and –CO are generated and exuded, which acidify soil and activate heavy metals. Several kinds of gel material are also excreted which facilitate complexion and chelation of metal ions. However, because of the relatively small amount and specific surface area compared with microbes, such improvement is neither notable nor stable. According to Yang et al. (Baker et al. 2000), after *Eisenia foetida* earthworm was inoculated, pH of a cock manure decreased by 0.7–0.9. However, if the inoculation occurred in an acidic red soil the pH value drops only by 0.03–0.18; if the inoculation happens in a sandy soil, no obvious decrease of pH is observed. Thus current studies imply that the effectiveness and efficiency of earthworm depend too much on outer conditions and may not be the optimum way of rapidly removing heavy metals. Further investigation in this field is needed.

4.2.3 Mechanism of Phytoremediation

4.2.3.1 Accumulation and Transport

In the rhizosphere of hyperaccumulator plants, protons are released by root to acidify the soil, which mobilize metal ions and increase metal bioavailability. This mechanism is supported by Crowley et al. (1991). However, due to metal ions' charge, lipohilic cellular membrane would be the first barrier of ions' entrance into cells. Fortunately, the following kinds of secretion can facilitate the transportation process.

1. Transporter proteins: Specific binding domain exists in such proteins. This binds to and transports metal ions from extracellular space into cells. Lasat et al. (1996) have found that hyperaccumulator *Thlaspi caerulescens* had bigger capacity for Zn^{2+} than its relative *T. arvense*. And such a gap of capacity is caused by different amounts of Zn transporter proteins (Mehra et al. 1991), which indicates that transporter proteins play a crucial role.
2. Nature chelators: As we know, chelators such as EDTA can bind to heavy metal ions and render them uncharged. An uncharged ion is of high mobility and is much easier to get through cellular membrane. In fact, plants can excrete nature chelators, which is much less toxic and more biodegradable as compared to EDTA (Table 4.1).

Among nature chelators, Phytochelatin (PC) and Metallothionein (MT) interest many scientists and is well studied. Metallothioneins (MT) are categorized into three classes: Class 1 MTs referred to polypeptides related to mammals, which contain 61 amino acids but lack aromatic amino acid or histidines; Class 2 MTs originally come from yeasts, and *Candida albicans* or cyanobacteria (Winge et al. 1985); A familiar chelator belonging to this class is *S. cerevisiae* MT, contributing to plants' high copper tolerance (Kagi 1991); Class 3 MTs is Phytochelatins (PC), which are composed of only three amino acid-Glu, Cys, and Gly, with Glu and Cys residues linked through a γ-carboxymide bond. In addition, Kagi (1991) has found

Table 4.1 Chelators that are commonly used

	Phytochelatin (PC)
	Metallothionein (MT)
Nature chelators	Organic acids
	EDTA (ethylene diamine tetra acetic acid)
	EGTA (ethylene glycol tetra acetic acid)
Synthetic chelator	DTPA(diethylenetriaminepentaacetic acid)

that heavy metals such as Cd, Zn, Hg, Ag, and Pb can induce the synthesis of MTs especially in animal and plant species. A recent study shows that the best activator is Cd followed by Ag, Bi, Pb, Zn, Cu, Hg, and Au (Kramer et al. 1996).

3. Organic acids: Several organic acids (e.g., malic acid and citrate) have been identified as positive bioreagents to accelerate the absorption of heavy metals by root. Such mechanisms are even more notable in the root–shoot transportation.

However, substantial achievements are lacked in the root–shoot transportation except two points: one is that root–shoot pathway is closely related to plants' transpiration efficiency; the other is one of chelator ligands (histidine) that is found in high levels in the xylem sap of a Ni-tolerant plant (*Alyssum lesbiacum*), and the coordination of Ni with histidine (Mehra et al. 1991) is substantiated by Kramer et al. (1996), which implies that chelation mechanism also works in the process of xylem transferring.

On the molecular level, accumulation and transport mechanism is partly clarified. Many transporters encoded by specific genes are investigated, and it is common that one kind of metal ion can be transported by different carriers (Zhao et al. 1999; Yang et al. 2007; Gleba et al. 1999; Geoffrey et al. 2006; Lugtenberg et al. 1991; Korshunova et al. 1999; Dubery et al. 2006; Yuebing et al. 2007) (Table 4.2).

4.2.3.2 Detoxification

As we know, some hazardous heavy metals exercise a detrimental influence on cells by binding to vital proteins, interfering with cellular activities, and inhibiting regulation of cells. Luckily, hyperaccumulator plants have evolved their own mechanisms to protect themselves from negative heavy metal stress. Several important detoxification mechanisms are explained as follows:

1. Chelation: Chelation plays a crucial role not only in the accumulation and transportation of heavy metals but also in the detoxification phase. Usually chelators have ligands (most commonly histidine and citrate) and can bind metal ions. Combined metal ions appear uncharged and inert when reacting with other substances, by which way heavy metals' damage toward cell are reduced significantly.
2. Vacuolar compartmentalization: Since the vacuole is widely considered as the main storage place of heavy metals in plant cells, vacuolar compartmentalization

Table 4.2 Metal ions and their transporters

ZNT protein(encoded by ZIP gene)	Zn^{2+}, Cd^{2+}
Nramp protein(encoded by AtNramp gene)	Cd^{2+}, Fe^{2+}
Protein encoded by NtCBP4	Pb^{2+}
Aquaglyceroprins	As^{3+}
Phosphate transporter	As^{5+}
IRT1 (iron-regulated transporter)	Fe^{2+}, Mn^{2+}, Zn^{2+}

is quite effective in controlling the distribution and concentration of metal ions. To compartmentalize the vacuole is to "arrest and imprison" hazardous metal ions, constricting them into a limited site. Thus other parts of the cell have no access to those dangerous metal ions and safety is of course ensured. This mechanism is proved to be true in the Cd detoxification and tolerance by Salt et al. (1995a, b), Kagi (1991): Cd induces the synthesis of PCs and then forms a Cd–PC molecule, which will be transferred into the vacuole by a Cd/H antiport and an ATP-dependent PC-transporter. In addition, Kramer et al. (2000) have reported that by "imprisoning" most of the intracellular Ni into the vacuole, metal tolerance of hyperaccumulator *T. goesingense* is greatly improved, which confirms the compartmentalization theory too.

3. Volatilization: By converting metal ions into a volatile state, some plant species avoid the lasting damage caused by accumulation and long-time stay of heavy metals. A representative example is the bioprocess of Hg, which is a worldwide volatile pollutant and which is able to accumulate in human bodies. However, not all the plants possess such ability and even among those innate Hg-resistant species, the relatively small amount of accumulation and their spatial distribution have greatly limited their wide cultivation. Thus scientists have employed genetic engineering and several transgenic plants have showed satisfactory performance to convert and volatilize metals. Transgenic species expressing organomercurial lyase (MerB) have much higher tolerance to organic Hg complex than wild type and can convert methylmercury to Hg(2), which is 100 times less toxic than the former one Huysen (2003). Furthermore, transgenic plants expressing both MerA (enzyme that reduces Hg(2) to Hg(0) and MerB have shown the highest tolerance to organic Hg(up to 10 μM) compared with MerB species' 5 μM and wild type's 0.25 μM Gisbert et al. (2003).

4.2.4 Mechanism of Microremediation

4.2.4.1 Metal-Binding Mechanism

Three substances should be mentioned for this mechanism: Metallothionein (MT), Phytochelatin (PC), and some novel metal-binding peptides. As we know from the former part of this article, MT and PC play a crucial part in plant–metal interaction. In fact, in the microbial world such interplay also exists. By binding to heavy metal ions MTs facilitate microbes' absorption or transportation of metal ions, and so do PCs, who are composed by only three amino acids (Gly, Cys, and Gly). That overexpression of PC synthase in microbes is effective in the accumulation and tolerance of metal ions has been reported by Sriprang et al. (2003). By expressing the *Arabidopsis thalina* gene encoding PC synthase, enhanced Cd accumulation is observed in *Mesorhizobium huakuii* subsp. rengei B3 and E. coli cells. In recent years, novel metal-binding peptides containing histidines or cysteines have been found and engineered. These peptides are usually of higher affinity, specificity and selectivity for a certain metal ion. Related and in-depth study, however, is scarce (Baker et al. 2000; Crowley et al. 1991; Bizily et al. 1999; Pence et al. 2000; Mehra et al. 1991; Winge et al. 1985; Kagi 1991; Loeffler et al. 1989; Kramer et al. 1996; Salt et al. 1995a, b; Kramer et al. 2000; Lasat et al. 1996; Bizily et al. 2000; Sriprang et al. 2003; Kashiwa et al. 2001; Fox et al. 1982).

4.2.4.2 Valence Transformation Mechanism

Metals of different valence vary in toxicity. By excreting special redox enzyme, plants skillfully convert hazard metals to a relatively less toxic state and decrease possible metal stress and damage. For example, reduction of Cr(6) to Cr(3) is widely studied, the latter one of which is both less mobile and less toxic. In addition, Kashiwa et al. (2001) has found that *Bacillus* sp. SF-1 was good at reducing high concentration of Se(6) into elemental Se. The most persuasive example of this mechanism is the mercury-resistant bacteria, in which organomercurial lyase (MerB) is produced. As we see from Fig. 4.2, methylmercury is converted to Hg (2), which is 100-fold less toxic than the former one (Yuebing et al. 2007; Begley et al. 1986; Rugh et al. 1998).

4.2.4.3 Volatilization Mechanism

By turning metal ions into volatile state, microbes escape possible negative effect that dangerous metal ions bring them. However, such approach is feasible for only a few metals such as Hg and metalloid Se. For the majority of most other metals which have no volatile state at natural conditions, this pathway is closed. To date, the way microbes deal with element Hg is relatively clear. In the cells of mercury-resistant bacteria there is a MerA enzyme, an enzyme that reduces Hg(2) to volatile form Hg(0) (Kashiwa et al. 2001; Fox et al. 1982) (Fig. 4.3).

4.2.4.4 Extracellular Chemical Precipitation Mechanism

Quite a number of binding substances were excreted by microbes, ranging from simple organic acid, alcohols to large polysaccharides, humic and fulvic acids. In fact, not only metals but also metal sulfides and oxides can be entrapped and absorbed by an extracellular mixture of polysaccharides, mucopolysaccharides,

Fig. 4.2 MerB converting organic Hg to Hg(2)

$$\text{R-CH}_2\text{-Hg}^+ + \text{H}^+ \xrightarrow{\text{MerB}} \text{R-CH}_3 + \text{Hg(II)}$$

$$\text{RSHg}^+ + \text{NADPH} \xrightarrow{\text{MerA}} \text{Hg(0)} + \text{RSH} + \text{NADP}^+$$

Fig. 4.3 MerA converting Hg(2) to Hg(0) and facilitating volatilization and proteins (Lugtenberg et al. 1991). Recent studies have found that peptidoglycan carboxyl groups are the main cation-binding sites for Gram-positive bacterial cell walls, whereas phosphate group for Gram-negative microbes and chitins are for fungi. No matter whether the precipitation happens in the outer surface of the cell wall or away from it, this mechanism is successful by keeping harmful metal ions out of the cytoplasm (Korshunova et al. 1999).

4.2.4.5 Symbiotic Mechanism

A big disadvantage of microremediation is that absorbed heavy metals would still stay in the soil, so symbiotic mechanism would be more effective by combining both microremediation and phytoremediation. Due to the symbiotic microbes' large amount and specific surface area, binding reagents such as MTs, PCs, and organic acid will be excreted more by symbiotic systems than by sole plants. Thus soil will be improved with better acidification, which ultimately leads to better solubility, mobility, and bioavailability of heavy metals. After heavy metal particles are activated, the subsequent process can be divided into two ways. One is that metal ions are accumulated by plants root, transported in the xylem and detoxified through chelation, vacuolar compartmentalization, and volatilization, just as normal phytoremediation does. The other way is heavy metals will be accumulated in rhizosphere and nodules. Rhizosphere bacteria's essential role in achieving optimum rates of selenium accumulation and volatilization has been proven by De Souza et al. (1998) in an Indian mustard experiment. Another symbiotic experiment was provided by Sriprang et al. (2002) by inserting the MTL4 gene into *M. huakuii* subsp. rengei B3. Data showed that the symbionts Cd^{2+} absorption increased by 2.3–6.6-fold in nodules, whereas obvious accumulation of Cu^{2+} was not observed.

4.3 Evaluation of Four Metal-Removing Methods

4.3.1 Summarized Disadvantages of Chemical/Physical Remediation

1. It is an expensive and labor-intensive method. Using large machine and synthesizing a great amount of chelators could be costly and the application of landfilling and leaching technology demands not only professional technicians but also many hours. It has been estimated that in order to reduce soil Pb concentration from 1.4 g/kg to 0.4 g/kg in 10 years, phytoremediation would cost only $27,900. Compared with landfilling method's $1,620,000 and soil-leaching method's $790,000, phytoremediation is very economical.
2. Natural soil's structure, texture and fertility can be impaired by the method itself and by the reagent added.

3. Excessive use of chelators would poison both plants and microbes. The most widely used chelator EDTA is both toxic and nonbiodegradable.
4. Chemical/physical method may lead to pollution by mobile heavy metal ions leaching into ground water, since the use of binding reagents makes metal ions more soluble and mobile.
5. Chelators such as EDTA usually lack selectivity and easily cause beneficial ions' loss (especially for Fe and Ca).

4.3.2 Summarized Disadvantages of Animal Remediation

(1) Animals such as earthworms are usually small in total number and specific surface area, which makes them inefficient in absorbing and accumulating heavy metals.
(2) Animals' rigorous demand for comfortable environment and excessive dependence on organic substance limit their wide use and decrease their practicality.
(3) Heavy metals are still in the soil if lack of a feasible method to collect all the earthworms.
(4) Clear mechanism of accumulating and detoxifying heavy metal in animals needs to be investigated by further studies.

4.4 Future Development and Opportunities of Bioremediation (Phyto- and Micro-remediation)

4.4.1 Application of Genetic Engineering or Cell Engineering

In recent years, genetic engineering technology seems increasingly necessary, because most nature hyperaccumulators are not satisfying(due to its slow growing, low biomass-production, and rigorous demand for growing conditions. Aided by this powerful tool, many heavy metal resistant genes have been introduced into plant cells (Wenzel et al. 2003) (Table 4.3). By overexpression of natural chelators (MTs, PCs, and organic acid), not only ions' entrance into cell but also translocation in xylem and other parts is facilitated by excreting certain transporter proteins; specific ions was bound and transferred by encoding special oxidoreductases such as MerA and MerB. Heavy metals' valence is changed into a less toxic one or a more volatile one by horizontal transfer of plasmid. Resistant genes are exchanged among rhizosphere bacteria and this benefits bacteria's accumulating and transferring metal ions.

Cell engineering technology (such as cell fusion or somatic hybridization) also shows great power. Because polyploid plants are usually bigger in size and more active in transpiration, which does good to the transportation of heavy metals in root-to-shoot process, this method is of great significance (Geoffrey et al. 2006).

Table 4.3 Genes introduced into plants and their source and target

Gene	Product	Source	Target	Gene	Product	Source	Target
merA	Hg(II) reductase	G+	*Liriodendron tulipifera Nicotiana tabacum*	CGS	Cystathione-gamma-synthase	*A. thaliana*	*B. juncea*
merA	Hg(II) reductase	G–	*A. thaliana*	CSase	Cysteine synthase	*Spinach*	*N. tabacum*
merB	Organomercurial lyase	G–	*A. thaliana*	gsh2	GSH synthase	*E. coli*	*B. juncea*
merA	Hg(II) reductase	G–	*N. tabacum*	gsh1	γ-Glu-Cys synthase	*E. coli*	*B. juncea*
merB	Organomercurial lyase	G–	*N. tabacum*	gsh1	γ-Glu-Cys synthase	*E. coli*	*Populus tremula*
AtPCSI	PC	*A. thaliana*	*A. thaliana*	NtCBP4	Cation channel	*N. tabacum*	*N. tabacum*
APSI	ATP sulfurylase	*A. thaliana*	*B. juncea*	SL	Se-Cys lyase	*Mouse*	*A. thaliana*
MT-1	MT	Mouse	*N. tabacum*	TaPCSI	PC	*T. aesitivum*	*N. glauca*

G+ stands for Gram-positive bacteria and G– stands for G-negative ones

Hybrids have not only a bigger size than its parent *Brassica juncea* and *Thlaspi caerulescens* but also better hyperextraction in absorbing Pb, Ni, and Zn. The amount of Pb absorbed by hybrids is almost the sum of its parents (Sriprang et al. 2002; Wenzel et al. 2003; Zhao et al. 1999).

However, a big disadvantage coming with transgenic technique is the genetic pollution caused by, for example, the floating of transgenic pollen and the horizontal transferring of plasmids among microbes. In addition, because both pollen and plasmid are invisible to naked eyes, not until the horrible mutant species grow up and the severe aftermath can we realize their damage. Therefore biosafety aspects must be taken into account and application of transgenic species need domestic legislation.

Here the authors contribute three feasible suggestions that minimize the potential damage caused by genetic pollution:

1. Choosing self-pollinated plants: In a self-pollinated plant, pollen moves to the female part of the same flower or to another flower on the same individual plant. And this means fertilization is completed within a closed environment. Therefore, self-pollination would be an effective way to avoid the spreading of transgenic genes into other species.
2. Using cell engineering technology to create infertile polyploid species: Infertile plants cannot transfer any gene to the next generation and so potential spread of gene segment becomes impossible. Technically two species are feasible: one is allopolyploid (the combination of two heterogeneous sets of chromosomes) and the other is triploid. Both approaches are able to make plants infertile by lack of equation–division mechanism of chromosomes.
3. Among microbes, transgenic technique should be limited within the sphere of rhizosphere bacteria and symbiotic bacteria, because such kind of microbes mainly congregate around the root and show a very slight distribution outside the rhizosphere. In detail, the population of "endophytic bacteria" living within the plant tissues is easier controlled than that of rhizospheric bacteria living on or round the plant roots, because the former depend more on plants. Therefore by restricting the transgenic and foreign genes in a limited area and a finite amount, potential harm of genetic pollution is minimized.

4.4.2 The Development of Crop Hyperaccumulators

Another new viewpoint toward phytoremediation is that adequate attention must be paid to the development of crop hyperaccumulators, termed by the author as "crop accumulators", such as wheat, maize, and rice (Heaton et al. 1998; Bizily et al. 2000; Ruiz et al. 2003; Lee et al. 2003). In other words, investigations of other "noncropaccumulators" species would be of little significance in application. A large area of farmland in the world is contaminated by heavy metals and needs remediation. Considering the finite and precious farmland sources and big population, no country would stop growing crops and turn to noncrop hyperaccumulators, which do not provide grain. Furthermore, it usually takes several years to completely remove a certain heavy metal from soil. This makes the application of noncrop hyperaccumulators more unrealistic. After all, no country could afford zero harvest and even famine for several years.

To date, more than 400 plant species have been identified as metal hyperaccumulators and the best hyperaccumulator varies according to regions with different climate and soil types. To date, the most studied and promising species is within Cruciferae family such as *Brassica* genus, *Alyssuns* genus, and *Thlaspi* genus but 75% of them are Ni tolerant, which grow on ultramafic soils (Yang et al. 2007; Gleba et al. 1999; Geoffrey et al. 2006; Lugtenberg et al. 1991; Korshunova et al. 1999; Dubery et al. 2006). Unfortunately almost none of them can be used as crops in the farmland. Thus it is critical that we paid immediate attention to the "crop accumulators" field. We recommend two ways to develop this "crop accumulators" species:

By transgenic approach: Through the introduction of foreign resistant genes into crops, capability for accumulating, transporting, and detoxifying heavy metals can be significantly improved. Even if the improvement is not as notable as expected, it doesn't matter because "crop accumulators" can be grown every year as normal field crops and heavy metals are bound to be removed thoroughly after a long period of time. As long as heavy metals do not accumulate in the edible part such as grain, long-time growing of "crop accumulators" would not reduce the total crop outcome and impair the grain quality.

By symbiotic approach: Finding an appropriate kind of symbiotic bacteria is the key in this method. In a symbiotic system, heavy metals are more likely to be accumulated in nodules. This is good for us, because generally crops bear grains aboveground and the fear that heavy metals may accumulate in an edible part is eliminated. Another big advantage is that some microbes produce antibiotics to enhance plants immunity and some produce necessary nutrients and even plant growth hormones (Korshunova et al. 1999; Dubery et al. 2006; Yuebing et al. 2007; Begley et al. 1986; Rugh et al. 1998; Heaton et al. 1998; Bizily et al. 2000; Ruiz et al. 2003). Therefore the future of this method would be bright.

4.4.3 Universal Procedures of Evaluation before Large-Scale Commercialization

As we discussed in Sect. 4.3.2, "crop accumulators" created by genetic or cell engineering technology will prove to be of great importance and hope (Salt et al. 1995b). However, related literature concerning concrete steps of applying a hyperaccumulator to practice is seldom found. The author tentatively puts forward a new evaluation plan that tells us clearly what should be done before large-scale commercialization of a new hyperaccumulator species (Lee et al. 2003; Hwang et al. 1999; Wangeline et al. 2004; Misra 1989; Pan 1994; Huysen 2003; Noji 2001; Zhu 1999a, b; Arasi 2000; Arazi et al. 1999; Pilon et al. 2003; Gisbert et al. 2003;

Shi et al. 2009a, b; Yang et al. 2009; Li et al. 2009; Shao et al. 2008, 2009, 2010; Wang and Ma 2008; Bi et al. 2010; Wu et al. 2010) (Fig. 4.4).

The procedures of evaluation before large-scale commercialization are shown in Fig. 4.4. They comprise the following steps:

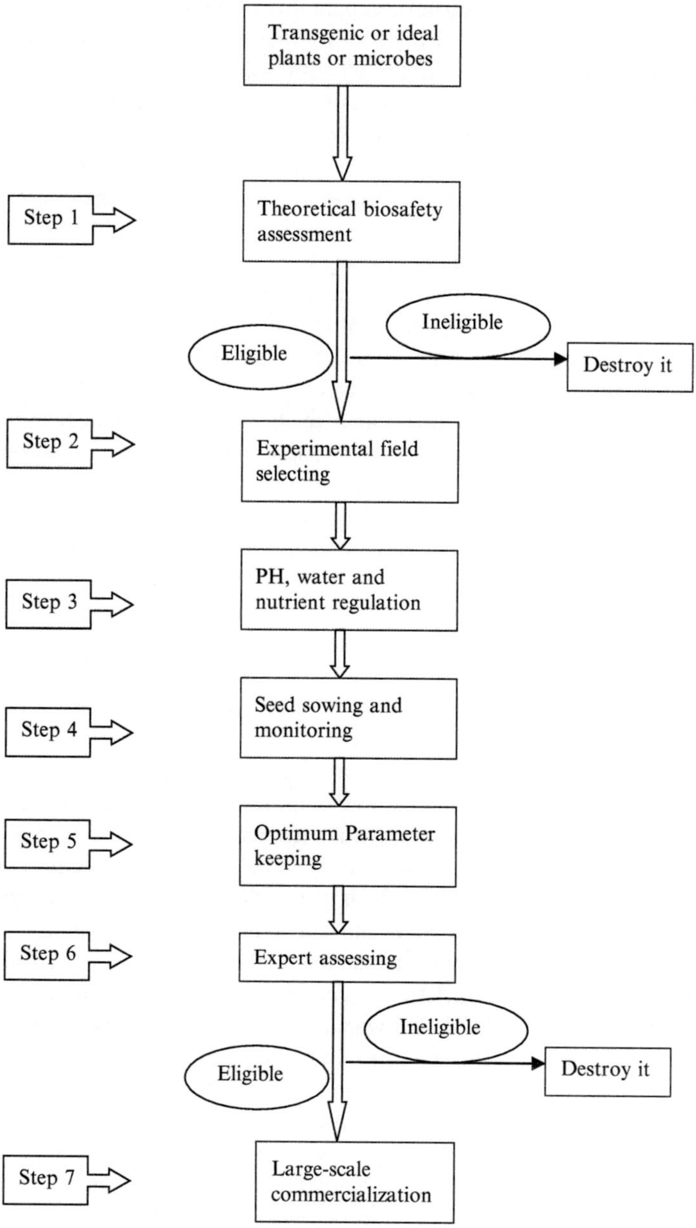

Fig. 4.4 Universal procedures of evaluation before large-scale commercialization

Step 1: After a transgenic or symbiotic species is created, theoretical biosafety assessment is indispensable. Since foreign genes would enter the ecological system automatically and cannot be easily removed once a seed is planted. Based on experience and hermetic experiments biosafety can be estimated. The following parameters are recommended: sexual compatibility of pollens, blossom, and spatial distribution of nearby plants, and the fertility of the next generation. If the new species is eligible and biosafe to the ecology, it is allowed to go to the next step. If not, simply destroy it (for the purpose of eliminating nonnative genes).

Step 2: Experimental field selected should be representative. A relatively closed environment is necessary to minimize gene-flow phenomenon. If possible, the selected field should be also easily regulated in water, pH, nutrient, and sampling for the subsequent steps.

Step 3: As we know, soil acidity is closely related to the activation and accumulation of heavy metal. Nutrient, however, also matters. Studies show that a proper amount of N and K promote the uptake of Pb but a large amount inhibits the Pb absorption. The addition of P doesn't improve the accumulation of Pb, which even counteracts the original absorbing (Begley et al. 1986). Water supplying relates with transpiration, which plays a great role in the transportation of the heavy metals. Every plant and microbe has its optimum temperature, pH, water content, and nutrient supplying for living, and satisfying these demands is important.

Step 4: We recommend a natural outdoor field with sunlight to sow the new test species and spring would be the best season for the most species. Although simulated lab field can provide similar conditions, it is not suitable for this evaluation process in many aspects especially in the assessment of genetic pollution where interaction with other plants is unavoidable. Besides, another important aim of Step 4 is to monitor how well the plant or symbiotic system survives while removing heavy metals. Several parameters are cited here, which can be categorized into two classes. One is a chemical method, which is stable, and the other one is biological parameters, which are less stable but more effective and sensitive. For example, chemical parameters can show the total amount of a certain metal, which may be useless because only soluble or biodegradable forms of metal ion can be absorbed. A government law (National Standard of the People's Republic of China) provides ways of heavy metal measurement, which is accurate and legal. Microparameters reflect not only heavy metal content and stress but also compatibility between bacteria and plants. Representative parameters include the ratio of microbial biomass *C* (Cmic)/*C* (Corg) and metabolic quotient (qCO2) in biochemical level and CLPP in community level. Usually when heavy metal concentration goes up, decrease in *C*/*C* ratio and increase in metabolic quotient can be observed.

Step 5: Keeping parameters at an optimum value can be achieved by several approaches. To decrease soil pH value, either H_2SO_4 or organic fertilizer can be used. To increase pH, lime or other alkaline substance is useful. By regulating water supply, both transpiration and Eh value (a redox index of soil) is controlled. The use of element P, N, and K can also affect those parameters. Through Step 5's regulation, we try to ensure that new hyperaccumulators work at a normal and relatively high efficiency.

Step 6: Different from Step 1, expert assessment is practical, because plants have matured. Maturity means pollens have been produced and spread. Therefore, at this time biosafety issue is quite realistic. A detailed and rigorous environmental risk assessment must be completed to avoid genetic pollution as much as possible. Other index should also be assessed such as biomass, total cost of the growing, and growth rate. Besides these, experts need to estimate how long bioremediation would take, for an excessively long duration is not suitable for process feasibility. If the plant is qualified in this step, it can go to further large-scale commercialization; if not, destroy it (for the purpose of eliminating nonnative genes).

Step 7: The performance of the last step needs a national legislation concerning transgenic species and its potential risk. If it is a novel symbiotic system using no transgenic technology, such procedure is not needed. In addition, when commercializing, different kinds of hyperaccumulators can be grown by turn, since one kind of hyperaccumulators is generally good at dealing with finite kinds of metal.

4.5 Conclusions

1. Bioremediation has more advantages and should be focused on more, especially on transgenic methods.
2. Future transgenic technology should focus on four promising targets: overexpression of natural chelators (MTs, PCs, and organic acid), transporter proteins, special oxidoreductase, and resistant gene in symbiotic microbe.
3. Several ways of minimizing genetic pollutions are put forward such as choosing self-pollinated plants, creating infertile polyploid species, and carefully selecting easy-controlled microbe species.
4. Crop hyperaccumulators should be paid more attention for studies on noncrop species are of limited use.
5. Universal procedures are put forward to evaluate hyperaccumulators' feasibility before large-scale commercialization.

Compared to chemico-physical method, bioremediation shows great advantage. Studies on this field are increasing and its mechanism is more and more clear (Shao et al. 2008, 2009, 2010; Shi et al. 2009a, b). Using transgenic technology is a tendency in the future to create an ideal species purposely. However, genetic pollution must be taken into consideration, which can be avoided or minimized by using methods mentioned above.

In the future crop hyperaccumulators will be a better choice due to its feasibility, in the field of which current emphasis is scarce. Microbes, in many cases, are more efficient in accumulating and absorbing heavy metals, because of their astronomical amount and specific surface area. Furthermore, technique of genetic engineering in microbes is easier and more mature than in plant cells. Therefore, using transgenic technology to create an optimum plant + soil + microbes combination would be a promising way in the future development (Arasi 2000; Arazi et al. 1999; Pilon et al.

2003; Gisbert et al. 2003; Shi et al. 2009a, b; Yang et al. 2009; Li et al. 2009; Shao et al. 2008, 2009, 2010; Wang and Ma 2008; Bi et al. 2010; Wu et al. 2010)

As to the estimation of a new species or symbiotic system, Fig. 4.4 offers a feasible risk assessment plan and may accelerate the application process.

Acknowledgements This work was jointly supported by the National Natural Science Foundation of China (No. 41001137), The Science and Technology Development Plan of Shandong Province (2010GSF10208), The Science and Technology Development Plan of Yantai City (20102450), One Hundred-Talent Plan of Chinese Academy of Sciences (CAS), the CAS/SAFEA International Partnership Program for Creative Research Teams, the Important Direction Project of CAS (KZCX2-YW-JC203) and CAS Young Scientists Fellowship (2009Y2B211).

References

Arazi T et al (1999) A tobacco plasma membrane calmodulin-binding transporter confers Ni^{2+} tolerance and Pb^{2+} hypersensitivity in transgenic plants. Plant J 20:171–182

Arasi ACM (2000) Responses to cadmium in leaves of transformed poplars overexpressing γ-glutamylcysteine synthetase. Physiol Plant 109:143–149

Baker AJM et al. (2000) Metal hyperaccumulator plants: a review of the ecology and physiology of a biological resource for phytoremediation of metal-polluted soils. In: Baker (ed) Phytoremediation of contaminated soil and water. CRC, Boca Raton, FL, pp 85–107

Begley TP et al (1986) Mechanistic studies of a protonolytic organomercurial cleaving enzyme: bacterial organomercurial lyase. Biochemistry 25:7192–7200

Bi RT, Bai ZK, Li H, Shao HB, Li WX, Ye BY (2010) Establishing a clean-quality indicator system for evaluating reclaimed land in the Antaibao opencast mine area, China. Clean – Soil Air Water 38:719–725

Bizily SP et al (2000) Phytodetoxification of hazardous organomercurials by genetically engineered plants. Nat Biotechnol 18:213–217

Bizily SP et al (1999) Phytoremediation of methylmercury pollution merB expression in *Arabidopsis thaliana* confers resistance to organomercurials. Proc Natl Acad Sci USA 96:6808–6813

Crowley DE et al (1991) Mechanisms of iron acquisition from siderophores by microorganisms and plants. Plant Soil 130:179–198

De Souza M et al (1998) Rate-limiting steps in selenium assimilation and volatilization by Indian mustard. Plant Physiol 117:1487–1494

Dubery RS et al (2006) Heavy metal uptake and detoxification mechanism in plants. Int J Agric Res 1:122–141

Fox B et al (1982) Mercuric reductase. Purification and characterization of a transposon-encoded flavoprotein containing an oxidation reduction active disulfide. J Biol Chem 257:2498–2503

Geoffrey M et al (2006) Microorganisms in toxic metal-polluted soils. In: Soil biology, vol 3, Microorganisms in soils: roles in genesis and functions. Springer, Berlin, pp 1–69

Gisbert C et al (2003) A plant genetically modified that accumulates Pb is especially promising for phytoremediation. Biochem Biophys Res Commun 303:440–445

Gleba D et al (1999) Use of plant roots for phytoremediation and molecular farming. Proc Natl Acad Sci USA 96:5973–5977

Heaton ACP et al (1998) Phytoremediation of mercury- and methylmercury-polluted soils using genetically engineered plants. J Soil Contam 7:497–509

Huysen V (2003) Overexpression of cystathionine-gamma-synthase enhances selenium volatilization in *Brasica juncea*. Planta 218:71–78

Hwang S et al (1999) Overexpression of ATP sulfurylase in Indian mustard leads to increased selenate uptake, reduction, and tolerance. Plant Physiol 119:123–132

Kagi JHR (1991) Overview of metallothionein. Methods Enzymol 205:613–626

Kashiwa M et al (2001) Removal of soluble selenium by a selenate-reducing bacterium *Bacillus* sp. *SF-1*. J Ferment Bioeng 83:517–522

Korshunova Y et al (1999) The IRT1 protein from *Arabidopsis thaliana*. Plant Mol Biol 40: 37–44

Kramer U et al (2000) Subcellular localization and speciation of nickel in hyperaccumulator and non-accumulator Thlaspi species. Plant Physiol 122:1343–1353

Kramer U et al (1996) Free histidine as a metal chelator in plants that accumulate nickel. Nature 379:635–638

Lasat MM, Baker AJM, Kochian LV (1996) Physiological characterization of root Zn^{2+} absorption and translocation to shoots in Zn hyperaccumulator and nonaccumulator species of *Thlaspi*. Plant Physiol 112(4):1715–1722

Lee S et al (2003) Overexpression of Arabidopsis phytochelatin synthase paradoxically leads to hypersensitivity to cadmium stress. Plant Physiol 131:656–663

Li H, Shi WY, Shao HB, Shao MA (2009) The remediation of the lead-polluted garden soil by natural zeolite. J Hazard Mater 169:1106–1111

Loeffler S et al (1989) Termination of the phytochelatin synthase reaction through sequestration of heavy metals by the reaction product. FEBS Lett 258:42–46

Lugtenberg BJJ et al (1991) Microbial stimulation of plant growth and protection from disease. Curr Opin Biotechnol 2:457–464

Mehra RK et al (1991) Metal ion resistance in fungi: molecular mechanisms and their regulated expression. J Cell Biochem 45:30–40

Misra S (1989) Heavy metal tolerant transgenic *Brassica napus* L. and *Nicotiana tabacum* L. plants. Theor Appl Genet 78:161–168

Noji M (2001) Cysteine synthase overexpression in tobacco confers tolerance to sulfur- containing environmental pollutants. Plant Physiol 126:973–980

Pan A (1994) Alpha-domain of human metallothionein I-A can bind to metals in transgenic tobacco plants. Mol Gen Genet 242:666–674

Pence NS et al (2000) The molecular physiology of heavy metal transport in the Zn/Cd hyperaccumulation *Thlaspi caerulescens*. Proc Natl Acad Sci USA 97:4956–4960

Pilon M et al (2003) Enhanced selenium tolerance and accumulation in transgenic Arabidopsis expressing a mouse selenocysteine lyase. Plant Physiol 131:1250–1257

Rugh CL et al (1998) Development of transgenic yellow poplar for mercury phytoremediation. Nat Biotechnol 16:925–928

Ruiz ON et al (2003) Phytoremediation of organomercurial compounds via chloroplast genetic engineering. Plant Physiol 132:1344–1352

Salt DE et al (1995a) Cadmium transport across tonoplast of vesicles from oat roots. Evidence for a Cd^{2+}/H^{+} antiport activity. J Biol Chem 268:12297–12302

Salt DE et al (1995b) MgATP-dependent transport of phytochelatins across the tonoplast of oat roots. Plant Physiol 107:1293–1301

Shao HB, Chu LY, Ruan CJ, Li H, Guo DG, Li WX (2010) Understanding molecular mechanisms for improving phytoremediation of heavy metal-contaminated soils. Crit Rev Biotechnol 30:23–30

Shao HB, Chu LY, Jaleel CA, Manivannan P, Panneerselvam R, Shao MA (2009) Understanding water deficit stress-induced changes in the basic metabolism of higher plants – biotechnologically and sustainably improving agriculture and the eco- environment in arid regions of the globe. Crit Rev Biotechnol 29:131–151

Shao HB, Chu LY, Lu ZH, Kang CM (2008) Primary antioxidant free radical scavenging and redox signaling pathways in higher plant cells. Int J Biol Sci 4:8–14

Shi WY, Shao HB, Li H, Shao MA, Du S (2009a) Co-remediation of the lead-polluted garden soil by exogenous natural zeolite and humic acids. J Hazard Mater 167:136–140

Shi WY, Shao HB, Li H et al (2009b) Progress in the remediation of hazardous heavy metal-polluted soils by natural zeolite. J Hazard Mater 170:1–6
Sriprang R et al (2003) Enhanced accumulation of Cd2+ by a *Mesorhizobium* sp. transformed with a gene from *Arabidopsis thaliana* coding for phytochelatin synthase. Appl Environ Microbiol 69:1791–1796
Sriprang R et al (2002) A novel bioremediation system for heavy metals using the symbiosis between leguminous plant and genetically engineered rhizobia. J Biotechnol 99:279–293
Wang XL, Ma XQ (2008) Advance in the research of phytoremediation in heavy metal contaminated soils. Subtrop Agric Res 4:44–49
Wangeline AL et al (2004) Overexpression of ATP sulfurylase in Indian mustard: effects on tolerance and accumulation of twelve metals. J Environ Qual 33:54–60
Wenzel WW et al (2003) Chelate-assisted phytoextraction using canola (*Brassica napus L*) in outdoors pot and lysismeter experiments. Plant Soil 249:83–96
Winge DR et al (1985) Yeast metallothionein: sequence and metal-binding properties. J Biol Chem 260:14464–14470
Wu G, Kang HB, Zhang XY, Shao HB, Chu LY, Ruan CJ (2010) A critical review on the bio-removal of hazardous heavy metals from contaminated soils: issues, progress, eco-environmental concerns and opportunities. J Hazard Mater 174:1–8
Yang PG, Mao RZ, Shao HB, Gao YF (2009) An investigation on the distribution of eight hazardous heavy metals in the suburban farmland of China. J Hazard Mater 167:1246–1251
Yang SG et al (2007) Experiment on *Eisenia foetida* for pre-compost of chook manure. Chin J Eco-agric 15:55–57
Yuebing S et al (2007) Phytoremediation and strengthening measures for soil contaminated by heavy metals. Chin J Environ Eng 1:23–28
Zhao M et al (1999) Rhizosphere bacteria enhance selenium accumulation and volatilization by Indian mustard. Plant Physiol 119:565–573
Zhu YL (1999a) Overexpression of gluthatione synthetase in Indian mustard enhances cadmium accumulation and tolerance. Plant Physiol 119:73–79
Zhu YL (1999b) Cadmium tolerance and accumulation in Indian mustard is enhanced by overexpressing γ-glutamilcysteine synthetase. Plant Physiol 121:1169–1177

Chapter 5
Plant Taxonomy and Metal Phytoremediation

Stanislaw W. Gawronski, Maria Greger, and Helena Gawronska

5.1 Introduction

Plants during 400 million years of evolution often were exposed to extreme environmental conditions. Plants, as organisms of sessile style of life, have developed unique defense mechanism(s) and strategies existing only in that group of organisms. As a result among all higher organisms only plants are able to survive in very polluted sites and tolerate high accumulation of toxic compounds in their tissues. They can thrive in soil, water and air contaminated to levels that are often orders of magnitude higher that can be accepted by other organisms. Nowadays pollution of soil, water and air environments is additionally raised as a side effect of antropopressure. At a certain point, the strength of the negative impact of human beings became so high that only the most resistant species are able to survive, very often at the cost of limited growth and development. Present state of knowledge and social awareness of negative anthropological impact on the environment lead to search for their elimination or at least lowering to permissible limits. Plants affect chemical, physical and biological processes in the environment and steer them in such a way as to change environmental conditions as close as possible to the optimum for plant growth. One of the spectacular example is the presence of the small crystals of pyromorphite in the rhizosphere of *Agrostis tenuis*, very tolerant to lead grass, nearby old lead mining site if soluble inorganic phosphate was present in the soil. Formation of the crystals in the surroundings of roots of *A. tenuis* indicates the plant's direct influence by release exudates or the impact on running this process by microbial community. Competence of plants for rock formation can be

S.W. Gawronski (✉) • H. Gawronska
Laboratory for Basic Research in Horticulture, Faculty of Horticulture and Landscape Architecture, Warsaw University of Life Sciences-SGGW, Warsaw, Poland
e-mail: stanislaw_gawronski@sggw.pl

M. Greger
Department of Botany, Stockholm University, Stockholm, Sweden

I. Sherameti and A. Varma (eds.), *Detoxification of Heavy Metals*, Soil Biology 30,
DOI 10.1007/978-3-642-21408-0_5, © Springer-Verlag Berlin Heidelberg 2011

recognized as a very intriguing phenomenon (Cotter-Howells and Caporn 1996). Lead, very common heavy metal polluting our environment, immobilization using phosphate amendments and shift from highly available to the most strongly insoluble form as pyromorphite is considered as worthy to work on and in some laboratories, research on the improvement and implementation of this process in practice take place (Miretzky and Fernandez-Cirelli 2008). This example very well highlights the plant's involvement in the process of "repairing of the environment" and underlines that plants were first in creating safe soil environment. Our duty is to find more species performing this activity or being more efficient.

Heavy metals (HM) are one of the most common pollutant and very dangerous for all living organisms. Toxic impact of the heavy metals on the plants and their responses was already recognized by Carl Linnaeus, the "Father" of taxonomy, who distinguished the leadwort plant family (Plumbaginaceae) represented by species very tolerant to lead. Sea Thrift (*Armeria maritima*), Cape Leadwort (*Plumbago auriculata*) and some species from genus Limonium are the most common for cultivation in lead polluted sites.

Plants respond to toxic metals in different ways starting from avoiding strategies such as, for example, of ions uptake but, if already taken up they exploit tolerance strategies comprising several other mechanisms. For example, detoxification and/or deposition of toxicants in various cells or cell compartments (vacuoles, cell walls) or secreting back to the environment. Plants possess two systems of heavy metals detoxification. One is based on metalothioneins (Goldsbrough 2000) that are common for all living organisms and the second one on the synthesis of phytochelatins (Cobbett 2000; Rauser 1995), substances typical for plants. Nowadays, our knowledge on plant accumulation and detoxification has been expanded mainly as a result of the development of molecular tools (Verbruggen et al. 2008). Among higher living organisms plants are one of the most tolerant to pollution with high capabilities for heavy metal uptake, detoxification and accumulation in easy harvestable organs; these makes them very useful in new emerging environmental biotechnology – phytoremediation.

Phytoremediation is typical in situ technology and still under development. The process of phytoremediation usually takes few years and becomes completed when contaminants no longer impose danger to human beings and environment.

Most advanced in development is soil phytoremediation, covering three main areas:

1. Post-industrial polluted sites (brown field), which are usually highly polluted but with one or few pollutants such as heavy metals and with others pollutants though usually at lower level such as polycyclic aromatic hydrocarbons.
2. Urban area polluted sites characterized by lower level of pollution but most often as a rule with several pollutants.
3. Agricultural soil with elevated levels of various pollutants such as heavy metals, arsenic, pesticides and salinity.

On the post-industrial sites, a branch of phytoremediation called phytoextraction will be the most successful. Plants take up heavy metals from soil using root system and transport them to the above-ground parts, thus enabling the removal of heavy metals from soil. Therefore, plants take responsibility for the most difficult part of soil remediation. Increasing interest in this technology emerges from its economic viability and final effect, as soil after heavy metals removal is brought to the natural state or to the acceptable level of pollution. Phytoextraction on the brown fields is based mainly on annual plants, tolerant to pollutants, with good ability to transfer pollutants to upper parts and whose annual cultivation provides huge biomass and therefore high "pollutants yield."

On these sites also phytostabilization can be considered, which is based on storage of heavy metals in plant tissues or in soil in the form of complex with limited solubility (Babula et al. 2008; Cotter-Howells and Caporn 1996).

Phytovolatilization as a variant of phytoremediation is limited to elements such as Hg, As and Se. In practice it is attenuating them in atmosphere. In the case of Se, since there is deficiency of this essential element, its release can be considered as a positive process, but volatilization of Hg and As will be similar to throwing the pollutants elsewhere.

Urban area polluted sites possess different character. The citizens are spending part of their daily life in these places, e.g., during driving to the working place or even working at such places. At the polluted sites of urban areas, phytoremediation should be maintained as a continuous process in order to remove the pollutants on regular basis. Although the pollutants are in much lower level but from several groups, so far their synergistic or additive effect is not well recognized yet. Transportation vehicles are major emitters of pollutants to air, soil and also water flowing from lanes and pavements. Vehicles are emitting heavy metals and other pollutants including particulate matter (PM).The latter pollutants are nucleus for condensation of HM and finally, after several days, are falling down polluting soil and water.

On the agricultural sites, metals and organics can be efficiently removed by growing energy crops such as *Salix viminalis*. This will decrease the metal content in post-growing crops. This has successfully been shown for the Cd content in wheat grains, which decreases by 25% if cultivating Salix for 2–3 years in advance (Greger and Landberg, in prep). As mentioned, organics such as pesticides will also be most efficiently degraded by phytodegradation, rhizodegradation and phytostimulation. The efficiency of depends on the level of pollution, which in general in agricultural sites is rather slightly elevated but for food production is to high . The lower concentration of pollutants is in the soil the shorter time it takes to remove them by phytoremediation.

In the last years, one of the major areas of research in phytoremediation is improvement of pollutant uptake by plants, in the case of heavy metals limited by their bioavailability. Synthetic chelators such as EDTA (ethylene-diamine-tetraacetic acid) or NTA (nitriloacetate) are questionable and never used on large scale. They are expensive; moreover, some environmentalists treat them as additional man-made pollutants which create risk of leaching heavy metals to the ground water during heavy rain just after application. The process of heavy metals chelation intensifies metal uptake in nature. Hence, plants use low molecular weight organic acids such as citric, ferulic, maleic, oxalic and tartaric acids; utilization of this phenomenon is always attractive for implementation (Strobel 2001; Evangelou et al. 2006). This can be recommended as a simple tool available in our hands. Lowering of soil pH, which would increase metal solubility, can be obtained by application of strong acid fertilizers such as sulfuric salts or even elemental sulfur. Elemental sulfur, being slowly oxidized by bacteria, increases acidity as well (Kayser et al. 2000).

The availability and uptake of heavy metals into the plants depend not only on direct uptake but also on rhizosphere. Plants' life is tightly associated with microorganisms surrounding the plants' roots. In this respect mycorrhizal fungi appear to play a central modulating role. Among them arbuscular mycorrhizal (AM) fungi intensify few crucial processes such as water and nutrients uptake in the plants but simultaneously heavy metals are taken up more efficiently as well. Mycorrhizal fungi alleviate heavy metal stress on plants by detoxification via greater metallothioneins synthesis (Gonzalez-Guerrero et al. 2007) and consequently hold HMs in their tissue thus performing the role of a buffer between soil environment and plants (Hildebrandt et al. 2007; Ferrol et al. 2009). Mycorrhizal fungi express metallothioneins genes, but further steps, i.e., heavy metal transport to and accumulation in roots and upper parts of the plants are controversial. This is because both lowering and increasing of the level of metal pollutants in the plant tissue were recorded (Toler et al. 2005; Sell et al. 2005). Nevertheless, plants inoculation with mycorrhiza, even if it does not support phytoextraction, would allow plants to grow on polluted sites at concentration(s) that otherwise would be eliminated (if not mycorrhized) (Adriaensen et al. 2006). Moreover, it can be considered as an example of phytostimulation.

The above shows that the ability of plants to heavy metal uptake, translocation and accumulation in particular organs depends on environmental factors. However, some of the species, families or even orders have capability higher than others for heavy metal tolerance and uptake.

5.2 Species for Phytoremediation

Plant species useful for phytoremediation need to fulfill some requirements: (a) fast growth; (b) produce high biomass; (c) and be tolerant to pollution. In case of phytoextraction, pollutants additionally need to be translocated and accumulated in the easy harvestable part of plants. If phytoremediation of urban areas is considered, species tolerating trimming are desired. Species recommended for phytodegradation, which usually takes place in the roots, should possess large and dense root system. Root systems with enough large surface are also preferred for microbial communities running together with the plant process of phytostimulation (Pilon-Smits 2005). Three elements, As, Se and Hg, can be removed from the soil environment by phytovolatilization (Greger et al. 2005). However, in case of Hg the picture is not clear yet and requires further studies.

The technology of phytoremediation was inspired by the existence of a group of plant species called hyperaccumulators (Brooks 1998). These plant species, in majority, belong to the following taxonomical orders: Poales, Malpighiales, Fabales, Rosales, Brassicales, Caryophyllales, Solanales and Asterales (Bremer et al. 2009; Chase and Reveal 2009; Haston et al. 2009).

In some parts of world, there are places polluted with radionuclear elements such as cesium, strontium and uranium. Plant species from two taxonomic families and orders, Asteraceae from Asterales and Betaceae from Caryophyllales, are recognized as good phytoremediants of radionuclides with the former showing better performance (Tang and Willey 2003).

The presence, on specific sites, of species that belong to genera, families or orders known to have high tolerance to particular heavy metal(s) that may play a role of bioindicators suggests pollution of this site with given heavy metal(s).

5.2.1 Order: Poales

Among monocotyledons this order is dominating in the number of families with species possessing high capabilities for heavy metal uptake. Most distinguished among them is the Poaceae family with quite a long list of cultivated species such as cereals and high biomass producing grasses that could be considered as good phytoremediants.

5.2.1.1 Family: Poaceae

Grass family is one of the most important for phytoremediation of heavy metals and organic pollutants such as PAHs and petroleum hydrocarbons. The advantage of plants from this family is that after cutting and drying, the plant material is not breakable. They possess a rather shallow root system, but as very dense and well penetrating soil these plants are very effective in phytoextraction of upper soil levels. Tolerance in this family to heavy metals is widely known, e.g., *Leersia hexandra,* is the hyperaccumulator of chromium with potential for phytoremediation of Cr-contaminated soil (Zhang et al. 2007). The biggest biomass producer from this family is maize (*Zea mays*) which is tolerant to heavy metals and petroleum. Sometimes, among cereals the older long-stem cultivars of wheat (*Triticum aestivum*) and barley (*Hordeum vulgare*) are recommended for phytoremediation. When polluted sites also contain high levels of salt, barley, as the most tolerant to salinity among plants from this family (Orcutt and Nilsen 2000), would be the first choice for phytoremediation.

Among other grasses very tolerant to several pollutants are two genera: Lolium and Festuca; both grow as wild on the polluted road sides and there is also quite long list of cultivars that can be used for phytoremediation. Although as typical pasture/loan grasses they produce lower biomass, they are cut 2–3 times during the vegetation season and as a sum they produce a reasonably good yield of polluted biomass. Besides, pollutants deposited on foliage are collected before they are fully washed off by rain and replace pollution into the soil. From the Lolium genus, in Europe, there are two common perennial species: *L. perenne* and *L. multiflorum*. In case of Festuca genus, *F. rubra, F. arundinacea* and *F. ovina* can be utilized in phytoremediation. In this family, there is a highly resistant group of grass belonging to the genus Agrostis with two species, *A. tenuis* and *A. alba*. These grasses are characterized by very high levels of tolerance to heavy metals. Because plants of these two species accumulate heavy metals mostly in the roots, they can be recommended rather for phytostabilization. In fact, earlier described case of *A. tenuis* forming crystals of pyromorphite in the rhizosphere on sites highly polluted with lead whenever soluble inorganic phosphate was present in the soil is also a form of phytostabilization.

On the terrestrial polluted places, at temper zone, *Elytrigia repens,* can be found very often. Plants of this species are very tolerant to salinity and heavy metals with the latter being in high amounts accumulated in the underground rhizomes. Rhizomes can easily be mechanically pulled out from the soil.

Two other grasses cultivated in Europe as ornamental plants also possess high phytoremediation capabilities: *Deschampsia cespitosa* and *Vetiveria zizanioides*.

Grasses may also be used for water and sediments phytoremediation but since for these purposes they are growing in water or wet stands the plants must also tolerate such conditions. As a number one species for this purpose *Phragmites australis* is listed as the basic species used in constructed wetlands, highly tolerant to a wide range of pollutants including high salinity. Other species such as *Phalaris arundinacea* and *Miscantus* sp. are also good candidates but as they require higher temperatures for growth, they can be recommended for rather warmer regions. A very attractive candidate for phytoremediation is the perennial species *Miscanthus gigantheus* which accumulates very high amounts of biomass when aged more than 2 years. This species is more tolerant to frost and survives in countries of Central Europe.

It is noteworthy that all grasses are easily colonized by a wide range of species of mycorrhizal fungi from the Glomus genus, with *G. intraradices* being the most tolerant to heavy metals.

5.2.1.2 Family: Cyperaceae

Several species highly tolerant to pollution belong to this family and many species from this family are growing on very wet sites, just on the border between water and land. Therefore, they are unique for the application in the above described, very difficult, for phytoremediation conditions. At the temper and cooler zones of Europe, several species of Carex were found on polluted sites (Pawlowska et al. 1996) and one of them *C. hirta* very often is recommended as an ornamental plant. For more warmer, southern part of the continent and for Mediterranean, subtropical zone *Cyperus alternifolius* and *Cyperus papyrus*, respectively, can be considered. All species are easily propagated vegetatively by dividing the mother plants.

Another species from this family, *Eriophorum angustifolium*, is very tolerant to a wide range of pH and is therefore very suitable on acidic mine tailing impoundments, which if untreated are very acidic, down to 2.3, and if limed, can have a pH up to 11. This plant is useful in phytostabilization of metals in mine tailings by its property to stabilize the pH around 6 and thus can decrease the bioavailability of metals up to 98% (Stoltz and Greger 2002).

Growing in the South Africa river, sedge species *Bolboschoenus maritimus* is considered as a good bioindicator of sediments pollution with metals (Shuping et al. 2011). In North America, *Scirpus californicus* was tested for constructed wetland with Zn-contaminated water (Gillespie et al. 1999).

5.2.1.3 Family: Typhaceae

From this family, species from Typha genus are very common not only in Europe but also in other parts of the world. Two species are recommended for phytoremediation in many countries: *Typha angustifolia* (Demirezen and Aksoy 2004) and *T. latifolia* (Carranza-Alvarez et al. 2007). Both species very well accumulate heavy metals and can be recommended as an important component of plant community in constructed wetland. Maintained on the bigger area of the shallow water, they not only uptake pollutants but also accumulate enough big biomass for "green" energy production (Ciria et al. 2005). Usually for producing bigger biomass *T. angustifolia* is of first choice, but for obtaining greater biodiversity both the species can be cultivated. Typha, similar to Phragmites, colonizes no running water. Both of them are known as allelopathically active against algae; however, they become effective when plants cover more than 25% of the water surface. Typha phytoremediates heavy metals and agricultural pollutants such as N and pesticides flowing into the water reservoir from intensive conventional agriculture. Typha plants translocate pollutants to the upper part of the plants better than those of Phragmites, but it is much less tolerant to salinity.

5.2.1.4 Family: Juncaceae

Some species from rashes family show potential for heavy metal uptake. But only few species from this family are used as ornamentals. The fact that they accumulate not so big biomass makes them less attractive for phytoremediation. The first species *Juncus lutea* was recognized as a hyperaccumulator of nickel more than 50 years ago (Pichi Sermolli 1948 after Brooks 1998). Plants of other species such as *J. effuses* (Marian et al. 2009) and *J. articulatus* (Vardanyan and Ingole 2006) were also found growing at the polluted sites accumulating heavy metals.

5.2.2 *Order: Malpighiales*

5.2.2.1 Family: Salicaceae

It is the most common family with woody plants that are used in phytoremediation. Plants species belonging to this family are very fast growing, accumulate high amounts of biomass and uptake wide range of pollutants. These mean that they meet the requirements to be one of the best phytoremediants. Usually very fast growing trees are short living, but it is not a problem for this technology because the woody plants still grow for a decade. The willows species are especially recommended for phytoextraction of heavy metals such as Cd, Pb and Zn.

Within the genus Salix there is a big variation in uptake, translocation to the shoot and tolerance to a certain metal. This variation is found to a higher extent between clones of a species than between Salix species (Landberg and Greger 1996; Greger and Landberg 1999). This makes Salix unique because it is then possible to choose a certain clone for a certain phytoremediation purpose, e.g., stabilizing soils with Zn, decreasing the leakage of Zn from the soil by a clone with low uptake of Zn, or by removal of Cd and Cu by a clone that would translocate high amounts of these metals to the shoot. However, at least in the species *Salix viminalis*; it is found that most clones are excluders of heavy metals and it seems to be the most common way of coping with heavy metal toxicity. It is also shown that heavy metal uptake, tolerance and translocation to the shoot in various Salix species are not evolved in areas with high levels of heavy metals (Landberg and Greger 1996).

The property to have high accumulation of heavy metals is not correlated with the property to have high biomass production (Greger et al. 2001). From Salix genus, several species are utilized. From Salix genus, several species are utilized. *Salix viminalis* has the leading position when dealing with heavy metals phytoextraction of heavy metals, which simultaneously evolve biomass which can be utilized as a source of renewable energy (Greger and Landberg 1999). Significant differences between genotypes of *S. viminalis* are observed; thus, in practice, cultivars already checked for their phytoremediation ability need to be cultivated. Nurseries and breeding companies offer special cultivars of *S. viminalis* for phytoremediation with high ability for metal uptake. The advantage of this willow species is that it can be very easily vegetatively propagated by wooden cuttings. Individuals received via this propagation are known to grow much better on moist soils and positively respond to irrigation. Plantations survive more than 10 years but during the first 2–3 years, before the full establishment, they require protection against weeds' competition.

S. caprea, a shrub or small tree, is also known for good phytoremediation abilities but is less common in application because it is propagated by seeds, which is less convenient and not all plants inherit the desired character. In the last few years, *S. burjatica* is recommended as a good candidate species for phytoremediation (Pulford et al. 2002); it is obvious that because of the origin of this species it is tolerant to low temperatures.

Willows are easy to cross between species and many commercially offered cultivars are hybrids (Pulford et al. 2002). One of the oldest cross is *S. viminalis* and *S. caprea* known as *S. smithiana*. This very fast grown hybrid, accumulates high biomass, very well uptakes heavy metals and is easily propagated by wooden cuttings. *S. viminalis, S. daphnoides* and *S. pupurea* are also cultivated as ornamental species and, in urban areas polluted by heavy traffic, can be employed for phytoremediation of pollutants from the air. Organic pollutants such as PAHs, PCBs and dioxins are accumulated in the waxes covering leaves and stems. Raking the leaf litter in the autumn and cutting stems in the spring lead to partial removal of these carcinogenic pollutants from the cities' environment.

The genus of Populus is more universal and can be applied for phytoremediation of both heavy metals and organic pollutants such as TCE, TNT explosive. These plants with very deep root system (up to 20 m) and capabilities to uptake up to 200 L of water per day (by 5-year-old tree) are superior for some of the contaminated sites. In temper zone, poplars are one of the fastest growing species and accumulate big biomass. They are also easily propagated by wooden cuttings. Among poplars, the most successful are not species but interspecies hybrids. Since the poplars are easy in vitro (Lonardo et al. 2011) and vegetatively propagated, the advantages obtained on the basis of heterosis characters can be continuously maintained. The most common are crosses between European and American species: *P. nigra* var. *italica x P. deltoides*, in literature also known as *P. euroamericana,* and others such as *P. trichocarpa x P. deltoides* or *P. trichocarpa x P. maximowiczi*. During application of poplar for phytoremediation it should be kept in mind that very intensive growth of these plants requires high water availability. Poplar hybrids, growing up to 3 m per year, are free of competition with weeds even during the beginning of plantation.

Species from both genera Salix and Poplar are easily colonized by mycorrhizal fungi and they respond with better efficiency of phytoremediation (Sell et al. 2005).

5.2.2.2 Family: Violaceae

This family has some historical meaning for phytoremediation because the endemic metalophyte *Viola calaminaria*, now named hyperaccumulator, was discovered in the nineteenth century on the Belgian/Germany border (Baumann 1885). More common in Europe, closely related to the above-mentioned species and also tolerant to heavy metals is *V. lutea* (Hildebrandt et al. 2007). Although both species accumulate low biomass during vegetation season, they still are interesting from the mechanisms involved from the point of view of tolerance to heavy metals. In China, another hyperaccumulator has been discovered from this family, i.e., *V. baoshanensis* producing high biomass and over there it is considered as a candidate for phytoextraction (Zhuang et al. 2007).

5.2.3 Order: Fabales

5.2.3.1 Family: Fabaceae

Species from this family are very good phytoremediants of heavy metals. They also trap organic pollutants in waxes covering leaves and bark. Moreover, plants from this family stimulate growth of soil microbial community, which are capable of degrading PAHs and PCBs. Additional and very important advantages that plants from this family possess are: (a) self-sufficiency for nitrogen; (b) high tolerance to drought; and (c) capacity of some plant species to survive on very poor soil. Heavy metal uptake by herbaceous species such as *Lupinus* sp. and *Vicia* sp., *Sesbania exaltata* (Miller et al. 2008) was studied and well documented in several laboratories. All common species of *L. albus, L. luteus, L. angustifolius* and *L. hispanicus* show similar growth and accumulation of Mn, Pb, Cr (III), Cr(VI) and CH_3Hg (Pilar et al. 2001). In urban areas, high tolerance is also observed in ornamental species such as *L. polyphyllus* (Gawronski own study, not published). Attention can be directed also to the capabilities of metals including mercury uptake by genus Vicia (Sierra et al. 2008). List of legume species tolerant to pollutants is much wider and include tolerance to heavy metals recorded in *Medicago sativa* (Poniedzialek et al. 2005, Sherifi et al. 2009), to the metalloids arsenate shown by *Cytisus striatus* (Bleeker et al. 2003) and other elements, e.g., to selenium found in *Melilotus indica* (Guo and Wu 1998).

It should be underlined that in this family several hyperaccumulators of selenium were found with *Astragalus bisulcatus* as the first. According to Brooks (1998) along time, the number of species on the list of those tolerant to selenium increased up to 13.

In urban areas, ornamental trees and shrubs from this family such as *Robinia pseudoaccacia, Caragana arborescens* and *Amorpha fruticosa* are commonly cultivated. All these species easily uptake heavy metals and capture organic pollutants in the waxes. *R. pseudoaccacia*, in addition to the above, also exude to the soil high amounts of flavonoids, chemicals with six carbon rings with high similarity to PAHs. This stimulates the development of specific rhizobial microorganism community, which when overpopulated would starve and then start to degrade PAHs and PCBs using carbon skeletons as a source of energy for their living processes.

The above listed trees and shrubs very well re-grow after spring time trimming, which allows removing polluted plant materials for controlled utilization.

5.2.4 Order: Rosales

This is one of the most important order in human life, cultivated mainly for food and ornamental purpose.

5.2.4.1 Family: Rosaceae

Rosaceae is the biggest family in this order, possessing above average tolerance to pollution several species but generally not so high such as some of those described earlier. Tolerant species from this family are mainly ornamentals and often cultivated in polluted urban areas. Roses, tolerant to the pollution, are very often cultivated on the median strips dividing traffic directions with *Rosa rugosa, Rosa rugotida* and *Rosa nitida* recommended for such purpose. Roses not only uptake heavy metals but also create a kind of natural barrier for trespassing. *R. rugosa* is good bioindicator of pollution (Calzoni et al. 2007). Similarly successful in the biomonitoring of Cd, Pb and Zn is other ornamental shrub *Pyracantha coccinea* (Akguc et al. 2008). In many cities, new very decorative two shrubs *Sorbaria sorbifolia* and *Physocarpus opulifolius* are evaluated and they also appear as very promising with respect to tolerance to the pollution.

5.2.4.2 Family: Cannabaceae

Industrial crop *Cannabis sativa* seems to be very promising in the new role as phytoremediant. It is short day species and as the length of the day increases the plant will continue to grow vegetatively, sometimes even up to 4 m height accumulating huge amounts of biomass. The industrial hemp grows fast, does not need to use herbicides because hemp by itself, as being highly allelopathic active, keeps weeds away, probably by excreting some substance. High levels of metals are mostly found not only in the shoots but also in the leaves (Linger et al. 2005, Khan et al. 2008). This with a small root mass, together with the fact that the root system is poor and shallow, makes *C. sativa* interesting for phytoextraction (Greger et al. in prep). As annual plants, all biomass of the above ground part is completely harvested in the autumn. High biomass yield require supplementing nitrogen, calcium (sensitive to shortage of Ca) and water. This species is also known as very good phytoremediant of heavy metals and radionuclides with significant differences in phytoremediation capabilities between cultivars recorded. Due to the strong similarity to another species, the *Cannabis indica* a source of narcotics, in countries with strong anti-drug law there are formal difficulties to cultivate this crop.

5.2.5 Order: Brassicales

5.2.5.1 Family: Brassicaceae

Plant species belonging to this family are among the best accumulators of heavy metals, with a long list of the hyperaccumulators and with record levels of heavy metal concentrations in plant tissues. Most of the discovered hyperaccumulators from this family belong to the genera Alysum and Thlaspi (Brooks 1998). For phytoremediation, most often, species with known agronomical practises are proposed and in this taxonomic group most common is the Indian mustard (*Brassica juncea*)(Kumar et al. 1995). Selected lines of this species uptake heavy metals in very high amounts. The fact that plants of this species can be cultivated twice every growing season makes them very attractive as heavy metals phytoremediants. White mustard (*Sinapis alba)* – species commonly cultivated in Europe for a green manure – possesses slightly lower capacities for heavy metals phytoremediation (Winska-Krysiak and Gawronski 2002). Very valuable was the discovery that to this family belongs the species *Iberis intermedia* with capability to hyperaccumulate thallium, a very toxic element (Leblanc et al. 1999). Recently, in many countries, programs of renewable energy have started and acreage of rapeseed cultivation for biodiesel production is increasing. Industrial utilization of the rapeseed (*Brassica napus* L. var. *napus)* oil allows cultivating this crop on the polluted sites; thus, besides oil production simultaneous process of phytoremediation can be run. However, it needs to be remembered that rapeseed requires better soil to obtain good yield; so not all fields can be used for the combined purposes of oil production and heavy metals phytoremediation in case of this species.

There are two disadvantages of using plants from this family for phytoremediation: (a) necessity of plant protection against insects and (b) leaf blades after drying are fragile and are dropped off the stems. Therefore, in order to avoid secondary emission, plants must be harvested before full maturity and the vegetative parts must be utilized in a controlled manner. Additional disadvantage is also the lack of colonization with symbiotic mycorrhizal fungi, which if takes place leads not only to increased heavy metals uptake but also to increased tolerance.

Technology of phytomining can be applied for the highly valuable metals such as gold using *Brassica juncea* (Anderson et al. 1998) and nickel with domesticated hyperaccumulator *Alyssum murale* (Chaney et al. 2005).

5.2.6 *Order: Caryophyllales*

To this order belong several families with species tolerant to heavy metals and most of them are also tolerant to salinity.

5.2.6.1 Family: Plumbaginaceae

In leadwort family (Plumbaginaceae) there are species very tolerant to lead and usually they also are tolerant to other heavy metals. Most common on lead polluted sites is *Armeria maritima*, a plant commonly cultivated in gardens but this pioneer plant is successfully surviving on postindustrial sites (Dahmani-Muller et al. 2000) and even colonizing places as heap (Olko et al. 2008). The shrub *Plumbago auriculata* is one of the most popular blue color flower plant in gardens at warmer parts of the world. Not many of us know that this is one of the most tolerant species to lead and that in earlier times it was used as an antidote against lead poisoning. From the genus Limonium several species are cultivated as ornamental plants.

5.2.6.2 Family: Betaceae

Plants from this family are known as highly tolerant to salinity and can be recommended for phytoremediation of salt polluted sites. They are also recognized as efficient in uptaking heavy metals (Rosso et al. 2005). The most common from Betaceae family sugar beet, crop producing very high yield, is not used in phytoremediation because of the problem with utilization of obtained biomass due to high water content. Two other cultivated species accumulating high biomass, *Atriplex hortensis* and *Kochia scoparia,* can be recommended for this technology. Both species are ornamental plants, easily propagated from seeds with wide selection of cultivars available offered by seed companies. Most species from this family are not colonized by mycorrhizal fungi.

5.2.6.3 Family: Amaranthaceae

Amaranth (*Amaranthus* ssp.) is cultivated as an ornamental plant and lately also as an alternative edible crop. *A. blitoides* was found on widely known toxic spill from Aznalcollar mine in Spain (Rio et al. 2002) and another weedy species *A. hybridus* was tested for phytoextraction (Puschenreiter et al. 2001). Morphologically, plants belonging to this genus are very much differentiated; most of them are interspecies crosses. They possess several interesting characters for phytoremediation such as very high tolerance to salinity and capabilities of heavy metal uptake. Very fast growing and accumulating high biomass species/cultivars (some can growth up 2 m) can be recommended for phytoremediation of salt, heavy metal or both of these pollutants simultaneously.

5.2.6.4 Family: Caryophylaceae

Plant species from this taxonomic family very often appear on salt or heavy metal polluted sites such as, for example, *Agrostemma githago* (Pichtel et al. 2000), *Dianthus carthusianirum* (Baranowska-Morek and Wierzbicka 2004) and *Silene vulgaris* (Baroni et al. 2000). Since, similar to other ornamental plant species from this family, they accumulate low biomass, they are not employed in phytoremediation but they are good indicators of pollution with these two groups of contaminants. During visit to polluted sites when decision on application of phytoremediation is under consideration, the presence of the plants of species from this family well indicates the presence of pollutants in the soil and probably on the level which allows to cultivate tolerant plant species.

5.2.6.5 Family: Polygonaceae

Most of the species from this family are known to uptake salt and heavy metals from the soil and they occupy the banks of salty ponds and lakes (Vardanyan and Ingole 2006). Usually these are species from Polygonum genus such as *P. amphibium* and grow on more dry sites such as *P. aviculare*. For phytoremediation, however, much more interesting are species from Fagopyron genus and in practice the following species are used: *F. esculentum*, *F. tataricum* and *F. cymosum*. In fact, utilization of *F. esculentum* for cleaning up soil from heavy metals has been patented in the US (Tamura et al. 2010) and Europe (Sato et al. 2007).

5.2.6.6 Family: Tamaricaceae

Species from this family are among the most tolerant to dryness and salinity. They are typical steppe, semi-desert and desert, being present on and therefore are considered as maintaining sandy dunes. There are around 50 *Tamarix* species known and two of them *T. symyrnesis* (Manousaki et al. 2008) and *T. tetrandra* (Kalmuckov et al. 2009) were already evaluated for their phytoextraction capabilities.

It is commonly known that species from Tamarix genera are salt and metal excluders; so in practice they can be used for temporary phytostabilization.

5.2.7 Order: Solanales

5.2.7.1 Family: Solanaceae

This plant family contains several genera that are very important for humans as a source of food such as potato, tomato or even as condiments such as tobacco. A majority of the species from that family uptake heavy metals from soil very well. In phytoremediation technology, most advanced is the application of *Solanum nigrum* by increasing its efficiency by multiple harvesting (Wei et al. 2006), inoculation with microorganism (*Paecilomyces lilacinus*), intensifying plant microorganism chelates synergy or application of citric acid (Gao et al. 2010). *Datura inoxia* demonstrates interesting capabilities for heavy metal sequestration (Lin and Rayson 1998). It is well known that *Solanum tabacum* easily take up heavy metals from soil (Evangelou et al. 2006) but not all cigarette smokers are so oriented that they introduce to their lung considerable amounts of heavy metals. Several *Nicotiana* spp. are used as ornamental and can be cultivated in polluted sites of urban areas. To lower cost and limit contact with polluted soil for cultivation in such places, *N. silvestris* the perennial species is recommended.

5.2.8 Order: Asterales

5.2.8.1 Family: Asteraceae

Plant species from that family are used for remediation of heavy metals and radionuclides such as Sr, Cs and U. Two crop species *Helianthus annus* and perennial *Helianthus tuberosus* are used for phytoremediation of industrial polluted sites. The *H. annus* – sunflower is a common annual crop plant, accumulating big biomass up to 100 tons of fresh matter. It is tolerant to drought, exhibits strong allelopathic activity and high competitiveness against weeds. It is the best species in European conditions for phytoremediation, because of its good uptake of heavy metals including antimony, which during the last few years has become an increasing threat to the environment (Tschan et al. 2008). *Helianthus tuberosus* is perennial crop which is advantageous in cultivation because of lower cost and the polluted soil is not tilled. Although most of the hyperaccumulators accumulate small biomass, one of the exception is *Berkheya coddi* (Robinson et al. 2003). This species is one of the best in nickel uptake and is considered as a candidate for phytomining of this element and of cobalt as well (Robinson et al. 1999). This family is rich in ornamental species and some of them which accumulate high amounts of biomass can be recommended for phytoremediation of urban areas. Species from three genera can be considered: Solidago, Tanacetum and Rudbeckia. Ornamental species usually greatly differ in many characters but for phytoremediation, tall cultivars with big biomass are preferred even if they would not be the best in terms of aesthetic characters. Some of the species from this family also demonstrates high tolerance to salt, for example *Artemisia vulgaris*.

Above-described orders, families, genera and species do not cover all and the list is not complete. Along with further discoveries and our increasing general knowledge on plant kingdom, the list will be extended. A good example of this is the discovery of hyperaccumulating arsenic fern *Pteris vitata,* which can be utilized for phytoremediation of this dangerous environment polluting elements (Ma et al. 2001). One year later, the paper describing further study revealed that other species from Pteris genera such as *P. cretica*, *P. longifolia,* and *P. umbrosa* hyperaccumulate arsenic to the same extent as the first one (Zhao et al. 2002).

5.3 Conclusions

Among plant kingdom there are several families represented by relatively high number of species tolerant to heavy metals including hyperaccumulators.

There are already species which are recognized as good candidates for phytoremediation. It can be assumed that along with our increasing knowledge on plant biology new species tolerant to heavy metals and candidate good phytoremediants will be found.

Candidates for phytoremediants should be tolerant to heavy metals, uptake them in reasonably big amounts and accumulate in easily harvestable of high biomass organs, thus ensuring high "yield of pollution."

Depending on the site of phytoremediation, i.e., brown fields, urban areas or agricultural polluted sites, different requirements should be fulfilled by considered species.

Simple and low-cost methods of evaluation of plant tolerance (and their progeny) to heavy metals are strongly desired.

References

Adriaensen K, Vangronsveld J, Colpaert JV (2006) Zinc-tolerant *Suillus bovinus* improves growth of Zn-exposed *Pinus sylvestris* seedlings. Mycorrhiza 16:553–558

Akguc N, Ozyigit II, Yarci C (2008) Pyracanyha coccinea Roem. (Rosaceae) as a biomonitor for Cd, Pb and Zn in Mugla Province (Turkey). Pak J Bot 40:1767–1776

Anderson CWN, Brooks RR, Stewart RB, Simcock R (1998) Harvesting a crop of gold in plants. Nature 395:553–554

Babula P, Adam V, Opatrilova R, Zehnalek J, Havel L, Kizek R (2008) Uncommon heavy metals, metalloids and their plant toxicity: a review. Environ Chem Lett 6:189–213

Baranowska-Morek A, Wierzbicka M (2004) Localization of lead in root tip of *Dianthus carthusianirum*. Acta Biol Cracov 46:45–56

Baroni F, Boscagli A, Protano G, Riccobono F (2000) Antimony accumulation in *Achillea ageratum, Plantago lanceolata* and *Silene vulgaris* growing in an old Sb-mining area. Environ Pollut 109:347–352

Baumann A (1885) Das Verhalten von Zinksalzen gegen Pflanzen und im Boden. Landwirtschaftliche Versuchsstation 31:1–53

Bleeker PM, Schat H, Vooijs R, Verkleij JAC, Ernst WHO (2003) Mechanisms of arsenate tolerance in *Cytisus striatus*. New Phytol 157:33–38

Bremer B, Bremer K, Chase MW, Fay MF, Reveal JL, Soltis DE, Soltis PS, Stevens PF (2009) An update of the Angiosperm Phylogeny Group classification for the orders and families of flowering plants: APG III. Bot J Linn Soc 161:105–121

Brooks RR (1998) Plants that hyperaccumulate heavy metals. CAB International, University Press, Cambridge, p 380

Calzoni GL, Antognoni F, Pari E, Fonti P, Gnes A, Speranza A (2007) Active biomonitoring of heavy metal pollution using *Rosa rugosa* plants. Environ Pollut 149:239–245

Carranza-Alvarez C, Alonso-Castro AJ, Alfaro De La Torre M, Garcia-De La Cruz RF (2007) Accumulation and distribution of heavy metals in *Scirpus americanus* nad *Typha latifolia* from an artificial lagoon in San Luis Potosi, Mexico. Water Air Soil Pollut 188(1–4):297–309

Chaney RL, Angle JS, Wang AS, McIntosh MS, Broadhurst L, Reeves RD (2005) Phytoextraction of soil Cd, Ni and Zn using hyperaccumulator plants to alleviate risks of metal contaminated soils requiring remediation. International Workshop. Current developments in remediation of contaminated lands, Pulawy, Poland, p 39

Chase MW, Reveal JL (2009) A phylogenic classification of the land plants to accompany APG III. Bot J Linn Soc 161:122–127

Ciria MP, Solano ML, Soriano P (2005) Role of macrophyte *Typha latifolia* in a constructed wetland for wastewater treatment and assessment of its potential as a biomass fuel. Biosyst Eng 92(4):535–544

Cobbett CS (2000) Phytochelatins and their role in heavy metal detoxification. Plant Physiol 123:825–832

Cotter-Howells J, Caporn S (1996) Remediation of contaminated land by formation of heavy metal phosphates. Appl Geochem 11:335–342

Dahmani-Muller H, van Oort F, Gelie B, Balabane M (2000) Strategies of heavy metal uptake by three plant species growing near a metal smelter. Environ Pollut 109:231–238

Demirezen D, Aksoy A (2004) Accumulation of heavy metals in *Typha angustifolia* (L.) and *Potamogeton pectinatus* (L.) living in Sultan Marsh (Kayseri, Turkey). Chemosphere 56:685–696

Evangelou MWH, Ebel M, Schaeffer A (2006) Evaluation of the effect of small organic acids on phytoextraction of Cu and Pb from soil with tobacco *Nicotiana tabacum*. Chemosphere 63:996–1004

Ferrol N, Gonzalez-Guerrero M, Valderas A, Benabdellah K, Azcon-Aguilar C (2009) Survival strategies of arbuscular mycorrhizal fungi in Cu-polluted environments. Phytochem Rev 8 (3):551–559

Gao Y, Miao C, Mao L, Zhou P, Jin Z, Shi W (2010) Improvement of phytoextraction and antioxidative defense in *Solanum nigrum* L. under cadmium stress by application of cadmium-resistant strain and citric acid. J Hazard Mater 181:771–777

Gillespie WB, Hawkins WB, Rodgers JH, Cano ML, Dorn PB (1999) Transfers and transformations of zinc in flow-through wetland microcosms. Ecotoxicol Environ Saf 43: 126–132

Goldsbrough P (2000) Metal tolerance in plants: the role of phytochelatins and metallothioneins. In: Terry N, Banuelos G (eds) Phytoremediation of contaminated soil and water. Lewis Publishers, Boca Raton, pp 221–235, pp 408

Gonzalez-Guerrero M, Cano C, Azcon-Aguilar C, Ferrol N (2007) *GintMT1* encodes a functional metallothionein in *Glomus intraradices* that responds to oxidative stress. Mycorrhiza 17: 327–335

Greger M, Landberg T (1999) Use of willow in phytoextraction. Int J Phytoremediation 1:115–123

Greger M, Landberg T, Berg B (2001) Salix clones with different properties to accumulate heavy metals for production of biomass. Report to STEM. Akademitryck AB, Edsbruk

Greger M, Wang Y, Neuschütz C (2005) Absence of Hg transpiration by shoot after Hg uptake by roots of six terrestrial plant species. Environ Pollut 134:201–208

Guo X, Wu L (1998) Distribution of free seleno-amino acids in plant tissue of *Melilotus indica* L. grown in selenium-laden soils. Ecotoxicol Environ Saf 39:207–214

Haston E, Richardson JE, Stevens PF, Chase MW, Harris DJ (2009) The Linear Angiosperm Phylogeny Group (LAPG) III: a linear sequence of the families in APG III. Bot J Linn Soc 161:128–131

Hildebrandt U, Regvar M, Bothe H (2007) Arbuscular mycorrhiza and heavy metal tolerance. Phytochemistry 68:139–146

Kalmuckov K, Alexnardova E, Georgiev GI (2009) Copper accumulation capacity of tamarisk (*Tamarix tetrandra* L.) and white mulberry (*Morus alba* L.) depending on soil type. Gen Appl Plant Physiol 35:179–187

Kayser A, Vegner K, Keller A, Attinger W, Felix HR, Gupta SK, Schulin R (2000) Enhancement of phytoextraction of Zn, Cd and Cu from calcareous soil: the use NTA and sulfur amendments. Environ Sci Technol 34:1778–1783

Khan MA, Wajid A, Noor S, Khattak FK, Akhter S, Rahman IU (2008) Effect of soil contamination on some heavy metals content of *Cannabis sativa*. J Chem Soc Pak 30(6):805–809

Kumar PBAN, Dushenkov V, Motto H, Raskin I (1995) Phytoextraction: the use of plants to remove heavy metals from soils. Environ Sci Technol 29:1232–1238

Landberg T, Greger M (1996) Differences in uptake and tolerance to heavy metals in *Salix* from unpolluted and polluted areas. Appl Geochem 11:175–180

Leblanc M, Petit D, Deram A, Robinson BH, Brooks RR (1999) The phytomining and environmental significance of hyperaccumulation of thallium by *Iberis intermedia* from Southern France. Econ Geol 94:109–114

Lin S, Rayson GD (1998) Impact of surface modification on binding affinity distributions of *Datura innoxia* biomass to metal ions. Environ Sci Technol 32:1488–1493

Linger P, Ostwald A, Haensler J (2005) *Cannabis sativa* L. growing on heavy metal contaminated soil: growth, cadmium uptake and photosynthesis. Biol Plantarum 49(4):567–576

Lonardo SD, Capuana M, Arnetoli M, Gabbrielli R, Gonnelli C (2011) Exploring the metal phytoremediation potential of three *Populus alba* L. clones using an in vitro screening. Environ Sci Pollut Res 18:82–90

Ma LQ, Komar KM, Tu C, Zhang W, Cai Y, Kennelley ED (2001) A fern that hyperaccumulates arsenic. Nature 409:579

Manousaki E, Kadukova J, Papadatonakis N, Kalograkis N (2008) Phytoextraction and phytoexcretion of Cd by the leaves of *Tamarix smyrnensis* growing on contaminated non-saline and saline soils. Environ Res 106:326–332

Marian M, Cozmuta LM, Varga C, Cozmuta AM, Nour E (2009) Vegetation dynamics depending on ecological particularities of Bozanta Mare (Maramures County – Romania) tailing pound: case study. Am J Environ Sci 5(1):116–123

Miller G, Begonia G, Begonia M, Ntoni J, Hundley O (2008) Assessment of efficacy of chelate – assisted phytoextraction of lead by coffeeweed (*Sesbania exaltata* Raf.). Int J Environ Res Public Health 5(5):428–435

Miretzky P, Fernandez-Cirelli A (2008) Phosphates for Pb immobilization in soils: a review. Environ Chem Lett 6:121–133

Olko A, Abratowska A, Żyłkowska J, Wierzbicka M, Tukendorf A (2008) *Armeria maritima* from calamine heap – initial studies on physiologic-metabolic adaptations to metal-enriched soil. Ecotoxicol Environ Saf 69:209–218

Orcutt DM, Nilsen ET (2000) The physiology of plants under stress. Wiley, New York, p 683

Pawlowska TE, Blaszkowski J, Ruhling A (1996) The mycorrhizal status of plants colonizing a calamine spoil mound in southern Poland. Mycorrhiza 6:499–505

Pichtel J, Kuroiwa K, Sawyerr HT (2000) Distribution of Pb, Cd and Ba in soils and plants of two contaminated sites. Environ Pollut 110:171–178

Pilar XE, Yolanda MA, Carmen C, Carmen C, Carmen B, Mercedes M (2001) Evaluation of lupinus species to accumulate heavy metals from waste waters. Int J Phytoremediation 4:369–379

Pilon-Smits E (2005) Phytoremediation. Annu Rev Plant Biol 56:15–39

Poniedzialek M, Sekara A, Ciura J, Jedrszczyk E (2005) Nickel and manganese accumulation and distribution in organs of nine crops. Folia Hort 17:11–22

Pulford ID, Riddell-Black D, Stewart C (2002) Heavy metal uptake by willow clones from sewage sludge-treated soil: the potential for phytoremediation. Int J Phytoremediation 4:59–72

Puschenreiter M, Stoger G, Lombi E, Horak O, Wenzel WW (2001) Phytoextraction of heavy metal contaminated soils with *Thlaspi goesingense* and *Amaranthus hybridus*: rhizosphere manipulation using EDTA and ammonium sulfate. J Plant Nutr Soil Sci 164:615–621

Rauser WE (1995) Phytochelatins and related peptides. Plant Physiol 109:1141–1149

Rio MD, Font R, Almela C, Velez D, Montoro R, Bailon DH (2002) Heavy metals and arsenic uptake by wild vegetation in the Guadiamar river area after the toxic spill of the Aznalcollar mine. J Biotechnol 98:125–137

Robinson BH, Brooks RR, Clothier BE (1999) Soil amendments affecting nickel and cobalt by *Berkheya coddii*: potential use for phytomining and phytoremediation. Ann Bot 84:689–694

Robinson BH, Lombi E, Zhao FJ, McGrath SP (2003) Uptake and distribution of nickel and other metals in the hyperaccumulator *Berkheya coddii*. New Phytol 158:279–285

Rosso PH, Pushnik JC, Lay M, Ustin SL (2005) Reflectance properties and physiological responses of *Salicornia virginica* to heavy metal and petroleum contamination. Environ Pollut 137:241–252

Sato T, Honda M, Tamura H (2007) Europatent no.: EP1795273 (A1)

Sell J, Kayser A, Schulin R, Brunner I (2005) Contribution of ectomycorrhizal fungi to cadmium uptake of poplar and willows from a heavily polluted soil. Plant Soil 277:245–253

Sherifi E, Bytyqi A, Lluga-Rizani K (2009) The concentration of Pb and Cd to *Medicago sativa* along Lipjan-Prizren Highway and their influence on biomass. J Eng Appl Sci 4(1):60–63

Shuping LS, Snyman RG, Odendaal JP, Ndakidemi PA (2011) Accumulation and distribution of metals in *Bolboschoenus maritimus* (Cyperaceae), from a South African river. Water Air Soil Pollut 216:319–328

Sierra MJ, Milan R, Esteban E, Cardona AI, Schmid T (2008) Evaluation of mercury uptake and distribution in *Vicia sativa* L. applying two different study scales: greenhouse condition and lysimeter experiments. J Geochem Explor 96:203–209

Stoltz E, Greger M (2002) Cottongrass effects on trace elements in submersed mine tailings. J Environ Qual 31:1477–1483

Strobel BW (2001) Influence of vegetation on low-molecular-weight carboxylic acids in soil solution – a review. Geoderma 99:169–198

Tamura H, Sato T, Honda M (2010) United States Patent no.: US 7,829,754 B2

Tang S, Willey NJ (2003) Uptake of ^{134}Cs by four species from Asteraceae and two varieties from the Chenopodiaceae grown in two types of Chinese soil. Plant Soil 250(1):75–81

Toler HD, Morton JB, Cumming JR (2005) Growth and metal accumulation of mycorrhizal sorghum exposed to elevated copper and zinc. Water Air Soil Pollut 164:155–172

Tschan M, Robinson B, Schulin R (2008) Antimony uptake by *Zea mays* (L.) and *Helianthus annus* (L.) from nutrient solution. Environ Geochem Health 30:187–191

Vardanyan LG, Ingole BS (2006) Studies on heavy metal accumulation in aquatic macrophytes from Sevan (Armenia) and Carambolim (India) lake system. Environ Int 32:208–218

Verbruggen N, Hermans C, Schat H (2008) Molecular mechanism of metal hyperaccumulation in plants. New Phytol 181(4):759–776

Wei S, Zhou Q, Koval PV (2006) Flowering stage characteristics of cadmium hyperaccumulator *Solanum nigrum* L. and their significance to phytoremediation. Sci Total Environ 369:441–446

Winska-Krysiak M, Gawronski SW (2002) Fizjologiczne aspekty tolerancji i hiperakumulacji ołowiu w wybranych genotypach *Brassica*. [Physiological aspects of lead tolerance and hyperaccumulationin *Brassica* species]. Zesz Prob Post Nauk Rol 481:605–613

Zhang X-H, Liu J, Huang H-T, Chen J, Zhu Y-N, Wang D-Q (2007) Chromium accumulation by the hyperaccumulator plant *Leersia hexandra* Swartz. Chemosphere 67:1138–1143

Zhao FJ, Dunham SJ, McGrath SP (2002) Arsenic hyperaccumulation by different fern species. New Phytol 156:27–31

Zhuang P, Yang QW, Wang HB, Shu WS (2007) Phytoextraction of heavy metals by eight plant species in the field. Water Air Soil Pollut 184:235–242

Chapter 6
Reclamation of Contaminated Mine Ponds Using Marble Wastes, Organic Amendments, and Phytoremediation

Dora M. Carmona, Raúl Zornoza, Ángel Faz, Silvia Martínez-Martínez, and Jose A. Acosta

6.1 Introduction

Mining activities in the Region of Murcia – SE Spain, which started 2,500 years ago and ended in the 1990s, generated several tailing ponds, which store residues from the extraction of lead and zinc (Pb/Zn). These tailing ponds contain materials of high Fe-oxyhydroxides, sulphates, and elevated contents of potentially leachable heavy metals (mainly Cd, Pb, Cu, and Zn) due to extreme acidic conditions. Since a long time those mine residues have been transported downstream during periods of high flow, erosion is evident in these areas, causing migration of pollutants into surface and ground water. These metal-contaminated soils also contribute to human

D.M. Carmona
Sustainable Use, Management, and Reclamation of Soil and Water Research Group. Department of Agrarian Science and Technology, Technical University of Cartagena, Paseo Alfonso XIII, 52, 30203 Cartagena, Murcia, Spain

Universidad Pontificia Bolivariana, Environmental Research Group, Circular 1ª #70-01, Medellín, Colombia

R. Zornoza • Á. Faz (✉) • S. Martínez-Martínez
Sustainable Use, Management, and Reclamation of Soil and Water Research Group. Department of Agrarian Science and Technology, Technical University of Cartagena, Paseo Alfonso XIII, 52, 30203 Cartagena, Murcia, Spain
e-mail: angel.fazcano@upct.es

J.A. Acosta
Sustainable Use, Management, and Reclamation of Soil and Water Research Group. Department of Agrarian Science and Technology, Technical University of Cartagena, Paseo Alfonso XIII, 52, 30203 Cartagena, Murcia, Spain

Institute for Biodiversity and Ecosystem Dynamics, University of Amsterdam, Nieuwe Achtergracht 166, 1018 WV Amsterdam, The Netherlands

I. Sherameti and A. Varma (eds.), *Detoxification of Heavy Metals*, Soil Biology 30,
DOI 10.1007/978-3-642-21408-0_6, © Springer-Verlag Berlin Heidelberg 2011

and animal metal exposure, through food chain transfer or inhalation of wind-blown dust (Pierzinski 1997). In terms of stabilizing metal contaminated mine sites, a lower metal concentration in vegetation shoot is preferred, in order to prevent metal from entering the ecosystem through food chain (Pichtel et al. 2000). Thus, efforts are needed for long-term reclamation of these contaminated areas to stabilize soil metals, create a structured soil, promote cover vegetation, and avoid health risks in the ecosystems.

In response to a growing concern for human health and environmental quality, many technologies have been developed to treat and remediate metal-contaminated soils. One of the remediation options gaining considerable interest over the last decade is the in situ immobilization of metals using metals immobilizing agents (Vangronsveld and Cunnungham 1998). Thus, the transformation of metals into harmless species or their removal in a suitable recycled mineral form such as carbonates using marble wastes or lime (Geebelen et al. 2003) is a possible solution for the remediation of a mining area. In addition, incorporation of organic amendments into contaminated mine soils has been proposed as feasible, inexpensive, and environmentally sound disposal practice, as generally such wastes can improve soil physical and chemical properties, and contain nutrients beneficial to initialize plant colonization (Barker 1997), favoring the reactivation of biogeochemical cycles and the natural establishment of vegetation. The increment in vegetation cover reduces or even prevents the dispersion of the contamination through wind and water erosion, and improves the aesthetic value of formerly bare areas (Vangronsveld and Cunnungham 1998). Besides, vegetation itself may contribute to metal immobilization processes through biological activities in the production of organic matter (Bouwman and Vangronsveld 2004), an emerging technology called phytostabilization.

Although there is a general consensus that efficiency of soil remediation also depends on the presence and activity of microorganisms, the long-term ecological consequences of inorganic and organic amendments for these features have received little attention (Mench et al. 2006). Biochemical properties may indicate the potential of a soil to sustain microbiological activity, which can be used to assess the effectiveness of a soil remediation process (Pérez de Mora et al. 2005). In this sense, soil enzymes have been reported to be highly sensitive to heavy metals, and, therefore, have been recommended as standard biochemical indicators to assess quality of heavy metal-polluted soils (Hinojosa et al. 2004).

The main objective of this study is to evaluate of the long-term effectiveness of different inorganic and organic amendments for remediation of contaminated mine soils by means of (1) monitoring the evolution of some physicochemical properties and availability of heavy metals, (2) determination of soil quality using biochemical properties as indicators, and (3) assessing the establishment of spontaneous vegetation and bioaccumulation of heavy metals in plants in order to avoid the risk of mobility in the food chain.

6.2 Material and Methods

6.2.1 Study Site

The study was conducted in the Region of Murcia (SE Spain), in the Cartagena-La Unión Mining District, which covers an area of ~50 km^2 with an elevation range from 0 to 110 m asl (Fig. 6.1). Great mining activity has been carried out for more than 2,500 years, the activity being stopped in the nineties. The climate of the area is semiarid Mediterranean, with annual average temperature of 18°C and mean annual rainfall of 200–300 mm. Two tailing ponds generated by mining activities were selected: El Lirio (L) and Brunita (B), representative of the 85 rest of existent tailing ponds in this Mining District. This mining area was an important center for the extraction of mineral ores such as sphalerite [(Zn,Fe)S], galena (PbS), and pyrite ($Fe^{2+}S_2$).

6.2.2 Field Experimental Set-Up

Twenty field plots (4 m^2) were established in 2004 in a completely randomized design to evaluate influence of the combined additions of industrial and organic wastes in soil development in tailing ponds. Two different organic amendments were used to reclaim the soils, pig manure (P), and sewage sludge (S). In addition, three different doses per amendment were applied. Thus, the treatments were: Untreated contaminated soil (C), soil treated with pig manure at dose 1 (P1), dose

Fig. 6.1 Study site and images of El Lirio tailing pond during the application of amendments (*above*) and 5 years after application (*below*)

2 (P2), and dose 3 (P3); and soil treated with sewage sludge at dose 1 (S1), dose 2 (S2), and dose 3 (S3). For pig manure, doses were 2.5, 5, and 10 kg per plot, respectively. For sewage sludge, doses were 1.99, 3.98, and 7.97 kg per plot, respectively. Marble waste was added at 22 kg per plot in all plots except for the control. Calcium carbonate from marble waste was applied to correct the acidity in the mine soils. The particle size distribution in marble waste was 26% <2 mm and the rest from 2 to 5 mm in size. Marble waste came from marble industry located in Cehegín (NE of Murcia Region). The anaerobically digested sewage sludge was collected from the drying process of Cartagena wastewater treatment plant, and pig manure came from a pig farm in the Cuevas de Reillo (SE of Murcia).

Dose of organic amendments (dry-weight) were calculated on the bases of European and Spanish legislation regarding the addition of N to soil (Directive 91/676/EEC 1991; Real Decreto 261/1996). Organic amendments were applied with the purpose of increasing soil pH to immobilize metals and create better conditions for microbial and plant development. Dose of marble waste was determined based on Eq $CaCO_3$ required to neutralize the acidity and reach soil pH ~7, according to the Sobek method (Sobek et al. 1978).

Amendments were manually applied. First, we added marble wastes and let it dry for 24 h. Then, we mixed the materials with the soil to a depth of 0–15 cm. Following this, we applied the organic amendments using the same procedure. After the addition of soil amendments, plots were exposed to the semiarid climatic conditions in the study area for long-term observations.

6.2.3 Soil and Vegetation Sampling and Analytical Methods

The soil samplings were carried out previously to the application of amendments (time 0), and at 6 months, 1, 2, and 5 years after application of amendments. Composite soil samples from five sub-samples were taken from the 0–15 cm layer for each treatment plot. Samples were air-dried for 7 days, passed through a 2-mm sieve and stored at room temperature prior to laboratory analyses.

Although the vegetation in the tailing pond of study was absent, the application of amendments in the plots conducted to spontaneous colonization of vegetation by the surrounding environment. Thus, at the same time of the last soil sampling (5 years after application), the identification of all plants species present in the plots was carried out (richness), as well as the percentage of vegetation cover. In addition, shoots of the most dominant species in P3 (the treatment with highest vegetation cover and richness) were collected in each of the three replicated plots, making a composite sample, for analyses of metals concentrations. These species were *Piptatherum miliaceum*, *Zygophylum fabago*, *Dactilis glomerata*, and *Brassica fruticulosa*.

Soil pH and electrical conductivity (EC) were measured in distilled water (1:1 and 1:5 w/v, respectively) (Peech 1965). Soil organic carbon (SOC) was determined by chemical oxidation using dichromate solution (Walkley and Black 1934), while total nitrogen (N_t) was determined according to Duchaufour (1970). Equivalent calcium carbonate (inorganic carbon) was estimated using a Bernard calcimeter. Total metals content were determined by acid digestion with conc. $HNO_3/HClO_4$ at 210°C for 1.5 h and addition in HCl 0.1 N (Risser and Baker 1990). Soluble metals were extracted by water (soil:distilled water ratio = 1:2) (Buurman et al. 1996), and available metals extracted using DTPA (for soils with pH > 6) or EDTA (for soils with pH < 6) (soil:DTPA = 1:2; soil:EDTA = 1:5) (Lindsay and Norvell 1978). For plant metals concentration, 1 g of dried plant sample was ashed in a muffle furnace at 450°C for 24 h. After that, the digested material was dissolved in HNO_3 and filtered. Measurements of metals (Cd, Cu, Pb, Zn) were carried out using flame atomic absorption spectrophotometer (AAnalyst 800, Perkin Elmer).

Microbial biomass carbon (MBC) was determined using the fumigation–extraction procedure (Vance et al. 1987); basal soil respiration (BSR) was determined according to Anderson (1982); β-glucosidase activity was measured following the method of Tabatabai (1982); arylesterase activity was established according to Zornoza et al. (2009); acid phosphatase activity was determined according to Tabatabai and Bremner (1969); phosphodiesterase was measured following the method of Browman and Tabatabai (1978); arylsulphatase activity was measured by the method of Tabatabai and Bremner (1970). In addition, the metabolic quotient qCO_2 (BSR/MBC) was calculated.

In order to assess the efficiency of plants for phytostabilization, the bioaccumulation factor (BF) was also calculated as $[\text{metal}]_{\text{shoot}}/[\text{bioavailable metal}]_{\text{soil}}$ (Kumar et al. 1995). Ideally this value would be <<1, but it should not exceed a ratio of 1, which would indicate that the plant is useful for phytoextraction (accumulation of metals in shoot tissue) but should not be used in phytostabilization (Brooks 1998).

6.2.4 Statistical Analyses

The fitting of the data to a normal distribution for all properties measured was checked with the Kolmogorov–Smirnov test. The data were submitted to one-way ANOVA to assess the differences among treatment and doses. The separation of means was made according to Tukey's verified significant difference at $P < 0.05$. Relationships among properties were studied using Pearson correlations. Soil chemical and biochemical properties were subjected to principal components analysis (PCA) to elucidate major variation patterns in terms of amendments and doses. Statistical analyses were performed with the software SPSS for Windows, Version 17.0.

6.3 Results

6.3.1 *General Physicochemical Properties of Tailing Ponds and Amendments*

Mine soils samples from El Lirio and Brunita can be classified as Anthropic Spolic Regosol according to WRB (2007), and Haplic Torriarent according to USDA (2010). Both tailing ponds El Lirio and Brunita present a similar particle size distribution corresponding to sandy loam textural class (sand: 83%, silt: 4%, clay: 13%).

Selected chemical properties of the studied tailing ponds at time 0 showed that lowest pH (2.6) and moderate salinity was measured in Brunita, while El Lirio had the highest pH (6.7) and high salinity (Table 6.1). Both SOC and N_t contents were absent, what indicate inhospitable conditions for vegetation growth. Tailing ponds had higher total contents of Pb and Zn compared to Cu and Cd, being higher in El Lirio. Both El Lirio and Brunita can be considered very contaminated, exceeding limit values of the European legislation for soil contamination (Zn: 300, Pb: 300, Cu: 140, Cd: 30, in mg/kg). These levels indicated high toxicity in mine soils which could adversely affect the biological activity (Nwachukwu and Pulford 2010).

With regards to amendments, pig manure and sewage sludge had high pH that favors increases in mine soil pH, and high levels of total N and C, needed to promote the activation of biochemical cycles. In terms of heavy metals, sewage sludge showed higher contents of Cd, Cu, and Zn than the studied tailing ponds. In the case of pig manure, only total Cu was high.

Table 6.1 Selected properties of amendments and mine soils

Parameter	S	P	MW	Soil L	Soil B
pH	7.57	8.58	7.88	6.72	2.65
EC (dS/m)	2.49	9.00	2.18	11.46	4.17
$CaCO_3$ (%)	4.8	19.6	97.9	0.49	0.33
N_t (%)	5.05	2.17	–	0.01	0.00
SOC (%)	34.0	32.0	–	0.0	0.0
C/N	7	14	–	–	–
Moisture (%)	75.0	40.0	2.1	–	–
Total metals					
Cd (mg/kg)	192	1.51	0.94	23.6	1.44
Cu (mg/kg)	357	832	4.98	93.4	39.7
Pb (mg/kg)	39.6	26.6	11.6	13973	1539
Zn (mg/kg)	8659	261	1.36	2351	1157
Bioavailable metals					
Cd (mg/kg)	–	–	–	10.46	0.10
Cu (mg/kg)	–	–	–	1.52	0.14
Pb (mg/kg)	–	–	–	436.9	5.57
Zn (mg/kg)	–	–	–	236.4	30.4

Marble waste (MW), Sewage sludge (S), Pig manure (P), Lirio (L), Brunita (B) on dry-weight basis. Soil samples (0–15 cm)*EC* electrical conductivity, N_t total nitrogen, *SOC* soil organic carbon

6.3.2 Evolution of Soil Properties in Field-Plots Trial

The evolution of soil properties and available metals for 5 years are represented in Figs. 6.2 (El Lirio) and 6.3 (Brunita). After the application of amendments, we observed an increase in pH in both ponds, being higher in Brunita, since the initial

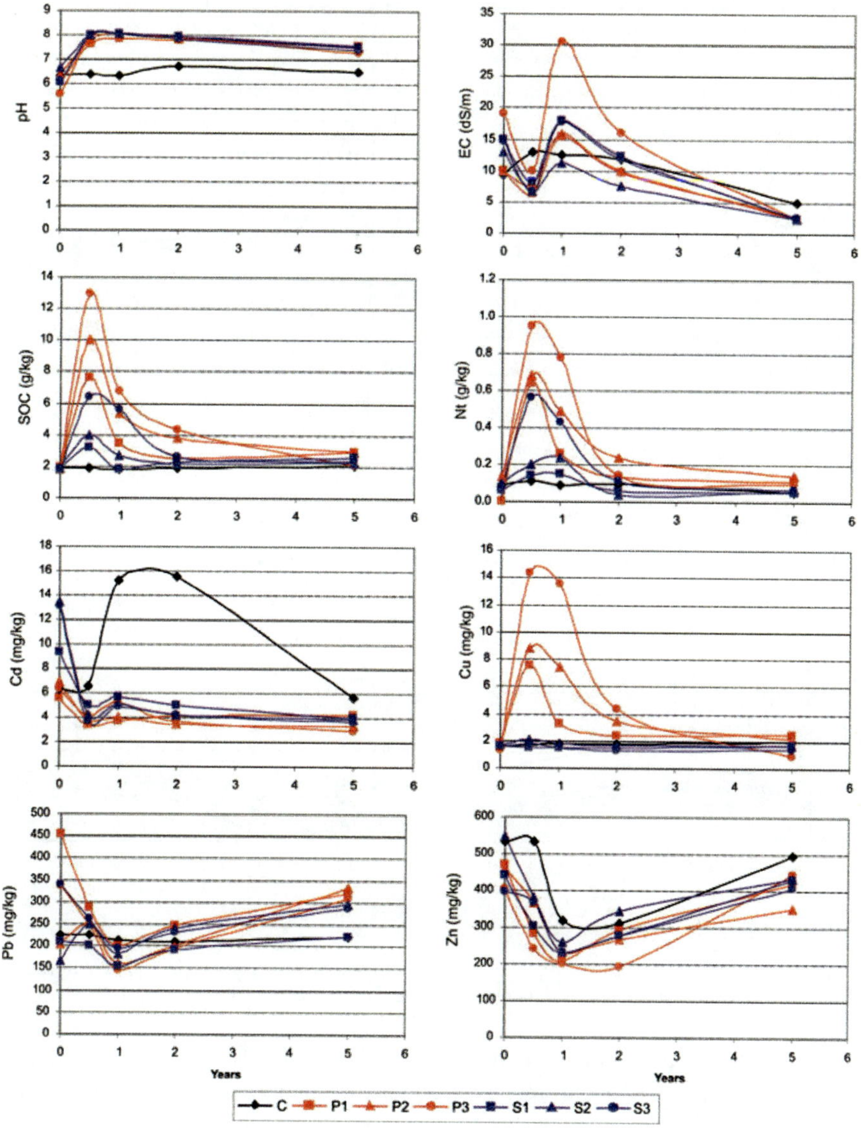

Fig. 6.2 Evolution of pH, electrical conductivity (EC), soil organic carbon (SOC), total nitrogen (N_t) and bioavailable metals in El Lirio plots (see the text for the meaning of plots abbreviations)

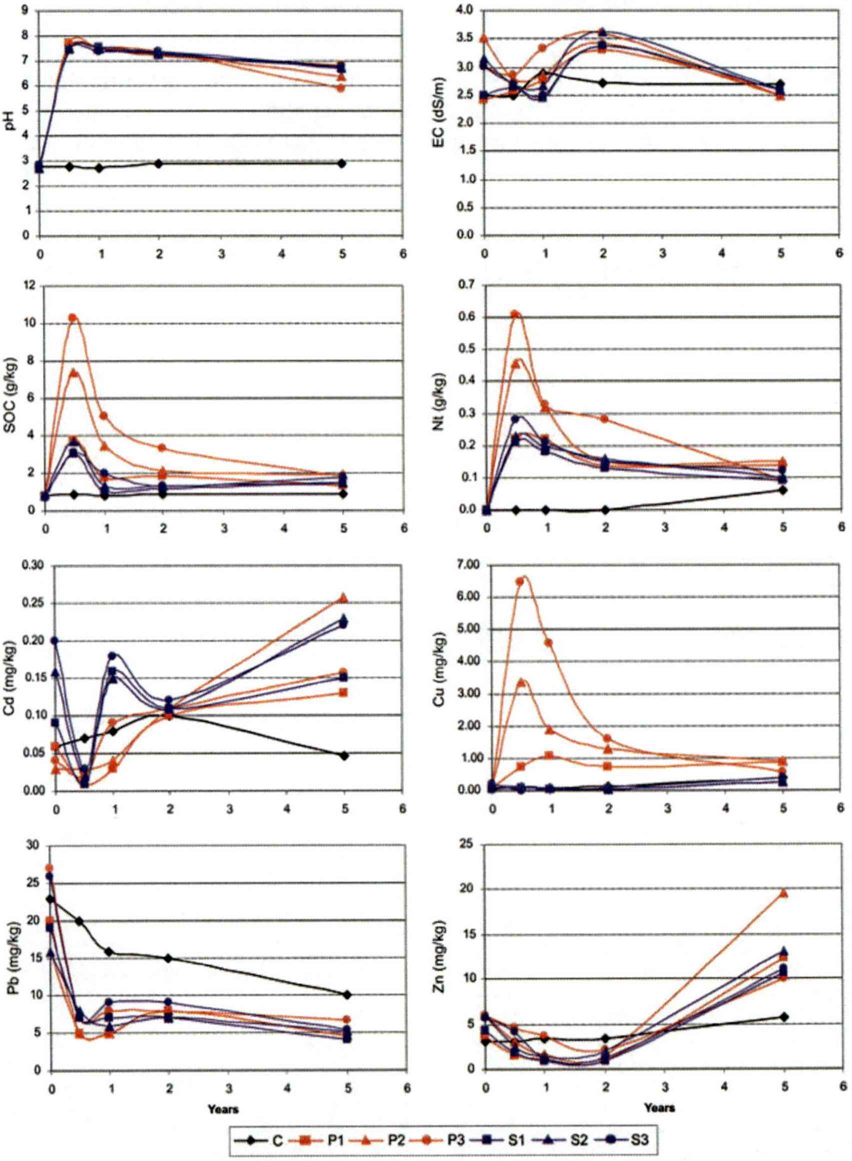

Fig. 6.3 Evolution of pH, electrical conductivity (EC), soil organic carbon (SOC), total nitrogen (N_t) and bioavailable metals in Brunita plots (see the text for the meaning of plots abbreviations)

pH value in this pond was extremely low (<3.0). The pH remained practically stable with time, without differences among treatments, although it tended to decrease after 5 years, mainly due to decreases in carbonates content (data not shown). With regards to EC, we observed an increment after 1 year of amendments application, owing to the high quantity of salts provided by the organic amendments and the solution of carbonates. After 2 years, there is a decreasing trend in EC, due to leaching of sulphates (easily soluble) from mine soils and soluble ions from organic amendments. SOC and N_t initially increased with the application of amendments, mainly in P plots. This increase was in general terms related with the dose of application. However, the values of these two properties decreased owing to leaching and mineralization, shifting down after 5 years of applications to values slightly higher to control.

Regarding total metals, Cd, Cu, Zn, and Pb were above European legislation thresholds, and did not change with time. Bioavailable metals decreased as general pattern in the amended plots in both tailing ponds. However, we detected slight increments in Pb and Zn in El Lirio, and in Cd and Zn in Brunita in amended plots after 5 years of monitoring. The possible explanation for this behavior is difficult owing to the different factors implied in the mobility of heavy metals and their interactions with soil properties. The detected slight decreases in pH and solubilization of carbonates may have likely contributed to increments in bioavailability of some metals. Moreover, the fact that we have also detected increments in the control plot in some metals could be indicating changes in bioavailability of these metals owing to water and wind erosion of surface particles that migrate to other zones, thereby exposing subsurface soils. In addition, the parent material is rich in Cd, Pb, Cu, and Zn sulphides. Thus, oxidation processes of these sulphides and dissolution of secondary sulphates may have also released some metals to the soil. On the other hand, decreases in soil organic matter could have had an important effect, since the application of organic amendments initially immobilized metals by complexation (Zanuzzi 2007). Biovailable Pb and Zn in El Lirio were positively correlated with plant richness and vegetation cover, whilst Cd and Zn were also positively correlated with plant richness in Brunita. This could indicate that the spontaneous establishment of vegetation could be influencing the availability of these metals. In fact, plant roots are known to exude organic compounds capable of complexing metals, which can increment the metals availability in the rhizosphere, and this process differs among different plant species (Jones 1998; Almeida et al. 2006). Plants release some labile compounds to soil to promote the availability and uptake of nutrients, provoking also the availability and uptake of heavy metals (Séguin et al. 2004).

6.3.3 Biochemical Properties

Results of the different biochemical properties determined in the plots emplaced in El Lirio 5 years after the application of the amendments are shown in Figs. 6.4 and 6.5. According to the general trends, all biochemical properties were higher in treated soils than in control, despite the fact that SOC and N_t were similar amongst the treatments after 5 years of the application in El Lirio plots (Fig. 6.2).

The highest increases with respect to control were for MBC in P plots (100%), β-glucosidase in P plots (250%), phosphodiesterase in P plots (210%), and

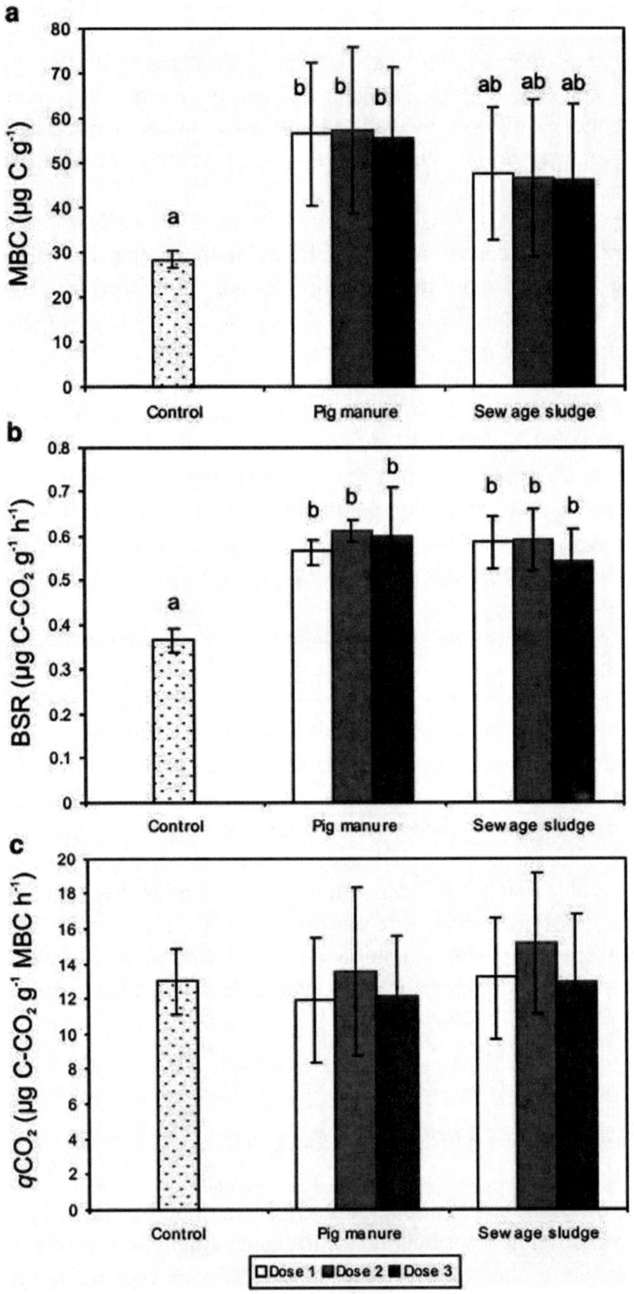

Fig. 6.4 Microbial biomass carbon, soil respiration and metabolic quotient (qCO_2) of the control soil and remediated plots with different organic amendments at three different doses. Different letters indicate mean values significantly different after Tukey's honestly significant difference at $P < 0.05$

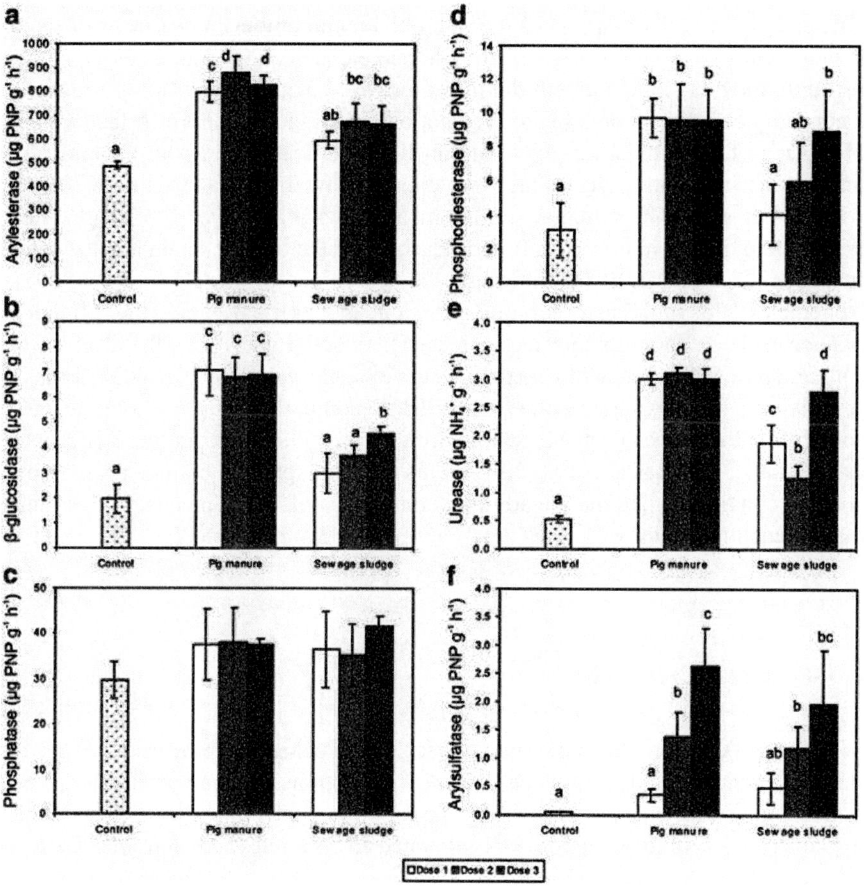

Fig. 6.5 Enzyme activities of the control soil and remediated plots with different organic amendments at three different doses. Different letters indicate mean values significantly different after Tukey's honestly significant difference at $P < 0.05$

arylsulfatase in P3 (4,000%). This confirms the high sensitivity of biochemical properties to evaluate soil quality (Nannipieri et al. 1990), as undetected shifts occurred with other chemical properties. In addition, also as general pattern, biochemical properties showed higher values after application of pig manure than after application of sewage sludge. These results are promising in an area like Murcia province where more than 10% of pig production in Spain is located. These industries generate a large volume of pig slurry that continuously increases with high demands for pork, and consequently creates disposal problem for many pig producers. However, doses did not have a great effect, being only significant for β-glucosidase, phosphodiesterase, and arylsulfatase.

Metabolic quotient has often been used as an indicator of efficiency in carbon mineralization (Insam and Haselwandter 1989). These authors postulated that the efficiency in the use of carbon of the microbial communities increases as the ecosystems succession progresses, resulting in decreases in qCO_2. Nonetheless, in this research, the metabolic quotient did not show this expected trend (lower values in amended plots, which have higher vegetation cover and richness). In fact, values of qCO_2 in this study are high, what can likely indicate a stressful situation for microorganisms, maybe due to high contents in heavy metals and salinity.

The fact that SOC remains similar in all plots after 5 years of application (Fig. 6.2), indicates a mineralization or leaching of the organic amendments, since treated plots had initially significantly higher values of SOC (Zanuzzi 2007). Nonetheless, this initial incorporation of organic matter has triggered the activation of microbial populations which has increased their activity, favoring the recovery of soils and the establishment of vegetation. However, the values of microbial biomass, respiration, and enzyme activities are still low comparing with noncontaminated soils from other zones from SE Spain with the same climatic conditions (Zornoza et al. 2006, 2007; Bastida et al. 2008). This can be explained by the still extreme edaphic conditions, like the already moderate levels of heavy metals, low organic matter, and high salinity.

6.3.4 Factor Analyses

With the PCA performed on the soil chemical and biochemical properties, 70.4% of the total variance could be explained by the first two principal components (Fig. 6.6). Soil samples were clearly clustered by the first principal component (PC1), which explained 48.7% of the variation. PC1 separated all P samples and S3 from the rest of

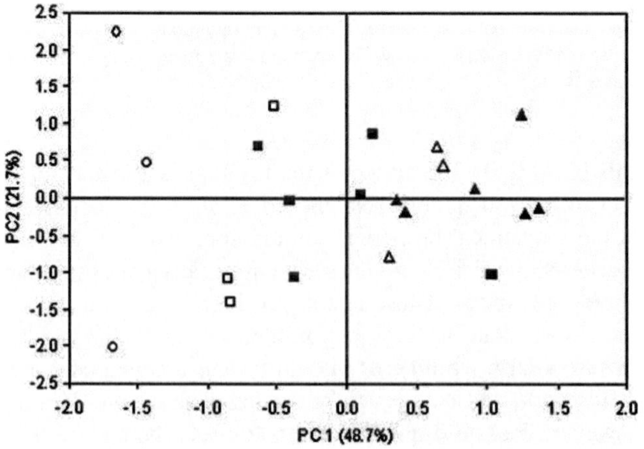

Fig. 6.6 PCA factor scores from chemical and biochemical properties for all treatments and doses. Treatments: control (*circles*), pig manure (*triangles*), sewage sludge (*squares*). Doses: dose 1 (*open symbols*), dose 2 (*gray symbols*), dose 3 (*black symbols*)

samples. Additionally, PC1 also separated control soils from S1 and S2, which clustered together. Samples were not separated by doses by any principal component; solely S3 was separated from S1 and S2 by PC1. This component was associated with all enzyme activities (except for acid phosphatase) and BSR. Second principal component was associated with bioavailable Zn, Pb, and Cd and acid phosphatase.

In general, the factor analyses performed with all samples showed a slight improvement in soil quality in amended soils, mainly in plots amended with pig manure (independently of the dose) and plots amended with the highest dose of sewage sludge. Thus, the use of organic amendments has been proved as a suitable and beneficial procedure to restore mine soils with high contents of heavy metals. The soil properties associated with the separation of treatments were the enzyme and metabolic activities, reinforcing the evidence that biochemical properties are the most sensitive to assess changes in soil quality after soil mine reclamation. Hence, the overall improvement in soil quality of mine sites should be evaluated based not only on soil chemical properties, but also with additional biochemical or biological assays that measure restoration of habitat functions (Hinojosa et al. 2008; Alvarenga et al. 2008).

6.3.5 Vegetation

The untreated plots for both sites remained without vegetation, while natural plant species spontaneously colonized the amended plots (Fig. 6.1). The vegetation cover increased with the dose as a general trend, although plots amended with pig manure showed highest vegetation cover (Table 6.2). Richness also increased with the application dose of amendments, with highest values in plots amended with pig manure. In fact, vegetation cover and richness were significantly positively correlated in both zones, El Lirio ($r = 0.78$; $P < 0.001$) and Brunita ($r = 0.67$; $P < 0.01$).

Thus, the initial and unique incorporation of organic matter has triggered the establishment of vegetation, which remains after 5 years of amendments application. The maintenance of this vegetation cover is essential for true landscape reclamation, activating nutrient cycles and microbial activity (Bouwman and Vangronsveld 2004).

Accumulation and distribution of heavy metals in plant tissues are important aspects to evaluate the role of plant in remediation of metalliferous soils (Friedland 1989). The present results showed that the contents of the different metals in shoots were similar in the most dominant plant species in the P3 plots, except for *Zygophylum fabago*, which had significantly higher values of Cd, Cu, and Zn (Table 6.3).

The BF showed that Cd, Cu, and Zn bioaccumulation occurred in most species (with values of BF $>$ 1), while for Pb no accumulation was observed (Table 6.4). Moreover, *P. miliaceum* only showed bioaccumulation for Cu, being the plant species with lower BF factors for the rest of metals. It is important to highlight

Table 6.2 Natural colonization of plant species on the plots

Treatment	Vegetation cover (%)	Richness	Plant species
El Lirio			
Control	0	0	–
LP1	43	4	*Zigophyllum fabago, Piptatherum miliaceum, Dittrichia viscosa, Phragmites australis*
LP2	45	4	*Zigophyllum fabago, Piptatherum miliaceum, Helichrysum decumbens, Sonchus tenerrimus*
LP3	60	5	*Zigophyllum fabago, Piptatherum miliaceum, Helichrysum decumbens, Dittrichia viscosa, Phragmites australis*
LS1	13	2	*Zigophyllum fabago, Piptatherum miliaceum,*
LS2	27	3	*Zigophyllum fabago, Piptatherum miliaceum, Helichrysum decumbens,*
LS3	32	4	*Zigophyllum fabago, Piptatherum miliaceum, Helichrysum decumbens, Sonchus tenerrimus*
Brunita			
Control	0	0	–
BP1	23	5	*Dactylis glomerata, Brassica fruticulosa, Piptatherum miliaceum, Bromus rubens, Helichrysum decumbens*
BP2	30	8	*Bromus rubens, Brassica fruticulosa, Helichrysum decumbens, Sonchus tenerrimus, Phagnalon saxalite, Dactylis glomerata, Zigophyllum fabago, Spergularia bocconei*
BP3	47	7	*Dactylis glomerata, Bromus rubens, Helichrysum decumbens, Dittrichia viscose, Phagnalon saxalite, Phallaris canariensis, Sonchus tenerrimus*
BS1	19	5	*Dactylis glomerata, Brassica fruticulosa, Bromus rubens, Phagnalon saxalite, Phallaris canariensis*
BS2	25	6	*Brassica fruticulosa, Phallaris canariensis, Bromus rubens, Sedum sediforme, Dactylis glomerata, Piptatherum miliaceum*
BS3	26	7	*Dactylis glomerata, Brassica fruticulosa, Sonchus tenerrimus, Bromus rubens, Helichrysum decumbens, Phallaris canariensis, Spergularia bocconei*

Table 6.3 Metal concentrations in shoots for the most dominant plant species in P3 plots

Metal in shoots (mg/kg)	Plant species				
	Dactilis glomerata	*Piptatherum miliaceum*	*Brassica fruticulosa*	*Zygophylum fabago*	*F* value[a]
Cd	4.2 ± 0.0a	4.4 ± 0.1a	4.2 ± 0.1a	13.2 ± 2.0 b	61.9***
Cu	3.4 ± 0.4a	3.8 ± 0.4a	4.4 ± 0.7a	12.0 ± 0.9 b	116.2***
Pb	23.7 ± 6.2 ab	13.1 ± 6.6a	49.9 ± 18.5 b	21.6 ± 4.8a	6.7*
Zn	21.8 ± 5.7a	90.9 ± 27.0 b	49.0 ± 8.4 ab	288.2 ± 30.2 c	100.0***
Fe	171 ± 53	325 ± 116	225 ± 119	180 ± 38	1.8 ns
Mn	46.1 ± 10.0a	28.0 ± 2.0 b	31.0 ± 6.9 ab	30.4 ± 4.5 ab	4.8*

Values are mean ± standard deviation ($n = 3$)

[a]Significant at: *$P < 0.05$, **$P < 0.01$, ***$P < 0.001$ after one-way ANOVA; *ns* not significant ($P > 0.05$). Different letters indicate significant differences ($P < 0.05$) among means in each location after Tukey's honestly significant difference

Table 6.4 Bioaccumulation factors (BF) of each metal in the most dominant plant species in P3 plots

	Plant species				
BF	*Dactilis glomerata*	*Piptatherum miliaceum*	*Brassica fruticulosa*	*Zygophylum fabago*	*F* value
Cd	18.7 ± 7.8[a]	1.0 ± 0.1 b	18.8 ± 6.7[a]	3.1 ±0.9 b	4.9*
Cu	6.2 ± 0.8[a]	5.9 ± 3.7[a]	8.0 ± 0.9 ab	12.7 ± 1.5 b	6.8*
Pb	0.4 ± 0.1[a]	0.1 ± 0.0 b	0.7 ± 0.2 c	0.1 ± 0.0 b	23.4***
Zn	3.2 ± 1.1 ab	0.3 ± 0.1[a]	5.0 ± 1.5 b	1.0 ± 0.3[a]	10.1**

Values are mean ± standard deviation ($n = 3$)
[a]Significant at: *$P < 0.05$, **$P < 0.01$, ***$P < 0.001$ after one-way ANOVA; ns: not significant ($P > 0.05$). Different letters indicate significant differences ($P < 0.05$) among means in each location after Tukey's honestly significant difference

the high values of BF for Cd in *B. fruticulosa* and *D. glomerata*, indicating the high bioaccumulation of this metal in these concrete species. Thus, these species may be more suitable for phytoextraction technique, rather than phytostabilization, since the most suitable species are those that show mechanisms for protecting themselves against uptake of metals and restricting their transport within the plant (Lefèvre et al. 2005). However, most species presents in these plots are not eaten by herbivores (Zanuzzi 2007), acting like a sink for metals and preventing it from becoming available to other organisms.

6.4 Conclusion

1. The application of pig manure and sewage sludge together with marble wastes has proved to be effective for long-term decrease in the bioavailability of most toxic heavy metals present in two tailing ponds from SE Spain, besides maintaining pH close to neutrality. Despite the initial decrease in SOC and N_t, mineralization and leaching have led to levels of organic matter only slightly higher in comparison to control plots. Since increments in some bioavailable metals have been monitored, future studies are needed to determine the causes, or mitigate this trend by new applications of organic amendments, until succession of vegetation progresses to provide enough litter to increase and maintain soil organic matter.
2. After 5 years of applications of amendments, plots with pig manure presented the best effects on microbial biomass and activity. However, the values of the biochemical properties are still low even for a semiarid environment. Besides, pig manure plots have proved to be more effective to initialize natural spontaneous vegetation colonization, richness, and vegetation cover. *Zygophylum fabago* accumulated moderated quantities of metals, not observed in the other plant species. However, BF was high for all plants except for *P. miliaceum*. Thus, even though most species grew in this study are refused by herbivores, a better

through selection of the most suitable species to continue with phytostabilization progress and mine soils remediation in SE Spain should be developed in the immediate future, focusing on reduction of erosion, tolerance to metals and salinity, nitrogen fixation and low accumulations of metals so that risks in the food chain are minimized.

Acknowledgements This work has been funded by the Project Nr.: CP-IP 213968–2 IRIS, funded by the European Union FP7. R. Zornoza acknowledges a "Juan de la Cierva" contract from the Ministry of Science and Innovation of the Government of Spain. J.A. Acosta acknowledges a grant from Fundación Séneca of Comunidad Autónoma de Murcia (Spain).

References

Almeida CMR, Mucha AP, Vasconcelos MTSD (2006) Comparison of the role of the sea club-rush *Scirpus maritimus* and the sea rush *Juncus maritimus* in terms of concentration, speciation and bioaccumulation of metals in estuarine sediment. Environ Pollut 142:151–159

Alvarenga P, Palma P, Gonçalves AP, Baião N, Fernandes RM, de Varennes A, Vallini G, Duarte E, Cunha-Queda AC (2008) Assessment of chemical, biochemical and ecotoxicological aspects in a mine soil amended with sludge of either urban or industrial origin. Chemosphere 72:1774–1781

Anderson JPE (1982) Soil respiration. In: Page AL, Miller RH, Keeney DR (eds) Methods of soil analysis, part 2, 2nd edn. American Society of Agronomy and Soil Science Society of America, Madison, WI, USA, pp 837–871

Barker AV (1997) Composition and uses of compost. In: Rechling JE, Mackimon HC (eds) Agricultural uses of by-products and wastes, vol 10. American Chemical Society, Washington, DC, pp 140–162, ACS Symposium Series No. 668

Bastida F, Barberá GG, García C, Hernández T (2008) Influence of orientation, vegetation and season on soil microbial and biochemical characteristics under semiarid conditions. Soil Biol Biochem 38:3463–3473

Bouwman LA, Vangronsveld J (2004) Rehabilitation of the nematode fauna in a phytoestabilized heavily zinc-contaminated, sandy soil. J Soils Sediment 4:17–23

Brooks RR (1998) Plants that hyperaccumulate heavy metals: their role in phytoremediation, microbiology, archaeology, mineral exploration and phytomining. CAB International, Wallingford, UK

Browman MG, Tabatabai MA (1978) Phosphodiesterase activity of soils. Soil Sci Soc Am J 42:284–290

Buurman P, van Lagen B, Veltorst EJ (1996) Manual for soil and water analysis, Technical report. Backhuys Publishers, Leiden, Netherlands

Directive 91/676/EEC (1991) Protection of waters against pollution caused by nitrates from agricultural sources. Official J L 375, pp 1–8, 31-12-1991

Duchaufour P (1970) Précis de Pedologie. Masson and Cie, París

Friedland AJ (1989) The movement of metals through soils and ecosystems. In: Shaw AJ (ed) Heavy metal tolerance in plants: evolutionary aspects. CRC Press, Inc., Boca Raton, pp 7–20

Geebelen W, Adriano DC, van der Lelie D, Mench M, Carleer R, Clijsters H, Vangronsveld J (2003) Selected biovailability assays to test the effect of amendment-induced immobilization of lead in soils. Plant Soil 249:217–228

Hinojosa MB, García-Ruíz R, Viñegla B, Carreira JA (2004) Microbiological rates and enzyme activities as indicators of functionality in soils affected by the Aznalcóllar toxic spill. Soil Biol Biochem 36:1637–1644

Hinojosa MB, Carreira JA, Rodríguez-Maroto JM, García-Ruíz R (2008) Effects of pyrite sludge pollution on soil enzyme activities: ecological dose-response model. Sci Total Environ 396:89–99

Insam H, Haselwandter K (1989) Metabolic quotient of the soil microflora in relation to plant succession. Oecologia 79:174–178

Jones DL (1998) Organic acids in the rhizosphere – a critical review. Plant Soil 205:25–44

Kumar PBAN, Dushenkov V, Motto H, Raskin I (1995) Phytoextraction: the use of plants to remove heavy metal from soils. Environ Sci Technol 29(5):1232–1238

Lefèvre I, Correal E, Lutts S (2005) Cadmium tolerance and accumulation in the noxious wed *Zygophylum fabago*. Can J Bot 83:1655–1662

Lindsay W, Norvell W (1978) Development of a DTPA soil test for Zn, Fe, Mn, and Cu. Soil Sci Soc Am J 42:421–428

Mench M, Renella G, Gelsomino A, Landi L, Nannipieri P (2006) Biochemical parameters and bacterial species richness in soils contaminated by sludge-borne metals and remediated with inorganic soil amendments, Methods of Soil Analysis, Part 2, 2nd edit. Environ Pollut 144:24–31

Nannipieri P, Greco S, Ceccanti B (1990) Ecological significance of the biological activity in soil. In: Bollag JM, Stotzky G (eds) Soil biochemistry, vol 6. Marcel Dekker Inc, New York, pp 293–355

Nwachukwu OI, Pulford ID (2010) Microbial respiration as an indication of metal toxicity in contaminated organic materials and soil. J Hazard Mater 185(2–3):1140–7

Peech M (1965) Hydrogen-ion activity. In: Black CA (ed) Methods of soil analysis, vol 2. American Society of Agronomy, Madison, WI, USA, pp 914–916

Pérez de Mora A, Ortega-Calvo JJ, Cabrera F, Madejón E (2005) Changes in enzyme activities and microbial biomass after "in situ" remediation of a heavy metal-contaminated soil. Appl Soil Ecol 28:125–137

Pichtel J, Kuroiwa K, Sawyerr HT (2000) Distribution of Pb, Cd and Ba in soils and plants of two contaminated sites. Environ Pollut 110:171–178

Pierzinski GM (1997) Strategies for remediating trace-element contaminated sites. In: Iskandar IK, Adriano DC (eds) Remediation of soils contaminated with metals. Science Reviews, Northwood, UK, pp 67–88

Real Decreto 261/1996 Protección de las aguas contra la contaminación producida por los nitratos procedentes de fuentes agrarias. BOE No. 61, de 11 de marzo de 1996

Risser JA, Baker DE (1990) Testing soils for toxic metals. In: Westerman RL (ed) Soil testing and plant analysis. Soil Science Society of America Special Publication, Madison, pp 275–298

Séguin V, Gagnon C, Courchesne F (2004) Changes in water extractable metals, pH and organic carbon sequestrations at the soil-root interface of forested soils. Plant Soil 260:1–17

Sobek AA, Schuller WA, Freeman JR, Smith RM (1978) Field and laboratory methods applicable to overburdens and minesoils. EPA-600/2-78-054

Tabatabai MA (1982) Soil enzymes. In: Page AL, Miller RH, Keeney DR (eds) Methods of soil analysis, part 2, 2nd edn. American Society of Agronomy and Soil Science Society of America, Madison, WI, USA, pp 501–538

Tabatabai MA, Bremner JM (1969) Use of p-nitrophenyl phosphate in assay of soil phosphatase activity. Soil Biol Biochem 1:301–307

Tabatabai MA, Bremner JM (1970) Arylsulphatase activity of soils. Soil Sci Soc Am J 34:225–229

USDA (2010) US Department of agriculture. Keys to soil taxonomy, 10th edn. Unites States Department of Agriculture-NRCS, Washington, USA

Vance ED, Brookes PC, Jenkinson DS (1987) An extraction method for measuring soil microbial biomass C. Soil Biol Biochem 19:703–707

Vangronsveld J, Cunnungham S (1998) Introduction to concepts. In: Vangronsveld J, Cunnungham S (eds) In situ inactivation and phytorrestoration of metal contaminated soils. Springer-, Berlin, pp 1–15, and RG Landes Company Georgetown

Walkley A, Black IA (1934) An examination of the digestion method for determining soil organic matter and a proposed modification of the chromic acid titration method. Soil Sci 37:29–38
WRB (2007) World reference base for soil resources 2006, first update 2007. World soil resources reports No 103, FAO, Rome
Zanuzzi A (2007) Reducción de impactos ambientales asociados a suelos contaminados por metales pesados mediante el uso de residuos antropogénicos. PhD Dissertation. Departamento de Ciencia y Tecnología Agraria. Universidad Politécnica de Cartagena
Zornoza R, Guerrero C, Mataix-Solera J, Arcenegui V, García-Orenes F, Mataix-Beneyto J (2006) Assessing air-drying and rewetting pretreatment effect on some soil enzyme activities under semiarid Mediterranean conditions. Soil Biol Biochem 38:2125–2134
Zornoza R, Guerrero C, Mataix-Solera J, Arcenegui V, García-Orenes F, Mataix-Beneyto J (2007) Assessing the effects of air-drying and rewetting pre-treatment on soil microbial biomass, basal respiration, metabolic quotient and soluble carbon under Mediterranean conditions. Eur J Soil Biol 43:120–129
Zornoza R, Landi L, Nannipieri P, Renella G (2009) A protocol for the assay of arylestease activity in soil. Soil Biol Biochem 41:659–662

Chapter 7
The Role of Membrane Transport in the Detoxification and Accumulation of Zinc in Plants

Miki Kawachi, Yoshihiro Kobae, Rie Tomioka, and Masayoshi Maeshima

7.1 Introduction

Plants incorporate inorganic elements and molecules, water, and organic compounds from the soil. The average content of Zn in the earth's crust is 80 mg/kg. Soils contain from 5 to 800 mg of Zn/kg of dry weight with an average content of 60 mg/kg. The most cultivated soils contain 70–110 mg of Zn/kg. Seawater has 50 mg/L and freshwater has less than 50 μg Zn/L (0.8 μM) (Greger 2004). The amount of Zn in soil, rivers, and lakes increases by human activities such as metal mining, refining, industry, and fertilization of crops. Automobile tires contain a significant amount of ZnO as a vulcanizing agent and the debris causes Zn contamination of soils on roadsides (Smolders and Degryse 2002).

Zn is a nutrient essential for the growth, differentiation, and survival of cells, and it affects the growth of all living organisms. The human body contains 2–4 g of Zn, and the recommended daily intake is 8–11 mg. The adequate concentration of Zn in plants is considered to be around 1.3 μg/g of fresh weight (20 μM). Zn-deficiency symptoms appear when a plant contains less than 20 μg Zn/g dry weight (Marschner 1995). Symptoms of Zn deficiency include reduction in growth of internodes and leaves, a puckered appearance along leaf margin, a decrease in stress resistance, and appearance of chlorotic spots in leaves (Broadley et al. 2007; Clemens 2010). Zn-deficient conditions tend to generate reactive oxygen species (ROS), which strongly

M. Kawachi
Department of Biology, Graduate School of Science, Kobe University, Rokkodai 1-1, Nada-ku, Kobe 657-8501, Japan

Y. Kobae • R. Tomioka
Graduate School of Bioagricultural Sciences, Nagoya University, Nagoya 464-8601, Japan

M. Maeshima (✉)
Laboratory of Cell Dynamics, Graduate School of Bioagricultural Sciences, Nagoya University, Nagoya 464-8601, Japan
e-mail: maeshima@agr.nagoya-u.ac.jp

I. Sherameti and A. Varma (eds.), *Detoxification of Heavy Metals*, Soil Biology 30,
DOI 10.1007/978-3-642-21408-0_7, © Springer-Verlag Berlin Heidelberg 2011

inhibit the growth of plants (Cakmak 2000). Nearly 30% of the world's cultivated soil is deficient in Zn. The problem is serious in several countries, especially India, Turkey, China, and Australia. Application of Zn is an efficient way to improve crop yield in such Zn-deficient soils. Widespread Zn deficiency in humans is also notable in such areas.

Even though Zn is an essential cofactor for many enzymes and regulatory proteins, excessive amounts of Zn(II) (hereafter referred to as Zn) disturb the balance of metal elements in cells and is harmful to the plants. Excessive absorption of Zn suppresses absorption of Fe and Cu absorption by plant roots (Rogers et al. 2000; Kawachi et al. 2009). Excess Zn concentrations cause chlorosis and growth disorders in plants. Damage due to excess Zn appears when Zn content in plants is more than 0.4 mg/g of dry weight (Beckett and Davis 1977). In cells, Zn replaces Fe, Mg, or Mn in metal-containing proteins and enzymes and suppresses their functions. For example, an increase in Zn concentration reduces photosynthesis, because the metal substitution of Mg with Zn suppresses the affinity of Rubisco for CO_2 and the activity of chlorophyll (Monnet et al. 2005). Furthermore, ROS is generated in mitochondria not only when Zn is lacking, but also when Zn is present in excess. Excess Zn inhibits or interferes with the respiratory chain, which might trigger ROS production in mitochondria (Dineley et al. 2003).

There are many excellent articles on recent findings on Zn in plants (Alkorta et al. 2004; Broadley et al. 2007; Clemens 2010; Eide 2006; Krämer et al. 2007; Krämer 2010; Palmer and Guerinot 2009; Palmgren et al. 2008). These articles provide comprehensive, key information for experimental research and for improvement of crops, soils, and the environment. Membrane transport systems are generally crucial for cell growth through uptake, sequestration, and export of nutrients and signaling substances (Groeneveld et al. 2009). Here we discuss the importance of Zn in living cells including signaling; we focus on Zn transporters, which are involved in the homeostasis of Zn in plant cells.

7.2 Biological Importance of Zn as a Structural Element and a Signal

7.2.1 Zn as a Cofactor of Enzymes and Transcription Factors

Zn acts as a Lewis acid and is able to react with a Lewis base to form an additional product (adduct). In cells, Zn exists as a coordination compound with proteins or organic acids. Zn tends to form a tetrahedral complex. Histidine (His) and cysteine (Cys) residues in polypeptides have high affinity for Zn. The Zn finger and RING (really interesting new gene) finger are known as Zn-binding motifs in proteins. The Zn finger motif in transcription factors is formed with two His and two Cys residues (Cys_2/His_2 type) and is involved in the stabilization of the DNA-binding domains. RING finger motif is defined as $CX_2CX_{9-27}CX_{1-3}HX_{2-3}CX_2CX_{6-17}CX_2C$ ($Cys_3HisCys_4$ type), which binds two Zn cations (Gamsjaeger et al. 2007). Other motifs containing Cys and His also bind Zn in cells. These Zn-binding proteins function in gene transcription, translation, mRNA trafficking, cytoskeleton organization, protein folding, chromatin remodeling, Zn transport, and Zn sensing in various organisms.

Plant cells contain many Zn-containing enzymes, for example, alcohol dehydrogenase, glutamate dehydrogenase, carbonic anhydrase, Cu,Zn-superoxide dismutase (Cu,Zn-SOD), ribonuclease, and DNA polymerase (Addepalli and Hunt 2008; Balasubrahmanyam and Hunt 2008; Clemens 2010; Gamsjaeger et al. 2007; Vallee and Auld 1990). Zn plays a key role in the assisting and structural stabilization of catalytic function in these enzymes. In carbonic anhydrase, the Zn that interacts with the water molecule attracts an electron in H_2O, depolarizes, and stabilizes the intermediate of the reaction, $H_2O \rightarrow H^+ + OH^-$; and finally completes the reaction, $CO_2 + H_2O \rightarrow HCO_3^- + H^+$(Lilsjas et al. 1994). Therefore, a deficiency in Zn causes dysfunctions in formation of functional enzymes and transcription factors and disrupts metabolism, cell proliferation (via DNA polymerase), synthesis of RNAs and ribosomes (RNA polymerase), and defense against ROS (Cu,Zn-SOD).

7.2.2 *Zn as a Signal*

Zn serves as a structural element for many proteins and plays a key role in the catalytic action of Zn-containing enzymes through attracting their substrates. Zn was recently highlighted as a dynamic signaling ion, especially in animal cells (Sensi et al. 2009). Zn acts as a neurotransmitter and mimics the actions of hormones and growth factors (Yamasaki et al. 2007). Zn binds to numerous proteins in cells, where it serves as a structural and functional element of proteins involved in various cell activities. For example, the Zn-sensitive protein tyrosine phosphatases, protein kinases, and transcription factors are thought to be transmitters of Zn signaling. Zn signaling may be involved in cell proliferation, cell differentiation, immune system, and senescence in animals (Sensi et al. 2009; Haase and Rink 2009).

In addition to cytoplasmic and nuclear proteins, Zn specifically binds to many membrane receptors, transporters, and channels, and modulates their activity (Fig. 7.1). A kind of voltage-gated K^+ channel requires Zn to form an active tetramer as an essential element (Strang et al. 2003). In human cells, intracellular Zn reversibly activates large-conductance Ca^{2+}-activated K^+ channels (Hou et al. 2010). These ion channels are possibly positive and direct effectors of the Zn signal in the cells. Zn has a negative effect on some types of voltage-dependent Ca^{2+} channels through binding to the extracellular metal-binding sites of the channels (Kang et al. 2010). In other cases, extracellular Zn activates a Na^+/H^+ exchanger mediated by a G protein-coupled, Zn-sensing receptor (Azriel-Tamir et al. 2004).

Zn has been found to act as a signal in plants. In wheat, Zn stimulates translation of several genes through enhancement of dimer formation of a translation initiation factor and a poly(A)-binding protein (Cheng et al. 2008). Proteins with the Zn-binding motif are present in numerous protein families in eukaryotes and the

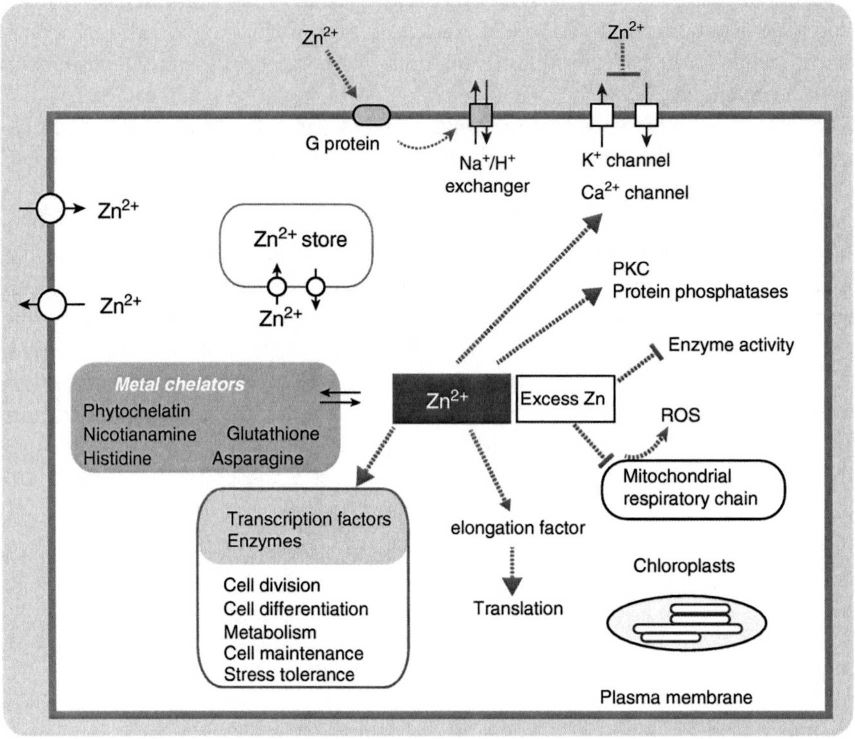

Fig. 7.1 *Zn-dependent cellular functions and homeostasis of Zn in plant cell.* Zn in the extracellular space (cell wall space) can regulate ion channels and ion transporters directly or indirectly. In the cell Zn is an essential element of many proteins, which are involved in various cell activities. Zn exists as a free ion, is stored in the organelles, or is chelated with organic compounds. Excess Zn damages many enzymes and stimulates the generation of reactive oxygen species (ROS). *G protein* guanine nucleotide binding protein, *PKC* protein kinase C

number of such proteins has been estimated to be more than 2,000 in plants (Gamsjaeger et al. 2007; Broadley et al. 2007). Several extracellular matrix metalloproteinases are included in this group (Flinn 2008). Some of these Zn-binding proteins and enzymes are targets of Zn signaling. Cell-imaging techniques reveal the homeostasis of Zn in the cytoplasm, vacuole, and nucleus, as shown for Ca in plant cells (Mazar et al. 2009). Zn signaling requires the maintenance of an adequate concentration of Zn in the cytoplasm, which is strictly regulated through the function of Zn transporters and Zn chelators. Loss-of-function mutants of Zn transporters in plants are key to understanding the function of Zn as a signal.

New experimental tools, including specific fluorescent probes (Zinpyr-1, etc.) coupled with transgenic plant lines that lack the Zn transporters, have revealed the dynamics of Zn in plant cells (Hanikenne et al. 2008; Kawachi et al. 2009). The function of many transcription factors and enzymes in the cytoplasm, nucleus, and other organelles is changed in response to the Zn level (Fig. 7.1). Membrane transporters are involved in the regulation of Zn levels in the cytoplasm and cell organelles such as the vacuole (Krämer 2010; Martinoia et al. 2007).

7.3 Acquisition and Distribution of Zn

Zn exists in rocks, sands, and soils in several forms including ZnS, $ZnSO_4$, ZnO, $ZnCO_3$, $Zn_3(PO_4)_2$, and Zn_2SiO_4 (Broadley et al. 2007). Fertilizers and agrochemicals provided to crop fields also contain Zn. Generally, a small part of the Zn in soils exists as free cations, the concentration of which increases at low soil pH and is available for plants. Concentrations of water-soluble Zn in the bulk soil solution have been estimated to be between 0.4 nM and 4 μM (Barber 1995). The Zn absorption properties of various plants suggest that the Michaelis–Menten parameter K_m for Zn has a submicromolar value (probably, <0.1 μM), close to the actual soil values (Broadley et al. 2007). Therefore, plant roots have a high potential to absorb Zn from the soil under normal conditions.

Zn absorbed by the epidermis including root hairs is transferred to cortical cells and then to the root xylem via the apoplast pathway, the transmembrane pathway, and the symplast pathway (xylem loading). The endodermis, which is lined between the cortex and the pericycle, has a Casparian strip that blocks the apoplast pathway. For this reason, the epidermal and endodermal cells might contain efficient Zn transporters in their plasma membranes. Zn is transported from the roots to shoots through the xylem and is redistributed to cells in stems and leaves by the apoplast and symplast pathways (xylem unloading).

Zn is compartmentalized in vacuoles, ER, Golgi apparatus, mitochondria, chloroplast, and the cell wall space (Fig. 7.2). There are numerous Zn-enzymes, Zn-proteins, and metal chelators with high and low affinity for Zn in the cytoplasm and cell organelles. Most Zn is required in the cytoplasm, nucleus, mitochondria, and chloroplasts, because there are many Zn-binding proteins in these compartments. Zn deficiency and toxicity occur at concentrations lower than 15–20 and higher than 100–300 μg/g of dry biomass, respectively (Krämer 2010). The adequate content of Zn in plants is 20–100 μg/g. However, the cytoplasmic Zn cation concentration ($[Zn]_{cyt}$) estimated for animal cells (<100 nM) (Sensi et al. 2009) and for *E. coli* (femtomolar level) (Outten and O'Halloran 2001) is very low. The $[Zn]_{cyt}$ might also be extremely low in plants (Broadley et al. 2007), although this value is not available because it is technically difficult to quantify. Most Zn is compartmentalized in cell organelles, especially vacuoles, and the cell wall space.

In mammalian and yeast cells, Zn-containing organelles called zincosomes have been identified. Zn-containing, vacuole-like small organelles have been shown to appear under conditions of excess Zn (Kawachi et al. 2009). It is still unknown what transporters of Zn exist in the organelles. Furthermore, ER and Golgi apparatus in plant cells may also function as a Zn store as shown in mammals (Ellis et al. 2005; Sensi et al. 2009).

Metal chelators with relatively high affinity to Zn include phytochelatin, glutathione, nicotianamine, histidine, and asparagine (Sharma and Dietz 2006). These

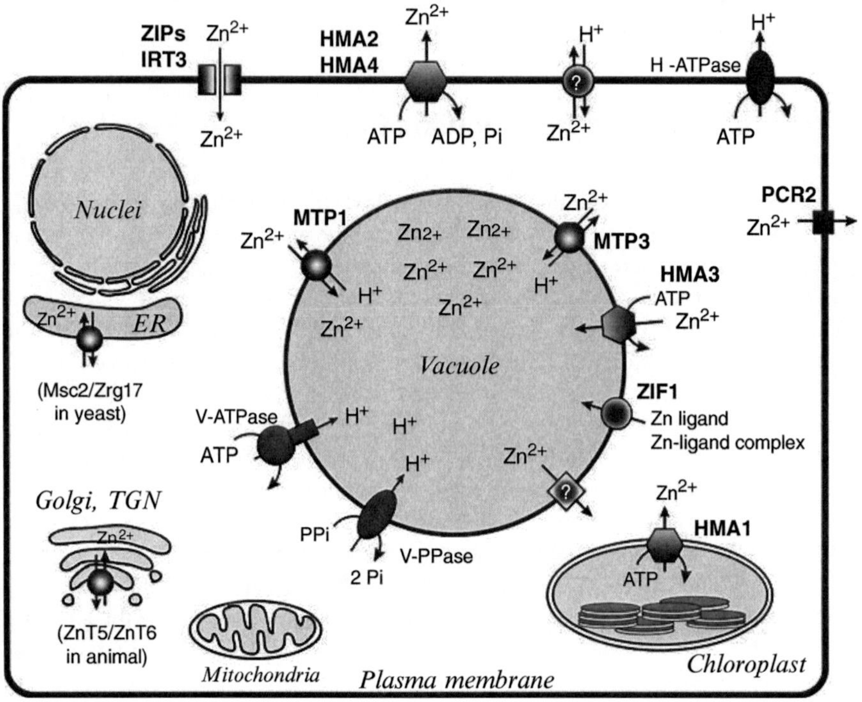

Fig. 7.2 *Membrane transport systems involved in homeostasis of Zn in the plant cell.* Various types of Zn transporters are involved in Zn homeostasis at the plasma membrane, vacuole, and chloroplast. *HMA* heavy metal ATPase, *IRT* iron-regulated transporter, *MTP* metal tolerance protein, *PCR* plant cadmium resistance, *TGN trans*-Golgi network, *V-ATPase* vacuolar H^+-ATPase, *V-PPase* vacuolar H^+-pyrophosphatase, *ZIF* zinc induced facilitator, *ZIP* ZRT1/IRT1-like protein, *ZnT* Zn transporter, *?* unknown proteins proposed to mediate transport of Zn

chelators immobilize Zn and play a role as a Zn pool in the cells. Zn is released from the chelators when $[Zn]_{cyt}$ decreases to the level of K_d of the chelators for Zn. Some phytochelatin is incorporated into vacuoles by an ATP-binding cassette (ABC) transporter and the vacuolar phytochelatin confers tolerance to Cd, but not Zn, in *Schizosaccharomyces pombe* (Mendoza-Cozatl et al. 2010). The plant homologue of the transporter remains to be identified.

7.4 Transport of Zn Across Membranes

7.4.1 Diversity of Zn Transporters

Plant cells contain several types of membrane transporters of Zn: metal tolerance protein (MTP), ZRT1/IRT1-like protein (ZIP) (also known as zinc–iron permease), and heavy-metal ATPase (HMA) (P_{1B} subgroup of P-type ATPase) (Krämer et al. 2007). These are different in their structure and transport mechanism. ZIP is a passive transporter that facilitates the "downhill" flow of Zn until its electrical and/or concentration gradient is dissipated. MTP is a subfamily of the cation diffusion facilitator (CDF) and belongs to the active transporters. MTP utilizes a pH gradient across the membrane. HMA actively transports Zn using energy provided by ATP hydrolysis. Therefore, MTP and HMA pump Zn in a mode of thermodynamically "uphill" transport. It is unclear whether ABC transporters are involved in the transport of Zn or Zn-compound complexes in plants.

Recently, a zinc-induced facilitator 1 (ZIF1) has been reported to influence the detoxification and accumulation of Zn in *Arabidopsis thaliana* and to be different from the known Zn transporters, which transport free Zn cation (Haydon and Cobbett 2007). Carboxylic organic acids are effective ligands for chelation of Zn in plant cells. Therefore, ZIF1 is thought to transport low-molecular-mass Zn-ligands, such as organic acids, and/or Zn-ligand complexes into the vacuoles.

7.4.2 Zn Transporters in the Plasma Membrane

ZIP mediates electrodiffusion of Zn across the plasma membrane. ZIP members including ZIP1, ZIP2, ZIP3, ZIP9, and ZIP10 and IRT3 are thought to be responsible for Zn uptake into cells at the plasma membrane (Grotz et al. 1998; Talke et al. 2006). *ZIP4* gene expression is strongly induced in roots and shoots upon shortage of Zn supply and *ZIP4* transcription factors have been identified (Assunçao et al. 2010). ZIP members take up Zn using a pH gradient generated by plasma membrane H^+-ATPase (Palmgren et al. 2008). At present, it is unclear which member of the ZIP family is involved in Zn uptake from the soil at the epidermal cells and root hairs. The *IRT3* gene is also induced under Zn-deficient conditions in *A. thaliana* and is a candidate for a Zn uptake transporter (Krämer et al. 2007; Lin et al. 2009; van de Mortel et al. 2006). However, its ion selectivity remains to be investigated.

Active Zn pumps, HMA2 and HMA4, export excess Zn out of the cytosol. This process is essential to the release of Zn from the root symplasm into the apoplastic xylem vessels, which provide the primary pathway for the movement of Zn from the root to the shoot (Hanikenne et al. 2008; Palmgren et al. 2008). Indeed, HMA4 is highly expressed in the xylem parenchyma and the cambium of leaves in *A. thaliana* and *Arabidopsis halleri*, a metal hyperaccumulator. In other words, HMA4 is a key transporter in the process of loading Zn at roots and unloading Zn at leaves. AtHMA4 has eight predicted transmembrane domains with a long cytoplasmic loop and an extremely long C-terminal domain. This C-terminal tail, which is exposed to the cytoplasm, is involved in ion selectivity and transport activity (Mills et al. 2010).

Plant cadmium resistance 2 (PCR2), a new transporter, is also involved in Zn export at the plasma membrane in *A. thaliana* (Song et al. 2010). PCR2 might be essential for Zn distribution in plant organs and exclusion from the roots. The functional mode of this transporter, which has only two transmembrane domains, is under investigation.

7.4.3 Zn Transporters in the Vacuolar Membrane

There is new information on the vacuolar Zn transporters that supports the vacuolar function as a Zn store in plant cells. MTPs function as Zn^{2+}/H^+ exchanger and belong to the CDF family as do mammalian Zn transporters (ZnTs) and bacterial Zn transporters (YiiB and ZitB) (Lu et al. 2009; Ohana et al. 2009). MTP1 is localized in the vacuolar membrane and takes charge of active incorporation of cytoplasmic Zn into the vacuole using a proton gradient (Krämer 2010; Kawachi et al. 2008), which is generated by H^+-ATPase and H^+-pyrophosphatase (Maeshima 2001; Martinoia et al. 2007) (Fig. 7.2). A loss-of-function mutant of *AtMTP1* is sensitive to Zn as low as 0.4 mM (Kobae et al. 2004; Desbrosses-Fonrouge et al. 2005) and is defective in Zn accumulation in roots (Kawachi et al. 2009). A Zn hyperaccumulator *A. halleri* has been reported to have five gene copies of MTP1; its transcript levels of MTP1-A1, MTP1-A2, and MTP1-B are extremely high under normal and excess Zn conditions (Shahzad et al. 2010). Another Zn hyperaccumulator *Thlaspi goesingense*, which has three allelic variants of TgMTP1 (Kim et al. 2009), accumulates TgMTP1s at high levels in the vacuolar membranes in shoots (Gustin et al. 2009). Furthermore, overexpression of TgMTP1 leads to increased tolerance to Zn in transgenic *A. thaliana* (Gustin et al. 2009). Therefore, MTP1 plays a crucial role in the tolerance to Zn and accumulation of Zn.

The loss-of-function *atmtp1* plants of *A. thaliana* have a low Zn content. Roots of the wild-type and *atmtp1* seedlings grown under normal medium conditions contain Zn of approximately 207 and 116 μg/g dry weight, respectively (Kawachi et al. 2009). MTP1 contributes to a Zn accumulation of about 91 μg Zn/g dry weight. Dry weight is 8.4% of the fresh weight. The contribution of MTP1 on Zn uptake into vacuoles can be calculated to be 120 μM on the assumption that the volume of the vacuole accounts for 75% of root tissue.

MTP1 proteins of *A. thaliana, A. halleri,* and *T. goesingense* have a histidine (His)-rich loop between the fourth and fifth transmembrane domains. The His-rich region is absent in *E. coli* YiiP (Lu et al. 2009). Deletion of this region from AtMTP1 enhances the Zn transport velocity 11-fold. The His-rich loop might absorb Zn in the cytoplasm and transfer cations to the Zn transport pore of AtMTP1 (Kawachi et al. 2008). This Zn transfer process requires energy and lowers the transport rate. Structural and functional changes may occur during saturation of the loop with Zn. X-ray crystallography of YiiP, which lacks a His-rich loop (Lu et al. 2009), and further structural analyses of MTP1 proteins might shed light on the regulatory role and mechanism of the His-rich region.

MTP3 is also located in the vacuolar membrane and accumulates at a relatively high level in epidermal and cortex cell layers of the root hair zone when they are exposed to excess Zn (>30 μM) (Arrivault et al. 2006), although MTP1 is expressed in the meristematic and elongating zones of the main and emerging lateral roots even under normal conditions. MTP3 might be involved in the compartmentalization and detoxification of Zn together with MTP1 under excess Zn conditions. Indeed, overexpression of MTP3 resulted in high accumulation of Zn in *A. thaliana* and conferred Zn tolerance compared with the wild-type plants (Arrivault et al. 2006). MTP3 has a short His-rich loop and an N-terminal region different from MTP1 (for sequence alignment, see Kawachi et al. 2008).

The vacuolar membrane contains HMA3, a P-type ATPase, which incorporates Zn, Cd, Co, and Pb ions into vacuoles (Morel et al. 2009). The gene is highly expressed in vascular tissues and the root apex. Overexpressing the gene enhances Zn tolerance in *A. thaliana*. AtHMA3 has been reported to be involved in the detoxification of biological (Zn) and other nonbiological heavy metals (Cd, Co, and Pb) by vacuolar sequestration. In rice, OsHMA3 of a *japonica* cultivar (Nipponbare), but not of an *indica* cultivar (Anjana Dhan), functions as a Cd transporter in the root vacuoles (Ueno et al. 2010). Cd is accumulated in the roots and the Cd content in the rice grains is kept low. The relationship between ion selectivity and specific amino acids is key to understanding the molecular evolution and physiological meaning of various metal transporters including MTP and HMA.

Zn exists in the form of free ions and complexes in vacuoles as well as the cytoplasm. In tobacco, almost 90% of Zn is sequestered by high levels of citrate in the vacuole while malate and oxalate hardly contribute to Zn accumulation in vacuoles (Wang et al. 1992). The stored Zn might be released from the vacuole and other compartments when the $[Zn]_{cyt}$ decreases. The transporter or channel for release of Zn from plant organelles remains to be identified.

7.4.4 Zn Transporters in Other Organelles

Recently, HMA1 has been identified in the chloroplast envelope of *A. thaliana* (Kim et al. 2009; Seigneurin-Berny et al. 2006). *AtHMA1* knockout plants are highly sensitive to high Zn levels and show increased accumulation of Zn in the shoot. In general, Zn can substitute easily for Mg and Fe because of similarity in ionic radius. Thus, many chloroplast proteins including plastoquinone, ferredoxin, and photosystems I and II are highly sensitive to excess Zn. AtHMA1 detoxifies Zn by transporting Zn from the chloroplast (plastid) stroma, which is rich in ATP, into the cytoplasm (Kim et al. 2009).

Mitochondria contain several Zn enzymes and require a minimum level of Zn in the matrix. Furthermore, mitochondria also have many Mg- and Fe-containing enzymes, such as aconitase and respiratory chain complexes, which might be sensitive to excess level of Zn. Therefore, Zn-uptake and Zn-export transporters may exist in mitochondria, although the transporters have not been identified.

In *Saccharomyces cerevisiae*, Zrg17p and Msc2p may form a heterodimer and transport Zn into the ER (Ellis et al. 2005). In mammals, ZnT5 and ZnT6, homologues of Msc2p and Zrg17, have been proposed to form a heterodimer and transport Zn into the Golgi apparatus (Ellis et al. 2005). These four transporter proteins belong to the CDF family. Some members of the mammalian ZnT family are involved in Zn transport in the Golgi apparatus, lysosome, and synaptic vesicle (Sensi et al. 2009). Particular Zn transporters probably exist in the ER and Golgi apparatus membranes and regulate Zn homeostasis in these organelles in plants. The MTP members are plant homologues of mammalian ZnT and include candidates for proteins in the ER or Golgi apparatus. For example, proteins with accession numbers AC007231 and BT014889 have relatively high sequence similarity with human ZnT5.

7.4.5 Regulation of Zn-Related Genes

Quantitative regulation of Zn transporters is critical for plants to adapt to both deficiency and excess of Zn. A set of proteins responsive to Zn under Zn excess conditions was identified by proteomics techniques (Fukao et al. 2009). In yeast, Zn-responsive transcription factor Zap1 functions as a Zn sensor, activates its own transcription in Zn-limited conditions, and binds the Zn-responsive element of more than 100 genes. Under Zn-deficient conditions, Zap1 upregulates the plasma membrane Zn transporter gene *ZRT1* and the vacuolar membrane Zn transporter gene *ZRC1* and suppresses the genes for ADH1 and ADH3, which are abundant Zn-binding proteins in cells (Eide 2006). No plant homologues of Zap1 have been identified. In *A. thaliana*, two transcription factors, bZIP19 and bZIP23, have been identified (Assunçao et al. 2010). Both transcription factors have a leucine zipper motif and associate with the promoter regions of the genes encoding the proteins that constitute the primary response to zinc deficiency. The genes *ZIP1*, *ZIP3*, *ZIP4*, *ZIP5*, *ZIP9*, *ZIP12*, and *IRT3* are under regulation of bZIP19 and bZIP23. In wheat, Zn promotes the interaction between translation initiation factor (eIF4B) and poly (A)-binding protein (PABP) (Cheng et al. 2008). Formation of the complex is essential at the initiation of translation. Thus, Zn-related genes are regulated directly by Zn and indirectly through Zn-sensing transcription factors.

7.5 Conclusion

Zn functions as an essential metal for numerous enzymes and regulatory proteins. Recent studies highlight the role of Zn as a cell signal in various organisms through functional regulation of enzymes, membrane transport systems, and proteins involved in transcription and translation. Zn transporters, Zn-chelate compounds, and intracellular Zn-storage organelles take part in the homeostasis of Zn and in the function of Zn as a signal. In plants, excess Zn is removed from the cytoplasm and chloroplasts to the extracellular space or the intracellular stores, especially vacuoles. Excessive Zn in the cytoplasm is incorporated in the vacuoles and Zn is supplied to the cytoplasm under Zn-deficient conditions. When the Zn concentration in cells increases beyond the capacity for Zn homeostasis, symptoms of excessive Zn appear. Zn accumulators resist excessive Zn by excluding Zn from the cytoplasm using Zn transporters or by detoxifying the high level of Zn by chelators.

Recent advances in genome database, metal quantification, and imaging techniques for Zn and Zn-related proteins have contributed knowledge on Zn homeostasis and on cross-homeostasis with other metals in plants. Studies on the structure–function relationship, functional regulation, and genetic regulation of Zn-related enzymes and transporters will enable us to fine-tune the homeostasis of Zn and grow plants adapted to the level of Zn in the environment. Zn signaling is an expanding area in plant science. The appearance of Zn-tolerant plants and high-Zn crops will improve the state of Zn nutrition in humans and livestock and enable phytoremediation of Zn-contaminated soil.

References

Addepalli B, Hunt AG (2008) Ribonuclease activity is a common property of *Arabidopsis* CCCH-containing zinc-finger proteins. FEBS Lett 582:2577–2582

Alkorta I, Hernandez-Allica J, Becerrisl JM, Amezaga I, Albizu I, Garbisu C (2004) Recent findings on the phytoremediation of soils contaminated with environmentally toxic heavy metals and metalloids such as zinc, cadmium, lead, and arsenic. Rev Environ Sci Biotechnol 3:71–90

Arrivault S, Senger T, Krämer U (2006) The *Arabidopsis* metal tolerance protein AtMTP3 maintains metal homeostasis by mediating Zn exclusion from the shoot under Fe deficiency and Zn oversupply. Plant J 46:861–879

Assunçao AGL, Herrero E, Lin YF et al (2010) *Arabidopsis thaliana* transcription factors bZIP19 and bZIP23 regulate the adaptation to zinc deficiency. Proc Natl Acad Sci USA 107: 10296–10301

Azriel-Tamir H, Sharir H, Schwartz B, Hershfinkel M (2004) Extracellular zinc triggers ERK-dependent activation of Na^+/H^+ exchange in colonocytes mediated by the zinc-sensing receptor. J Biol Chem 279:51804–51816

Balasubrahmanyam A, Hunt AG (2008) Ribonuclease activity is a common property of Arabidopsis CCCH-containing zinc-finger proteins. FEBS Lett 582:2577–2582

Barber SA (1995) Soil nutrient bioavailability, 2nd edn. Wiley, New York

Beckett PHT, Davis RD (1977) Upper critical levels of toxic elements in plants. New Phytol 79:95–106

Broadley MR, White PJ, Hammond JP, Zelko I, Lux A (2007) Zinc in plants. New Phytol 173:677–702

Cakmak I (2000) Possible roles of zinc in protecting plant cells from damage by reactive oxygen species. New Phytol 146:185–205

Cheng S, Sultana S, Goss DJ, Gallie DR (2008) Translation initiation factor 4B homodimerization, RNA binding, and interaction with poly(A)-binding protein are enhanced by zinc. J Biol Chem 283:36140–36153

Clemens S (2010) Zn: a versatile player in plant cell biology. In: Hell R, Mendel RR (eds) Cell biology of metals and nutrients. Springer, Berlin

Desbrosses-Fonrouge AG, Voigt K, Schroeder A, Arrivault S, Thomine S, Krämer U (2005) *Arabidopsis thaliana* MTP1 is a Zn transporter in the vacuolar membrane which mediates Zn detoxification and drives leaf Zn accumulation. FEBS Lett 579:4165–4174

Dineley KE, Votyakova TV, Reynolds IJ (2003) Zinc inhibition of cellular energy production: implications for mitochondria and neurodegeneration. J Neurochem 85:53–570

Eide DJ (2006) Zinc transporters and the cellular trafficking of zinc. Biochim Biophys Acta 1763:711–722

Ellis CD, MacDiarmid CW, Eide DJ (2005) Heteromeric protein complexes mediate zinc transport into the secretory pathway of eukaryotic cells. J Biol Chem 280:28811–28818

Flinn BS (2008) Plant extracellular matrix metalloproteinases. Funct Plant Biol 35:1183–1193

Fukao Y, Ferjani A, Fujiwara M, Nishimori Y, Ohtsu I (2009) Identification of zinc-responsive proteins in the roots of *Arabidopsis thaliana* using a highly improved method of two-dimensional electrophoresis. Plant Cell Physiol 50:2234–2239

Gamsjaeger R, Liew CK, Loughlin FE, Crossley M, Mackay JP (2007) Sticky fingers: zinc-fingers as protein-recognition motifs. Trends Biochem Sci 32:63–70

Greger M (2004) Metal availability, uptake, transport and accumulation in plants. In: Prasad MNV (ed) Heavy metal stress in plants: from biomolecules to ecosystems, 2nd edn. Springer, Berlin

Groeneveld P, Stouthamer AH, Westerhoff HV (2009) Super life: how and why "cell selection" leads to the fastest-growing eukaryote. FEBS J 276:254–270

Grotz N, Fox T, Connolly E, Park W, Guerinot ML, Eide D (1998) Identification of a family of zinc transporter genes from *Arabidopsis* that respond to zinc deficiency. Proc Natl Acad Sci USA 95:7220–7224

Gustin JL, Loureiro ME, Kim D, Na G, Tikhonova M, Salt DE (2009) MTP1-dependent Zn sequestration into shoot vacuoles suggests dual roles in Zn tolerance and accumulation in Zn-hyperaccumulating plants. Plant J 57:1116–1127

Haase H, Rink L (2009) Functional significance of zinc-related signaling pathways in immune cells. Annu Rev Nutr 29:133–152

Hanikenne M, Talke IN, Haydon MJ et al (2008) Evolution of metal hyperaccumulation required *cis*-regulatory changes and triplication of *HMA4*. Nature 453:391–396

Haydon MJ, Cobbett CS (2007) A novel major facilitator superfamily protein at the tonoplast influences zinc tolerance and accumulation in *Arabidopsis*. Plant Physiol 143: 1705–1719

Hou S, Vigeland LE, Zhang G et al (2010) Zn^{2+} activates large conductance Ca^{2+}-activated K^+ channels via an intracellular domain. J Biol Chem 285:6434–6442

Kang HW, Vitko I, Lee SS, Perez-Reyes E, Lee JH (2010) Structural determinants of the high affinity extracellular zinc binding site on $Ca_v3.2$ T-type calcium channels. J Biol Chem 285:3271–3281

Kawachi M, Kobae Y, Mimura T, Maeshima M (2008) Deletion of a histidine-rich loop of AtMTP1, a vacuolar Zn^{2+}/H^+ antiporter of *Arabidopsis thaliana*, stimulates the transport activity. J Biol Chem 283:8374–8383

Kawachi M, Kobae Y, Mori H, Tomioka R, Lee Y, Maeshima M (2009) A mutant strain *Arabidopsis thaliana* that lacks vacuolar membrane zinc transporter MTP1 revealed the latent tolerance to excessive zinc. Plant Cell Physiol 50:1156–1170

Kim Y-Y, Choi H, Segami S et al (2009) AtHMA1 contributes to detoxification of excess Zn(II) in *Arabidopsis*. Plant J 58:737–753

Kobae Y, Uemura T, Sato MH, Ohnishi M, Mimura T, Maeshima M (2004) Zinc transporter of *Arabidopsis thaliana* AtMTP1 is localized to vacuolar membranes and implicated in zinc homeostasis. Plant Cell Physiol 45:1749–1758

Krämer U (2010) Metal hyperaccumulation in plants. Annu Rev Plant Biol 61:517–534

Krämer U, Talke IN, Hanikenne M (2007) Transition metal transport. FEBS Lett 581: 2263–2272

Lilsjas A, Hikansson K, Jonsson BH, Xue Y (1994) Inhibition and catalysis of carbonic anhydrase-recent crystallographic analyses. Eur J Biochem 219:1–10

Lin YF, Liang HM, Yang SY et al (2009) Arabidopsis IRT3 is a zinc-regulated and plasma membrane localized zinc/iron transporter. New Phytol 182:392–404

Lu M, Chai J, Fu D (2009) Structural basis for autoregulation of the zinc transporter YiiP. Nat Struct Biol 16:1063–1067

Maeshima M (2001) Tonoplast transporters: organization and function. Annu Rev Plant Physiol Plant Mol Biol 52:469–497

Marschner H (1995) Mineral nutrition of higher plants, 2nd edn. Academic, London

Martinoia E, Maeshima M, Neuhaus E (2007) Vacuolar transporters and their essential role in plant metabolism. J Exp Bot 58:83–102

Mazar C, Bourque S, Mithofer A, Pugin A, Ranjeva R (2009) Calcium homeostasis in plant cell nuclei. New Phytol 181:261–274

Mendoza-Cozatl DG, Zhai Z, Jobe TO et al (2010) Tonoplast-localized Abc2 transporter mediates phytochelatin accumulation in vacuoles and confers cadmium tolerance. J Biol Chem 285: 40416–40426

Mills RF, Valdes B, Duke M et al (2010) Functional significance of AtHMA4 C-terminal domain *in planta*. PLoS One 5:e13388

Monnet F, Vaillant N, Hitmi A, Sallanon H (2005) Photosynthetic activity of *Lolium perenne* as a function of endophyte status and zinc nutrition. Funct Plant Biol 32:121–139

Morel M, Crouzet J, Gravot A et al (2009) AtHMA3, a P-1B-ATPase allowing Cd/Zn/Co/Pb vacuolar storage in Arabidopsis. Plant Physiol 149:894–904

Ohana E, Hoch E, Keasar C, Kambe T et al (2009) Identification of the Zn^{2+} binding site and mode of operation of a mammalian Zn^{2+} transporter. J Biol Chem 284:17677–17686

Outten CE, O'Halloran TV (2001) Femtomolar sensitivity of metalloregulatory proteins controlling zinc homeostasis. Science 292:2488–2492

Palmer CM, Guerinot ML (2009) Facing the challenges of Cu, Fe and Zn homeostasis in plants. Nat Chem Biol 5:333–339

Palmgren MG, Clemens S, Williams LE et al (2008) Zinc biofortification of cereals; problems and solutions. Trends Plant Sci 13:464–473

Rogers EE, Eide DJ, Guerinot ML (2000) Altered selectivity in an Arabidopsis metal transporter. Proc Natl Acad Sci USA 97:12356–12360

Seigneurin-Berny D, Bravot A, Auroy P et al (2006) HMA1, anew Cu-ATPase of the chloroplast envelope, is essential for growth under adverse light conditions. J Biol Chem 281: 2882–2892

Sensi SL, Paoletti P, Bush AI, Sekler I (2009) Zinc in the physiology and pathology of the CNS. Nat Rev Neurosci 10:780–791

Shahzad Z, Gostil F, Ferot H et al (2010) The five AhMTP1 zinc transporters undergo different evolutionary fates towards adaptive evolution to zinc tolerance in *Arabidopsis halleri*. PLoS Genet 6:e1000911

Sharma SS, Dietz KJ (2006) The significance of amino acids and amino acid-derived molecules in plant responses and adaptation to heavy metal stress. J Exp Bot 57:711–726

Smolders and Degryse (2002) Fate and effect of zinc from tire debris in soil. Environ Sci Technol 36:3706–3710

Song WY, Choi KS, Kim DY et al (2010) *Arabidopsis* PCR2 is a zinc exporter involved in both zinc extrusion and long-distance zinc transport. Plant Cell 22:2237–2252

Strang C, Kunjilwar K, DeRubeis D, Peterson D, Pfaffinger PJ (2003) The role of Zn^{2+} in Shal voltage-gated potassium channel formation. J Biol Chem 278:31361–31371

Talke IN, Hanikenne M, Krämer U (2006) Zinc-dependent global transcriptional control, transcriptional deregulation, and higher gene copy number for genes in metal homeostasis of the hyperaccumulator *Arabidopsis halleri*. Plant Physiol 142:148–167

Ueno D, Yamaji N, Kono I et al (2010) Gene limiting cadmium accumulation in rice. Proc Natl Acad Sci USA 107:16500–16505

Vallee BL, Auld DS (1990) Zinc coordination, function, and structure of zinc enzymes and other proteins. Biochemistry 29:5647–5659 van de Mortel JE, Villanueva LA, Schat H et al (2006) Large expression differences in genes for iron and zinc homeostasis stress response, and lignin biosynthesis distinguish roots of *Arabidopsis thaliana* and the related metal hyperaccumulator *Thlaspi caerulescens*. Plant Physiol 142:1127–1147

Wang J, Evangelou BP, Nielsen MT, Wagner GJ (1992) Computer, simulated evaluation of possible mechanisms for metal ion activity in plant vacuoles: II. Zinc. Plant Physiol 99: 621–626

Yamasaki S, Sakata-Sagawa K, Hasegawa A et al (2007) Zinc is a novel intracellular second messenger. J Cell Biol 177:637–645

Chapter 8
Initial Steps of Copper Detoxification: Outside and Inside of the Plant Cell

Valentina P. Kholodova, Elena M. Ivanova, and Vladimir V. Kuznetsov

8.1 Introduction

Increasingly high technogenic load on the environment raises a problem of strategies and mechanisms of plant adaptation to abiotic stressors, among which heavy metals (HM) occupy a specific place. Contamination of vast territories with HM acquires more and more threatening character.

According to their toxicity, all HM are arranged in the so-called Irving–Williams series, where Cu takes the first place corresponding to the highest level of toxicity (Krämer et al. 2007; Yruela 2009). Copper (Cu), being the most toxic metal, at the same time belongs to so-called essential elements. This term designates the group of HM, which trace amounts are required for plant metabolism, growth, and development, whereas their high concentrations are toxic (Hall and Williams 2003).

Under physiological conditions, copper occurs in the cell in reduced (Cu^{+}) and oxidized (Cu^{2+}) states. Being a protein cofactor, copper is involved in the processes of photosynthesis and respiration, in the perception of the ethylene signal, in plant protection against oxidative stress, in the biogenesis of molybdenum cofactor, and also in the cell wall metabolism (Yruela 2009). However, the range of Cu concentrations suitable for the optimum cell metabolism and plant development is rather narrow. It is believed that even twofold exceeding of these concentrations could exert harmful effects, whereas the higher Cu concentrations induce a toxicity syndrome (chlorosis and necrosis, stunting, inhibition of root and shoot growth) and in most cases result in plant death. An exclusion from this rule is metallophytes highly tolerant to Cu, but their number is small.

At the cellular level, Cu and other HM toxicity is determined by (1) binding to SH groups in proteins, thereby inhibiting enzyme activity or protein function;

V.P. Kholodova (✉) • E.M. Ivanova • V.V. Kuznetsov
Laboratory of Physiological and Molecular Mechanisms of Adaptation, Timiryazev Institute of Plant Physiology RAS, Botanicheskaya ul., 35, Moscow 127276, Russia
e-mail: v-p-kholodova@ippras.ru

I. Sherameti and A. Varma (eds.), *Detoxification of Heavy Metals*, Soil Biology 30,
DOI 10.1007/978-3-642-21408-0_8, © Springer-Verlag Berlin Heidelberg 2011

(2) induction of deficiency of other essential ions; (3) impaired cell transport processes; and (4) oxidative damage (Yruela 2009; Cohu and Pilon 2010).

Soil contamination with Cu occurs mainly on territories adjacent to the zones of copper ore deposits close to the surface of the earth, in its mining and processing sites, and near numerous industrial plants. Another source of large-scale soil contamination is agricultural industry: copper entries into soil with fertilizers or plant defense preparations. Such soil contamination is especially characteristic for vineyard where multiyear plant treatment with Cu-containing preparations results in a sharp increase in its content in soil. At the usual average Cu concentration in soil of 20–30 mg/kg, its content in contaminated soil could exceed these values in tens and hundreds times and attain 250–1,000 mg/kg of soil (Krämer and Clemens 2006; Yruela 2009).

In this connection, the question raises how plants control Cu homeostasis under conditions of its excess in environment, e.g., how they can provide the cells of all tissues and organs with Cu micro amounts required to sustain life and, on the other hand, prevent its toxic effects resulting in plant death.

It should be noted that Cu is really present in all cellular structures: nucleus, mitochondria, chloroplasts, Golgi apparatus, and endoplasmic reticulum. This is achieved due to functioning of several families of more or less specialized membrane transporters localized in the plasma membrane and all other cell membranes. However, at a great Cu excess in environment, its transport into the cells is performed also by unspecialized transporters, which transport other two- and/or monovalent ions across the membranes as well. Fine regulation of transporter and metallochaperon functioning provides all organism cells with Cu micro amounts.

Our task was to study the mechanisms underlying plant protection against toxicity of high Cu concentrations in soil. Furthermore, it is established that the intracellular concentration of free Cu ions does not exceed 10^{-21}–10^{-18} M, i.e., one ion per cell (Changela et al. 2003). This indicates that a rather efficient system of Cu ion chelating operates in the cell cytosol. In this process, an important role belongs to highly specialized chelators, phytochelatins and metallothioneins, which function in coordination with another group of proteins, metallochaperons. Metallochaperons provide for delivering of Cu–ligand complexes to the sites of Cu destination. It is evidently true for both transport of Cu micro amounts to the sites of their functioning in corresponding macromolecules and transfer of excessive chelated Cu ions penetrated into the cytosol to the sites of their detoxification. Thus, the main components of the intracellular defense system against Cu excess are chelators, metallochaperons, and, naturally, membrane transporters (Lee et al. 2006; Mari and Lebrun 2006; Yruela 2009). In spite of rather wide-ranging studying of HM excess chelating in the cytosol of plant cells, the contribution of chelators and metallothioneins in Cu detoxification remains rather debatable (Gonzalez-Mendoza et al. 2007; Guo et al. 2008; Yruela 2009).

The problem of Cu ion detoxification in the apoplast is poorly studied, although just the apoplast is the first site of the plant cell contact with Cu excess in environment and evidently it could be considered one of the important barriers on the route of excessive HM penetration into the cytosol. We would call the apoplast all compartments beyond the plasma membrane (Sattlemacher 2001), although the main attention would be paid to the interfibrillar and intermicellar spaces of the cell walls. However, before discussing the role of the apoplast per se in Cu detoxification, we would recede still farther from the tissue symplast pool of Cu and analyze briefly the role of arbuscular mycorrhizal fungi (AMF), which colonize the root cortex of most higher plants and produce an extraradical mycelium (Joner et al. 2000; Ferrol et al. 2009), i.e., we consider events occurring in the rhizosphere and appraise the role of these structures in Cu ion immobilization.

8.2 Role of Arbuscular Mycorrhiza

AMF are obligate biotrophs of higher plants. They colonize the root cortex of most higher plants and develop an extraradical mycelium, which grow in soil around the roots (Ferrol et al. 2009). Mycorrhizal fungi constitute the only group of microorganisms that are capable of transporting mineral elements from the soil solution to plants (Joner et al. 2000). By now, it is well established that symbiotic combination of metallophytes with AMF could enhance plant ability to grow on highly contaminated soils (see review by Hildebrandt et al. 2007). One of the main reasons for activation of plant growth in such symbiotic systems is believed to be the improved plant supply with phosphorus and nitrogen (Ferrol et al. 2009).

May be this was the reason for a protective effect of AMF *Glomus mosseae*, which was clearly manifested in three from four tested plant species (Chen et al. 2007). In *Trifolium repens, Coreopsis drummondi*, and *Pteris vitata* (fern), the root length and the root and shoot dry weights were substantially increased at 14–18% root colonization with *G. mosseae*. AMF did not essentially affect the concentration of Cu in the roots but markedly reduced it in the shoots. However, since the shoot biomass was strongly increased, total Cu amount in shoots, as calculated per plant (or vessel), was increased by 4.7 times in *T. repens*, 4.9 times in *C. drummondii*, and 8.13 times in *P. vitata*. The authors did not analyze the reasons for such AMF effects on plants, considering a "dilution effect" as the key mechanism of AMF-mediated diminishing of Cu (and Cd) concentration in aboveground organs.

At the same time, another hypothesis of AMF protective action is more popular. According to this hypothesis, AMF, which colonize plant roots, considerably reduce the uptake of HM into plant cells, and this may be one of the mechanisms of metallophyte tolerance to HM (Hildebrandt et al. 2007). Nevertheless, the realization of this mechanism is evidently species specific.

This suggestion was supported by the experiments of Sudova et al. (2008), who failed to establish a synergism between *Agrostis capillaries* and AMF *Glomus intraradices*, although these authors used various clones and isolates originated either from contaminated or uncontaminated sites. It turned out that AMF did not confer significant additional Cu tolerance to either Cu-tolerant or Cu-sensitive host plants grown on the contaminated substrate, even in the case of highly Cu-sensitive clones. Nevertheless, the authors did not exclude a possibility that beneficial interaction between the two organisms might be manifested for other combinations of plant species and fungal species/isolates.

Another situation was observed at inoculation of the salt marsh plant *Aster tripolium* with inoculates of AMF *Glomus geosporum* originated from polluted salt marshes (PL isolates) or from nonpolluted sites (NP isolates) (Carvalho et al. 2006). In the presence of high Cu concentrations in soil (up to 2.0 mM), tolerance of nonmycorrhizal (NM) plants was relatively high: the dry weight of their roots and shoots exceeded corresponding indices for PL and NP plants. At the same time, the content of Cu in the roots of PL and NP plants was twice higher that in the roots of NM plants, whereas, in contrast, the shoots of NM plants contained the highest amount of Cu. The authors concluded that AMF enhanced Cu uptake and accumulation in the root system but retarded its translocation to shoots, as compared to NM plants. They called this effect "toxic metal trapping."

Oryza sativa colonization by the extraradical mycelium of AMF *G. mosseae* resulted in the stronger fungal effects (Zhang et al. 2009). When colonized plants (GM plants) were grown in the presence of 5–100 μM $CuCl_2$, Cu contents in their roots and especially in their shoots were significantly reduced. Thus, the content of Cu in the shoots was only 48% of that in NM plants. In this case, more than 60% of Cu present in the GM plants was retained in the root cell walls, whereas the corresponding value in NM plants was only 25%. The authors supposed that the reason for this difference is some changes they observed in the composition of the GM plant cell walls and, thus, in their Cu-binding capacity, i.e., cation exchange capacity (CEC).

Gonzalez-Chavez et al. (2002) evaluated directly the role of CEC for extraradical mycelium of three *Glomus* species colonizing *Sorgum vulgare*. Isolates of all species were obtained from polluted soil and one of them from nonpolluted soil as well. Observation performed with the usage of TEM and SEM linked to an energy dispersive X-ray spectrometer (EDAX) showed that AMF from polluted soil were able to accumulate Cu in different zones of their extraradical mycelium, in mucilaginous outer hyphal wall zone, cell walls, and inside the hyphal cytoplasm. The difference between the three AMF in the content of Cu accumulated in the extraradical mycelium was evidently primarily (might be partly) related to their difference in CEC. Metal ion filtering during their uptake by the roots might be one of important protective AMF functions in Cu-contaminated sites.

Such AMF role was confirmed in several researches. Thus, in the work of González-Guerrero et al. (2008), it was shown using EDXS that Cu accumulated mainly in the fungal cell walls and in the electron-dense granules in the vacuoles of fungal (*Glomus intraradices*) spores but not in the cytoplasm. A comparison of CEC of *Trifolium subterraneum* roots and extraradical hyphae of several *Glomus* species (Joner et al. 2000) showed that fungal hyphae manifested approximately tenfold higher CEC than roots, and there was no difference between different fungal species.

Thus, investigations on the effects of excessive Cu concentrations on the symbiotic plant-AMF systems showed clearly a diversity of mechanisms used at plant fungal colonization for diminishing Cu excess toxicity. Along with sometimes observed dilution, such a mechanism may be Cu entrapping or filtering resulting in retaining the great part of Cu excess in the mycelium and, as a consequence, in observed decrease in the Cu content within the plant and its translocation from roots to shoots. It is quite possible that the main reason for this effect is a high CEC value of AMF mycelium (toward Cu), which exceeds markedly CEC of colonized plant roots. In this connection, the data concerning the involvement of insoluble glycoprotein, glomalin, in Cu detoxification are of a great interest. In the experiments performed in vitro, this protein could sequester up to 28 mg Cu/g of protein (González-Chávez et al. 2004). For the appraisal of the cell wall protective role, the fact (established with monoclonal antibody MAb32B11) that glomalin is mainly located in the cell walls of AMF hyphae is of especial importance (Purin and Rillig 2008; Ferrol et al. 2009).

At the same time, data concerning changes in the cell wall composition of roots colonized by AMF (Zhang et al. 2009) deserve an earnest consideration. Such changes may be the reason for Cu retaining in the root tissues; they indicate a possibility of direct influence of fungal mycelium on the metabolic processes in the host plant tissues. This suggestion is confirmed by studies performed in the laboratory of Hildebrandt (see review Hildebrandt et al. 2007); it was shown that, in the symbiotic system, the HM, including Cu, excess not only activated markedly the defense system of the mycelium but also induced complex and ambiguous changes in expression of plant genes encoding products putatively involved in HM detoxification.

8.3 Apoplast Involvement in Copper Detoxification

Copper penetrates into the plant mainly as a bivalent cation Cu^{2+}, although it was reported that sometimes it is reduced near the root surface and penetrates into the root cells as a monovalent cation (Krämer and Clemens 2006; Cohu and Pilon 2010) or even as a free metal (see Sect. 8.5.2). Since the cell wall is a negatively charged ion exchanger, it is quite clear that a part of Cu ions is absorbed by the cell wall on the path of their movement into the cell. Moreover, an additional copper amount may be released into the apoplast from symplast because transporters of the P_{1b}-ATPase family, denoted as heavy metal ATPase 5 (HMA5), are located in the plasma membrane and provide for the efflux into the apoplast of excess Cu ions, which have previously penetrated inside the cell. The important role of apoplast in Cu detoxification was supported by experiments of Andres-Colas et al. (2006), who demonstrated that the *hma5* T-DNA insertion mutant exhibited Cu hypersensitivity, which was especially dramatic in roots where *HMA5* was mostly expressed. These data allow a consideration of HMA5 P-type ATPase functioning as a potential mechanism for improving Cu detoxification under Cu excess.

Cell walls of different plant species differ substantially in their composition (Carpita et al. 2001). Negatively charged groups of the polymeric matrix play a key role in binding of HM, including Cu ions. The high content of pectic compounds containing unetherified carboxyl groups of polygalacturonic acid plays a crucial role in this process. The content and qualitative composition of phenolic compounds and lignin also exert substantial influence.

Strong differences in the composition and properties of the cell walls in different plant species are evident; they depend on plant age, environmental cues, and other factors. CEC of polymers in the cell wall matrix is one of the main parameters determining ion exchange in them (Meychik and Yermakov 2001; Wehr et al. 2010). This parameter is frequently assessed on isolated, chemically purified and, thus, to some degree modified cell walls. This raises a question as to whether degree the obtained results correspond to the natural state of the cell walls before their isolation. Microscopic methods offer a closer approach to the evaluation of the natural cell wall capacity for HM immobilization. However, the methods of direct Cu visualization with the usage of specific reagents are not essentially developed, excluding few studies (Arru et al. 2002; Ranathunge et al. 2005). The methods of electron microscopy present great possibilities; the combination of TEM or SEM with modern physical methods, EDX or EELS, makes a reality the quantification of Cu intracellular localization in plants.

8.3.1 Microscopic Methods of Cu Localization in the Cell Wall and Periplasmatic Space

Using TEM, Cu was detected in the roots of *Zea mays* (Ouzounidou et al. 1995) and *Oreganum vulgare*, a plant of the Mediterranean aromatic flora (Panou-Filotheou and Bosabalidis 2004) grown under Cu great excess in medium. On TEM micrographs, Cu was seen as electron-dense globular inclusions, near the cell walls in particular. Cu was deposited in the cell walls, including their middle-lamellar regions. In addition, dense compact material was located behind the plasmalemma, adjacent to the walls in cortical and stellar parenchyma cells.

Copper localization in leaf and stem tissues of *Cannabis sativa* plants, grown for 10 days on medium supplemented with 1 mM $CuSO_4$ (100-fold above normal concentration) but without any visible toxicity symptoms, was studied after Cu immobilization with Na_2S (Arru et al. 2002). In these TEM experiments, electron-dense deposits were observed in vacuoles and also as very fine precipitates at the cell junctions, mainly on the middle lamella. The presence of Cu in these structures was confirmed by EDX-ray microanalysis in the epidermis and in leaf palisade layers. The authors concluded that, in *C. sativa* leaves, Cu ions followed symplastic as well as apoplastic route.

Structural changes, induced by the excess of Cu (10 μM) in the photosynthesizing suspension cell culture of *Glycine max,* were studied (Bernal et al. 2006). Using LM, numerous dark rounded deposits were observed attached at the outer surface of the cell wall. These extracellular dark deposits of different sizes and shapes (round and ellipsoid), contacting with the cell walls, contained Cu, and this was confirmed by EDX-ray microanalysis. The authors noted that such structures were present only in

Cu-stressed cells (21 days on medium with 10 μM $CuSO_4$) but not in control treatment or after long-term acclimation (after 22 subculturings in the presence of 10 μM Cu).

A comprehensive study was performed on *Armeria maritima* plants growing on the copper-rich soil, which accumulated from 2,000 (leaves) to 4,000 (roots) higher amounts of Cu as compared with its standard concentrations in plants (Neumann et al. 1995). As measured by TEM-EDX analysis, a great part of Cu in roots and leaves was retained in vacuoles of idioblasts ("tannin cells"). Similar osmiophilic precipitates (with high Cu concentrations) were shown between the cell wall and plasmalemma and in the cell wall of root cortical parenchyma. Copper contents in cell walls of the exodermis (in roots) and idioblast/bundle (in leaves) of *A. maritima* were especially high. EELS spectra of cell-wall-bound Cu showed that Cu was evidently bound in the protein–copper complexes. Two copper-binding proteins were extracted from the cell walls.

Another species, *Elsholtzia splendens*, a native Chinese Cu-tolerant and Cu-accumulating plant, was studied by Peng et al. (2005). Using TEM, these authors observed in the leaves large amounts of Cu, which were intensively deposited both within the cells (in the chloroplast membranes) and in the cell walls. Still larger and numerous big dark Cu particles were noted in the root cells, especially at the high $CuSO_4$ concentration in medium. These particles were deposited near the root cell walls and plasma membrane. Numerous Cu deposits separated the cytoplasm and the cell wall. It should be noted that such numerous large Cu deposits were found in this work in spite of the very high Fe–EDTA concentration (100 μM) in medium and 20-min root washing with 5 mM $Pb(NO_3)_2$ before fixation. These researchers also presented quantitative Cu estimates for isolated cell walls obtained by the gradient centrifugation technique (see Sect. 8.3.3).

Quantification of Cu distribution between root and leaf tissues of *Avicennia marina* plants, a facultative halophyte and a typical mangrove species, was performed using SEM X-ray microanalysis in the work of MacFarlane and Burchett (2000). Because of method limitations, a very high $CuSO_4$ concentration (4 g/l) in the substrate was applied, but the time of exposure was shortened to 4 days to minimize the onset of Cu toxicity responses. In the root zone of developed endodermis with Casparian bands, the relative Cu concentration was similar from epidermal through parenchyma to the endodermal cell walls, the lowest concentration being in the phloem. Cell wall Cu concentrations were higher than cytoplasmic ones; the difference was small in epidermis and parenchyma cells and significantly higher (by ~30%) in the endodermal cell walls. In the leaf tissues, the highest concentration was found in the xylem cell walls; it decreased gradually with an increased distance from the xylem; in the mesophyll cells, Cu content in the cell walls was higher than in the cytoplasm.

As distinct from above data, in experiments with *Allium sativum* plants grown in the presence of high Cu concentrations (10 and 100 μM) in nutrient medium (and 1 μM Cu in control treatment), only trace amounts of Cu were detected in the cell walls after 9-day-long exposure (Liu and Kottke 2004). The TEM-EELS analysis showed that the main amount of Cu was precipitated in the cell walls of cortical cells. The authors believe that the difference of their results from those of other researchers is determined by species specificity. One more cause may be a rather high EDTA concentration (10 μM Fe–EDTA) in medium, which could limit Cu availability for the roots. A decrease in the free copper ion concentration in the near-root medium can be a reason for the diminished Cu ion absorption by the cell walls.

Thus, the results of microscopic investigations are mainly qualitative. Nevertheless, in various plant species belonging to various ecological groups, Cu precipitates, usually numerous globular structures, were observed in the cell walls, mainly on the middle lamellae and also in the closed intercellular space restricted by the cell walls. It should be emphasized that Cu-containing globules were also detected near the cell wall, in the periplasmatic space, between the cell wall and plasma membrane (Neumann et al. 1995; Panou-Filotheou and Bosabalidis 2004; Peng et al. 2005; Bernal et al. 2006). The presence of Cu in this space in its insoluble form (small precipitates, clumps, globules, etc.) could be evidently explained by the ways of its immobilization in this small volume: it seems likely that Cu was retained in the periplasmatic space by other compounds than the components of the cell wall.

8.3.2 Some Approaches for Distinction Between Apoplastic and Symplastic Cu Pools

To quantify Cu distribution between its main tissue pools, i.e., between the apoplast and symplast, it is of importance to measure the CEC of the cell walls. This parameter characterizes the process of cation exchange and a possibility of Cu immobilization by the cell walls (Meychik et al. 2010). The main way to estimate CEC is a correct performing of metal (Cu) desorption using Cu-displacing agents, which could replace it from the cell wall matrix, with subsequent determination of Cu content by any one of methods, most often spectrophotometrically. This approach is now applied most frequently, although, regretfully, the common methodological version of the assessment is not elaborated. Only few works analyze the correctness of the approaches applied.

One of the first studies of copper distribution between the apoplast and symplast was that of Harrison et al. (1979). The authors screened several metal cations on their relative capacity for Cu desorption from the free space of *Hordeum* roots. Such desorption is difficult because of especially high stability of copper complexes with charged sites of the cell wall matrix. After testing the collection of metals (Pb, Ca, Cd, Co, Mg, and Ni), the authors established that Pb was most suitable as a Cu-displacing agent. The efficiency of 5 mM $Pb(NO_3)_2$ (60 min at 0°C) relative the wide range of Cu concentrations was confirmed by the authors in both time-course and kinetic studies.

Rather few early investigations using $Pb(NO_3)_2$ as a desorbent were reviewed by Ernst et al. (1992). The authors concluded that the amount of metals bound to the cell wall was usually less than 10% of its total cellular amount. As to the role of cell walls in HM detoxification, both positive correlation between HM tolerance and cell wall binding capacity and the absence of such correlation were observed.

Later, Pb was repeatedly used as an effective agent for apoplastic Cu displacing in the roots and leaves of many plant species (De Vos et al. 1991; Yang et al. 2002; Llugany et al. 2003; Peng et al. 2005; Russo et al. 2008; and others). In most cases, the role of this cellular Cu pool was analyzed.

In recent years, short washing with 3 mM EDTA came to be applied to remove free cations. This approach was, for example, used for soybean suspension culture and leaves (Bernal et al. 2006, 2007) and Arabidopsis plants (Andres-Colas et al. 2006). Branquinho et al. (1997) compared carefully several ways of Cu desorption, including EDTA and Pb application, for several lichen species. They showed that elution procedure with 20 mM Na_2–EDTA (a chelating agent) (pH 4.5, sequentially for 40 and 20 min) was as efficient as that with 20 mM $Pb(NO_3)_2$. In contrast, the usage of $NiCl_2$ for Cu desorption resulted in the underestimation of the apoplastic Cu pool. It is of importance that EDTA and Pb did not disturb plasma membrane integrity, as evident from the absence of the effects of these agents on the content of intracellular potassium. In authors' opinion, this might be a cause for great errors. The efficiency of Na_2–EDTA usage for determination of the apoplastic Cu pool was confirmed in some studies (Monnet et al. 2005; Russo et al. 2008; see also the review of Bačkor and Loppi 2009).

However, in many studies, researchers applied only poorly controlled washing with distilled water. Such approach did not allow the assessment of Cu concentration within the cells, in the symplast. Application of some surfactants, such as Alconox or Triton in combination with EDTA, and lauryl sulfate were used mainly for cleansing the surface of aboveground organs from Cu contamination (Faucon et al. 2007, 2009). Only few publications comprise quantitative data concerning Cu content in the apoplast and its proportion in the total Cu content in tissues.

Branquinho et al. (1997) and Monnet et al. (2005) presented quantitative data obtained for intact lichens. Branquinho et al. (1997) used Na_2-EDTA as a displacing agent to assess extracellular bound Cu (see above) in *Usnea* spp. and *Ramalina fastigiata*. It was established that, in the presence of the highest Cu concentration in medium (15.8 mM $CuSO_4$), extracellular binding of Cu reached equilibrium after 10–20 min, and after 2 h, almost all Cu was immobilized by the cell walls, and not more than 2% in *Usnea* and about 5% in *R. fastigiata* was localized within the cells. For both species, the concentration dependence of extracellular Cu binding showed saturation kinetics, and the maximum capacity to bind extracellular Cu was approximately fivefold higher for *Usnea* spp. than for *R. fastigiata*.

Monnet et al. (2005) also used Na_2–EDTA as an extracellular Cu extractant from *Dermatocarpon luridum* thalli. They were able to determine and quantify the cellular location of Cu in lichens in the presence of 0.25–1.00 mM Cu in medium. At all Cu concentrations in medium, the extracellular Cu concentration reached a maximum value after 3–6 h, but a significant decrease was then observed in the presence of 0.25 or 0.50 mM $CuSO_4$. (It is worth mentioning that the first measurement was performed in this study only in 3 h after treatment.) Measurements performed for 48 h allowed the authors to conclude that the extracellular content of Cu represented more than 90% of total Cu content.

A specific approach for separation of apoplastic and symplastic Cu pools was applied in the study of Zhang et al. (2009) performed on *O. sativa* plants, and this is evidently the only work of this sort for Cu. After Cu desorption from the roots with 5 mM $CaCl_2$ (four times, 5 min each), the roots were subjected to fast freezing in liquid nitrogen with subsequent thawing and rinsing with water. Cu released after such treatments was considered symplastic Cu. The total content of Cu released in the 20-min desorption and Cu remained in the root after the freeze–thaw cycle was considered apoplastic Cu. In these experiments, the highest value of apoplastic Cu was about 25% of its total content in the *O. sativa* roots not colonized by *G. mosseae*, but this proportion increased to 60% under conditions of mycorrhizal colonization, mainly due to a decrease in the total Cu content in plants.

Regretfully, only several studies comprised quantitative data characterizing the Cu apoplastic pool in intact plant organs. The investigations in this direction must be expanded. Correct methods for separation of apoplastic and symplastic Cu pools in plant organs should be elaborated. In our opinion, the principle of the method developed by Harrison et al. (1979) could be used for the quantitative assessment of Cu partition between the root apoplast and symplast, maybe after some modification. We performed some work in this direction.

The main principles of the Harrison et al. (1979) method are: (1) Cu uptake by excised roots to exclude its translocation from roots to shoots and (2) low temperature during replacement of the apoplast pool to prevent Cu transport from the apoplast (cell walls and periplasmatic space) to the symplast. In our method modification, all procedures, Cu uptake and displacing, were performed at low temperature (lower than 4°C).

To apply these methods, it was of importance to find a correct criterion for distinction between the pools. Na_2–EDTA was chosen as a displacing agent, whereas EDTA concentration and the time of desorption were variable parameters. When the values of these two parameters were increased to a definite limit, a sharp jump in Cu leakage from the tissue was observed. We consider such a jump as a sign of plasma membrane damage resulting in the mixing of two divided pools. Therefore, for separate quantification of these pools, we use conditions preceding this jump.

When Cu uptake occurred at a temperature close to zero, we can obtain direct information about apoplast filling because transmembrane Cu transfer into symplast, which would be performed by protein transporters under sufficient level of energy supply, is blocked by low temperature.

It is quite clear that possibilities of the approach we elaborated are rather limited. Indeed, for each plant species and experimental conditions, the limiting values of variable parameters and their correctness should be thoroughly checked. Undoubtedly, some additional modifications are required for wide application of this method.

The analysis of the Cu apoplastic pool, its physiological functions, and possible involvement in metal detoxification will be presented below together with the results obtained for isolated cell walls.

8.3.3 *Isolated Cell Walls as a Model for Studying Cu Immobilization in the Apoplast*

In some studies, CEC relative to Cu was studied on cell walls isolated from roots and/or leaves of plants, in particular plants differing in their tolerance to excessive Cu concentrations. The methods used for cell wall isolation differ substantially, and this hinders generalization of published works. Nevertheless, the basic information concerning this problem was obtained just on isolated cell walls.

In most works, cell walls were isolated using buffer solutions (pH 7.0–7.5) with corresponding additions for the removal of low-molecular soluble organic compounds (Nishizono et al. 1987; Konno et al. 2005; Peng et al. 2005; Fritioff and Greger 2006; Ke et al. 2007; Wehr et al. 2010). In the works of Lou et al. (2004), Shi et al. (2008), and Wei et al. (2008), before cell wall isolation, plant material was soaked in the mixture of methanol and chloroform (2:1, v/v). Zhang et al. (2009) applied a very specific method for cell wall isolation but also with the usage of organic solvents.

It turned out that a great part of studies were performed on isolated cell walls from Cu-tolerant and Cu-accumulating plants.

Several works were performed on the two species of Elsholtzia. Important information concerning the dynamics of apoplastic Cu pool accumulation was obtained for *Elsholtzia haichowensis*, a native Cu-tolerant Cu-accumulating plant species growing on copper mining deposits in China (Lou et al. 2004). The saturation of sites for Cu adsorption was achieved after 72 h exposure to the 100 μM $CuSO_4$ solution. During the first 24 h, most of the increased Cu in roots was found in the cell wall fraction. By the end of experiment (over 120 h), Cu bound to the cell walls accounted for 68% of the total Cu content in roots.

Ultrastructural Cu distribution in both the roots and aboveground organs was examined for another highly tolerant species, *E. splendens* (Peng et al. 2005; Shi et al. 2008). In these plants exposed to 500 μM Cu during 8 days, total Cu concentration in the leaves increased fivefold, up to 250 mg Cu/kg, whereas in the roots an increase was much greater, up to 8,000 mg Cu/kg (Peng et al. 2005). The authors calculated that 37% of total Cu in the leaves and 56% in the roots were accumulated in the cell walls. On this basis, it was concluded that in *E. splendens*, a Cu-tolerant and Cu-accumulating plant species, the plant cell wall was the main Cu location site both in the root and leaf cells. It could mainly account for the high detoxification of Cu in the plant.

In another work performed on the same plant species, *E. splendens,* subcellular distribution of Cu was analyzed by Cu K-edge XANES method (Shi et al. 2008). It was established that 30–39% of total Cu in the roots and stems and 20–44% in the leaves were localized in the cell walls. It turned out that, in the interval between 10- and 60-day-long plant exposures to 300 μM $CuSO_4$, Cu concentration in the roots increased only by 17%, whereas in the leaves, by 50%. Cu redistribution was manifested much more profoundly in the leaf cell walls (from 44 to 20%) as compared with the root cell walls (from 39 to 31%). At the same time, the content of Cu bound with histidine-like intracellular ligands was increased.

One of the earliest studies of intracellular Cu localization was that by Nishizono et al. (1987) performed on the fern *Athyrium yokoscense*, which is known in Japan as a highly HM-tolerant plant. The concentration of Cu in the roots of this plant growing on metalliferous soils exceeded this index in plants from nonmetalliferous habitats by many times (2,602–3,846 vs. 26–70 μg/g DW of roots). At the same time, the difference between relative Cu content in the root cell walls of these plants from different habitats was not too strong: 85–92% of its total content in plants growing on metalliferous soils vs. 72–76% in plants on nonmetalliferous soil.

The values of CEC for Cu, determined in the isolated *A. yokoscense* cell walls, were by 5–7 times higher than this index measured for several plant species growing in copper-contaminated areas. The CEC value was well correlated with the Cu concentration accumulated in *A. yokoscense* roots. The high values of cell wall CEC in *A. yokoscense* roots did not depend on the extent of habitat contamination. However, a difference between *A. yokoscense* and other fern species, nontolerant to Cu excess, was insignificant. These results and the data about much higher Cu concentration in the *A. yokoscense* root cell cytoplasm as compared with the concentration characteristic of other ferns allowed the authors to conclude that, in addition to the increased CEC of their root cell walls, some specialized intracellular mechanisms of Cu detoxification were responsible for a high *A. yokoscense* tolerance to Cu.

In studies of specific features of adaptation of one more highly tolerant Japan fern *Lygoduim japonicum* (Konno et al. 2005), substantial changes in matrix polysaccharides of the cell walls in fern prothallium were detected. It turned out that 20-day-long growth on medium containing 0.4 mM $CuSO_4$ resulted in the accumulation of 17.2 μmol Cu/g DW of the cell wall preparation. Cu excess in medium induced strong changes in the content and fraction composition of the basic cell wall components. The content of solubilized pectin was reduced by 53%, solubilized hemicellulose, by 82% of control, and the content of uronic acid increased markedly. With the usage of *endo*-pectate-lyase, it was also established that most of the Cu accumulated into the *L. japonicum* prothallium was tightly bound to galacturonic acids of homogalactouronan of the cell wall pectin.

In laboratory experiments performed on the aquatic plant *Potamogeton natans* (pondweed), the inhabitant of eutrophic lakes with increased HM concentrations, a distribution of Cu (and also Cd, Pb, and Zn) was studied (Fritioff and Greger 2006). The highest content of Cu was always found in the roots. The analysis of cell walls isolated from leaves and stems showed that 40–42% of Cu in leaves and 33–51% in stems were bound to the cell walls. The authors noted that CEC for Cd (48 $\pm$ 6 and 51 $\pm$ 5 meqv 100 g^{-1} $\pm$ se for leaves and stems, respectively) was not correlated with the organic matter content in these plant organs, as distinct from usual state of affairs. The data for Cu are absent from this work.

A comparison of plant species differing in their tolerance to Cu in relation to the possible role of their cell walls was performed by Wei et al. (2008). *Sorghum sudanense* plants more tolerant to Cu in medium and less tolerant *Chrysanthemum coronarium* plants were grown in hydroponic culture in the presence of Cu excess (up to 50 μM). It turned out that, in *C. coronarium* shoots and their cell walls, total

Cu concentrations were much higher than in *S. sudanense*. However, in *C. coronarium*, the percentage of cell wall bound Cu was much lower (and the percentage of water-soluble Cu was higher). In contrast, in *S. sudanense*, substantially greater Cu amounts remained in the roots, and the percentage of cell wall bound Cu was also higher (76% of total content in the cell walls vs. 60% in *C. coronarium*). This was correlated with a strong increase in the content of uronic acids in *S. sudanense* induced by Cu excess and with their higher concentration as compared to *C. coronarium* (85 μg/g FW against 55 μg/g). The authors concluded that all these peculiarities could provide for a higher *S. sudanense* tolerance. Among the reasons for improved *S. sudanense* tolerance, it was also indicated that the 2.5-fold lower Cu concentration was present in this plant shoots, especially taking into account that more than its 15% (vs. 11% in *C. coronarium*) was retained by the cell walls. As a result, Cu load on *S. sudanense* protoplast turned out to be much lower than in *C. coronarium.*

Using cell walls isolated from *O. sativa* roots (including those inoculated with AMF *G. mosseae*), Zhang et al. (2009) demonstrated that Cu absorption was rapid, with saturation of binding sites being achieved within 10 min. The concentration dependence of Cu binding in isolated cell walls showed its high capacity and saturation at around 1.5 mM Cu. Along with significantly increased value of Cu-binding capacity of cell walls isolated from AMF-inoculated roots in comparison with noninoculated roots, a substantial difference was also observed during successive cell wall fractionation. It was shown that a significantly higher (approximately by 15%) Cu binding capacity (CEC equivalent) of the cell walls isolated from inoculated roots was largely determined by the increased contribution of the pectin-containing fraction into this characteristic.

Ke et al. (2007) compared two natural populations of *Daucus carota*, grown in soil culture from seeds collected from Cu contaminated site (CS) and uncontaminated site (UCS). In the presence of high Cu concentration (>400 mg/kg of soil), biomass accumulation in UCS population was reduced to 68–60% of control, but growth of CS population plants was not affected. Total and cell wall Cu contents were increased similarly (by 4.5 times) in leaves and roots of both plant populations. In the leaves, where Cu concentration was higher than in the roots, 51–54% of the whole leaf Cu was localized in the cell wall in both CS and UCS plants (vs. only 42–43% in control plants). Thus, any reliable contribution of extracellular Cu retention to the improved tolerance of CS plants was not established. In contrast, the higher level of Cu localization in vacuoles of CS population plants (14.2 vs. 9.1% in UCS plants) indicated the probability of intracellular mechanism of Cu detoxification functioning in *D. carota* plants.

Thus, all available approaches indicated that Cu was always detected in the apoplast, especially at its excess in environment. It is believed that the volume of the apoplast comprises only about 5% of total volume of plant tissues, but it represents a zone of a direct contact between the root and external medium (soil, for example), which could contain a great Cu excess in contaminated areas. Keeping in mind that the apoplast manifests weak metabolic activity, it might be expected that just the apoplast could fulfill some protective functions against increased Cu concentrations toxic for plants. The materials presented confirm these expectations to some degree.

However, two circumstances interfere with the reliable assessment of the apoplast role in Cu detoxification. Firstly, appropriate methods for separation of apoplastic and symplastic Cu pools in intact organs (roots primarily) are absent. Therefore, reported basic indices, such as, for example, time of saturation, are very variable (from 4–6 h to 3 days); the data concerning a concentration dependence of apoplast pool filling are scarce, etc. As to the results obtained on isolated cell walls, there are great doubts relative to their equivalence to the initial cell wall state. Indeed, isolation of the cell polymeric matrix resulted frequently in the disturbance of its architecture and changes in chemical composition. Thus, lipid-free cell wall preparations were used in some experiments (Shi et al. 2008), ionically and tightly bound proteins and structural proteins were removed in other ones (Konno et al. 2005). However, direct data are available about the involvement of cell wall proteins in Cu immobilization. Neumann et al. (1995) demonstrated in studies of *A. maritima* EELS spectra that Cu binding to the cell walls is largely determined by the production of copper–protein complexes. The two copper-binding proteins were extracted from the cell walls. Remember also information concerning AMF-inoculated roots, e.g., a sharp increase in the CEC related to the presence of specific proteins in fungal hyphae. One of such protein is evidently a recently discovered glomalin. The confirmation of this result could open new possibility for plant genetic modifications.

Summing the material presented in this division, it may be concluded that, in spite of described difficulties, it is undoubtedly that the apoplastic Cu pool comprises a very substantial, sometimes dominating part of metal absorbed by the plant, especially at its excess in environment. Thus, the role of extracellular Cu pool must be taken into account when analyzing the problem of Cu detoxification.

8.4 Initial Steps of Intracellular Copper Detoxification

As was mentioned earlier, essentially all Cu ions are present in the cytosol in the bound form (Changela et al. 2003; Krämer et al. 2007). Thus, after Cu ions transfer across the plasma membrane with the help of membrane transporters, they are immediately subjected by chelating. The basic components of the defense system in the cytosol are S-containing ligands: phytochelatins (PC) and metallothioneins (MT). MT belong to the group of cysteine-rich proteins (8–10 kDa) encoded by the family of nuclear genes (Murphy and Taiz 1995; Cobbett and Goldsbrough 2002). As distinct from MT, PCs are oligomers of glutathione produced by the enzyme phytochelatin synthase (PCS) (Cobbett and Goldsbrough 2002).

Information concerning the role of MT and PC in detoxification of Cu excess is contradictory and sometimes mutually exclusive (Schäfer et al. 1997; Gonzalez-Mendoza et al. 2007; Guo et al. 2008).

We studied the involvement of MT and PC in Cu detoxification in *Brassica napus* L. plants grown in the presence of 10, 50, and 150 μM $CuSO_4$ for 10 days;

control plants were kept in the presence of 0.25 μM $CuSO_4$. In these experiments, we compared the time course of Cu accumulation in the intracellular Cu pool in the roots and leaves with the level of transcripts of *PCS* gene and two MT-encoding genes, *MT1* and *MT2*. Apoplastic Cu desorption was performed as described earlier (Ivanova et al. 2010; Kulikova et al. 2011); Cu concentration was measured with the atomic absorption spectrophotometer. Transcript content was assessed by routine methods (Kholodova et al. 2011). Plant material was fixed for analyses in 3, 6, and 24 h and also in 5 and 10 days of Cu treatment.

Figures 8.1 and 8.2 compare directly the Cu content in the intracellular (symplast) pool of the roots (Fig. 8.1) or its total content in the leaves (Fig. 8.2)

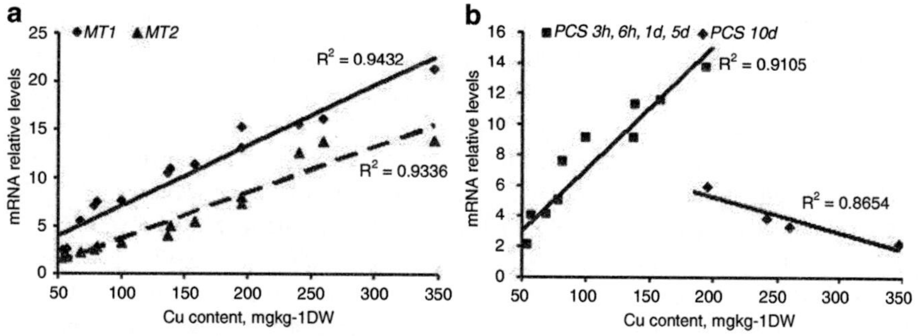

Fig. 8.1 Correlation between Cu content in the root symplast and the level of *MT1*, *MT2* (**a**) and *PCS* (**b**) mRNAs in *Brassica napus* plants. Plants were kept on the 10, 50, and 150 μM $CuSO_4$ solutions from 3 h to 10 days. For estimation of Cu content in the roots, they were firstly washed with a large volume of running water, then with 10 mM EDTA for 15 min with stirring, and at last with distilled water. Cu content was determined using an AAS Labist-400 (Labist, Russia). mRNA relative level is the ratio of each gene mRNA amount in experimental rapeseed plants to the same mRNA amount in control plants (grown on standard nutrient medium). 18S rRNA was used as an internal standard. Measurements were performed in triplicate

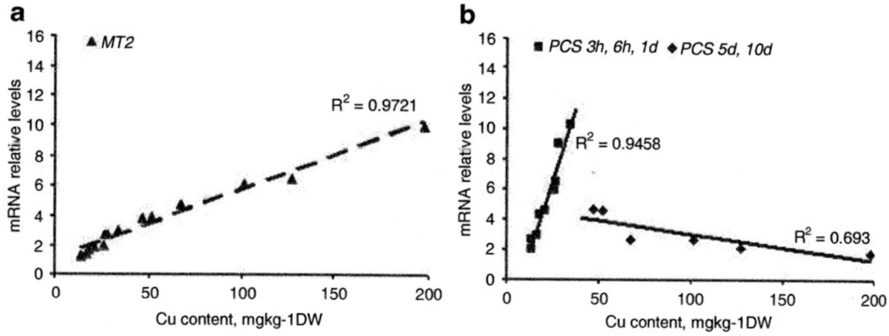

Fig. 8.2 Correlation between Cu content in the *Brassica napus* leaves and the level of *MT2* (A) and *PCS* (B) mRNAs. Plants were kept on the 10, 50, and 150 μM $CuSO_4$ solutions from 3 h to 10 days. Cu content was determined using an AAS Labist-400 (Labist, Russia). mRNA relative level is the ratio of each gene mRNA amount in experimental rapeseed plants to the same mRNA amount in control plants (grown on standard nutrient medium). 18S rRNA was used as an internal standard. Measurements were performed in triplicate

with the level of mRNAs of tested genes at all used $CuSO_4$ concentrations in medium.

Under these conditions, Cu accumulation in the root symplast occurred very actively. For the first 3 and 6 h, Cu accumulated to the level of 68.7 and 99.6 μg/g DW, respectively. It continued to grow during experiment, attaining the highest concentration of 347 μg/g DW; this value exceeded initial Cu concentration (39.7 μg/g DW) by 8.7 times.

Already from the beginning of Cu treatment, the amount of *MT1* gene transcripts increased by 21 times as compared to the initial value (Fig. 8.1a). Activation of *MT2* gene transcription was less profound; nevertheless, the level of its mRNA increased until the end of experiment. It was unexpected that Cu excess in the root symplast initiated transcription of *PCS* gene, which product catalyzes the synthesis of phytochelatins, i.e., S-containing peptides of another group of ligands (Fig. 8.1b). During initial intracellular Cu accumulation, the extent of *PCS* gene activation was comparable with that of *MT1* gene, which is root specific (Guo et al. 2008). The highest value in the *PCS* transcripts was 13.7-fold higher that its initial level at the Cu content in the root symplast of 159 μg/g DW. However, further Cu accumulation was accompanied by a strong decrease in the level of *PCS* mRNAs.

Under similar conditions, in the presence of 10–150 μM $CuSO_4$, the leaves accumulated Cu noticeably slower than the roots. In 3 h, the initial Cu level in leaves (12.1 μg/g DW) was exceeded only by 1.1–2.2 times; however, by the end of experiment, the leaves accumulated up to 198 μg Cu/g DW, i.e., 16-fold more than its initial level.

As distinct from the roots, expression of *MT1* gene was not detected in leaves, confirming its root specificity. A statistically significant increase in the level of *MT2* transcripts was observed only in 24 h of treatment, the highest level, exceeding control values by 6.1–9.9 times, was attained at Cu concentration in leaves of 127–198 μg/g DW to the end of the experiment. In contrast to this slow response of *MT2* gene transcription to the Cu excess in the leaves, the *PCS* gene transcription was the earliest response: the level of *PCS* transcripts exceeded the initial level by 2–4 times at the less than twice increased Cu concentration. However, like in the roots, the level of *PCS* mRNAs reduced markedly with continued Cu accumulation on the background of a stable increase in the *MT2* transcript level (Fig. 8.2).

The correlation analysis was performed for tested indices. The analysis of the bulk of information obtained for roots showed that the coefficients of correlation between Cu concentration ([Cu]) and the level of mRNA were equal to +0.971 for *MT1* and +0.966 for *MT2* gene, for leaves, $r = +0.986$ for *MT2* gene. These data make very probable that the transcription levels of these genes were tightly connected with Cu concentration in the symplast of corresponding organs. In spite of the complex type of dependence between [Cu] and the level of *PCS* mRNA, the coefficients of correlation for each of the branches of this dependence were rather high: $r = +0.957$ and $r = -0.930$ for roots and $r = +0.972$ and $r = -0.832$ for leaves. It is unquestionable that the role of PCS in Cu detoxification must be studied in more detail.

Thus, as judged from the levels of tested gene transcription, in the roots, where Cu ion absorption by the cells was very intense, both groups of main S-containing ligands were evidently involved in the early steps of Cu detoxification. The products of two genes encoding MT could directly participate in this process. Activation of *PCS* gene expression could activate phytochelatin synthesis via PCS enzyme accumulation. As to the leaves, chelating Cu excess in them is seemingly organized rather rational. Indeed, since Cu penetration into the leaves was somewhat delayed, its early small excess was not evidently sufficient for *MT2* gene activation. In contrast, it was quite sufficient for very active transcription of the *PCS* gene. It is characteristic that *PCS* transcripts accumulated in the leaves even more actively than in the roots experiencing the large-scale Cu penetration into the cells. When the Cu content in leaves attained 70 μg/g DW, the relative levels of *PCS* and *MT2* mRNAs became equal, and then the level of *MT2* mRNAs exceeded that of *PCS* mRNAs. Such behavior of two ligand groups at excessive Cu concentrations in the rapeseed leaves could be considered as their coordinated functioning in Cu excess detoxification, which was manifested at the earliest step of Cu action on rapeseed plants. Similar interaction between phytochelatins and metallothioneins in Cu detoxification, realized under different conditions, has been earlier observed in black mangrove *Avicennia germinans* (Gonzalez-Mendoza et al. 2007) and Arabidopsis (Guo et al. 2008) plants.

8.5 Nanoparticles of Metallic Copper in Plants

Recent years are marked by an intense development of nanotechnologies, which start to be used not only in diverse fields of industry but also in medicine and plant studies. In this connection, the interaction between nanoparticles (NPs) and plants was focused in recent studies (Nowack and Bucheli 2007; Ma et al. 2010). Below, we consider briefly two aspects related closely to the issue.

8.5.1 Production of Copper Nanoparticles by Plants

A possibility of the presence within the plant cell of metallic Cu usually entering plants, like other metals, as ions does not seem impossible or exotic. Cu^{2+} is reduced to Cu^{+} in soil in the near-root environment, and compounds comprising Cu^{+} were detected in plant roots (Naftel et al. 2007). But most important is that plant cells comprise some oxido-reductively labile metabolites capable of metal ion reduction to the metallic forms. Among them, ascorbic acid evidently plays a crucial role as a mild reductant (Manceau et al. 2008).

In recent years, it was established that the extracts from plant tissues could be used for reduction of some metal ions. It is especially interesting that, as a result of this reduction, metals produce nanoparticles (NPs) because the sizes of individual structures produced are less than 100 nm in more than one dimension (which is characteristic for nanostructures) (Nowack and Bucheli 2007).

For obtaining metallic Ag NPs from $AgNO_3$, Jha et al. (2009) used 50% ethanolic tissue extracts of various ecological plant groups: xerophytes *Bryophyllum* sp., mesophytes *Cyperus* sp., and hydrophytes *Hydrilla* sp. With the usage of X-ray diffraction (XRD) spectra technique, it was detected the formation of cubic Ag NPs. With the usage of TEM micrographs, the sizes of particles were found to be in the range of 2–5 nm. The authors believe that soluble flavones, quinines in plant tissue extracts were involved in the reduction of silver ions at 40°C, which lasted for several hours.

Haverkamp and Marshall (2009) studied the mechanism of metallic NP formation in plants as exemplified by several Ag salts absorbed by *Brassica juncea*. The upper limit for reduced Ag metal NPs was established as 0.35% Ag of plant dry weight, and it depended on the reducing capacity of the plant under experimental conditions. It became clear why, along with Ag, only Au and Cu NPs with the reduction potential of at least 0.04 V were detected in plants. From a review of the electrochemical potential observed, it was proposed that metal NP formation in plants is restricted to those elements, whose salts have a potential for reduction to metal above about 0 V. This indicates that the plant metal NP production will be limited to the precious and semiprecious metals. Along with already mentioned Ag (up to 0.80 V), Au (1.0 V), and Cu (0.35 V), Pd (0.64 V) and Pt (0.74 V) belong to this group.

It should be noted that, in recent years, the interest to the production of NPs with the help of plants increased markedly, in particular in connection with their usage in medicine. Indeed, the biological formation provides for the production of low-cost, energy-efficient, and nontoxic metallic NPs (Thakkar et al. 2010). A possibility exists of NP usage for plant life control, including for nanogenetic crop manipulations (Nair et al. 2010).

8.5.2 Action of Cu^0 Nanoparticles on Plants

As is seen from the material presented in the previous division (Sect. 8.5.1), metal ions, Cu in particular, may be reduced to the metallic form in the living plant (Haverkamp and Marshall 2009), although the extent of this process is rather limited. In contrast, natural and engineered sources of NPs are rather sizeable; among the natural sources, volcanoes are of importance. As to the production of engineered nanoparticles (ENPs), according to the evidence from The Royal Society and Royal Academy of Engineering 2004, presented by Navarro et al. (2008), in 2004 the annual global production of ENPs was of the order of 10^3 tons and it is expected to increase to 10^4–10^5 tons per year after 2010. In this connection, the effects of ENPs on plants and the role of the cell walls in defense against a potential NP threat should be closely considered.

The cell wall is a barrier on the route of ENP entry into the plant cell. The diameter of pores across the cell wall is within the range of 5–20 nm, and this determines its sieving properties (Fleischer et al. 1999). Such sizes of pores would limit penetration of large NPs. At the same time, it is known that newly synthesized cell walls manifest the higher permeability; in addition ENPs themselves might induce the formation of new, bigger than usual pores. Like for macromolecules, endocytosis can serve as a main way for translocation across the plasma membrane (Ovecka et al. 2005). Thus, there are no principal obstacles for NP penetration into and spreading over the plant. In this connection the question arises as to the nature of NP and especially ENP action on plants and a possible role of the cell wall in this interaction. Below, we present scarce available data concerning Cu ENPs.

When studying ENP action on germination of lettuce seeds, Shah and Belozerova (2009) introduced Cu in soil as copper nanosize activated powder to a final concentration of 0.013 or 0.066%. The shoot/root ratio was used as an index of possible NP action. In 11 days at the highest Cu concentration, this ratio increased significantly from 1.92 to 2.70, which evidently reflected the stronger inhibitory effect on root than shoot growth. This effect was manifested only after a 15-day-long soil pretreatment with NPs; the authors believe that this fact indicates a possibility of indirect Cu NP action on lettuce seedlings.

In more thorough investigations performed by Stampoulis et al. (2009) on *Cucurbita pepo*, the effects of Cu ENPs and bulk Cu powder were compared. Any effect on germination was not found. However, 15-day-long growth in hydroponic solution revealed the stronger inhibitory action of Cu ENPs on root length and biomass as compared with bulk Cu powder. Thus, root length was reduced by 77% in treatment with Cu ENPs, but only by 64% in treatment with bulk Cu powder. A decrease in biomass attained 90% in treatment with Cu ENPs, but only 69% in treatment with bulk Cu powder. An additional verification showed that a possible ionization of a small part of Cu during experiment did not change substantially a conclusion about the higher phytotoxicity of Cu ENPs as compared with that of bulk Cu powder.

Similar conclusion was made by Lee et al. (2008), who used *Phaseolus radiatus* and *Triticum aestivum* seed germination as a toxicity test for Cu ENPs. The seeds were germinated in Petri dishes on dual agar media for homogeneous exposure of plant roots to NPs. In *P. radiatus*, 50% growth inhibition of seedlings was observed at the Cu NP median effective concentration of 335 mg/l, whereas in *T. aestivum*, at 570 mg/l. It was also checked that the concentration of copper ions released during sonication process was very low and the apparent toxicity clearly resulted from Cu NPs. For a comparison of various plant species, a bioaccumulation factor, [Cu] in plants (mg/kg dry tissue)/[Cu] in media (mg/l), was used. Its values were 8 and 32 l/kg for *P. radiata* and *T. aestivim*, respectively, which reflected, in particular, specific root morphology of these plant species. The application of TEM for monitoring Cu localization showed the presence of individual and aggregated NPs in the cell walls and inside of the root cells of *P. radiata* and *T. aestivum*. Bigger deposits were found at the higher concentrations of Cu NPs. Using TEM-EDS and EDS-scanning technique, the high Cu concentration in deposits was demonstrated, although only a small portion of NPs could be transported within the plant from roots to shoots.

In the field experiments, the common wetland plants, *Phragmites australis* and *Iris pseudoacorus,* were grown in Cu^{2+}-contaminated soil (Manceau et al. 2008). Using synchrotron microanalysis, Cu grains 5–20 μm in size were observed in the rhizosphere of *P. australis* on the root surface in the sites of its contact with the fungal mycelium and in the cortical parenchyma, but not in the central vascular cylinder of the root. In *I. pseudoacorus,* Cu spots were detected by XRD in the zone of root hairs and mycorrhizal hyphae and, using EXAF spectroscopy, were identified as Cu^0. X-ray diffraction confirmed that Cu grains were the aggregates of NPs; some of them were represented by individual NPs. The size of Cu^0 NPs was evidently within the range of 1.0–15.0 nm. The authors concluded that copper was reduced biotically, evidently with the involvement of organic molecules as templates to control the shape and size of metallic nanoparticles. They believe that this newly identified mode of copper biomineralization by plant roots (and mycorrhizal hyphae) under copper stress would prevent copper from entering the food chain.

Thus, Cu^0 NPs in plants could arise in the process of Cu^{2+} or Cu^+ reduction within the plants or penetrate from environment. In spite of the scarcity of experimental data, an increased toxicity of Cu^0 ENPs (in comparison with Cu ions) seems rather likely. In the connection with predicted global production of ENPs, the expanded and deepened analysis of their action on plants is required.

8.6 Conclusion

The analysis of presented data concerning the mechanisms of copper detoxification and the maintenance of its homeostasis in plant cells at its excessive content in environment allows some generalizations.

It should be first noted that a possibility of metallic copper (Cu^0) presence in the plant was until now essentially outside researcher attention. In recent years, it became evident that Cu^{2+} reduction to Cu^0 in the plant cells is not only allowable theoretically but, in some cases, experimentally proven, e.g., metallic copper is detected in plant cells. It is also established that Cu^0 particles have nanosizes. It is of especial importance that copper nanoparticles were observed not only in the apoplast but also in the cytoplasm. A rather limited material obtained in this field did not allow unambiguous answer to the question whether the formation of copper nanoparticles is a novel way of excess Cu detoxification or, in contrast, these particles represent a real threat to plant life. It is undoubtedly that this field of research will develop violently in the nearest years.

The second important aspect of the problem of Cu detoxification and the maintenance of its homeostasis is an essential absence of free Cu ions in the plant cell cytoplasm (Changela et al. 2003). This is a consequence of the functioning in plants of very efficient systems of its chelating and compartmentation. Currently, a great progress is achieved in studying Cu detoxification mechanisms operating within the cells (in symplast). Indeed, well-grounded experimental publications and comprehensive reviews destined to this problem appear regularly. Our experiments showed that, at early steps of Cu excess detoxification in the plant cell cytosol, both metallothioneins and phytochelatins evidently play an important role. A good coordination between functioning of these two groups of Cu high-molecular ligands was also demonstrated.

Of considerable value is also the notion that, along symplastic, its apoplastic pool, which is characteristic only for plants, is also involved in Cu detoxification. These two pools differ fundamentally in the mechanisms used for the restriction of Cu ion bioavailability, which primarily follows from differences in the types of Cu chemical bonds dominating in these two compartments. In recent years, due to the development of new methods for Cu quantification in various cell compartments, including quantitative microscopy, vast information is accumulated concerning the apoplast participation in Cu immobilization, which allows a discussion of a possible role of this phenomenon in Cu detoxification.

The role of the apoplast in Cu excess detoxification attracted the attention of researchers since the 1970s. The idea was put forward that HM association with plant cell walls is a primary mechanism of tolerance because Cu accumulation in the cell wall hinders its penetration into more sensitive cell metabolic sites in the root. Despite the alternative view is known (Ernst et al. 1992), the materials presented earlier show that, in some plants, up to 60–92% of total copper of the root cells was localized in the apoplast. This shows a necessity of further intense investigations of the cell wall role in Cu retention.

It should be noted that even under ordinary Cu content in medium, cell walls immobilize a great amount of Cu (up to 43–47% of total Cu content in leaves and roots). Plant growth in the presence of Cu excess results not only in the increased proportion of Cu in the cell walls but also in the substantial increase (up to 100-fold) in the Cu concentration (Nishizono et al. 1987). The important conclusion is that usually a potential capacity of cell walls relative to Cu absorption is used only insignificantly. It is also worth mentioning that a comparison of two plant species differing in their tolerance to Cu showed that Cu retention by the cell walls in the roots and leaves of more tolerant species was higher that in less tolerant species.

In all likelihood, improved tolerance due to enhanced Cu immobilization in the apoplast could be realized at the initial stage of growing root penetration into Cu-enriched soil. This response disappears later, which could indicate a gradual transfer of Cu, initially held by the cell wall, inside the cell.

It is not excluded that plant growth in Cu-enriched medium could initiate an increase in the cell wall CEC. This is indicated by some researchers detecting changes in the composition and structure of the cell walls as a result of long-term action of Cu excess. In particular, it was established that Cu induced an increase in the contents of pectin and hemicelluloses in the cell walls, changes in their fraction composition, accumulation of uronic acids, which occurred only in species tolerant to Cu, and also an increased content of lignin and suberin (Llugany et al. 2003; Konno et al. 2005; McKenna et al. 2010).

At the same time, in some cases an improved plant tolerance to excess Cu in medium may be provided otherwise than by the expansion of its apoplastic pool.

Thus, at the evaluation of the ways of Cu detoxification under natural conditions, the contribution of AMF was clearly undervalued. An increased Cu tolerance of the plants colonized by AMF is explained by the effects of Cu entrapping or filtration. This important positive AMF effect is evidently determined by the much higher cell wall CEC values characteristic of fungal hyphae in comparison with plant cell walls. Accepting the urgency of the problem of purification of HM-contaminated territories and keeping in mind the absence of plants promising for phytoremediation of Cu, more attention should be paid to cultivation of AMF-colonized plants on Cu-enriched soils.

Thus, the apoplastic pool of copper includes a great, sometimes dominating part of Cu absorbed by the plant, especially at its excess in environment. Chemical and structural changes occurring in the cell wall under the influence of Cu excess (and also under the influence of AMF) allow consideration of the cell wall as a substantial component in the system of plant adaptation to HM action.

Acknowledgments The authors are grateful to Professor Nella L. Klaychko from the Institute of Plant Physiology (Moscow, Russia) for valuable discussions and interesting ideas. This work was partially supported by the grants of the Russian Foundation for Basic Research (No. 10-04-00799_a and No. 10-04-90456-ukr_a) and by the Program of the Presidium of RAS "Molecular and Cell Biology."

References

Andres-Colas N, Sancenon V, Rodriguez-Navarro S, Mayo S, Thiele DJ, Ecker JR, Puig S, Penarrubia L (2006) The Arabidopsis heavy metal P-type ATPase HMA5 interacts with metallochaperones and functions in root copper detoxification. Plant J 45:225–236

Arru L, Regnoni S, Baroncini M, Bonatti PM, Perata P (2002) Copper localization in *Cannabis sativa* L. grown in copper-rich solution. Euphytica 140:33–38

Bačkor M, Loppi S (2009) Interactions of lichens with heavy metals. Biol Plant 53:214–222

Bernal M, Sanches-Testillano P, Risueno MD, Yruela I (2006) Excess copper induces structural changes in cultured photosynthetic soybean cells. Funct Plant Biol 33:1001–1012

Bernal M, Cases R, Picorel R, Yruela I (2007) Foliar and root supply affect differentially Fe- and Zn- uptake and photosynthetic activity in soybean plants. Environ Exp Bot 60:145–150

Branquinho C, Brown DH, Catarino F (1997) The cellular location of Cu in lichens and its effects on membrane integrity and chlorophyll fluorescence. Environ Exp Bot 38:165–179

Carpita N, Tierney M, Campbell M (2001) Molecular biology of the plant cell wall: searching for the genes that define structure, architecture and dynamics. Plant Mol Biol 47:1–5

Carvalho LM, Cacador I, Martins-Loucao MA (2006) Arbuscular mycorrhizal fungi enhance root cadmium and copper accumulation in the roots of the salt marsh plant *Aster tripolium* L. Plant Soil 285:161–169

Changela A, Chen K, Xue Y, Holschen J, Outten CE, O'Halloran TV, Mondragorn A (2003) Molecular basis of metal-ion selectivity and zeptomolar sensitivity by CueR. Science 301: 1383–1387

Chen BD, Zhu Y-G, Duan J, Xiao XY, Smith SE (2007) Effects of the arbuscular mycorrhizal fungus *Glomus mosseae* on growth and metal uptake by four plant species in copper mine tailings. Environ Pollut 147:374–380

Cobbett C, Goldsbrough P (2002) Phytochelatins and metallothioneins: roles in heavy metal detoxification and homeostasis. Annu Rev Plant Biol 53:159–182

Cohu CM, Pilon M (2010) Cell biology of copper. In: Hell R, Mendel RR (eds) Cell biology of metals and nutrients, vol 17, Plant cell monographs. Heidelberg, Berlin

De Vos CHR, Schat H, De Waal MAM, Vooijs R, Ernst WHO (1991) Increased resistance to copper-induced damage of the root cell plasmalemma in copper tolerant *Silene cucubalis*. Physiol Plant 82:523–528

Ernst WHO, Verkleij JAC, Schat H (1992) Metal tolerance in plants. Acta Bot Neerl 41:229–248

Faucon M-P, Shutcha MN, Meerts P (2007) Revising copper and cobalt concentrations in supposed hyperaccumulators from SC Africa: influence of washing and metal concentrations in soil. Plant Soil 301:29–36

Faucon M-P, Colinet G, Mahy G, Ngongo M, Verbruggen N, Meerts P (2009) Soil influence on Cu and Co uptake and plant size in the cuprophytes *Crepidorhopalon perennis* and *C. tenius* (Scrophulariaceae) in SC Africa. Plant Soil 317:201–212

Ferrol N, Gonzalez-Guerrero M, Valderas A, Benabdellah K, Azcon-Aguilar C (2009) Survival strategies of arbuscular mycorrhizal fungi in Cu-polluted environments. Phytochem Rev 8:551–559

Fleischer A, O'Neill MA, Ehwald R (1999) The pore size of non-graminaceous plant cell walls is rapidly decreased by borate ester cross-linking in the pectic polysaccharide rhamnogalacturonan II. Plant Physiol 121:829–838

Fritioff A, Greger M (2006) Uptake and distribution of Zn, Cu, Cd, and Pb in an aquatic plant *Pomatogeton natans*. Chemosphere 63:220–227

Gonzalez-Chavez C, D'Haen J, Vangronveld J, Dodd JC (2002) Copper sorption and accumulation by the extraradical mycelium of different Glomus spp. (arbuscular mycorrhizal fungi) isolated from the same polluted soil. Plant Soil 240:287–297

González-Chávez MC, Carrillo-González R, Wright SF, Nichols KA (2004) The role of glomalin, a protein produced by arbuscular mycorrhizal fungi, in sequestering potentially toxic elements. Environ Pollut 130:317–323

González-Guerrero M, Melville L-H, Ferrol N, Azcon-Aguilar C, Peterson R-L (2008) Ultrastructural localization of heavy metals in the extraradical mycelium and spores of the arbuscular mycorrhizal fungus *Glomus intraradices*. Can J Microbiol 54:103–110

Gonzalez-Mendoza D, Moreno AQ, Zapata-Perez O (2007) Coordinated responses of phytochelatin synthase and metallothionein genes in black mangrove, *Avicennia germinans*, exposed to cadmium and copper. Aquat Toxicol 83:306–314

Guo W-J, Meetam M, Goldsbrough PB (2008) Examining the specific contributions of individual Arabidopsis metallothioneins to copper distribution and metal tolerance. Plant Physiol 146:1697–1706

Hall JL, Williams E (2003) Transition metal transporters in plants. J Exp Bot 54:2601–2613

Harrison SJ, Lepp NW, Phipps DA (1979) Uptake of copper by excised roots. II. Copper desorption from the free space. Z Pflanzenphysiol 49:27–34

Haverkamp RG, Marshall AT (2009) The mechanism of metal nanoparticle formation in plants: limits on accumulation. J Nanopart Res 11:1453–1463

Hildebrandt U, Regvar M, Bothe H (2007) Arbuscular mycorrhiza and heavy metal tolerance. Phytochemistry 68:139–146

Ivanova EM, Kholodova VP, VlV K (2010) Biological effects of high copper and zinc concentrations and their interaction in rapeseed plants. Russ J Plant Physiol 57:864–873

Jha AK, Prasad K, Prasad K, Kulkarni AR (2009) Plant system: nature's nanofactory. Colloids Surf B Biointerfaces 73:219–223

Joner EJ, Briones R, Leyval C (2000) Metal-binding capacity of arbuscular mycorrhizal mycelium. Plant Soil 226:227–234

Ke W, Xiong Z, Xie M, Luo Q (2007) Accumulation, subcellular localization and ecophysiological responses to copper stress in two *Daucus carota* L. populations. Plant Soil 292: 291–304

Kholodova VP, Volkov KS, Abdeeva AR, Kusnetzov VlV (2011) Water relations in *Mesembryanthemum crystallinum* as affected by heavy metal stress. Environ Exp Bot 71:382–389

Konno H, Nakato T, Nakashima S, Katoh K (2005) *Lygodium japonicum* fern accumulates copper in the cell wall pectin. J Exp Bot 56:1923–1931

Krämer U, Clemens S (2006) Functions and homeostasis of zinc, copper, and nickel in plants. In: Tamás MJ, Martinoia E (eds) Molecular biology of metal homeostasis and detoxification, vol 14, Topics in current genetics. Heidelberg, Berlin

Krämer U, Talkea IN, Hanikenneb M (2007) Transition metal transport. FEBS Lett 581: 2263–2272

Kulikova AL, Kuznetsova NA, Kholodova VP (2011) Copper excesses influence to root cell vitality and root morphology of soybean plants. Russ J Plant Physiol 58:836–843

Lee J, Adle D, Kim H (2006) Molecular mechanisms of copper homeostasis in yeast. In: Tamás MJ, Martinoia E (eds) Molecular biology of metal homeostasis and detoxification, vol 14, Topics in current genetics. Heidelberg, Berlin, pp 1–36

Lee W-M, An Y-J, Yoon H, Kweon H-S (2008) Toxicity and bioavailability of copper nanoparticles to the terrestrial plants mung bean (*Phaseolus radiatus*) and wheat (*Triticum aestivum*): plant agar test for water-insoluble nanoparticles. Environ Technol Chem 27:1915–1921

Liu DH, Kottke I (2004) Subcellular localization of copper in the root cells of *Allium sativum* by electron energy loss spectroscopy (EELS). Bioresour Technol 94:53–158

Llugany M, Lombini A, Poschenrieder C, Dinelli E, Barcelo J (2003) Different mechanisms account for enhanced copper resistance in *Silene armeria* ecotypes from mine spoil and serpentine sites. Plant Soil 251:55–63

Lou L-q, Shen Z-g, Li X-d (2004) The copper tolerance mechanisms of *Elsholtzia haichowensis*, a plant from copper-enriched soils. Environ Exp Bot 51:111–120

Ma X, Geiser-Lee J, Deng Y, Kolmakov A (2010) Interactions between engineered nanoparticles (ENPs) and plants: phytotoxicity, uptake and accumulation. Sci Total Environ 408:3053–3061

MacFarlane GR, Burchett MD (2000) Cellular distribution of copper, lead and zinc in the grey mangrove *Avicennia marina* (Forks.). Vierh Aquat Bot 68:45–69

Manceau A, Nagy KL, Marcus MA, Lanson M, Geoffroy N, Jacquet T, Kirpichtchikova T (2008) Formation of metallic copper nanoparticles at the soil-root interface. Environ Sci Technol 42:1766–1772

Mari S, Lebrun M (2006) Metal immobilization: where and how? In: Tamás MJ, Martinoia E (eds) Molecular biology of metal homeostasis and detoxification, vol 14, Topics in current genetics. Heidelberg, Berlin, pp 273–298

McKenna BA, Kopittke PM, Wehr B, Blamey FPC, Menzies NW (2010) Metal ion effects on hydraulic conductivity of bacterial cellulose-pectin composites used as plant cell wall analogs. Physiol Plant 138:205–214

Meychik NR, Yermakov IP (2001) Ion exchange properties of plant root cell walls. Plant Soil 234:181–193

Meychik NR, Yermakov IP, Khonarmand SD, Nikolaeva YuI (2010) Ion-exchange properties of cell walls in chickpea cultivars with different sensitivities to salinity. Russ J Plant Physiol 57:620–630

Monnet F, Bordas F, Deluchat V, Chatenet P, Botineau M, Baudu M (2005) Use of the aquatic lichen *Dermatocarpon luridum* as bioindicator of copper pollution: accumulation and cellular distribution tests. Environ Pollut 138:455–461

Murphy A, Taiz L (1995) Comparison of metallothionein gene expression and nonprotein thiols in ten Arabidopsis ecotypes correlation with copper tolerance. Plant Physiol 109:945–954

Naftel SJ, Martin RR, Macfie SM, Courchesne F, Seguin V (2007) An investigation of metals at the soil/root interface using synchrotron radiation analysis. Can J Anal Sci Spect 52:18–24

Nair R, Varghese SH, Nair BG, Maekawa T, Yoshida Y, Kumar DS (2010) Nanoparticulate material delivery to plants. Plant Sci 179:154–163

Navarro E, Baun A, Behra R, Hartmann NB, Filser J, Miao A-J, Quigg A, Santschi PH, Sigg L (2008) Environmental behavior and ecotoxicity of engineered nanoparticles to algae, plants, and fungi. Ecotoxicology 17:372–386

Neumann D, zur Nieden U, Lichtenberger O, Leopold I (1995) How does *Armeria maritima* tolerate high heavy metal concentrations? J Plant Physiol 146:704–717
Nishizono H, Ichikawa H, Suziki S, Ishii F (1987) The role of root cell wall in the heavy metal tolerance of *Athrium yokoscense*. Plant Soil 101:15–20
Nowack B, Bucheli TD (2007) Occurrence, behavior and effects of nanoparticles in the environment. Environ Pollut 150:5–22
Ouzounidou G, Chiamporova M, Moustakas M, Karatsglis S (1995) Responses of maize (*Zea mays* L.) plants to copper stress. I. Growth, mineral content and ultrastructure of roots. Environ Exp Bot 35:167–176
Ovecka M, Lang I, Baluska F, Ismail A, Illes P, Lichtscheidl IK (2005) Endocytosis and vesicle trafficking during tip growth of root hairs. Protoplasma 226:39–54
Panou-Filotheou H, Bosabalidis AM (2004) Root structural aspects associated with copper toxicity in oregano (*Oreganum vulgare* sudsp. *hirtum*). Plant Sci 166:1497–1504
Peng H-Y, Yang XE, Tian SK (2005) Accumulation and ultrastructural distribution of copper in *Elsholzia splendens*. J Zhejiang Univ 6B:311–318
Purin S, Rillig MS (2008) Immunolocalization of glomalin in the mycelium of the arbuscular mycorrhizal fungus *Glomus intraradices*. Soil Biol Biochem 40:1000–1003
Ranathunge K, Steudle E, Lafitte R (2005) A new precipitation technique provides evidence for the permeability of Casparian bands to ions in young roots of corn (*Zea mays* L.) and rice (*Oryza sativa* L.). Plant Cell Environ 28:1450–1462
Russo M, Sgherri C, Izzo R, Navari-Izzo F (2008) *Brassica napus* subjected to copper excess: phospholipases C and D and glutathione system in signaling. Environ Exp Bot 62:238–246
Sattlemacher B (2001) The apoplast and its significance for plant mineral nutrition. New Phytol 149:167–192
Schäfer HJ, Greiner S, Rausch T, Haag-Kerwer A (1997) In seedlings of the heavy metal accumulator *Brassica juncea* Cu^{2+} deferentially affects transcript amounts for γ-glutamylcysteine synthetase (γ-ECS) and metallothionein (MT2). FEBS Lett 404:216–220
Shah V, Belozerova I (2009) Influence of metal nanoparticles on the soil microbial community and germination of lettuce seeds. Water Soil Pollut 192:143–148
Shi JY, Wu B, Yuan XF, Cao YY, Chen XC, Chen YX, Hu TD (2008) An X-ray absorption spectroscopy investigation of speciation and biotransformation of copper in *Elsholtzia splendens*. Plant Soil 302:163–174
Stampoulis D, Sinha SK, White JC (2009) Assay-dependent phytotoxicity of nanoparticles to plants. Environ Sci Technol 43:9473–9479
Sudova R, Doubkova P, Vosatka M (2008) Mycorrhizal association of *Agrostis capillaries* and *Glomus intraradices* under heavy metal stress: combination of plant clones and fungal isolates from contaminated and uncontaminated substrates. Appl Soil Ecol 40:19–29
Thakkar KN, Mhatre SS, Parikh RY (2010) Biological synthesis of metallic nanoparticles. Nanomedicine 6:257–262
Wehr JB, Blamey FPC, Menzies NW (2010) Comparison between methods using copper, lanthanum, and colorimetry for the determination of the cation exchange capacity of plant cell walls. J Agric Food Chem 58:4554–4559
Wei L, Luo C, Li X, Shen Z (2008) Copper accumulation and tolerance in *Chrysanthemum coronarium* L. and *Sorghum sudanense* L. Arch Environ Contam Toxicol 55:230–246
Yang MJ, Yang XE, Roemheld V (2002) Growth and nutrient composition of *Elsholtia splendens* Nakai under copper toxicity. J Plant Nutr 25:1359–1375
Yruela I (2009) Copper in plants: acquisition, transport and interactions. Funct Plant Biol 36: 409–430
Zhang X-H, Lin A-J, Gao Y-L, Reid RJ, Wong M-H, Zhu Y-G (2009) Arbuscular mycorrhizal colonization increases copper binding capacity of root cell walls of *Oryza sativa* L. and reduces copper uptake. Soil Biol Biochem 41:930–993

Chapter 9
Arsenic Tolerance and Detoxification Mechanisms in Plants

Dharmendra K. Gupta, Sudhakar Srivastava, H.G. Huang, Maria C. Romero-Puertas, and Luisa M. Sandalio

9.1 Introduction

Arsenic (As) is a widely distributed metalloid in the earth's crust with an average concentration of 2 mg kg^{-1}. Inorganic arsenic is a class 1 carcinogen. There is widespread chronic As poisoning in regions of Asia and elsewhere, due to the consumption of drinking water with geo-genically elevated As content, and the situation is worst in the densely populated flood plains and deltas of South and Southeast Asia (Tripathi et al. 2007; Zhao et al. 2010a). In addition, intake of As from food, mainly rice, which is the most efficient crop plants accumulating As in grains (Zhao et al. 2009, 2010a) and constitutes the major portion of diet for people in Southeast Asia.

Arsenic can exist in four valence states: −3, 0, +3, and +5. Arsenic salts exhibit a wide range of solubility depending on pH and ionic environment. Under reducing conditions, arsenite (As^{III}) is the dominant form, while in oxygenated environments arsenate (As^{V}) is generally the stable form. Arsenate is an analogue of phosphate and thus enters the plant system through phosphate transporters, which has been experimentally proven through various physiological and electrophysiological

D.K. Gupta (✉) • M.C. Romero-Puertas • L.M. Sandalio
Departamento de Bioquimica, Biologia Cellular y Molecular de Plantas, Estacion Experimental Del Zaidin, CSIC, Apartado 419, Granada 18008, Spain
e-mail: guptadk1971@gmail.com

S. Srivastava
Nuclear Agriculture and Biotechnology Division, Bhabha Atomic Research Centre, Mumbai 400085, India

H.G. Huang
Ministry of Education Key Laboratory of Environmental Remediation and Ecosystem Health, Zhejiang University, Hangzhou 310029, China

I. Sherameti and A. Varma (eds.), *Detoxification of Heavy Metals*, Soil Biology 30, DOI 10.1007/978-3-642-21408-0_9, © Springer-Verlag Berlin Heidelberg 2011 studies demonstrating inhibition of arsenate uptake in the presence of phosphate (Abedin et al. 2002; Srivastava et al. 2007) and is strengthened by recent reports demonstrating that *A. thaliana* mutants defective in phosphate transport are more tolerant to arsenate (Caterecha et al. 2007). In contrast, As^{III} exists as a neutral molecule [$As(OH)_3$] at the prevailing pH and redox conditions of the environment, and hence, its uptake occurs via aquaglyceroporin channels of nodulin 26-like intrinsic protein family, which transport various neutral molecules including salicylic acid and boric acid (Isayenkov and Maathuis 2008; Zhao et al. 2009). Inside the plant cell, As^{V} interferes with essential cellular processes, such as oxidative phosphorylation and ATP synthesis in mitochondria by replacing phosphate moiety, while the toxicity of As^{III} is due to its propensity to bind to sulfhydryl groups, with consequent detrimental effects on general protein functioning (REF actual). Arsenic toxicity is also mediated by oxidative stress by overproducing reactive oxygen species (ROS) and altering antioxidant defenses in plant tissues, which can be detected by oxidative modification of lipids and DNA (Singh et al. 2006; Lin et al. 2008a). Exogenous supply of nitric oxide prevented the arsenic-dependent oxidative stress in tall fescue leaves (Jin et al. 2010). Overproduction of antioxidants, CuZn-SOD, and ascorbate peroxidase in transgenic tall fescue plants confers tolerance to arsenic exposure and some other metals, which demonstrates the relevance of oxidative stress in the toxicity mechanisms imposed by these metals (Lee et al. 2007).

Pteris vittata and *Oryza sati*va are two interesting cases as they are much more efficient in As accumulation than non-accumulator ferns and cereals, respectively (Williams et al. 2007; Xie et al. 2009). The relatively high As accumulation in rice occurs due to the fact that anaerobic conditions empowered by submerged paddy soils keep As predominantly in the more bioavailable form As^{III} and that rice is one of the most efficient accumulators of Selenium, whose transporters of NIP (nodulin intrinsic protein) family also transport As(III) (Ma et al. 2008).

Pteris possesses efficient uptake systems for As^{V}/As^{III} uptake, for their transport to shoots, and for their final sequestration in the vacuoles in addition to the fact that it is hypertolerant to As (Su et al. 2008; Xie et al. 2009; Indriolo et al. 2010). For this reason, *Pteris* has been used as a model plant for analyzing hyperaccumulation mechanisms for prospective generation of high transgenic biomass, while rice has been used to understand how to repress the As accumulation into grains for prospective development of safe rice cultivar in near future (Tripathi et al. 2007). In other front, *Holcus lanatus* is a classical example of a hypertolerant plant, which can grow in mining areas at highly elevated As concentrations through lowering the uptake of arsenate due to suppression of high-affinity phosphate/arsenate uptake systems (Meharg and Macnair 1992; Hartley-Whitaker et al. 2001; Meharg and Hartley-Whitaker 2002).

In addition, various other aquatic and terrestrial plants are being analyzed to comprehensively understand the mechanistic details of As uptake and detoxification to achieve well-funded and extremely charming scenarios for the betterment of our lives.

9.2 Mechanisms of Arsenic Tolerance and Accumulation

To date, knowledge about the mechanisms of As tolerance/accumulation can be put into the following categories:

1. **Lowering the rate of uptake of As**:
   One way of achieving tolerance to As is through decreasing the rate of As uptake infiltration. Several plants, which have adapted to grow preferentially in heavily As-contaminated soils, such as *Holcus lanatus* and *Cytisus striatus*, achieve As hypertolerance by constitutive suppression of high-affinity phosphate/As^{V} transport (Meharg and Macnair 1992; Bleeker et al. 2003). Such a constitutive suppression of the activity of high-affinity phosphate transporters leads to significant decrease in the rate of As uptake as compared to non-adapted plants. The phosphate/arsenate transporter has a higher affinity for phosphate, and if external phosphate status is high, phosphate will be taken up more effectively compared to arsenate. Hence, non-resistant plants can increase their resistance to arsenate by raising their phosphorus status, which in turn leads to lower levels of arsenate accumulation through suppression of phosphate/arsenate uptake (Meharg et al. 1994). A correlation between phosphate transporters and the level of As accumulation was also established by analysis of *Arabidopsis thaliana pho2* mutants, which accumulates high P concentrations in shoots (Delhaize and Randall 1995), flowered earlier than wild-type plants and uptake more in its reproductive biomass. Same type of result was also reported for *H. lanatus* (Fitter et al. 1998). *PHO2* encodes an ubiquitin-conjugating E2 enzyme 24, which acts as a negative regulator of phosphate-starvation-induced genes including phosphate transporters. Hence, in *pho2* mutant, expression of phosphate transporters remains higher (Lin et al. 2008b) which gives higher As accumulation than in wild-type plants (Quaghebeur and Rengel 2004). Screening 108 recombinant inbred lines of rice Bala × Azucena mapping population revealed the presence of a major gene, AsTol, on chromosome 6. This AsTol gene was found to have similarity with the phosphorus uptake QTL in another population of rice, providing circumstantial evidence for a mechanism involving the behaviour of arsenate as a phosphate analogue (Dasgupta et al. 2004).
2. **Transformation and volatilization of accumulated As:**
   As mentioned before, several studies have provided a link between arsenate resistance phenotype and reduced/suppressed phosphate uptake. However, such a mechanism of lowered uptake rate has not yet been discovered in case of As^{III}. Nevertheless, other mechanisms of reducing total As accumulation by effluxing/volatilizing As has been found to occur in bacteria and yeast and recent reports suggest towards existence of these mechanisms in plants as well. S-adenosylmethionine-dependent methyltransferase is used by various bacteria, fungi, and mammals to convert arsenite to the gaseous trimethylarsine (TMA) (Messens and Silver 2006). Qin et al. (2009) isolated two genes encoding As(III) methylases (CmarsM7 and CmarsM8) from the eukaryotic alga *Cyanidioschyzon merolae* and demonstrated the capacity of the purified enzymes to convert As^{III} to TMA.

Furthermore, the expression of CmarsM genes, as well as its homologue arsM from soil bacterium *Rhodopseudomonas palustris*, in arsenite hypersensitive *E. coli* resulted in production of TMA and improved tolerance to As (Qin et al. 2006, 2009). Natural occurrence of methylated arsenic species in rice and other plants has been reported previously (Zhu et al. 2008; Williams et al. 2005). Recently, Norton et al. (2008) found that putative As(III) methylase gene gets up regulated in response to arsenate in rice during its transcriptomics analysis. In submerged plants collected from Moira River and Moira Lake, Canada, a variety of organic arsenic compounds, including simple methylated compounds (methylarsonic acid and dimethylarsinic acid), trimethylarsine oxide, and tetramethylarsonium cation were detected at trace levels. An unknown compound, most probably an arsenosugar, was also detected in two submerged plants, *Ceratophyllum demersum* and *Elatine triandra* (Zheng et al. 2003). These studies positively supported the idea that As methylation may be a common mechanism of tolerance against inorganic As since organic methylated As species are generally less toxic than inorganic ones. However, if As methylation in plants leads to final volatilization or not is yet to be clearly demonstrated.

3. **Efflux of accumulated As.**
The efflux of accumulated As back to the medium in the form of As^{III} is another strategy employed by the plants to avoid toxic responses (Logoteta et al. 2009; Zhao et al. 2010b; Liu et al. 2010b). In *Lycopersicon esculentum* and *Oryza sativa*, Xu et al. (2007) analyzed As^{V} reduction in roots and As^{III} efflux to the medium and found that roots of both plants rapidly reduced arsenate to arsenite and effluxed some part of arsenite to the medium. Similarly, both non-tolerant and tolerant ecotypes of *H. lanatus* have almost similar capacity to efflux 80–100% of the arsenate taken up to the medium as arsenite within 2–24 h (Logoteta et al. 2009). Hence, As^{III} efflux is not adaptively enhanced in the tolerant phenotype *H. lanatus*, but it appears to be a basal tolerance mechanism to greatly decrease cellular As burden in both phenotypes. Zhao et al. (2010b) studied the role of the rice silicon transporters, Lsi1 (OsNIP2;1), which is a major route for As^{III} entry into rice roots, but can also participate in As efflux. Since aquaporins are bidirectional channels mediating transport for metal/metabolite in concentration gradient manner. It was found that during 24 h exposure to 5 μM As^{V}, rice roots extruded As^{III} to the external medium rapidly, accounting for 60–90% of the As^{V} uptake, whereas a *lsi1* mutant extruded significantly less arsenite than the wild-type rice (Zhao et al. 2010b). Investigation of tolerance mechanisms in aquatic fern *Azolla* demonstrated operation of similar strategies (Zhang et al. 2008). Comparative evaluation of the highest (*Azolla caroliniana*) and lowest (*Azolla filiculoides*) As-accumulating ferns suggested that variable accumulation was due to differential rate of arsenate influx rate and due to this *A. filiculoides* was more resistant to external arsenate. Both strains, however, showed a similar degree of tolerance to internal As and effluxed more arsenate than arsenite, and the amount of As efflux was proportional to the amount of As accumulation.

4. **Effective reduction of arsenate to arsenite.**
For the above-discussed mechanisms (volatilization and effluxing) and following complexation mechanism, reduction of arsenate to arsenite is one of the basic and important steps. Thus, differential capacity for reduction of As^{V} to As^{III} may also affect plants' potential to tolerate the As^{V} stress. This point was experimentally proven by Bleeker et al. (2006) who suggested that next to decreased uptake rate of As^{V}, enhanced expression of arsenate reductase might act as an additional determinant of As^{V} hypertolerance and As transport in tolerant ecotypes of *H. lanatus.* They also demonstrated that enhanced based sequestration in As^{V} hypertolerant *H. lanatus* (Hartley-Whitaker et al. 2002) is not due to enhanced capacity of phytochelatin synthesis but due to increased As^{V} reductase activity (Bleeker et al. 2006). The overexpression of arsenate reductase in *Arabidopsis* and Holcus was found to increase, though variably, the tolerance of t plants to As^{V}.

9.3 Mechanisms of Arsenic Detoxification

9.3.1 Chelation of Arsenite with Glutathione and Phytochelatins

One of the most studied mechanisms of As detoxification is its chelation as As^{III} with glutathione and PCs. Only a few plants are known which can suppress phosphate transport systems and there are even few candidates, where As methylation and volatilization would occur to significant levels. The efflux is also recently discovered and it is still uncertain to what extent it is prevalent among plants. However, the synthesis of metal or metalloid binding peptides like GSH and PCs is observed in almost all plants studied to date and leads to immediate detoxification (in terms of loss of reactivity in chelated form) of an specific ion. Plants exposed to As substantially increase the synthesis of GSH and PCs (Gupta et al. 2004; Grill et al. 2006; Srivastava et al. 2007) and augmented PC synthesis has been observed in non-tolerant, and non-accumulator, as well as in hypertolerant and hyperaccumulator plants (Grill et al. 2006; Tripathi et al. 2007; Gupta et al. 2008). Importance of sulphur metabolism in As detoxification is also demonstrated by the fact that As exposure leads to stress on sulphur metabolism. In *Hydrilla,* an increase in sulphur supply enhances accumulation of As and tolerance (Srivastava and D'Souza 2009, 2010). The complexation of As with GSH and PCs has been experimentally demonstrated in various plants, such as *Cicer arietinum*, *Rauvolfia serpentina*, *Holcus Lanatus*, *Pteris Cretica*, *Helianthus annuus*, and *Brassica juncea* (Gupta et al. 2004, 2008; Tripathi et al. 2007). Tolerance to As is enhanced by increased thiol synthesis in transgenic plants overexpressing the genes of thiol (cysteine, GSH, or PC) biosynthesis pathway (Tripathi et al. 2007; Gasic and Korban 2007; Wojas et al. 2008, 2010). An arsenic-accumulating, hypertolerant *Brassica*, *Isatis capadocica* (Karimi et al. 2009), collected from Iranian arsenic-contaminated mine had PC-based tolerance ($>50\%$ As complexed with PCs) rather than through suppression of high-affinity phosphate/arsenate root transport. These findings provide conclusive evidence that thiols, particularly PCs, play a crucial role in As detoxification. However, although studies point to the essential role of PCs in both constitutive and adaptive tolerance to As (Schat et al. 2002), essentially PCs do not contribute significantly towards As tolerance phenotype in hypertolerant (*H. lanatus* and *Silene paradoxa*) and hyperaccumulator plants (*P. vittata* and *P. cretica*), since the amount of PCs synthesized is too low in these cases (Raab et al. 2004; Arnetoli et al. 2008). Other studies also confirm that non-hyper accumulator plants like *Helianthus annuus, Ceratophyllum demersum*, and *Hydrilla verticillata* (Raab et al. 2005; Srivastava et al. 2007, 2010; Mishra et al. 2008) rely more on PC-based detoxification mechanisms.

9.3.2 Restricting Translocation of Arsenic from Root to Shoot Through Efficient Chelation in Roots

Complexation of arsenite [As(III)] with phytochelatins (PCs) is an important mechanism employed by plants to detoxify As; how this complexation affects As mobility is yet little known. In a recent study, high ICP-MS and ESI-MS coupled to HPLC were employed to identify and quantify As(III)–thiol complexes and free thiol compounds in *Arabidopsis* exposed to As^{V} (Liu et al. 2010). In wild-type plants roots, 69% of As was complexed as As(III)-PC4, As(III)-PC3, and As(III)-(PC2)2, while both the GSH-deficient mutant cad2-1 and the PC-deficient mutant cad1-3 were more sensitive to As^{V} than the wild type, and complexation of As was only 25 and 8%, respectively. The two mutants had a greater As mobility from root to shoot and higher As^{III} efflux from roots to the medium than the wild-type plants. Furthermore, when wild-type plants were exposed to L-buthionine sulfoximine (GSH biosynthesis inhibitor) or deprived of sulphur, both As^{III} efflux and root-to-shoot translocation were enhanced. The authors suggested that complexation of As^{III} with PCs in *Arabidopsis* roots decreases both the efflux to the external medium and the translocation to root to shoot (Liu et al. 2010).

9.3.3 Sequestration of Arsenic into Root or Shoot Vacuoles Either in the Form of Complex or as Ions

The complexes As-GSH-PCs are supposed to be sequestered in the vacuoles (Bleeker et al. 2006), although there is, as yet, no in vivo evidence in plants. Recent research on hyperaccumulating and tolerant species showed that vacuoles of *P. vittata, P. cretica*, and *H. lanatus* contain large concentrations of inorganic As^{III} (Raab et al. 2004). This suggests that the majority of stored As is not complexed, and that plants might have vacuolar As^{III} transporters, possibly with similarities to

Table 9.1 Arsenic hypertolerant and hyperaccumulator plants identified to date

Name of the plant	References
Hyperaccumulators	
Pteris vittata	Ma et al. (2001)
Pteris umbrosa	Koller et al. (2007)
Pteris aspericaulis, P. cretica var. nervosa, *P. fauriei, P. multifida, P. multifida, P. serrulata, P. oshimensis*	Wang et al. (2007)
Pityrogramma calomelanos	Francesconi et al. (2002)
Pteris biaurita L., *P. quadriaurita, P.ryukyuensis*	Srivastava et al. (2006)
Pteris cretica	Meharg (2003)
Myriophyllum propinquum, Eigeria densa, Ceratophyllum demersum, Rorippa nasturtium-aquaticum	Robinson et al. (2006) and references therein
Hypertolerant/Arsenic excluder	
Bidens pilosa	Sun et al. (2009)
Andropogon scoparius, Agrostis castellana, A. delicatula, A. capillaris, Deschampsia cespitosa, Silene vulgaris, Calluna vulgaris, Plantago lanceolata	See references in Meharg and Hartley-Whitaker (2002)
Cytisus striatus	Bleeker et al. (2003)
Silene paradoxa	Arnetoli et al. (2008)
Isatis capadocica	Karimi et al. (2009)
Holcus lanatus	Macnair and Cumbes (1987)

bacterial As^{III} extrusion pumps. This point has been recently proven in *P. vittata*, where Indriolo et al. (2010) isolated and characterized ACR3 and ACR3;1 genes, which encode proteins similar to the ACR3 arsenite effluxes of yeast in vacuolar membrane. PvACR3 is able to rescue the As-sensitive phenotypes of yeast deficient for ACR3, while knocking down the expression of ACR3 in the gametophyte resulted in an As^{III}-sensitive phenotype, indicating that ACR3 plays a necessary role in arsenic tolerance in the gametophyte. However, this transporter is missing in flowering plants, which is why most of the As hyperaccumulators discovered to date belong to Pteridaceae (lower plants) (Table 9.1).

9.4 Conclusion

In conclusion, the As tolerance and detoxification mechanisms may be summarized in a sequential flow as follows: (1) Restrict As inflow at very first stage from medium to the roots, whatever amount of As enters the roots. (2) Efflux the most of this accumulated As back to the medium, whatever of As remains in root cells. (3) Restrict translocation of the metalloid from root to shoot through its complexation and sequestration in vacuoles, either complexing it there with GSH and PCs or transforming inorganic As species to other non-toxic organic species and volatilize it into the atmosphere. A complete diagrammatic representation of As detoxification mechanism has been given in Fig. 9.1.

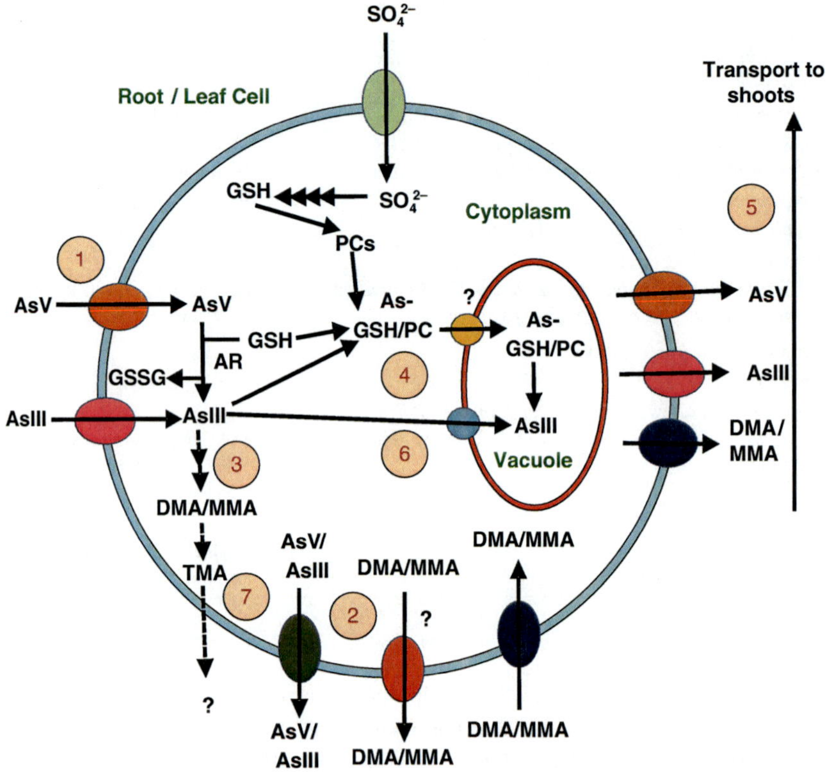

Fig. 9.1 Known and postulated mechanisms of arsenic detoxification in plants. In general, plants possess only a few of the available mechanisms for achieving arsenic tolerance; however, they may also employ other mechanisms to some extent. Here those mechanisms are proposed to occur in following sequence: **1**. rate of arsenate uptake can be reduced through constitutive suppression of high-affinity phosphate/arsenate uptake systems; however, no such mechanism of restricted uptake is known for arsenite, DMA, and MMA; **2**. whatever amount of arsenic enters the cells, it can be effluxed either directly as arsenate or after reduction as arsenite; DMA and MMA are known to enter root cells and probably they may also be effluxed but that is still not known; whatever amount of arsenic remains in the root cells, **3**. it can either be converted to less toxic organic arsenic forms, or **4**. transported to vacuoles as arsenite itself or as arsenite–glutathione–phytochelatin complexes so that its translocation to shoots is prevented; **5**. the level of arsenic transport to shoots can also be reduced through decreasing its entry into xylem stream; whatever amount of arsenic reached leaves can **6**. either be sequestered in vacuoles as in the roots or **7**. can be removed to atmosphere in the volatile organic form through a series of reactions. The still unknown and only hypothetical features have been marked with a *question mark* (?). Abbreviations used: *AsV* arsenate; *AsIII* arsenite; *AR* arsenate reductase; *DMA* dimethylarsinic acid; *MMA* monomethylarsinic acid; *GSH* reduced glutathione; *GSSG* oxidized glutathione; *PCs* phytochelatins

References

Abedin MJ, Feldmann J, Meharg AA (2002) Uptake kinetics of arsenic species in rice plants. Plant Physiol 128:1120–1128

Arnetoli M, Vooijs R, ten Bookum W, Galardi F, Gonnelli C, Gabbrielli R, Schat H, Verkleij JAC (2008) Arsenate tolerance in *Silene paradoxa* does not rely on phytochelatin- dependent sequestration. Environ Pollut 152:585–591

Bleeker PM, Schat H, Vooijs R, Verkleij JAC, Ernst WHO (2003) Mechanisms of arsenate tolerance in *Cytisus striatus*. New Phytol 157:33–38

Bleeker PM, Hakvoort HWJ, Bliek M, Souer E, Schat H (2006) Enhanced arsenate reduction by a CDC25-like tyrosine phosphatase explains increased phytochelatin accumulation in arsenate-tolerant *Holcus lanatus*. Plant J 45:917–929

Caterecha P, Segura MD, Franco-Zorrilla JM, Garcia-Ponce B, Lanza M, Solano R, Paz-Ares J, Leyva A (2007) A mutant of the *Arabidopsis* phosphate transporter PHT1;1 displays enhanced arsenic accumulation. Plant Cell 19:1123–1133

Dasgupta T, Hossain SA, Meharg AA, Price AH (2004) An arsenate tolerance gene on chromosome 6 of rice. New Phytol 163:45–49

Delhaize E, Randall PJ (1995) Characterization of a phosphate-accumulator mutant of *Arabidopsis thaliana*. Plant Physiol 107:207–213

Fitter AH, Wright WJ, Williamson L, Belshaw M, Fairclough J, Meharg AA (1998) The phosphorus nutrition of wild plants and the paradox of arsenate tolerance: does leaf phosphate concentration control flowering? In: Lynch JP, Deikman J (eds) Phosphorus in plant biology: regulatory roles in molecular, cellular, organismic and ecosystem processes. American Society of Plant Physiologists, Washington, DC, pp 39–51

Francesconi K, Visootiviseth P, Sridokchan W, Goessler W (2002) Arsenic species in an arsenic hyperaccumulating fern, *Pityrogramma calomelanos:* a potential phytoremediator of arsenic-contaminated soils. Sci Total Environ 284:27–35

Gasic K, Korban SS (2007) Transgenic Indian mustard (*Brassica juncea*) plants expressing an *Arabidopsis* phytochelatin synthase (AtPCS1) exhibit enhanced As and Cd tolerance. Plant Mol Biol 64:361–369

Grill E, Mishra S, Srivastava S, Tripathi RD (2006) Role of phytochelatins in phytoremediation of heavy metals. In: Singh SN, Tripathi RD (eds) Environmental bioremediation technologies. Springer, Heidelberg, pp 101–146

Gupta DK, Tohoyama H, Joho M, Inouhe M (2004) Changes in the levels of phytochelatins and related metal binding peptides in chickpea seedlings exposed to arsenic and different heavy metal ions. J Plant Res 117:253–256

Gupta DK, Tripathi RD, Mishra S, Srivastava S, Dwivedi S, Rai UN, Yang XE, Huang H, Inouhe M (2008) Arsenic accumulation in roots and shoots vis-à-vis its effects on growth and level of phytochelatins in seedlings of *Cicer arietinum* L. J Environ Biol 29:281–286

Hartley-Whitaker J, Ainsworth G, Vooijs R, Ten Bookum W, Schat H, Meharg AA (2001) Phytochelatins are involved in differential arsenate tolerance in *Holcus lanatus*. Plant Physiol 126:299–306

Hartley-Whitaker J, Woods C, Meharg AA (2002) Is differential phytochelatin production related to decreased arsenate influx in arsenate tolerant *Holcus lanatus*? New Phytol 155:219–225

Indriolo E, Na G, Ellis D, Salt DE, Banks JA (2010) A vacuolar arsenite transporter necessary for arsenic tolerance in the arsenic hyperaccumulation fern *Pteris vittata* is missing in flowering plants. Plant Cell 22:2045–2057

Isayenkov SV, Maathuis FJM (2008) The *Arabidopsis thaliana* aquaglyceroporin AtNIP7; 1 is a pathway for arsenite uptake. FEBS Lett 582:1625–1628

Jin JW, Xu YF, Huang YF (2010) Protective effect of nitric oxide against arsenic-induced oxidative damage in tall fescue leaves. Afr J Biotechnol 9:1619–1627

Karimi N, Ghaderian SM, Raab A, Feldmann J, Meharg AA (2009) An arsenic accumulating, hyper tolerant brassica, *Isatis capadocica*. New Phytol 184:41–47

Koller CE, Patrick JWP, Rose RJ, Offler CE, MacFarlane GR (2007) *Pteris umbrosa* R. Br. as an arsenic hyperaccumulator: accumulation, partitioning and comparison with the established As hyperaccumulator *Pteris vittata*. Chemosphere 66:1256–1263

Lee SH, Ahsan N, Lee KW, Kim DH, Lee DG, Kwak SS, Kwon SY, Kim TH, Lee BH (2007) Simultaneous overexpression of both CuZn superoxide dismutase and ascorbate peroxidase in transgenic tall fescue plants confers tolerance to a wide range of abiotic stress. J Plant Physiol 164:1626–1638

Lin A, Zhang X, Zhu YG, Zhao FJ (2008a) Arsenate-induced toxicity: effects on antioxidative enzymes and DNA damage in *Vicia faba*. Environ Toxicol Chem 27:413–419

Lin SI, Chiang SF, Lin WY, Chen JW, Tseng CY, Wu PC, Chiou TJ (2008b) Regulatory network of microRNA399 and *PHO2* by systemic signalling. Plant Physiol 147:732–746

Liu WJ, Wood BA, Raab A, McGrath SP, Zhao FJ, Feldmann J (2010) Complexation of arsenite with phytochelatins reduces arsenite efflux and translocation from roots to shoots in *Arabidopsis*. Plant Physiol 152:2211–2221

Logoteta B, Xu XY, Macnair MR, McGrath SP, Zhao FJ (2009) Arsenite efflux is not enhanced in the arsenate-tolerant phenotype of *Holcus lanatus*. New Phytol 183:340–348

Ma LQ, Komar KM, Tu C, Zhang W, Cai Y, Kennelley ED (2001) A fern that hyperaccumulates arsenic. Nature 409(6820):579

Ma JF, Yamaji N, Mitani N, Xu XY, Su YH, McGrath SP, Zhao FJ (2008) Transporters of arsenite in rice and their role in arsenic accumulation in rice grain. Proc Natl Acad Sci USA 105:9931–9935

Macnair MR, Cumbes Q (1987) Evidence that arsenic tolerance in *Holcus lanatus* L. is caused by an altered phosphate uptake system. New Phytol 107:387–394

Meharg AA (2003) Variation in arsenic accumulation-hyperaccumulation in ferns and their allies. New Phytol 157:25–31

Meharg AA, Hartley-Whitaker J (2002) Arsenic uptake and metabolism in arsenic resistant and non-resistant plant species. New Phytol 154:29–43

Meharg AA, Macnair MR (1992) Suppression of the high-affinity phosphate uptake system: a mechanism of arsenate tolerance in *Holcus lanatus* L. J Exp Bot 43:519–524

Meharg AA, Naylor J, Macnair MR (1994) Phosphorus nutrition of arsenate tolerant and non tolerant phenotypes of velvet grass. J Environ Qual 23:234–238

Messens J, Silver S (2006) Arsenate reduction: thiol cascade chemistry with convergent evolution. J Mol Biol 362:1–17

Mishra S, Srivastava S, Tripathi RD, Trivedi PK (2008) Thiol metabolism and antioxidant system complement each other during arsenate detoxification in *Ceratophyllum demersum* L. Aquat Toxicol 86:205–215

Norton GJ, Lou-Hing DE, Meharg AA, Price AH (2008) Rice–arsenate interactions in hydroponics: whole genome transcriptional analysis. J Exp Bot 59:2267–2276

Qin J, Rosen BP, Zhang Y, Wang GJ, Franke S, Rensing C (2006) Arsenic detoxification and evolution of trimethylarsine gas by a microbial arsenite S-adenosylmethionine methyltransferase. Proc Natl Acad Sci USA 103:2075–2080

Qin J, Lehr CR, Yuan CG, Le XC, McDermott TR, Rosen BP (2009) Biotransformation of arsenic by a Yellowstone thermo acidophilic eukaryotic alga. Proc Nat Acad Sci USA 106:5213–5217

Quaghebeur M, Rengel Z (2004) Arsenic uptake, translocation and speciation in pho1 and pho2 mutants of *Arabidopsis thaliana*. Physiol Plant 120:280–286

Raab A, Feldmann J, Meharg AA (2004) The nature of arsenic-phytochelatin complexes in *Holcus lanatus* and *Pteris cretica*. Plant Physiol 134:1113–1122

Raab A, Schat H, Meharg AA, Feldmann J (2005) Uptake, translocation and transformation of arsenate and arsenite in sunflower (*Helianthus annuus*): formation of arsenic-phytochelatin complexes during exposure to high arsenic concentrations. New Phytol 168:551–558

Robinson B, Kim N, Marchetti M, Moni C, Schroeter L, van den Dijssel C, Milne G, Clothier B (2006) Arsenic hyperaccumulation by aquatic macrophytes in the Taupo volcanic zone New Zealand. Environ Exp Bot 58:206–215

Schat H, Llugany M, Vooijs R, Hartley-Whitaker J, Bleeker PM (2002) The role of phytochelatins in constitutive and adaptive heavy metal tolerances in hyperaccumulator and non-hyperaccumulator metallophytes. J Exp Bot 53:2381–2392

Singh N, Ma LQ, Srivastava M, Rathisasabapathi B (2006) Metabolic adaptations to arsenic-induced oxidative stress in *Pteris vittata* L. and *Pteris ensiformis* L. Plant Sci 170:274–282
Srivastava S, D'Souza SF (2009) Increasing sulfur supply enhances tolerance to arsenic and its accumulation in *Hydrilla verticillata* (L.f.) Royle. Environ Sci Technol 43:6308–6313
Srivastava S, D'Souza SF (2010) Effect of variable sulfur supply on arsenic tolerance and antioxidant responses in *Hydrilla verticillata* (L.f.) Royle. Ecotoxicol Environ Saf 73(6):1314–1322
Srivastava M, Ma LQ, Santos JAG (2006) Three new arsenic hyperaccumulating ferns. Sci Total Environ 364:24–31
Srivastava S, Mishra S, Tripathi RD, Dwivedi S, Trivedi PK, Tandon PK (2007) Phytochelatins and antioxidant systems respond differentially during arsenite and arsenate stress in *Hydrilla verticillata* (L.f.) Royle. Environ Sci Technol 41:2930–2936
Srivastava S, Mishra S, Dwivedi S, Tripathi RD (2010) Role of thiol metabolism in arsenic detoxification in *Hydrilla verticillata* (L.f.) Royle. Water Air Soil Pollut 212:155–165
Su YH, McGrath SP, Zhu YG, Zhao FJ (2008) Highly efficient xylem transport of arsenite in the arsenic hyperaccumulator *Pteris vittata*. New Phytol 180:434–441
Sun YB, Zhou QX, Liu WT, An J, Xu ZQ, Wang L (2009) Joint effects of arsenic and cadmium on plant growth and metal bioaccumulation: a potential Cd-hyperaccumulator and As-excluder *Bidens pilosa* L. J Hazard Mater 165:1023–1028
Tripathi RD, Srivastava S, Mishra S, Singh N, Tuli R, Gupta DK, Maathuis FJM (2007) Arsenic hazards: strategies for tolerance and remediation by plants. Trends Biotechnol 25:158–165
Wang HB, Wong MH, Lan CY, Baker AJM, Qin YR, Shu WS, Chen GZ, Ye ZH (2007) Uptake and accumulation of arsenic by 11 *Pteris* taxa from southern China. Environ Pollut 145:25–233
Williams PN, Price AH, Raab A, Hossain SA, Feldmann J, Meharg AA (2005) Variation in arsenic speciation and concentration in paddy rice related to dietary exposure. Environ Sci Technol 39:5531–5540
Williams PN, Villada A, Deacon C, Raab A, Figuerola J, Greens AJ, Feldmann J, Meharg AA (2007) Greatly enhanced arsenic shoot assimilation in rice leads to elevated grain levels compared to wheat and barley. Environ Sci Technol 41:6854–6859
Wojas S, Clemens S, Hennig J, Sklodowska A, Kopera E, Schat H, Bal W, Antosiewicz DM (2008) Overexpression of phytochelatin synthase in tobacco: distinctive effects of AtPCS1 and CePCS genes on plant response to cadmium. J Exp Bot 59:2205–2219
Wojas S, Clemens S, Sklodowska A, Antosiewicz DM (2010) Arsenic response of *AtPCS1*- and *CePCS*-expressing plants-effects of external As(V) concentration on As-accumulation pattern and NPT metabolism. J Plant Physiol 167:169–175
Xie QE, Yan XL, Liao XY, Li X (2009) The arsenic hyperaccumulator fern *Pteris vittata* L. Environ Sci Technol 43:8488–8495
Xu XY, McGrath SP, Zhao FJ (2007) Rapid reduction of arsenate in the medium mediated by plant roots. New Phytol 176:590–599
Zhang X, Lin AJ, Zhao FJ, Xu GZ, Duan GL, Zhu YG (2008) Arsenic accumulation by the aquatic fern Azolla: comparison of arsenate uptake, speciation and efflux by *A. caroliniana* and *A. filiculoides*. Environ Pollut 156:1149–1155
Zhao FJ, Ma JF, Meharg AA, McGrath SP (2009) Arsenic uptake and metabolism in plants. New Phytol 181:777–794
Zhao FJ, McGrath SP, Meharg AA (2010a) Arsenic as a food chain contaminant: mechanisms of plant uptake and metabolism and mitigation strategies. Annu Rev Plant Biol 61:7.1–7.25
Zhao FJ, Ago Y, Mitani N, Li RY, Su YH, Yamaji N, McGrath SP, Ma JF (2010b) The role of the rice aquaporin Lsi1 in arsenite efflux from roots. New Phytol 186:392–399
Zheng J, Hintelmann H, Dimock B, Dzurko MS (2003) Speciation of arsenic in water, sediment, and plants of the Moira watershed, Canada, using HPLC coupled to high resolution ICP-MS. Anal Bioanal Chem 377:14–24
Zhu YG, Sun GX, Lei M, Teng M, Liu YX, Chen NC, Wang LH, Carey AM, Deacon C, Raab A, Meharg AA, Williams PN (2008) High percentage inorganic arsenic content of mining impacted and non impacted Chinese rice. Environ Sci Technol 42:5008–5013

Chapter 10
Cadmium Metal Detoxification and Hyperaccumulators

Kavita Shah

10.1 Introduction

Increasing anthropological practices have led to the addition of a highly toxic metal, cadmium, into the natural environment that affects the overall growth and metabolism of plants (Shah and Dubey 1998; Agrawal and Sharma 2006; Shah and Nongkynrih 2007). Cadmium toxicity can elicit a variety of adaptive responses that have been comprehensively reviewed by Toppi di Sanita and Gabbrielli (1999). Bioaccumulation of such toxic metals in the plants poses a risk to human and animal health (Wang et al. 2003). Removal of excess of metal ions from the contaminated site is brought about by chemical as well as biological means. In recent years, the reclamation of metal-contaminated soils through remediation measures has attracted considerable attention. In this context, hyperaccumulator plants can be used for phytoremediation of Cd-polluted soils, as they are able to extract and concentrate metals in their upper parts that can be harvested (Chaney 1983; Brooks and Robinson 1998).

The genetic basis, the physiological pathways for Cd in particular, and the adaptive significance of metal hyperaccumulation are not well understood (Pollard 2000; Pollard et al. 2002; Roosen et al. 2003). Of the four main hypotheses, namely, (1) Drought hypothesis, (2) Interference hypothesis (3), Defense hypothesis, and (4) Tolerance/Disposal hypothesis postulated by Boyd (2004) to explain the need for hyperaccumulation, the tolerance/disposal hypothesis seems to be attractive as it suggests that hyperaccumulation is a mechanism for increased tolerance that allows sequestration of metal in tissues (tolerance) or their elimination from the plant body by shedding (disposal) (Antonovics et al. 1971; Baker 1981). The mechanism of hyperaccumulation would enable plants to tolerate metals through

K. Shah (✉)
Environmental Science, Institute of Environment and Sustainable Development, Banaras Hindu University, Varanasi 221005, U.P, India
e-mail: kavitashah@bhu.ac.in

I. Sherameti and A. Varma (eds.), *Detoxification of Heavy Metals*, Soil Biology 30,
DOI 10.1007/978-3-642-21408-0_10, © Springer-Verlag Berlin Heidelberg 2011

internal detoxification (Basic et al. 2005; Shah and Nongkynrih 2007; Roelofs et al. 2008). Therefore, it should be kept in mind that metal hyperaccumulation and metal tolerance are not equivalent. There is little evidence linking hyperaccumulation with metal tolerance/disposal (Jiang et al. 2005), yet under controlled conditions, Roosen et al. (2003) and Casio (2004) have found a decrease in hyperaccumulator biomass production with increasing Cd concentration in solution and shoots. This suggests that metal hyperaccumulation though presupposes tolerance of plants toward metals is not necessarily associated with tolerance to the metal burden in the soil. The stress responses of plants to environmental pollutant Cd are widely studied since 1980. This chapter focuses primarily on the accumulation and detoxification of Cd metal in plants and gives an overview of plants that have potential as Cd hyperaccumulators and are recently reported in the literature. The chapter also deals with genetic and molecular basis of Cd metal detoxification and bioaccumulation in plants.

10.2 Cadmium in the Environment

Cadmium, an element identified in 1817 (Schwarz 1974), is present in nature in low concentrations and is normally bound to Zn, Pb, or Cu. High levels of cadmium are associated with sources of industrial emission observed as a steep increase in contamination by this metal during 1980s and beyond. Normal levels of atmospheric Cd range from 1 ng/m^3 in uninhabited areas to 40 ng/m^3 in urban centers; near industrial complexes the atmospheric concentration of Cd may reach 5 mg/m^3 (FAO-WHO 1986; Robards and Worsfold 1991; Seiler and Sigel 1988; Shah and Dubey 1995; Cabrera et al. 1998). The extent of environmental cadmium concentration and the risk this element poses to human health have stimulated interest among scientists to study the factors that contribute to cadmium contamination, chemistry of toxicity, intake, and removal. The ecotoxicity of Cd is determined mainly by the fact that this element and the compounds that contain it are not biodegradable. Indiscriminate use of phosphate containing fertilizers (contain >20 mg Cd/kg, but repeated applications over years cause accumulation) further increases the levels of atmospheric Cd, which is absorbed or retained by the soil particles and gets deposited in the earth's crust (Jackson and Alloway 1991; reviewed extensively by Cabrera et al. 1998). Dissolved Cd is readily absorbed by plants and leaches rapidly into subsoil, contaminating deep and surface waters. Cd persists in the environment and has a long biological half-life (Goyer 1988; Shah and Dubey 1998). The element Cd can remain suspended in the soil for 1–3 year; in coastal sediments its estimated half-life is 2 years and in oceanic water it can persist for 7,000 years or longer Robards and Worsfold (1991).

An excess of Cd leads to toxic symptoms in plants, such as growth retardation, inhibition of photosynthesis, induction and inhibition of enzymes, altered stomatal action and water relations, efflux of cations, and generation of free radicals (Shah and Dubey 1995; Prasad 1995; Shah et al. 2001; Nahakpam and Shah 2011).

In freshwater, Cd is available as free hydrated or carbonated ions, in seawater it exists as chlorine ion complexes, and in animal tissues it appears as Cd–metallothionein (Concon 1988; Seiler and Sigel 1988; Robards and Worsfold 1991).

Cadmium reaches foodstuff mainly through soil and plant and its levels are further enhanced by methods of production, processing, additives, packaging, cooking utensils, etc. (Reilly 1980). According to FAO-WHO standards (1986) in food of plant origin, Cd levels range from 5 to 100 ng/g, the maximum levels being reported in vegetables and grains (Concon 1988). Cd contamination is also reported in beverages, Cd-based pigments used for glazing, and recycled paper (for details, see Haguenoer and Furon 1981; Tahvonen 1996).

10.3 Cadmium Hyperaccumulator Plants

The annual worldwide release of cadmium has reached 22,000 metric tons (Singh and Jain 2003). Cadmium is considered to be carcinogenic to humans by International Agency for Research on Cancer (Beyersman and Hartwig 2008). The removal and recovery of metals from contaminated sites are of great importance in order to protect the environment as well as prevent Cd phytotoxicity (Kim et al. 2004; Shah and Nongkynrih 2007). Although Cd is generally extremely toxic for plants (Shaw et al. 2004), some plant ecotypes have developed the ability to survive in highly Cd-contaminated soils and to accumulate large amounts of Cd in their tissues (Vogel-Mikus et al. 2010). Such plants are termed as Cd hyperaccumulator. A review by Baker and Brooks (1989) listed 26 plant families that contain hyperaccumulator of different metals. Recently, Environment Canada has developed a database PHYTOREM of 775 plants with capabilities to accumulate or hyperaccumulate one or several of 19 key metallic elements (McIntyre 2003). The first hyperaccumulators characterized were members of the Brassicaceae and Fabaceae families, and to date nearly 500 plant species are reported to serve as hyperaccumulators (Reeves and Brooks 1983; Kramer 2010). Numerous studies have focused on the identification of hyperaccumulator plants and their growth, tolerance, and hyperaccumulation on highly metal-contaminated soils. A hyperaccumulator plant is defined as one that can accumulate 0.1% of dry mass of elements such as Ni, Co, or Pb, 1% dry mass of Zn and Mn, and 0.01% of dry mass of Cd (Baker and Brooks 1989; Shah and Nongkynrih 2007). The ecological role of metal hyperaccumulators is still unclear; however, it is suggested that perhaps metal accumulation provides protection against fungal and insect infection (Boyd et al. 1994).

The first field trial for Cd phytoextraction was conducted together with Zn by Baker et al. 1991, and now presently, this technology is receiving considerable attention for cleanup of soil contaminated with heavy metals (USEPA 2000). Important aspects of the novel green technology (phytoremediation) have been discussed in several comprehensive reviews that focus on the physiological and molecular basis of phytoremediation (Salt et al. 1998; Dzantor and Beauchamp

2002; Shah and Nongkynrih 2007; Padmavathiamma and Loretta 2007; Kramer 2010; Sarma 2011). *Thlaspi caerulescens* (Brassicaceae) is a Cd hyperaccumulator (defined as accumulating >100 mg Cd/kg) in shoots in the natural habitat (Brown et al. 1995; Baker et al. 2000; Lu et al. 2008); however, *Arabidopsis halleri* (Brassicaceae) and *Sedum alfredii* (Crassulaceae) are also shown to be capable of hyperaccumulating Cd under hydroponic condition (Küpper et al. 2000). Recently, there are reports of several plant species that have potential as Cd hyperaccumulators and have been listed in Table 10.1. Hongli and Wei (2009) studied 23 *Amaranth* cultivars for Cd hyperaccumulation from different ecological regions under hydroponic culture conditions and reported that *Amaranthus* cv. *Tianxigmi* is a typical Cd hyperaccumulator. Similarly, *Rorippa globosa* and *Rorippa islandica* behaved as Cd accumulating plants (Sun et al. 2000). Desert plant *Prosopis laevigata* is also shown to have a bioaccumulation factor greater than 100 for Cd, suggesting it to be a promising candidate for phytoremediation purposes (Buendia-Gongalez et al. 2010). *Thlaspi praecox* can accumulate up to 7,430 μg Cd/g dry weight in shoots without suffering toxic effects (Lombi et al. 2002; Vogel-Mikus et al. 2006, 2010). The phytoremediation potential of *Hibiscus cannabinus* L. for Cd removal has also been carried out (Banuelos et al. 1997; Bada and Raji 2010). Study by Zhu (2010) on phytoremediation of Cd-polluted soil by *Brassica chinensis* showed an increased Cd concentration inside the plant which further enhanced with the application of a chelator to the soil, implying that *B. chinensis* is a promising crop for removal of Cd from contaminated soil. All these studies infer that metal hyperaccumulators belong to a diverse range of plant families that are distributed

Table 10.1 List of Cd hyperaccumulators with a potential for phytoremediation

Plant species	References
Raphanus sativus	Baker et al. (1991)
Rorippa globosa	Sun et al. (2000)
Rorippa islandica	Sun et al. (2000)
	Küpper et al. (2000)
Arabidopsis halleri	Berts and Meerts (2003)
Pelargonium sp.	Dan et al. (2002)
Arabis gemmifera	Kubota and Takenka (2003)
Pistia stratiotes	Odjegba and Fasidi (2004)
Sedum alfredii	Xiong et al. (2004)
Thlaspi caerulescens	Bansova and Horak (2008)
Tamarix smyrnensis	Manousaki et al. (2008)
Brassica napus	Selvam and Wong (2008)
Atriplex halimus subsp. *schweinfurthii*	Nedjimi and Daoud (2009)
Arabidopsis thaliana	Saraswat and Rai (2009)
Amaranthus mangostanus	HongLi and Wei (2009)
Hibiscus cannabinus	Bada and Raji (2010)
Thlaspi praecox	Vogel-Mikus et al. (2010)
Brassica chinensis	Zhu (2010)
Prosopis laevigata	Buendia-Gongalez et al. (2010)

over areas equally diverse geographically, but possessing a common characteristic of natural enrichments for some specific element(s) (Sarma 2011).

Transgenic metal hyperaccumulator *Brassica juncea* overexpressing ATP sulphurylase has been shown to have higher tolerance to increased Cd levels (Pilon Smits et al. 1999, Van Huysen et al. 2004). Bioengineered tobacco plants tolerant to the presence of toxic levels of Cd are also known (Kawashima et al. 2004). *Arabidopsis* plants transformed with an *E. coli* gene *Znt A* that encodes for Cd/Zn transporter were observed to have improved resistance to Cd (Lee et al. 2003).

10.4 Uptake and Bioaccumulation of Cadmium

The uptake of dissolved Cd by plants is influenced by pH, soil characteristics, the type of plant species involved, the presence of other elements in the vicinity (total metal ion concentration in soil), cation-exchange capacity, organic matter content, soil texture, and interaction with other metals (Bosque et al. 1990; Jackson and Alloway 1991). Table 10.2 shows bioaccumulation potential of some Cd hyperaccumulators. Cd concentration in root of cucumber was found higher (Moreno-Caselles et al. 2000) than for other heavy metal such as Ni and Cr (Moral et al. 1994). Specificity of accumulation is also reflected by different plant families (Baker 1981). Low Cd accumulations have been observed in Fabaceae, moderate in Poaceae, Cucurbitaceae, and Apiaceae, and high in Chenopodiaceae, Brassicaceae, Solanaceae, and Asteraceae. Hydroponic culture studies have demonstrated the ability of *Brassica juncea* to accumulate significant amounts of cadmium in its tissues, with bioaccumulation coefficients (ratio of metal concentration in plant to concentration of metal in soil) of up to 175 in shoots and 20,574 in the roots when

Table 10.2 Bioaccumulation potential of some Cd hyperaccumulators

Plants	Bioaccumulation	References
Thlaspi caerulescens	522 mg/kg	Basic et al. (2005)
Myriophyllum heterophyllum	21.46 μg/g	Sivaci et al. (2008)
Potamogeton crispus	49.09 μg/g	Sivaci et al. (2008)
Arabis paniculata	1,127 mg/kg (in shoots)	Zeng et al. (2009)
Atriplex halimus sudsp. *schweinfurthii*	606.51 μg/g DW	Nedjimi and Daoud (2009)
Sedum alfredii	3,100 mg/kg DW (in roots)	Yang et al. (2004), Zhou and Qiu (2005)
Rorippa globosa	218.9 μg/g DW	Sun et al. (2000)
Thlaspi praecox	71,000 μg/g DW (in seeds)	Vogel-Mikus et al. (2010)
Amaranthus mangostanus	260 mg/kg (in shoots)	HongLi and Wei (2009)
Prosopis laevigata	8,176 mg/kg (in shoots) 21,437 mg/kg (in roots)	Buendia-Gongalez et al. (2010)

grown at a nonphytotoxic Cd concentration (Duskenhov et al. 1995; Salt et al. 1995). Dicots have been found to absorb more Cd metals than monocots (Sharma and Agrawal 2005). In this review, Sharma and Agrawal (2005) also proposed that metals accumulate more in reproductive than in vegetative stage. Studies on Cd hyperaccumulation and reproductive traits in natural hyperaccumulator *Thlaspi caerulescens* populations also suggest that Cd hyperaccumulation induces a better nutrient uptake and transfer, leading thereby to more shoot, fruits, and heavier seeds (Basic et al. 2005).

10.4.1 Cadmium Uptake by Absorption

The absorption of cadmium by plants is dependent upon the mechanism of soil contamination, its subsequent distribution in different profiles, the extension of the root system and structure, physiological and metabolic characteristics of the plant (interactions with other ions), and the relations between cadmium and micro-organisms in the soil water (McLaughlin et al. 1996). It is reported that some microorganisms are able to form organic Cd (Yannai and Berdicevsky 1995). Presence of phosphate or calcium and processes, namely, exudation, volatization, leaf drop, or leaching can alter absorption (Martin and Coughtrey 1982; Concon 1988). Biotic factors such as temperature, salinity, and drought also influence the bioavailability of Cd (Shah and Nongkynrih 2007; Shah and Nahakpam 2011). Cadmium absorption by plants is also partially regulated by the soil Cd/Zn ratio (Xue and Harrison 1991).

10.4.2 Transporter Mediated Uptake of Cd in Plants

Once the metal ions enter the roots, they can either be stored or exposed to the shoot. Metal transport to the shoot primarily takes place through the xylem. Cd loading into the xylem sap of *Brassica juncea* displays biphasic saturation kinetics (Salt et al. 1995), suggesting that xylem loading of metal ions is facilitated by specialized membrane transport processes. Movement of metal ions, particularly Cd, in xylem vessels appears to be mainly dependent on transpiration-driven mass flow (Salt et al. 1995).

In the last decade, our understanding of metal ion transporters in plant cells has been greatly advanced. Some of the known Cd metal transporters have been listed in Table 10.3. It is believed that Cd uptake by non-accumulator plants represents opportunistic transport via cation channels for Ca and Mg or via a carrier for other divalent cations such as Zn, Cu, or Fe (Welch and Norvell 1999). In the hyperaccumulator *T. caerulescens*, Cd was proposed to enter the cell via either a high-affinity uptake system for Fe (Lombi et al. 2002; Vert et al. 2002) or a low-affinity system for Ca or Zn uptake (Baker et al. 1994; Escarré et al. 2000;

Table 10.3 Some Cd-metal ion transporters in plants

Localization	Transporter	Metal ions
	ZIP1-4	Zn, Cd
	ZNT1	Zn, Cd
	IRT1	Fe, Mn, Zn, Cd
	AtVramp1/3/4	Fe, Cd
Cytosol	LCT1	Cd, Ca
	At MRP	Cd-PC
	HMT1	Cd-PC
Vacuole	CAX2	Cd

Roosen et al. 2003; Casio 2004; Molitor et al. 2005; Zha et al. 2004), depending on the population studied. The differences in Cd uptake found between populations suggested a high-affinity transporter for Cd in specific populations (Lombi et al. 2001; Zha et al. 2004). It is often assumed that Cd and other heavy metals without a biological function are taken up by transporters for essential elements because of a lack of specificity (Lombi et al. 2002). Dan and coworkers (2000) showed that a multiple metal hyperaccumulator *Pelargonium* sp. can tolerate and accumulate multiple metals including Cd and still maintain normal metabolic process. The multiple metal accumulation and metal transfer factors from soil (TFS) have been reported in three wild macrophytic species, namely, *Ipomea* sp., *Eclipta* sp., and *Marsilea* sp. where Gupta et al. (2008) recorded *Ipomea* sp. and *Marsilea* sp. to be associated with TFS for Cd metal. There are reports that Cd activates the signaling pathways, namely, the mitogen activating protein kinase (MAPK) and jasmonic acid pathways in plants (Yeh et al. 2004) involving transcription factors bZIP, MYB, and MYC involved in metal transport (Roelofs et al. 2007).

Various plant metal transporters are known including plasma membrane localized ZIP1-4, ZNT1, IRT1, COPT1, AtVramp1/3/4, and LCT1, and vacuolar AtMRP, HMT1, CAX2, ABC type, ZAT, and RAN1 located in Golgi bodies. Manipulations of these transporters to achieve removal of metal ions from the cell hold great potential (Tong et al. 2004; Shah et al. 2011).

IRT1 is isolated from *Arabidopsis* and its transcription is induced in *Arabidopsis* roots by iron starvation, which makes this transporter a likely candidate for mediating Fe(II) uptake from soil. IRT1 shows a broad substrate range and also transport Mn, Zn, and possibly Cd (Korshunova et al. 1999). LAC1 mediates uptake of Cd in plants (Clemens et al. 1999). ZIP transporters 1–3 confer Zn uptake activity (Grotz et al. 1988; Guerinot and Eide 1999). Characterization of new metal transporters in *Medicago truncatula* with high similarities with the ZIP family (MtZIP) revealed their functions as metal transporters (Lopez-Millan et al. 2004).

The natural resistance associated macrophage proteins (Nramp) family of transporters has been recently characterized from rice and Arabidopsis. Based on sequence comparison, the family is divided into two classes of transporters. One consists of AtNramp1 and OsNramp1 and the other of AtNramp2-5 and OsNramp2. AtNramp3 is involved in Cd uptake. Disruption of this gene enhanced Cd tolerance, whereas its overexpression led to Cd hypersensitivity in the above plants

(Curie et al. 2001). The expression of another transporter LCT1 that enhances uptake of Cd elevates the protective action of calcium against Cd toxicity in tobacco (Clemens et al. 1998; Antosiewicz and Henning 2004). The HMT1 is a protein with similarity to ABC-type transporters in *Schizosaccharomyces pombe* which is localized in the vacuolar membrane and mediates Mg-ATP, vanadate-inhibitable transport of PC–Cd complexes, and apo-PC (Ortiz et al. 1995). Similar activities have been observed in the tonoplast of the oat roots, indicating the operation of an HMT1-like mechanism in plant cells (Salt and Rauser 1995). Since very few HMT1 sequences are known in plants, AtMRPs are considered as probable transporters of PC–Cd across the tonoplast (Rea et al. 1998).

YCF1 is a MgATP-energized vacuolar transporter responsible for sequestration of compounds after their S-conjugation with glutathione from *S. cerevisiae* (Tommasini et al. 1998). Overexpression of the YCF1 gene in *A. thaliana* exhibited a fourfold higher rate of glutathione–Cd uptake in YCF-1 transgenics than those of wild-type plants, indicating that its expression strongly increases Cd transport (Song et al. 2003).

The cation diffusion facilitator (CDF) family of transporters reported in plants (Paulsen and Saier 1997; Van der Zaal et al. 1999) has Zn and Cd ions as their substrates. The protein ZRC1 from *S. cerevisiae* when overexpressed confers Zn/Cd tolerance in these plants (Kamizono et al. 1989; Conklin et al. 1992). ZRC1 is located in the tonoplast, suggesting its role in metal sequestration (Li and Kaplan 1998). CAX2 transporter characterized from *Arabidopsis* is considered to be a high-affinity, high-capacity H^+/metal cation antiporter and can transport Cd (Hirschi et al. 1996, 2000). Howden et al. (1995) isolated a cad1 mutant of *Arabidopsis thaliana* sensitive to Cd ions and deficient in its ability to form Cd–PC complexes. In view of the fact that *Arabidopsis* has only single pathway for PC synthesis, the finding of the Cd-sensitive mutant (impaired in PC synthesis) when challenged to Cd confirmed the role of PC in Cd detoxification and the authors concluded that *CAD1* gene is likely the structural gene for PC synthase (Howden et al. 1995).

10.5 Biomass Production and Biochemical Responses of Cd Hyperaccumulator Plants Toward Elevated Cd Levels

Cadmium is a non-redox metal that is strongly phytotoxic and causes growth inhibition and even plant death (Shah and Dubey 1998). An altered lipid profile, enzyme activities associated with membranes (such as enhanced lipid peroxidation, H^+-ATPase), and production of reactive oxygen species (ROS) are some of the very common biochemical responses reported under Cd stress (Ouariti et al. 1997; Fodor et al. 1995; Shah et al. 2001, 2004). Cd also affects the photosynthetic apparatus (Siedlecka and Baszynski 1993), decreases chlorophyll content, and inhibits the stomatal regulations (Barcelo and Poschenreider 2002). Cd-induced reduction of photosynthetic pigment was recorded in two species, namely, *M. heterophyllum* and

P. crispus. Cd hyperaccumulator *Atriplex halimus* subsp. *schweinfurthii* was sensitive to high Cd and resulted in reduction of chlorophyll pigments, stomatal transpiration, and root hydraulic conductivity (Nedjimi and Daoud 2009).

A study on the ROS metabolism during Cd hyperaccumulation at 400–600 μmol/L $CdCl_2$ by Zhang and Qiu (2007) in hyperaccumulator plant *Sedum alfredii* suggested that the accumulation of ROS induced by Cd treatment might be involved in Cd hyperaccumulation. Similar to published work of Shah et al. 2001 and Olmos et al. 2003, Cd enhanced superoxide anion generation and H_2O_2 accumulation. The group concluded that the ROS scavenging enzyme peroxidase plays an important role in *S. alfredii* during the process of Cd hyperaccumulation. ROS accumulated at a certain concentration could function as a signal mediating adaptive response to Cd stress such as enhancing the expression of genes responsible for encoding antioxidant enzymes, namely, *CAT1* (catalase), *cAPX* (sscorbate peroxidase), and *GR1* (glutathione reductase). Studies on the effect of cadmium on molybdate containing hydroxylases in *Phragmites australis* suggested activation of aldehyde oxidase and xanthine dehydrogenase enzyme so as to confer Cd resistance to these plants. It must be noted that *P. australis* is adapted to specific environments including the presence of metal contaminants as Cd (Van 1991) and is considered a plant with high detoxification and phytoremediation potential (Jean and De 1997).

In Cd hyperaccumulators, the biomass production depends on the concentration of Cd and duration of exposures. Selvam and Wong (2008) showed that the biomass had a negative correlation with Cd concentration in Cd hyperaccumulator *Brassica napus*. Suitable levels of Cd have been shown to stimulate biomass production in *A. panniculata* (Tang et al. 2009) and *Oryza sativa* (Shah and Dubey 1995).

10.6 Chelation and Detoxification of Cadmium in Plants

Plants respond to Cd metal toxicity via a number of mechanisms (Cobbett 2000). Figure 10.1 shows the general response of plants to cadmium metal, oxidative burst, stress-recognition, signal transduction, and the altered biological events involved therein. One of the most documented aspects of plant response to Cd metal is the synthesis of certain compounds that help in sequestering Cd ions or metal detoxification. Diverse strategies are involved in detoxification process. Cadmium metal detoxification by buffering cytosolic metal concentrations is brought about by chelators, whereas chaperones specifically deliver metal ions to organelles and metal requiring proteins (see review by Shah and Nongkynrih 2007). In plants, the principal classes of metal chelators and sequesters include phytochelatins, metallothioneins, organic acids, and amino acids. Chelate-assisted transport of metal to shoots appears to occur in the xylem via the transpiration stream. The metal moves to shoots as a metal–chelate complex (as discussed in Sect. 10.4) where water evaporates and the metal–chelate complex remains. In this way, after chelate-assisted induction the plant becomes a wick, which drives chelated metal from the soil solution into the leaves. The operation of the wick

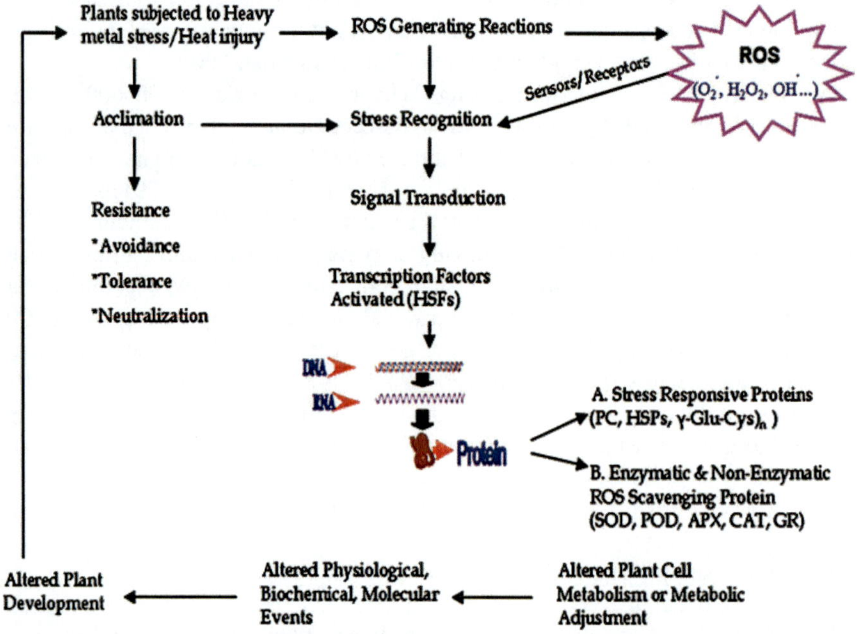

Fig. 10.1 General response of plants to cadmium metal, oxidative burst, stress-recognition, signal transduction, and the altered biological events involved therein

relies on the high-surface area collection system provided by the roots and by the efficient capillary plumbing system inside the plant (Salt et al. 1998). Plasma membrane also plays a prominent role either by reducing the heavy metal Cd uptake or stimulates efflux pumping of Cd metal that enters the cytosol (Hall 2002). Ectomycorrhiza are also effective in reducing the metal burden in plants (Marschner 1995).

10.6.1 Metallothioneines

Metallothioneines (MTs) are gene-encoded, low-molecular-weight Cys-rich polypeptides (Robinson et al. 1993) which are induced by and have high affinity for copper. The presence of large number of cystein residue in MTs binds a variety of heavy metals through mercaptide bonds (Cobbett and Goldsbrough 2002). A large number of MT genes and proteins have been identified and reported in plants (Cobbett and Goldsbrough 2002). MTs provide protection against Cd toxicity in animals (Klassen et al. 1999); however, their role in plants still remains unclear. Shah and Dubey (1998) reported an 18 kDa cadmium-binding protein complex in rice, which was rich in Cys and could have implications in Cd accumulation and tolerance in rice; nevertheless, targeted gene disruption studies are needed for confirmatory results.

10.6.2 Phytochelatins

Phytochelatins (PCs) are small metal-binding peptides found in plants and are well documented in the literature (Grill et al. 1986a; Mehra and Winge 1988; Meuwly et al. 1995; Klapheck et al. 1994; Chen et al. 1997). PC formation uses glutathione (Grill et al. 1989), homoglutathione, hydroxymethyl-glutathione (Klapheck et al. 1995), or γ-glutamylcysteine (Hayashi et al. 1991). It is catalyzed by phytochelatin synthase (PCS), a constitutive enzyme requiring posttranslational activation by heavy metals and/or metalloids, in particular Cd, Ag, Pb, Cu, Hg, Zn, Sn, As, and Au both in vivo and in vitro (Grill et al. 1987; Grill et al. 1989; De Knecht et al. 1995; Klapheck et al. 1995; Maitani et al. 1999; Chen et al. 1997; Wojcik and Tukiendorf 2005). Iso-PCs, a series of PC-like homologous chelating peptides, are reported with varying terminal amino acids such as alanine (Grill et al. 1986b), serine (Klapheck et al. 1994), glutamic acid (Meuwly et al. 1995), and glutamine (glu; Maitani et al. 1999) and have a C-terminal modified residue other than glycine (gly; Gekeler et al. 1989).

Peptides lacking the C-terminal amino acid are reported by Bernand and Kagi (1987) in *Zea mays*. These have the formula (γ-glu-cysteine)*n* and are called desGly-PC. The PC and iso-PC molecules form complexes with heavy metals such as Cd. Low-molecular-mass cytosolic PC–metal complexes are then transported into the vacuole where high-molecular complexes are formed with incorporation of sulfur (Maitani et al. 1999; Shah and Dubey 1998).

In vitro experiments have shown that a series of metal-sensitive plant enzymes can tolerate a 10- to 1,000-fold concentration of Cd in the form of a PC complex than as free radical ion (Kneer and Zenk 1992). In *B. juncea*, it has been observed that the accumulation of Cd is followed by synthesis of PC, sufficient enough to chelate Cd (Heiss et al. 1999). PCs reactivate metal-poisoned plant enzymes such as nitrate reductase up to 1,000-fold better than chelators. Several review papers have concluded that PCs played an important role in Cd tolerance in higher plants (Rauser 1990; Cobbett 2000); however, some differ (De Knecht et al. 1992; Schat et al. 2002). Zhang et al. (2010) proposed that in hyperaccumulator *S. alfredii* more than 60% Cd localized in the leaf cell wall and only 5% Cd entered the cytoplasm, activated PCS to synthesize PCs, which then acted as the major intracellular detoxification mechanism.

10.6.3 Organic Acids

Organic acids as amino acids and carboxylic acids are suggestive potential ligands for chelation, owing to the capacity of metal ions to react with S, N, and O (Clemens

2001). Citrate, malate, and oxalate have been implicated in a range of processes, including differential metal tolerance, metal transport through xylem, and vacuolar metal sequestration (reviewed in Rauser 1995). Citric acid has been hypothesized to be a major Cd ligand at low Cd concentrations (Wagner 1993). Xylem cell walls have a high cation exchange capacity; therefore, they are expected to retard severely the upward movement of metal cation. Therefore, noncationic metal chelate complexes, such as Cd citrate, should be transported more efficiently in the transpiration stream (Senden et al. 1990). Extended X-ray absorbance fine structure (EXAFS) analysis showed that Cd in the xylem sap of *B. juncea* was chelated by oxygen or nitrogen atoms, suggesting the involvement of organic acids in Cd translocation (Salt et al. 1995). These workers reported that there appeared no evidence for sulfur coordination of Cd, confirming that phytochelatins and other thiol-containing ligands play no direct role in Cd transport in the xylem. In contrast, a recent study using EXAFS on complexation of Cd in seeds and vegetative tissues of Cd hyperaccumulator *Thlaspi praecox* conducted by Vogel-Mikus and group (2010) revealed that in intact seeds and isolated embryos, almost two thirds of the Cd ligands were thiol groups (Cd–S–C–). In addition, there was coordination to phosphate groups via bridging oxygen (Cd–O–P–). The phosphate ligand was absent in vegetative tissues. In roots and shoots, 80% of the Cd ligands were oxygen ligand provided by the cell walls and by organic acids stored in vacuoles. Iron chelate nicotianamine and the free amino acid histidine have also been implicated in chaperoning metals during the transport process in *A. halleri* (Becher et al. 2004; Weber et al. 2004).

10.6.4 Ectomycorrhiza

Ectomycorrhizal symbionts associated with plant roots play specific role in modulating metal stress in plants (Schutzendubel and Polle 2002). Mycorrhiza usually adopt the principle of metal exclusion like absorption of metal by fungal hyphae, non-accessibility to apoplast, and chelation, thereby restricting the movement of metals into the roots. Ectomycorrhizal fungi *Pisolothus tincotrius* (Tam 1995) and *Paxillus involutus* (Blaudez et al. 2000) have been shown to tolerate Cd and play a role in Cd-metal amelioration.

10.7 Genes Involved in Cd Metal Hyperaccumulation and Detoxification in Plants

It has been demonstrated that tolerance to high levels of metals by hyperaccumulators is under genetic control (McNair 1993), which allows the plant to produce specific molecules as discussed above that react with metals to form complexes which can then be stored away from sensitive tissues (Dzantor and Beauchamp 2002). From all the above discussions, it can be hypothesized that the genes associated with metal tolerance and detoxification in hyperaccumulator plants would be classified according to the type of molecule that they employ for metal hyperaccumulation or detoxification. Therefore, for the convenience of the readers, the genes related to the above phenomenon in plants have been categorized into genes involved in:

1. Induction of metal transporters
2. Formation of metal-detoxifying chelators (PCs and MTs)
3. Sulfur metabolism

Table 10.4 lists the genes reported in the literature that encode for Cd-metal detoxification and hypertolerance processes in plants. Quantitative mRNA in situ hybridization (QISH) in *Thlaspi caerulescens* shows that transporter gene expression changes during cadmium hyperaccumulation (Küpper and Kochian 2010). Cd induced changes in cellular expression of ZIP family of transition metal transporters ZNT1, ANT5, and MTP1, suggesting that they are possibly a part of acclimatization process in plants for Cd toxicity (Sarma 2011). Synthesis of

Table 10.4 Genes encoding for Cd-metal detoxification and hypertolerance processes in plants

Gene	Plant	References
(i) Cd-metal transporter genes *ZnT1*	*T. caerulescens*	Pence et al. (2000) Lombi et al. (2002) Becher et al. (2004)
Nicotianamine synthase (*NAS 2&3*) genes	*A. halleri*	Weber et al. (2004)
IRT1 (Iron transporter genes)	*T. caerulescens*	Bernard et al. (2004)
FRO2	*T. caerulescens*	Bernard et al. (2004)
HMA4 (heavy metal ATPase gene)	*T. caerulescens*	Bernard et al. (2004)
Znt A (encodes for Cd/Zn transporter)	*Arabidopsis*	Lee et al. (2003)
Ospdr9 (encodes for (PDR)-type ABC transporter)	*O. sativa*	Moons (2003)
(ii) Synthesis of metal-detoxifying chelators		
ATP-PRT1 gene involved in histidine biosynthesis	*A. thaliana*	Ingle et al. (2005)
(iii) Over-expression of sulfur metabolism enzymes		
Serine-acetyltransferase	*Thlaspi*	Freeman et al. (2004) Pilon Smits et al. (1999)
ATP sulphurylase	*Brassica juncea*	Van Huysen et al. (2004)
(iv) PC biosynthesis genes		
Gsh1(encodes for GCS/GSH biosynthesis)	*S. pombe*	Cobbett et al. (1998)
CAD2(encodes for GCS/GSH biosynthesis)	*Arabidopsis*	Cobbett et al. (1998)
Gsh2 (encodes for GS/GSH biosynthesis)	*S. pombe*	Cobbett et al. (1998)
CAD1(encodes for PC synthase/PC biosynthesis)	*Arabidopsis*	Howden et al. (1995) Ha et al. (1999)
PCS1 (encodes for PC synthase/PC biosynthesis)	*S. pombe*	Howden et al. (1995) Ha et al. (1999)

metal-detoxifying chelators encoded by and overexpression of *ATP-PRT1* gene involved in histidine biosynthesis has been shown in A. *thaliana* and Alyssum (Ingle et al. 2005).The gene for the enzyme PC synthase was first identified genetically in *Arabidopsis*. Expression of Arabidopsis and wheat cDNA libraries in *S. cerevisae* were used to identify genes conferring Cd resistance, namely, *AtPCS1* and *TAPCS1* (Clemens 2001; Vatanamuik et al. 1999). The processes of sulfur uptake, assimilation, and sequential sulfur metabolism in plant respond to Cd stress (Heiss et al. 1999; Lee and Leustek 1999). The expression of sulfur transporters with varied affinity was changed in different ways under Cd stress, and the high expression of ATP sulfurylase (APS) and adenosine 5′ phosphosulfate reductase (APR) (Pilon Smits et al. 1999; Van Huysen et al. 2004) may help to keep the supply of S for cysteine (Cys) synthesis. The efficiency of Cys synthesis may function in Cd detoxification, and the upregulated expression of Ser acetyltransferase (SAT) (Freeman et al. 2004) and O-acetyl-ser (thiol)-lyase (OASTL) has been found in some Cd-treated plants. Expression of genes responsible for glutathione synthesis, namely, gamma-glutamylcysteine synthetase (GCS) and glutathione synthetase (GS) was enhanced upon treatment with Cd (Xiang and Oliver 1998). In transgenic *B. juncea* in which either GCS or GS was overexpressed, the PC biosynthesis and Cd tolerance was enhanced (Zhu et al. 1999, Yong et al. 1999); however, dissimilar results were obtained when GCS was overexpressed in poplar (Noctor et al. 1998). In *A. halleri*, nicotianamine synthase (*NAS*) genes overexpressed in both roots and shoots under metal stress (Becher et al. 2004; Weber et al. 2004). A cDNA encoding a peptide with homology to the C-terminal part of heavy metal ATPase *HMA4* from *T. caerulescens* conferred Cd tolerance in yeast and could be involved in Cd transport and Cd hyperaccumulation character in this species (Bernard et al. 2004). *Ospdr9* which encodes a pleiotropic drug resistance (PDR)-type ATP-binding (ABC) transporter protein has also been shown to be induced by 20 μM Cd in rice roots (Moons 2003).

10.8 Conclusion

The present situation of Cd pollution in the ecosystem is alarming. The understanding of Cd-metal detoxification processes supplemented by investigations in model systems such as *Arabidopsis, Brassica*, *Sedum*, and *Thlaspi* in the near future will allow us to explore the mechanisms by which some species are capable of hyperaccumulation of metals like Cd and how they can be best used in green technology. In the long term, we aim to understand Cd-metal detoxification at the whole plant level, to extrapolate, as well as to exploit this knowledge to develop genetically manipulated plant that stands fair chance in bioremediation studies. Use of in vitro culture systems to investigate the mechanism of Cd-metal hyperaccumulation as used in *Sedum alfredii* by Zhao and coworkers (2009) could lead to better insights into metal homeostasis in higher plants. Wide species variations are recorded for the accumulative efficiency of Cd metal; however, correlation with

Cd-metal tolerance remains unclear. Correlating studies in Cd accumulation and metal tolerance is therefore needed in the near future. Identification of new metal transporters and elucidation of their functions will help to develop the knowledge of Cd uptake and translocation by plants. With the number of studies on plant genomics focused on the effects of metal stress, new genes are likely to be identified and annotated that confer resistance/tolerance to Cd and other metals in higher plants. More comparative transcriptomic studies employing hyperaccumulator (*T. caerulescens*) and non-accumulator (*T. arvense*) species would be helpful in understanding many genes involved in homeostasis and open response in plants. Screening of mutagenized Arabidopsis populations could be used to identify mutants tolerant to Cd. Although most of the mechanistic knowledge about Cd tolerance and sequestration comes from model species *A. thaliana*, the study of species living under natural but extreme conditions of Cd stress, e.g., *Thlaspi caerulescens* might add to our knowledge that otherwise could not be attained from mesophilic representatives. Gene discovery programs in such naturally tolerant models should essentially focus on expression sequence tags (EST) sequencing of stressed and unstressed libraries that are underrepresented in the current genomic databases related to metal stress. Use of Bioinformatics tools to compare stress-tolerant species from varied evolutionary lineages would help to identify the universal gene complement underlying metal stress tolerance, detoxification, and hyperaccumulation in plants. Biotechnological approaches to the production of high-biomass metal hyperaccumulators could also be considered. Modern genetics can be used to transfer hyperaccumulating genes to non-accumulating plants.

Acknowledgments I wish to acknowledge the contributions of the members of Environmental Science, Biochemical and Bioinformatics Laboratory, IESD, Banaras Hindu University, and valuable collaboration with colleagues. The work of my laboratory is supported by the Department of Science and Technology, Government of India, New Delhi. My apologies to all those researchers whose work could not be included in this chapter due to space limitation.

References

Agrawal V, Sharma K (2006) Phytotoxic effects of Cu, Zn, Cd and Pb on *in vitro* regeneration and concomitant protein changes in *Holarrhena antidysentrica*. Biol Plantarum 50:307–310

Antonovics J, Bradshaw AD, Turner RG (1971) Heavy metal tolerance in plants. Adv Ecol Res 7:1–85

Antosiewicz DM, Henning J (2004) Over-expression of LCT1 in tobacco enhances the protective action of calcium against cadmium toxicity. Environ Pollut 129:237–245

Bada BS, Raji KA (2010) Phytoremediation potential of kenaf (*Hibiscus cannabinus* L.) grown in different soil textures and cadmium concentration. Afr J Environ Sci Technol 4:250–255

Baker AJM (1981) Accumulators and excluders – strategies in the response of plants to heavy metals. J Plant Nutr 3:643–654

Baker AJM, Brooks RR (1989) Terrestrial higher plants which hyperaccumulate metallic elements – a review of their distribution, ecology and phytochemistry. Biorecovery 1:81–126

Baker AJM, Reeves RD, McGrath SP (1991) In situ decontamination of heavy metal polluted soils using crops of metal accumulating plants – a feasibility study. In: Hinchee RE, Olfenbuttel RF (eds) Bioreclamation. Wiley, New York, pp 539–544

Baker AJM, McGrath SP, Sidoli CMD, Reeves RD (1994) The possibility of in situ heavy metal decontamination of polluted soils using crops of metal-accumulating plants. Resour Conservat Recycl 11:41–49

Baker AJM, McGrath SP, Reeves RD, Smith JAC (2000) Metal hyperaccumulator plants: a review of the ecology and physiology of a biochemical resource for phytoremediation of metal polluted soil. In: Terry N, Baneulos G (eds) Phytoremediation of contaminated soil and water. Lewis Publications, Boca Raton, pp 85–107

Bansova V, Horak O (2008) Heavy metal content in *Thlaspi caerulescens* J. Et C. Presl growing on metalliferous and non metalliferous soils in central Slovakia. Int J Environ Pollut 33:133–145

Banuelos GS, Ajwa HA, Mackey B, Akohoue S (1997) Evaluation of different plant species used for phytoremediation of high soil selenium. J Environ Qual 26:639–646

Barcelo J, Poschenreider C (2002) Fast root growth responses, root exudates and internal detoxification as clue to mechanism of aluminium toxicity and resistance. Environ Exp Bot 48:75–92

Basic N, Keller C, Fontanillas P, Vittoz P, Besnard G, Galland N (2005) Cadmium hyperaccumulation and reproductive traits in natural *Thlaspi caerulescens* population. Plant Biol 8:64–72

Becher M, Talke IN, Krall L, Kramer U (2004) Cross species microarray transcript profiling reveals high constitutive expression of metal homeostasis gene in shoots of the zinc hyperaccumulator *Arabidopsis halleri*. Plant J 37:251–268

Bernand WR, Kagi HR (1987) Purification and characterization of a typical cadmium binding peptides from *Zea mays*. Experimentia 52:309–315

Bernard C, Roosens N, Czernic P, Lebrum M, Verbruggen N (2004) A novel CPx-ATPase from the cadmium hyperaccumulator *Thlaspi caerulescens*. FEBS Lett 569:140–148

Berts V, Meerts P (2003) Genetic basis of Cd tolerence and hyperaccumulation in *Arabidopsis halleri*. Plant Soil 249:9–18

Beyersman D, Hartwig A (2008) Carcinogenic metal compounds: recent insight into molecular and cellular mechanism. Arch Toxicol 82:493–512

Blaudez D, Botton B, Chalot M (2000) Cadmium uptake and subcellular compartmentalization in ectomycorrhizal fungus *Paxillus involvutus*. Microbiology 146:1109–1117

Bosque MA, Schuhmacher M, Domingo JL, Llobet JM (1990) Concentration of lead and cadmium in edible vegetables from Tarragona province, Spain. Sci Total Environ 95:61–70

Boyd RS (2004) Ecology of metal hyperaccumulation. New Phytol 162:563–567

Boyd RS, Shaw JJ, Martens SN (1994) Nickel hyperaccumulation defends *Streptanthus polygaloids* (Brassicaseae) against pathogens. Am J Bot 81:294–300

Brooks RR, Robinson BH (1998) The potential use of hyperaccumulator and other plants for phytomining. In: Brooks RR (ed) Plants that hyperaccumulate heavy metals: their role in phytoremediation, microbiology, archeology, mineral exploration and phytomining. CAB International, Wallingford, UK, pp 327–356

Brown SL, Chaney RL, Angle JS, Baker AJM (1995) Zinc and cadmium uptake by hyperaccumulator *Thlaspi caerulescens* and metal tolerant *Silene vulgaris* grown on sludge-amended soils. Environ Sci Technol 29:1581–1585

Buendia-Gongalez L, Orozco-Villafuerte J, Cruz-Sosa F, Barrera-Diaz CE, Vernon-Carter EJ (2010) *Prosopis laevigata* a potential chromium (VI) and cadmium (II) hyperaccumulator desert plant. Bioresour Technol 101:5862–5867

Cabrera C, Ortege E, Lorenzo ML, Lopez MDC (1998) Cadmium contamination of vegetable crops, farmlands and irrigation waters. Rev Environ Contam Toxicol 154:55–81

Casio C (2004) Phytoextraction of heavy metal by hyperaccumulation and non hyperaccumulating plants – comparison of cadmium uptake and storage mechanism in the plants. PhD thesis, ENAC, EPFL, Switzerland, No 2937

Chaney RL (1983) Plant uptake of inorganic waste constituents. In: Parr JF, Marsh PB, Kla JS (eds) Land treatment of hazardous waste. Noyes Data Crop, Park Ridge, NJ, pp 50–76

Chen J, Zhao J, Goldsbrough PB (1997) Characterization of phytochelatin-synthase from tomato. Physiol Plant 101:165–172

Clemens S (2001) Molecular mechanism of plant metal tolerance and homeostasis. Planta 212:475–486

Clemens S, Antonsiewz DM, Ward JM, Schachtman DP, Schroder JJ (1998) The plant cDNA LCT1 mediates the uptake of calcium and cadmium in yeast. Proc Nat Acad Sci USA 95:12043–12048

Clemens S, Kim EJ, Neumann D, Schroeder JI (1999) Tolerance to toxic metals by a gene family of phytochelatin synthases from plants and yeast. EMBO J 18:3325–3333

Cobbett CS (2000) Phytochelatins and their role in heavy metal detoxification. Plant Physiol 123:825–832

Cobbett C, Goldsbrough P (2002) Phytochelatins and metallothioneins: roles in heavy metal detoxification and homeostasis. Annu Rev Plant Biol 53:159–182

Cobbett CS, May MJ, Howden R, Rolls B (1998) The glutathione-deficient, cadmium-sensitive mutant, cad2-1, of *Arabidopsis thaliana* is deficient in γ-glutamylcysteine synthetase. Plant J 16:73–78

Concon JM (1988) Food toxicology: contamination and additives. Marcel Dekker, New York

Conklin DS, MacMasters JA, Culbertson MR, Kung C (1992) *COT1* genes involved in cobalt accumulation in *Saccharomyces cerevisiae*. Mol Cell Biol 12:3678–3688

Curie C, Panaviene Z, Loulerguech C, Delaporta SL, Briat F, Walker EL (2001) Maize yellow stripe encodes a membrane protein directly involved in Fe (III) uptake. Nature 409:346–349

Dan TV, Krishna S, Saxena PK (2000) Metal tolerance of scented germanium (*Pelargonium* sp. *Fresham*): effects of cadmium and nickel on chlorophyll fluorescence kinetics. Int J Phytoremediation 2:91–104

Dan TV, Krishna S, Saxena PK (2002) Cadmium and nickel uptake and accumulation in scented geranium (*Pelargonium* sp. *fresham*). Water Air Soil Pollut 137:355–364

De Knecht JA, Koevoets PLM, Verkleji JAC, Ernst WHO (1992) Evidence against a role of phytochelatins in naturally selected increased cadmium tolerance in *Silene vulgaris* (Moench) *Garcke*. New Phytol 122:681–688

De Knecht JA, Van Baren N, Ten Bookum WM, Wong F, Sang HW, Koevoet PLM, Schat H, Verkleij JAC (1995) Synthesis and degradation of phytochelatins in cadmium sensitive and cadmium-tolerant *Silene vulgaris*. Plant Sci 106:9–18

Duskenhov V, Kumar PBA, Motto H, Raskin I (1995) Rhizofiltration: the use of plants to remove heavy metals from aqueous streams. Environ Sci Technol 29:1239–1245

Dzantor EK, Beauchamp RG (2002) Phytoremediation, part I: fundamental basis for the use of plants in remediation of organic and metal concentration. Environ Pract 4:77–487

Escarré J, Lefèbvre C, Gruber W, Leblanc M, Lepart J, Rivière Y, Delay B (2000) Zinc and cadmium hyperaccumulation by *Thlaspi caerulescens* from metalliferous and nonmetalliferous sites in the Mediterranean area: implications for phytoremediation. New Phytol 145:429–437

FAO-WHO (1986) Toxicological evaluation of certain food activities and contaminants, WHO food additives series. World Health Organization, Rome

Fodor E, Szabo-Nagy A, Erdei L (1995) The effects of cadmium on the fluidity and H^+-ATPase activity of plasma membrane from sunflower and wheat roots. J Plant Physiol 147:87–92

Freeman JL, Pearson MW, Nieman K, Albrecht C, Peer W, Pickering IJ, Salt DE (2004) Increased glutathione biosynthesis plays a role in nickel tolerance in *Thlaspi* nickel hyperaccumulators. Plant Cell 16:2176–2191

Gekeler W, Grill E, Winnacker E-L, Zenk MH (1989) Survey of the plant kingdom for the ability to bind metals through phytochelatins. Z Naturforsch 44:361–369

Goyer RF (1988) Lead. In: Seiler HG, Sigel H (eds) Handbook of toxicity of inorganic compounds. Marcel Dekker, New York

Grill E, Winnacker E-L, Zenk MH (1986a) Homo-phytochelatins as heavy metal-binding peptides of homoglutathione containing *Fabales*. FEBS Lett 205:47–50

Grill E, Winnacker E-L, Zenk MH (1986b) Synthesis of seven homologous phytochelatins in metal-exposed *Schizosaccharomyces cerevisiae* cells. FEBS Lett 197:115–120
Grill E, Winnacker E-L, Zenk MH (1987) Phytochelatins the metal binding peptides from plants are functionally analogous to metallothioneins. Proc Nat Acad Sci USA 84:439–443
Grill E, Loffler S, Winnacker EL, Zenik MH (1989) Phytochelatins, the heavy metal binding peptides of plants are synthesized from glutathione by a specific Y-glutamyl cysteine dipeptidyl transpeptidase (phytochelatin synthase). Proc Natl Acad Sci USA 86:6838–6842
Grotz N, Fox T, Cannoly E, Park W, Guerinot ML, Eide D (1988) Identification of a family of zinc transporter from *Arabidopsis* that respond to zinc deficiency. Proc Nat Acad Sci USA 86:6838–6842
Guerinot ML, Eide D (1999) Zeroing in on zinc uptake in yeast and plants. Curr Opin Plant Biol 2:244–249
Gupta SS, Nayek S, Saha RN, Satpati S (2008) Assessment of heavy metal accumulation in macrophyte, agriculture soil and crop plants adjacent to discharge zone of sponge iron factory. Environ Geol 55:731–739
Ha SB, Smith AP, Howden R, Dietrich WM, Bugg SO, Connell MJ, Goldsbrough PB, Cobbett CS (1999) Phytochelatin synthase gene from *Arabidopsis* and the yeast, *Schizosaccharomyces pombe*. Plant Cell 11:1153–1164
Haguenoer JM, Furon D (1981) Toxicologie et hygiène Industrielles. Technique et Documentation, Paris
Hall JL (2002) Cellular mechanisms for heavy metal detoxification and tolerance. J Exp Bot 53:1–11
Hayashi Y, Nakagawa CW, Mutoh N, Isobe M, Goto T (1991) Two pathways in the biosynthesis of cadystins (γ EC)nG in the cell free system of fission yeast. Biochem Cell Biol 69:115–121
Heiss S, Schafer HJ, Haag-Kerwer A, Rausch T (1999) Sulphur assimilation genes of *Brassica juncea* L.: cadmium differentially affect the expression of a putative low affinity sulphate transporter and isoform of ATP suphurylase and APS reductase. Plant Mol Biol 39:847–857
Hirschi KD, Zhen RG, Cunningham KW, Rea PA, Fink GR (1996) CAX1 and H^+/Ca^{2+} antiporter from *Arabidopsis*. Proc Nat Acad Sci USA 93:8782–8786
Hirschi ED, Korenkey VD, Wilganewski NI, Wagner GI (2000) Expression of *Arabidopsis* CAX2 in tobacco. Altered metal accumulation and increased manganese tolerance. Plant Physiol 124:128–133
Hongli F, Wei Z (2009) Screening of amaranthus cultivars (*Amaranthus mangostanus* L.) for cadmium hyperaccumulation. Sci Agric Sin 42:1316–1324
Howden R, Goldsborough PB, Anderson CR, Cobbett CS (1995) Cadmium-sensitive, *cadl* mutants of *Arabidopsis thaliana* are phytochelatin deficient. Plant Physiol 107:1059–1066
Ingle RA, Mugford ST, Rees JD, Campbell MM, Smith JA (2005) Constitutively high expression of the histidine biosynthetic pathway contributes to nickel tolerance in hyperaccumulator plants. Plant Cell 17:2089–2106
Jackson AP, Alloway BJ (1991) The bioavailability of cadmium to lettuce and cabbage in soils previously treated with sewage sludge. Plant Soil 132:179–186
Jean L, De M (1997) Constructed wetlands for sludge dewatering. Water Sci Technol 35:279–285
Jiang RF, Ma DY, Zhao FJ, Mcgrath SP (2005) Cadmium hyperaccumulation protects *Thlaspi caerulescens* from leaf feeding damage by thrips (*Frankliniella occidentalis*). New Phytol 167:805–814
Kamizono A, Nishizawa M, Teranishi Y, Murata K, Kimura A (1989) Identification of a gene conferring resistance to zinc and cadmium in *Saccharomyces cerevisiae*. Mol Gen Genet 219:161–167
Kawashima CG, Noji M, Nakamura M, Ogra Y, Suzuki KT, Saito K (2004) Heavy metal tolerance of transgenic tobacco plants over-expressing cysteine synthase. Biotechnol Lett 26: 153–157
Kim JB, Kanmg JY, Kim SY (2004) Over expression of a transcription factor regulating ABA-responsive gene expression confers multiple stress tolerance. Plant Biotechnol J 2:459–466

Klapheck S, Fleigner W, Zimmer I (1994) Hydroxymethyl phytochelatins [(γ-glutamyl-cystine)*n* serine] are metal induced peptides of *Poaceae*. Plant Physiol 104:1325–1332

Klapheck S, Schlunz S, Bergmann L (1995) Synthesis of phytochelatins and homo phytochelatins in *Pisum sativum*. Plant Physiol 107:515–521

Klassen CD, Liu J, Choudhuri S (1999) Metallothionein: an intracellular protein to protect against cadmium toxicity. Annu Rev Pharmacol Toxicol 39:267–294

Kneer R, Zenk MH (1992) Phytochelatins protect plant enzymes from heavy metal poisoning. Phytochemistry 31:2663–2667

Korshunova YO, Eide D, Clark WG, Guerinot ML, Pakrasi HB (1999) The IRT1 protein from *Arabidopsis thaliana*, is a metal transporter with a broad substrate range. Plant Mol Biol 40:37–44

Kramer U (2010) Metal hyperaccumulation in plants. Annu Rev Plant Biol 61:517–534

Kubota H, Takenka C (2003) *Arabis gemmifera* is a hyperaccumulator of Cd and Zn. Int J Phytoremediation 5:197–220

Küpper H, Kochian L (2010) Transcriptional regulation of metal transport genes and mineral nutrition during acclimation to cadmium and zinc in the Cd/Zn hyperaccumulator, *Thlaspi caerulescens (Ganges population)*. New Phytol 185:114–129

Küpper H, Lombi E, Zhao FJ, McGrath SP (2000) Cellular compartmentation of cadmium and zinc in relation to other elements in the hyperaccumulator *Arabidopsis halleri*. Planta 212:75–84

Lee SM, Leustek T (1999) The effect of cadmium on sulphate assimilation enzyme in *Brassica juncea*. Plant Sci 141:201–207

Lee S, Moon JS, Petros D, Goldsbroug PB, Korban SS (2003) Over-expression of *Arabidopsis* phytochelatin synthase paradoxically leads to hypersensitivity to cadmium stress. Plant Physiol 131:656–663

Li L, Kaplan J (1998) Defect in yeast iron transport system result in increased metal hypersensitivity because of the increased expression of transporter with broad transition metal specificity. J Biol Chem 271:22181–22187

Lombi E, Zhao FJ, Dunham SJ, McGrath SP (2001) Phytoremediation of heavy metal – contaminated soils: natural hyperaccumulation versus chemically enhanced phytoextraction. J Environ Qual 30:1919–1926

Lombi E, Tearall KL, Howarth JR, Zhao FJ, Hawkesford MJ, McGrath SP (2002) Influence of iron status on Cadmium and Zinc uptake by different ecotypes of the hyperaccumulator *Thlaspi caerulescens*. Plant Physiol 128:1359–1367

Lopez-Millan AF, Ellis DR, Grusak MA (2004) Identification and characterization of several new members of the ZIP family of metal ion transporters in *Medicago truncatula*. Plant Mol Biol 54:583–596

Lu LL, SK Tian, Yang XE, Wang XC, Brown P, Li TQ, He ZL (2008) Enhanced root-to-shoot translocation of cadmium in the hyperaccumulating ecotype of *Sedum alfredii*. J Exp Bot 59:3203–3213

Maitani T, Kubota H, Sato K, Yamada T (1999) Phytochelatins (class III metallothioneins) and their desglycyl peptides induced by cadmium in root culture of *Rubia tinctorum* L. In: Klassen C (ed) Metallothionein, vol IV. Birkhauser Verlag, Basel, pp 201–205

Manousaki E, Kadukova J, Papadantonakis N, Kalogerakis N (2008) Phytoextraction and phytoexcretion of Cd by the leaves of *Tamarix smyrnensis* growing on contaminated non-saline and saline soils. Environ Res 106:326–332

Marschner H (1995) Mineral nutrition in higher plants, 2nd edn. Academic, London

Martin MH, Coughtrey PJ (1982) Biological monitoring of heavy metal pollution. Applied Science Publishers, New York

McIntyre T (2003) Phytoremediation of heavy metal from soil. In: Schepet T (ed) Advances in biochemical engineering/biotechnology, vol 78. Springer, Heidelberg

McLaughlin MJ, Tiller KG, Naidu R, Stevens DP (1996) Review: the behaviour and environmental impact of contaminants in fertilizers. Aust J Soil Res 34:1–54

McNair MR (1993) The genetics of metal tolerance in vascular plants. New Phytol 124: 541–559

Mehra RK, Winge DR (1988) Cu (I) binding to *Schizosaccaromyces pombe* γ-glutamyl transferase peptides varying in chain lengths. Arch Biochem Biophys 265:381–389

Meuwly P, Thibault P, Schwan AL, Rauser WE (1995) Three families of thiol peptides are induced by cadmium in maize. Plant J 7:391–400

Molitor M, Dechamps C, Gruber W, Meerts P (2005) *Thlaspi caerulescens* on nonmetalliferous soil in Luxembourg: ecological niche and genetic variation in mineral element composition. New Phytol 165:503–512

Moons A (2003) *Ospdr 9*, which encodes a PDR-type ABC transporter, is induced by heavy metals, hypoxic stress and redox perturbation in rice roots. FEBS Lett 553:370–376

Moral R, Gomez I, Navarro-Pedreno J, Mataix J (1994) Effect of Cd on nutrient distribution, yield and growth of tomato grown soilless culture. J Plant Nutr 17:953–962

Moreno-Caselles J, Moral R, Pere-Espinosa A, Marcia MD (2000) Cadmium accumulation and distribution in cucumber plants. J Plant Nutr 23:243–250

Nahakpam S, Shah K (2011) Expression of key antioxidant enzymes under combined effect of heat and cadmium toxicity in growing rice seedlings. Plant Growth Regul 63(1):23–35

Nedjimi B, Daoud Y (2009) Cadmium accumulation in *Atriplex halimus* subsp. *schweinfurthii* and its influence on growth, proline, root hydraulic conductivity and nutrient uptake. Flora Morphol Distrib Funct Ecol Plants 204:316–324

Noctor G, Arisi A, Jouanin L, Kunert K, Renenberg H, Foyer C (1998) Glutathione biosynthesis, metabolism and relationship to stress tolerance explored in transformed plants. J Exp Bot 49:623–647

Odjegba VJ, Fasidi IO (2004) Accumulation of trace elements by *Pistia stratiotes*: implications for phytoremediation. Ecotoxicology 13:637–646

Olmos E, Solano MJR, Piqueras A (2003) Early steps in the oxidative burst induced by cadmium in cultured tobacco cells (BY-2 line) [J]. J Exp Bot 54:291–301

Ortiz DF, Ruscitti T, McCuc KF, Ow DW (1995) Transport of metal binding peptides by HMT1, a fission yeast ABC-type B vacuolar membrane protein. J Biol Chem 270:4721–4728

Ouariti O, Boussama N, Zarrouk M, Cherif A, Ghorbal MH (1997) Cadmium- and copper-induced changes in tomato membrane lipids. Phytochemistry 45:1343–1350

Padmavathiamma PK, Loretta YL (2007) Phytoremediation technology: hyper-accumulation of metals in plants. Water Air Soil Pollut 184:105–126

Paulsen IT, Saier MH (1997) A novel family of ubiquitous heavy metal ion transport proteins. J Membr Biol 156:99–103

Pence NS, Larsen PB, Ebbs SD, Letham DLD, Lasat MM, Garvin DF, Eide D, Kochian LV (2000) The molecular physiology of heavy metal transport in Zinc/Cadmium hyperaccumulator *Thlaspi caerulescens*. Proc Natl Acad Sci USA 97:4956–4960

Pilon Smits E, Hwang S, Lytle M, Zhu Y, Tai JC, Bravo RC, Chen Y, Leustek T, Terry N (1999) Over-expression of ATP sulfurylase in *Brassica juncea* leads to increased selenate uptake, reduction and tolerance. Plant Physiol 119:123–132

Pollard AJ (2000) Metal hyperaccumulation. New Phytol 146:179–181

Pollard A, Powell K, Harper F, Smith J (2002) The genetic basis of metal hyperaccumulation in plants. Crit Rev Plant Sci 21:539–566

Prasad MNV (1995) Cadmium toxicity and tolerance in vascular plants. Environ Exp Bot 35: 525–540

Rauser WE (1990) Phytochelatins. Annu Rev Biochem 59:61–86

Rauser WE (1995) Phytochelatins and related peptides. Plant Physiol 109:1141–1149

Rea PA, Li ZS, Lu YP, Drosdowicz YM, Martinoia E (1998) From vacuolar GS-X pumps to multi-specific transporters. Annu Rev Plant Physiol Plant Mol Biol 49:727–760

Reeves RD, Brooks RR (1983) Hyperaccumulation of lead and zinc by two metallophytes from mining areas of Central Europe. Environ Pollut Ser A 31:277–285

Reilly C (1980) Metal contamination of food. Applied Science, London

Robards K, Worsfold P (1991) Cadmium: toxicology and analysis. A review. Analyst 116: 549–568

Robinson NJ, Tommey AM, Kuske C, Jackson PJ (1993) Plant metallothioneins. Biochem J 295:1–10

Roelofs D, Marien J, Van Straalen NM (2007) Differential gene expression profiles associated with heavy metal tolerance in the soil insect *Orchesella cincta*. Insect Biochem Mol Biol 37:287–295

Roelofs D, Aarts MGM, Schat H, Van Straalen NM (2008) Functional ecological genomics to demonstrate general and specific responses to abiotic stress. Funct Ecol 22:8–18

Roosen N, Verbruggen N, Meerts P, Ximenez-Embun P, Smith JAC (2003) Natural variation in Cadmium tolerance and its relationship to metal hyperaccumulation for seven populations of *Thlaspi caerulescens* from western Europe. Plant Cell Environ 26:1657–1672

Salt DE, Rauser WE (1995) Mg-ATP dependent transport of phytochelatins across the tonoplast of oat roots. Plant Physiol 107:1293–1301

Salt DE, Prince RC, Pickering IJ, Raskin I (1995) Mechanisms of cadmium mobility and accumulation in Indian mustard. Plant Physiol 109:1427–1433

Salt DE, Smith RD, Raskin I (1998) Phytoremediation. Annu Rev Plant Physiol Plant Mol Biol 49:643–648

Saraswat S, Rai JPN (2009) Phytoextraction potential of six plant species grown in multimetal contaminated soil. Chem Ecol 25:1–11

Sarma H (2011) Metal hyperaccumulation in plants: a review focusing on phytoremediation technology. J Environ Sci Technol 4:118–138

Schat H, Llugany M, Voojis R, Harley-Whitaker J, Bleeker PM (2002) The role of phytochelatins in constitutive and adaptive heavy metal tolerances in hyperaccumulator and non-hyperaccumulator metallophytes. J Exp Bot 53:2381–2392

Schutzendubel A, Polle A (2002) Plant response to abiotic stresses: heavy metal-induced oxidative stress and protection by mycorrhization. J Exp Bot 53:1351–1365

Schwarz Y (1974) Trace element metabolism in animals. University Park Press, Baltimore

Seiler H, Sigel H (1988) Handbook on toxicity of inorganic compounds. Dekker, New York

Selvam A, Wong JW (2008) Phytochelatin synthesis and cadmium uptake of *Brassica napus*. Environ Technol 29:65–773

Senden MHMN, Van Paassen FJM, Van Der Meer AJGM, Wolterbeek HTh (1990) Cadmium-citric acid-xylem cell wall interactions in tomato plants. Plant Cell Environ 15:71–79

Shah K, Dubey RS (1995) Effect of cadmium on RNA level as well as activity and molecular forms of ribonuclease in growing rice seedlings. Plant Physiol Biochem 33:577–584

Shah K, Dubey RS (1998) A 18 kDa cadmium inducible protein complex: its isolation and characterization from rice (*Oryza sativa* L.) seedlings. J Plant Physiol 152:448–454

Shah K, Nahakpam S (2011a) Expression of key antioxidant enzymes under combined effect of heat and cadmium toxicity in growing rice seedlings. Plant Growth Regul 63:23–35

Shah K, Nahakpam S (2011b) Heat stress and Cadmium toxicity in higher plants-an overview. In: Hemantranjan A (ed) Advances in plant physiology, vol 12. Scientific Publishers (India), Jodhpur, pp 441–453

Shah K, Nongkynrih JM (2007) Metal hyperaccumulators and bioremediation. Biol Plantarum 51:618–634

Shah K, Kumar RG, Verma S, Dubey RS (2001) Effect of Cadmium on lipid peroxidation, superoxide anion and activities of antioxidant enzymes in growing rice seedlings. Plant Sci 161:1135–1144

Shah K, Penel C, Gagnon X, Dunand C (2004) Purification and identification of a Ca^{2+}-pectate peroxidase from *Arabidopsis* leaves. Phytochemistry 65:307–312

Shah K, Raghuvanshi R, Singh I (2011). Phosphate Transporters in Symbiotic Arbuscular Endomycorrhizal Association: A Bioinformatics Approach. In: Sinha RP, Sharma NK and Rai AK, (eds) Advances in life sciences, IK International Publishers (India), New Delhi, pp 93–115

Sharma RK, Agrawal M (2005) Biological effects of heavy metals – an overview. J Environ Biol 26(2l):301–313
Shaw BP, Sahu SK, Mishra RK (2004) Heavy metal induced oxidative damage in terrestrial plants. In: Prasad MNV (ed) Heavy metal stress in plants. Springer, Heidelberg, pp 84–126
Siedlecka A, Baszynski T (1993) Inhibition of electron flow around photosystem I in chloroplasts of Cd-treated maize plants is due to Cd-induced iron deficiency. Physiol Plant 87:199–202
Singh OV, Jain RK (2003) Phytoremediation of toxic aromatic pollutants from soil. Appl Microbiol Biotechnol 63:128–135
Sivaci A, Elmas E, Gumu F, Sivaci ER (2008) Removal of Cadmium by *Myriophyllum heterophyllum michx* and *Potamogeton crispus* L. and its effect on pigments and total phenolic compounds. Arch Environ Contam Toxicol 54:612–618
Song W-Y, Sohn EJ, Martinoia E, Lee YJ, Yang YY, Jasinski M, Forestier C, Hwang I, Lee Y (2003) Engineering tolerance and accumulation of lead and cadmium in transgenic plants. Nat Biotechnol 21:914–919
Sun R, Zhou Q, Wei S (2000) Cadmium accumulation in relation to organic acids and non- protein thiols in leaves of the recently found Cd-hyperaccumulator *Rorippa globosa* and the Cd-accumulating plant *Rorippa islandica*. J Plant Growth Regul 30(1):83–91
Tahvonen R (1996) Contents of Lead and Cadmium in foods and diets. Food Rev Int 12:1–70
Tam PCF (1995) Heavy metal tolerance by ectomycorrhizal fungi and metal amelioration by *Pisolothus tinctorius*. Mycorrhiza 5:181–187
Tang YT, Qiu RL, Zeng XW, Ying RR, Yu FM, Zhou XY (2009) Lead, zinc, cadmium hyperaccumulation and growth stimulation in *Arabis paniculata* Franch. Environ Exp Bot 66:126–134
Tommasini R, Vogt E, Fromenteau M, Hortensteiner S, Matile P, Amrhein N, Martinoia E (1998) An ABC transporter of *Arabidopsis thaliana* has both glutathione conjugate and chlorophyll catabolite transport activity. Plant J 13:773–780
Tong YP, Kneer R, Zhu YG (2004) Vacuolar compartmentalization: a second-generation approach to engineering plants for phytoremediation. Trends Plant Sci 9:7–9
Toppi di Sanita L, Gabbrielli R (1999) Response to Cadmium in higher plants. Environ Exp Bot 41:105–130
USEPA (2000) Introduction to phytoremediation. United States Environmental Protection Agency, Washington DC, USA
Van DWM (1991) Coomon reed. In: Rozema J, Verkleij JAC (eds) Ecological responses to environmental stress. Kluwer Academic, The Netherlands, pp 172–182
Van der Zaal BJ, Neutboom LW, Pinars JE, Chardonnens AN, Schat H, Verkleji JA, Hooykaas PJ (1999) Over-expression of a novel *Arabidopsis* gene related to putative Zinc transport genes from animals can lead to enhanced Zinc resistance and accumulation. Plant Physiol 119:1047–1056
Van Huysen T, Terry N, Pilon-Smits EAH (2004) Exploring the selenium phytoremediation potential of transgenic Indian mustard over-expressing ATP sulfurylase or cystathionine-γ-synthase. Int J Phytoremediation 6:111–118
Vatanamuik OK, Mari S, Lu YP, Rea PA (1999) A phytochelatin synthase from *Arabidopsis*: isolation and in vitro reconstitution. Proc Natl Acad Sci USA 96:7110–7115
Vert G, Grotz N, Dedaldechamp F, Gaymard F, Guerinot ML, Briata JF, Curie C (2002) IRT1, an *Arabidopsis* transporter essential for iron uptake from the soil for plant growth. Plant Cell 14:1223–1233
Vogel-Mikus K, Pongrac P, Kump P, Necemer M, Regvar M (2006) Colonisation of a Zn, Cd and Pb hyperaccumulator *Thlaspi praecox Wulfen* with indigenous arbuscular mycorrhizal fungal mixture induces changes in heavy metal and nutrient uptake. Environ Pollut 139:362–371
Vogel-Mikus K, Arcon I, Kodre A (2010) Complexation of cadmium in seeds and vegetative tissues of the cadmium hyperaccumulator *Thlaspi praecox* as studied by X-ray absorption spectroscopy. Plant Soil 331:439–451

Wagner GJ (1993) Accumulation of cadmium in crop plants and its consequences to human health. Adv Agron 51:173–212

Wang QR, Cui YS, Liu XM, Dong YT, Christie P (2003) Soil contamination and uptake of heavy metals at polluted sites in China. J Environ Sci Health 38:823–838

Weber M, Harada E, Vess C, Roepenack-Lahaye EV, Clemens S (2004) Comparative microarray analysis of *Arabidopsis thaliana* and *Arabidopsis halleri* roots identifies nicotinamine synthase, a ZIP transporter and other genes as potential metal hyperaccumulation factor. Plant J 37:269–281

Welch RM, Norvell WA (1999) Mechanisms of Cadmium uptake, translocation and deposition in plants. In: McLaughlin MJ, Singh BR (eds) Cadmium in soils and plants. Kluwer Academic, Dordrecht, pp 125–150

Wojcik M, Tukiendorf A (2005) Cadmium uptake, localization and detoxification in Zea mays. Biol Plantarum 49:237–245

Xiang C, Oliver DJ (1998) Glutathione metabolic genes coordinately respond to heavy metals and jasmonic acid in Arabidopsis. Plant Cell 10:1539–1550

Xiong YH, Yang XE, Ye ZQ, He ZL (2004) Characteristics of cadmium uptake and accumulation by two contrasting ecotypes of Sedum alfredii Hance. J Environ Sci Health A Tox Hazard Subst Environ Eng 39:2925–2940

Xue Q, Harrison HC (1991) Effect of soil Zinc, pH and cultivar on Cadmium uptake in leaf lettuce (*Lactuca sativa* L. var. crispa). Commun Soil Sci Plant Anal 22:975–991

Yang XE, Long XX, Ye HB (2004) Cadmium tolerance and hyperaccumulation in a new Zn-hyperaccumulating plant species (*Sedum alfredii* Hance) [J]. Plant Soil 259:181–189

Yannai S, Berdicevsky I (1995) Formation of organic cadmium by microorganisms. Eco-toxicol Environ Saf 32:209–214

Yeh CM, Hsiao LJ, Huang HJ (2004) Cadmium activates a mitogen-activated protein kinase gene and MBP kinase in rice [J]. Plant Cell Physiol 45:1306–1312

Yong LZ, Pilon-Smits EAH, Tarun AS, Weber SU, Jouanin L, Terry N (1999) Cadmium tolerance and accumulation in Indian mustard is enhanced by over-expressing γ glutamyl cysteine synthetase. Plant Physiol 121:1169–1177

Zeng X, Ma LQ, Qiu R, Tang Y (2009) Responses of nonprotein thiols to Cd exposure in Cd hyperaccumulator Arabis paniculata Franch. Environ Exp Bot 66:242–248

Zha HG, Jiang RF, Zhao FJ, Vooijs R, Schat H, Barker JHA, McGrath SP (2004) Co-segregation analysis of Cadmium and Zinc accumulation in Thlaspi caerulescens interecotypic crosses. New Phytol 163:299–312

Zhang Z, Qiu B (2007) Reactive oxygen species metabolism during the cadmium hyper-accumulation of a new hyperaccumulator Sedum alfredii (Crassulaceae). J Environ Sci 19: 1311–1317

Zhang CZ, Chen BX, Qiu BS (2010) Phytochelatin synthesis plays a similar role in shoots of the cadmium hyperaccumulator Sedum alfredii as in non-resistant plants. Plant Cell Environ 33:1248–1255

Zhao S-J, Zhang Z-C, Gao X, Tohsun G, Qiu BS (2009) Plant regeneration of the mining ecotype Sedum alfredii and cadmium hyperaccumulation in regenerated plants. Plant Cell Tissue Organ Cult 99:9–16

Zhou WB, Qiu BS (2005) Effects of cadmium hyperaccumulation on physiological characteristics of Sedum alfredii Hance (Crassulaceae). Plant Sci 169:737–745

Zhu Y (2010) Phytoremediation of Cadmium polluted soil by *Brassica chinensis*. Guangdong Agric Sci DOI:CNKI:SUN:GDNY.0.2010-02-024

Zhu YL, Pilon-Smits EAH, Jouanin L, Terry N (1999) Over-expression of glutathione synthases in Indian Mustard enhances cadmium accumulation and tolerances. Plant Physiol 119:73–79

Chapter 11
Transport, Accumulation, and Physiological Effects of Vanadium

Dieter Rehder

11.1 Introduction

Vanadium is the 21st element in abundance in Earth's crust. Although more than 120 minerals are known, the main amount of vanadium is distributed in such a way that this element is ubiquitous and omnipresent. Most vanadium minerals have formed by geological processes, but a biogenic formation by several strains of bacteria, employing vanadium(V) as an electron acceptor in respiration, is also conceivable. Vanadium is further enriched in fossil "fuels," where it accumulates in the form of vanadyl porphyrins. Burning of these fuels, exposure to vanadium oxides in the course of mining and processing of vanadium ores, and re-mobilization of vanadate from vanadinite (a lead vanadate) accumulated in the scales of lead water pipes, are the main sources for potential hazardous vanadium exposure.

Vanadium compounds are classified as mutagenic and teratogenic, and potentially carcinogenic. Toxic levels of vanadium in humans are, except in case of direct exposure, not easily achieved. The main health problems appear to arise from the inhalation of vanadium pentoxide. The structural similarity of vanadate and phosphate, plus the stability (in contrast to phosphate) of penta-coordinated vanadium, give rise to a vanadate–phosphate antagonism and synergism which can be both beneficial (in the sense of vanadate as a regulator of the phosphate metabolism) and detrimental (by inhibition of phosphate-activated substrates and phosphate-dependent enzymes). The inhibition of a protein–tyrosine phosphatase by vanadate, lending vanadium compounds insulin-enhancing properties, is an example for an inhibition with benign consequences. Along with possible health hazards arising from the vanadate–phosphate antagonism, vanadium compound are involved in the production as well as the annihilation of reactive oxygen species, and thus,

D. Rehder (✉)
Institut für Anorganische und Angewandte Chemie, Martin-Luther-King-Platz 6, 20146 Hamburg, Germany
e-mail: rehder@chemie.uni-hamburg.de

I. Sherameti and A. Varma (eds.), *Detoxification of Heavy Metals*, Soil Biology 30,
DOI 10.1007/978-3-642-21408-0_11, © Springer-Verlag Berlin Heidelberg 2011

potentially, in the proliferation as well as in the slowdown of the growth of tumor cells. Trapping of alkyl radicals by vanadium compounds points into the same direction.

The essentiality of vanadium for life, in particular for human life, has not yet been established, although, based on the regulatory function of vanadate in the phosphate metabolism, vanadium is likely to be an essential trace element. Vanadium(V) (in the form of vanadate) is more toxic than vanadium(IV) (in the form of the vanadyl ion and its compounds). The lower toxicity of vanadium(IV) is a result of the insolubility of vanadyl hydroxide. V^V and V^{IV} readily interconvert under physiological conditions. Vanadium(III) does not appear to play a pivotal role.

Vanadium overloads, due to acute exposure to vanadium compounds, are comparatively rapidly removed from serum and tissues; here, half-lives amount to a few hours at the worse. Vanadate is built in into the hydroxyapatite lattice of the bones, a fact which increases the overall half-life of vanadium to a couple of days, corresponding to a residence time of about a month. Acute intoxication can be treated with reducing agents such as ascorbic acid, with medications working both as reductants and chelators, such as catechins (an oligophenol), and with efficient chelating agents, e.g., desferoxamine.

11.2 The Aqueous Chemistry of Vanadium

Under physiological conditions, vanadium can occur in the oxidation states +III, +IV, and +V, which are also the oxidation states of vanadium in terrestrial minerals (Sect. 11.3). Vanadium(II) is unstable under oxic conditions. V^{IV} and V^V are by far the most important oxidation states in aqueous media, and thus in living organisms. Under normal physiological conditions, V^{III}, when generated by sufficiently strong reductants or by disproportionation of V^{IV}, is short-lived and rapidly re-oxidized to higher oxidation states. An exception are the vanadocytes of ascidians, where V^{III} is stable in the highly acidic cytosol of these blood cells (Sect. 11.3).

Vanadium(V) may occur in cationic form (monooxidovanadium(V) VO^{3+} or dioxidovanadium(V) VO_2^+) and in anionic form as vanadate. While the cationic variants are soluble only when coordinated to – and thus "protected" (against hydrolysis) by – sufficiently strong ligands, the bare vanadates are readily soluble in water. Oxygen has a high affinity to V^V and V^{IV}; the cationic species are therefore almost always present in the form of these oxido ions. Eventually, deoxygenation of VO^{2+} can occur, e.g., by interaction with thiols such as provided by cysteine residues RSH in peptides and proteins (11.1). In blood serum, most of the vanadium in the +IV and +V states is coordinated to (apo)transferrin and, to a much lesser extent, to serum albumin.

$$VO^{2+} + 2RSH \rightarrow V^{4+} + 2RS^- + H_2O \tag{11.1}$$

Although “nonoxido” vanadium(IV) exists in nature (see amavadin; Sect. 11.3), the by far more common form of vanadium(IV) is VO^{2+}, embedded in the center of coordination compounds. “Free” VO^{2+} is hardly available under physiological conditions, i.e., in the pH range 5–8, because it forms almost insoluble oxido-vanadium hydroxides “$VO(OH)_2$,” commonly termed vanadyl hydroxides in the literature. The solubility of vanadyl hydroxide lies in the nanomolar range; the species present in solution is the anion $[VO(OH)_3]^-$. This tetrahedral vanadate(IV) anion is iso-structural with the simplest vanadate(V) anion $[VO_2(OH)_2]^- \equiv H_2VO_4^-$ (at pH 7), and phosphate $[PO_3(OH)]^{2-} \equiv HPO_4^{2-}$ (again at pH 7). The reason for the vanadate–phosphate antagonism, to be addressed in Sect. 11.3, roots in this structural similarity.

The speciation of vanadate(V) very much depends on the pH and the concentration. At pH 7 and concentrations in the micro and sub-micromolar range (and hence at physiological conditions), the dominant species is diprotonated monovanadate $H_2VO_4^-$. The pK_a for the protonation equilibrium $H_2VO_4^- \leftrightarrows HVO_4^{2-} + H^+$ at an ionic strength of 0.15 mM (blood serum) is 8.5 (Schmidt et al. 2001). At concentrations exceeding ca. 0.1 mM, divanadate $H_nV_2O_7^{(4-n)-}$ ($n = 1$, 2) and tetravanadate $V_4O_{12}^{4-}$ form in the pH range 5–9. At still higher concentrations, pentavanadate $V_5O_{15}^{5-}$ comes in, and at pH < 6.3, decavanadates $H_nV_{10}O_{28}^{(6-n)-}$ ($n = 1$–3) are present. At pH > 6.3, decavanadate is thermodynamically unstable; it is, however, kinetically stable to the extent that its decay into vanadates of lower nuclearity takes several hours. The speciation diagrams in Fig. 11.1 convey an impression of the vanadate species present in water as a function of pH and concentration. In the mixed vanadate–phosphate system, the labile mixed species $H_2VPO_7^{2-}$ is present (Andersson et al. 2005).

Vanadate and vanadyl are linked by the redox couple depicted in (11.2). The redox potential at pH 7, $E^{pH=7} = -0.34$ V, compares to that of NAD^+/NADH ($E^{pH=7} = -0.315$ V). Vanadate(V) is readily reduced to VO^{2+} under physiological conditions by reductants such as glutathion and NADH, while VO^{2+} in turn is readily oxidized in the presence of oxygen or reactive oxygen species (ROS).

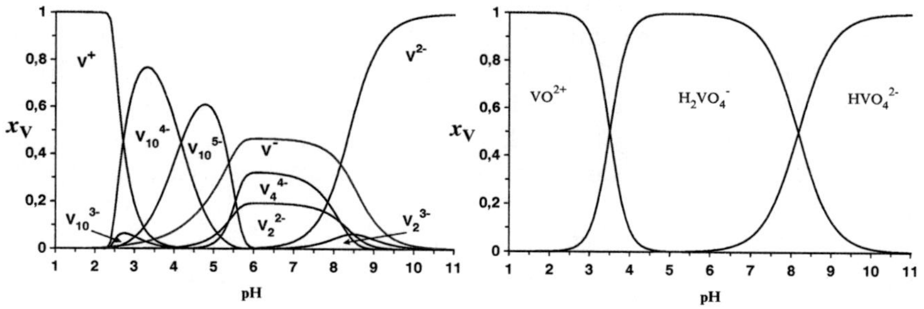

Fig. 11.1 Speciation diagrams of vanadate for c(V) = 1 mM (*left*) and 1 μM (*right*) at an ionic strength of 0.15 mM Na(Cl) and a temperature of 25°C; x_V is the mole fraction. $V^+ = [VO_2(H_2O)_4]^+$, V_{10} = decavanadate, V_4 = tetravanadate, V_2 = divanadate, $V^- = H_2VO_4^-$, $V^{2-} = HVO_4^{2-}$. Courtesy L. Pettersson, Umeå University (Sweden)

VO^{2+} by itself can be involved in the generation of ROS such as superoxide (11.3), and hydroxyl radicals (Sakurai 1994) (11.4).

$$H_2VO_4^- + 4H^+ + e^- \leftrightarrows VO^{2+} + 3H_2O \qquad (11.2)$$

$$VO^{2+} + O_2 \rightarrow VO^{3+} + O_2^- \qquad (11.3)$$

$$VO^{2+} + H_2O_2 \rightarrow VO_2^+ + HO\cdot + H^+ \qquad (11.4)$$

11.3 Occurrence and Biological Usage

The abundance of vanadium in Earth's crust, 0.013% w/w, compares to that of carbon (0.02%) and nitrogen (0.017%). In seawater, where vanadium is present in the form of ion pairs $Na^+H_2VO_4^-$ at a concentration of ca. 30 nM, it is the second-to-most abundant transition metal, only outplayed by molybdenum (100 nM, present in the form of molybdate). The comparatively high abundance of V and Mo in sea water and other oxic aqueous systems reflects the fact that these two *anionic* transition metal species are easily soluble in water. In volcanic regions, concentrations of vanadate in subsoil water can go up to 120 μg L^{-1} (2.4), exceeding the notification level, 15 μg vanadium per liter, set by, e.g., the California Office of Environmental Health Hazard Assessment.

In rock, vanadium is an omnipresent trace element. In vanadium-based minerals, vanadium occurs in the oxidation states +III, +IV, and +V. Representative examples are roscoelite $K(Al,V^{III})_2(OH,F)_2[AlSi_3O_{10}]$, patronite $V^{IV}(S_2)_2$ and vanadinite $PbCl_2{\cdot}3Pb_3[V^VO_4]_2$. In addition, vanadium(+II) has been retrieved in meteorites, such as in chondrules of the Vigarano meteorite, where octahedral magnesium sites in the mineral forsterite (Mg_2SiO_4) are partially replaced by V^{2+} (Giuli et al. 2006). Vanadium minerals are mainly formed in the course of geological processes, but biogenic formation of some minerals, based on oxidovanadium(IV) hydroxide, is an additional option (vide infra). Further, vanadium can accumulate in fossilized bio-mass, such as peat, coal, crude oil, bitumen, and oil-shales. Venezuelan crude oil can contain up to 0.12% vanadium. This enrichment reflects a secondary process: the extraction of vanadium from vanadium-bearing rock as oil passes through these rocks. Crude oil contains porphinogens, originating from hemes and chlorophyll-derived porphyrins of the fossilized organisms, and these porphyrinogenic systems are excellent ligands which extract vanadium from rock and firmly complex the VO^{2+} cation.

Certain bacteria, such as the soil bacterium *Shewanella oneidensis*, can use vanadate as an electron acceptor in respiration (Carpentier et al. 2005), Fig. 11.2. The reduction occurs at the outer cell wall, and the reduction product is vanadyl hydroxide of the approximate formula $VO(OH)_2$. Around pH 7, vanadyl hydroxide

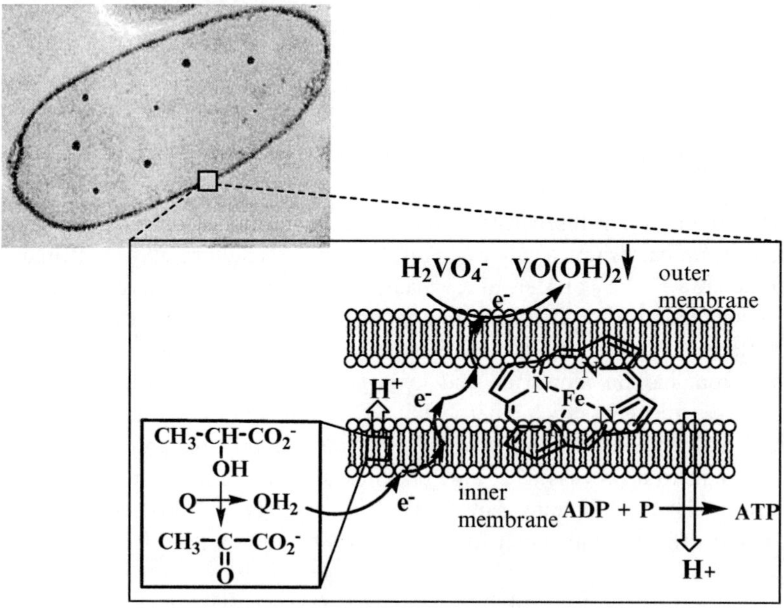

Fig. 11.2 Electron transfer pathway in the membrane of *Shewanella oneidensis* (*upper left*): Electrons and protons delivered by the oxidation of lactate to pyruvate (box at *lower left*) are picked up by a quinone (Q) and delivered to cytochrome-c type hemoproteins, which finally carry the reduction equivalents to the outer membrane, where V^V is reduced to V^{IV}. The reduction of vanadate is coupled H^+ influx into the cytosol and ATP synthesis at the inner membrane

is an almost insoluble compound, which eventually can mineralize and thus give rise to the formation of minerals resembling sherwoodite, a mixed-valence (V^{IV}/V^V) calcium–aluminum polyoxidovanadate. The electrons for the reduction are delivered via the oxidation of lactate to pyruvate in the cytosolic membrane. The electron transport across the periplasm is accomplished by oligomeric heme proteins, $\{Fe^{2+/3+}\}$; for the reaction sequence see (11.5). Other bacteria, e.g., *Pseudomonas isachekovii*, use molecular hydrogen and carbon monoxide as reductants (Antipov et al. 2000) (11.6).

The bacterium *Geobacter metallireducens* has been employed to effectively remove vanadate from groundwater contaminated by mining activities, employing acetate for electron delivery (Ortiz-Bernad et al. 2004) (11.7). The insoluble compound formed in this process resembles the mineral sincosite $CaV_2(OH)_4[PO_4]_2{\cdot}3H_2O$, symbolized by $\{VO(OH)_2\}$ in (11.7).

$$\text{lactate} \rightarrow \text{pyruvate} + 2H^+ + 2e^-$$
$$e^- \text{ transport chain}: \{Fe^{3+}\} + e^- \rightarrow \{Fe^{2+}\}$$
$$\{Fe^{2+}\} + H_2VO_4^- + 2H^+ \rightarrow \{Fe^{3+}\} + VO(OH)_2\downarrow + H_2O \quad (11.5)$$

$$H_2 \rightarrow 2H^+ + 2e^- \quad (11.6a)$$

$$CO + H_2O \rightarrow CO_2 + 2H^+ + 2e^- \quad (11.6b)$$

$$8H_2VO_4{}^- + CH_3CO_2H + 8H^+ \rightarrow 8\{VO(OH)_2\}\downarrow + 2CO_2 + 6H_2O \quad (11.7)$$

Other microorganisms, such as the proteobacterium *Azotobacter vinelandii* and the cyanobacterium *Anabaena variabilis* use vanadium as a constituent of the cofactor, a Fe_7VS_9 cluster, of a vanadium-dependent nitrogenase for the ATP-driven conversion of N_2 + H^+ to ammonium ions and molecular hydrogen (Eady 2003; Rehder 2008a) (11.8), and eventually for the production of alkanes and alkenes from carbon monoxide and hydrogen (Lee et al. 2010). For the latter reaction, see the non-stoichiometric (11.9). The reactions are analogous to the Haber–Bosch and Fischer–Tropsch syntheses, respectively, but proceed, in contrast to these industrial processes, at ambient pressure and temperature. *A. vinelandii* synthesizes the vanadophore azotochelin (Bellenger et al. 2007), **1** in Fig. 11.3, which is an efficient transporter for vanadium(V).

$$N_2 + 14H^+ + 12e^- \rightarrow 2NH_4{}^+ + 3H_2 \quad (11.8)$$

$$CO + H^+ + e^- \rightarrow C_2H_4, C_2H_6, \ldots \quad (11.9)$$

Another vanadium-based group of enzymes, the vanadate-dependent haloperoxidases, are present in marine algae, and in fungi and lichen. In these organisms, vanadate $H_2VO_4{}^-$ is coordinated to an imidazol nitrogen of a histidine in the active site protein pocket. The enzyme catalyzes the two-electron oxidation of halide, Hal^- (Cl^-, Br^-, I^-), to a Hal^+ species such as hypohalous acid, (11.10a) for Hal = Br, which then halogenates organic substrates RH, (11.10b). The enzymes also catalyze the oxidation of (prochiral) sulfides to (chiral) sulfoxides (11.11). Oxidant is hydrogen peroxide. The active center of these peroxidases, with vanadium(+V) in a trigonal–bipyramidal $\{O{=}VO(OH)_2N\}$ coordination

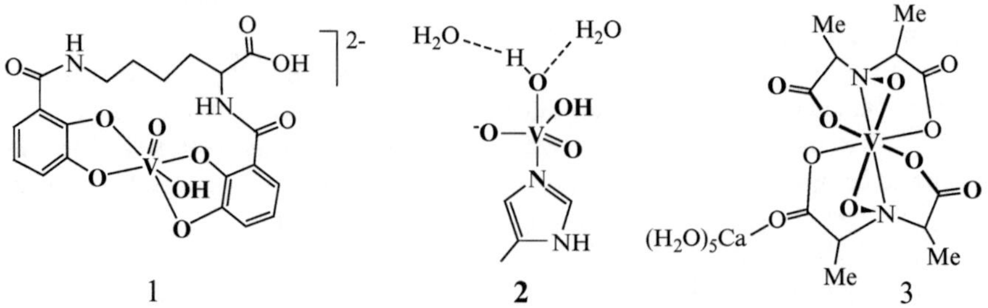

1 **2** 3

Fig. 11.3 Biogenic vanadium compounds. **1**: Transport of oxidovanadium(V) by the vanadophore azotochelin from *A. vinelandii*; **2**: The active center of vanadium-dependent haloperoxidases, and vanadate-inhibited acid phosphatase (from rat prostate); **3**: Amavadin in the mushroom *Amanita muscaria* environment (**2** in Fig. 11.3), is analogous to the respective center in vanadate-inhibited phosphatases, cf. Sect. 11.4.

$$Br^- + H^+ + H_2O_2 \rightarrow BrOH + H_2O \qquad (11.10a)$$

$$BrOH + RH \rightarrow RBr + H_2O \qquad (11.10b)$$

$$SR(R') + H_2O_2 \rightarrow O{=}SR(R') + H_2O \qquad (11.11)$$

Vanadium is further accumulated in mushrooms of the genus *Amanita*, e.g., in the toad stool (fly agaric) *Amanita muscaria*, and further by some ascidians (sea squirts) and fan worms. The non-oxido vanadium(+IV) compound isolated from *A. muscaria*, amavadin (**3** in Fig. 11.3), contains vanadium coordinated to (S,S)-*N*-hydroxyimino-2,2′-diisopropionic acid (Berry et al. 1999), an organic derivative of hydroxylamine. Given the versatility of vanadium in redox catalysis, amavadin may be a relic of a vanadium-dependent oxidase or oxygenase, used by the mushrooms in an earlier stage of their evolution (Rehder 2008a). Many ascidians take up vanadate from sea-water, reduce it to VO^{2+} by means of a reductase with NADPH as the cofactor, and bind VO^{2+} to lysine residues of a $(Cys)_2$-rich protein called vanabin (Ueki et al. 2009). The vanadyl cation is further reduced to vanadium(III) and stored, mainly in the form of $[VO(H_2O)_4(HSO_4)]^+$, in the highly acidic medium of special ascidian blood cells termed vanadocytes. The highest vanadium concentration, 0.35 M and hence an accumulation from sea water by a factor of 10^8, has been measured in *Ascidia gemmata*. The function of vanadium in these marine animals has remained elusive during the almost 100 years that have elapsed since its discovery in ascidians.

11.4 Physiological Effects and Toxicity

The total body pool of vanadium is low; depending on the source of information, it varies between ca. 0.1 and 1 mg (Wenning and Kirsch 1988; Léonard and Gerber 1994), corresponding to an approximate concentration of $c = 0.03$–0.3 μM. The average vanadium concentration in blood is 2.3 μg L^{-1} ($c = 45$ nM) (Leuschner et al. 1994). Vanadium contents in food are typically around 30 μg V kg^{-1} food (Wenning and Kirsch 1988), while drinking water commonly contains less than 10 μg L^{-1} ($c < 0.2$ μM) (Cohen 1996). In basaltic volcanic areas, the concentration can go up by an order of magnitude. Breathable air contains 0.25–75 ng V m^{-3} in rural areas, and 60–300 ng m^{-3} in urban settings (Cohen 1996). The daily overall intake, dominated by dietary sources, amounts to 0.01–2 mg vanadium (Cohen 1996; Scior et al. 2005). Less than 1% of this remains absorbed: vanadium, in particular in its oxidation state +IV, is effectively excreted through feces and urine. The no-effect level is an intake of about 10 mg kg^{-1} day^{-1}; hence, normal diet-related exposure to vanadium does not affect our health.

High exposure risks (>30 mg m^{-3} in the breathable air), coupled with potentially severe health problems, go along with the mining and milling of vanadium containing ores, the production of vanadium metal, vanadium oxides used in redox catalysis and batteries (Tracey et al. 2007), and fly ashes from oil firing. The latter problem arises from the sometimes high amounts of vanadium (present as vanadyl porphyrins) in crude oil; cf. Sect. 11.3. The worldwide industrial emission of vanadium into the atmosphere totaled 71,000 tons per year in 1995 (Nriagu and Pirrone 1998), which may be extrapolated to 85,000 tons in 2010. This compares to an estimated world-wide emission of vanadium from natural sources (continental dust, volcanic activity, forest fires) of approximately 10 tons per year.

The toxicity of vanadium compounds is generally low. Orally applied, pentavalent compounds (vanadates) are more toxic than tetravalent compounds (vanadyl) because the latter form almost insoluble $VO(OH)_2$ in the gastro-intestinal tract. In the US, vanadyl sulfate $VOSO_4$ is a common supplement, in daily doses up to 60 mg, for athletes for an alleged increase their muscle mass (Barceloux 1999). Vanadium pentoxide V_2O_5, the main aerial contaminant, is potentially mutagenic and teratogenic. In mice, V_2O_5 causes pulmonary inflammation and tumor promotion (Rondini et al. 2010). In rats, vanadium compounds cause DNA cleavage, likely as a results of the intermittent formation of reactive hydroxyl radicals (Sakurai 1994); (11.4) in Sect. 11.2. Vanadium oxides may therefore be considered "suspected carcinogenic compounds" for humans: In workers exposed to the inhalation of V_2O_5, its metabolites cause oxidation of DNA, affect DNA repair, and induce the formation of tumor-associated antigens (Ehrlich et al. 2008). On the other hand, vanadium compounds such as sodium vanadate, $VO(acac)_2$ (acac = acetylacetonate(1-)), and $VO(maltolate)_2$ (bis(maltolato-oxidovanadium(IV), BMOV) inhibit the proliferation of hepatoma cells to a higher extent than of hepatic cells, which is paralleled by lower levels of ROS in hepatoma than in hepatic cells (Wang et al. 2010).

Human vanadium poisoning symptoms are mainly restricted to the conjunctivae and respiratory system, renal and gastrointestinal irritation. Exposure can thus give rise to conjunctivitis, rhinitis, pulmonary inflammation resulting in bronchitis and asthma-like diseases, and dysfunctions of the digestive system. The limit value for immediate danger to health for an average human is about 7 mg V in the case of intravenous application, and 35 mg V m^{-3} in breathing air. The following compilation lists selected official exposure limits (MAC) and LD_{50}/LC_{50} values. MAC refers to the maximum allowable concentration at the workplace (40-h week, 8-h time-weighted average). LD_{50} and LC_{50} indicate the level of a harmful substance (in mg per kg body weight) causing the death of 50% of the test animals by oral (LD) or inhalative (LC) administration, respectively.

- V_2O_5: MAC: 0.05 mg m^{-3}
- V_2O_5: LD_{50} (rat, oral) 10 mg kg^{-1}
- V_2O_5: LD_{50} (mouse, oral): 5 mg kg^{-1}
- V_2O_5: LC_{50} (rat) 126 mg m^{-3}, 6 h exposure

A biological threshold limit of 50 μg V g^{-1} of creatinine in urine collected at the end of the work week has been adopted in the US. Vanadium is also toxic for aquatic organisms: LD_{50} values are 4.8 mg L^{-1} for soft water, and 30 mg^{-1} for hard water. The lower toxicity in hard water reflects the formation of sparingly soluble calcium vanadates.

Irritation of the conjunctivae and pulmonary systems either directly by vanadium oxides (V_2O_5, V_2O_4, V_2O_3) and/or the oxidovanadium moieties VO_2^+, VO^{3+}, and VO^{2+} formed by solubilization of the oxides in the physiological systems goes along with the generation of reactive oxygen species (H_2O_2, and the radicals $O_2{\bullet}^-$ and •OH), considered to be the actual agents responsible for tissue impairment, either by oxidative damage and/or intervention with the phosphorylation of signaling and transcription pathways (Rondini et al. 2010). In hepatic cells, *N*-acetylcysteine can ameliorate this (indirect) cytotoxicity (Wang et al. 2010). While oxidovanadium species are usually considered to generate ROS, they can also consume ROS, e.g., in the course of the oxidation of vanadyl to vanadate(V) reversal of (11.4) in Sect. 11.2. Further, trivanadate $V_3O_9^{3-}$, and VO_2^+, complexed to sylicylidenehydrazide ligands, have been shown to consume alkylating toxins (Hamilton et al. 2006; Fautch et al. 2009), thus preventing DNA alkylation and, concomitantly, cancer risk.

Physiological effects of vanadium applied in the form of vanadate or simple vanadium complexes also arise from its direct intervention with phosphatases, phosphorylation enzymes, kinases, ribonucleases, and the phosphate metabolism in general (Tracey et al. 2007; McLauchlan et al. 2010). These interventions go back to the structural similarity between vanadate $VO(OH)_3^-/H_2VO_4^-$ and phosphate HPO_4^{2-} on the one hand (cf. also Sect. 11.1), and the ability of the transition metal vanadium to enlarge its coordination sphere and thus to form stable penta- and hexa-coordinated coordination compounds on the other hand. In contrast, penta-coordinated phosphorus only exists as a transitory species in, e.g., phosphatase reactions. In Fig. 11.4, this is pictured for the pentavalent transition state formed in the course of the phosphoester cleavage by a phosphatase with histidine in the active center, and the inhibition of this hydrolysis through the build-in of vanadate into the active site of the enzyme. The awareness of the role of vanadate as a phosphatase inhibitor goes back to the discovery of the switch-off of the Na^+, K^+-pump by sodium vanadate (Cantley et al. 1977). Trace amounts of vanadium have since been proposed to be an essential regulatory nutrient for most if not all living beings. The omnipresence of oxidic vanadium in soil, and of vanadate in the aqueous medium, has likely provoked an adaption – in terms of beneficial use – to vanadium already in the primordial development of life (Rehder 2008b). This is also suggested by the similarity of the active centers of vanadate-dependent haloperoxidases and vanadate-inhibited phosphatases (**2** in Fig. 11.3). Interestingly, vanadate-inhibited phosphatases can exhibit some haloperoxidase activity (Tanaka et al. 2002).

The vanadate–phosphate analogy (Stankiewicz and Tracey 1995; Crans et al. 2004; Steens et al. 2009) is also the key to the potency, in the treatment of diabetes mellitus, of vanadate, vanadyl and simple vanadium compounds such as BMOV and its ethyl analog, *bis*(ethylmaltolato)oxidovanadium BEOV. Here, vanadium acts as a regulator of glucose homeostasis and inhibits free fatty acid release.

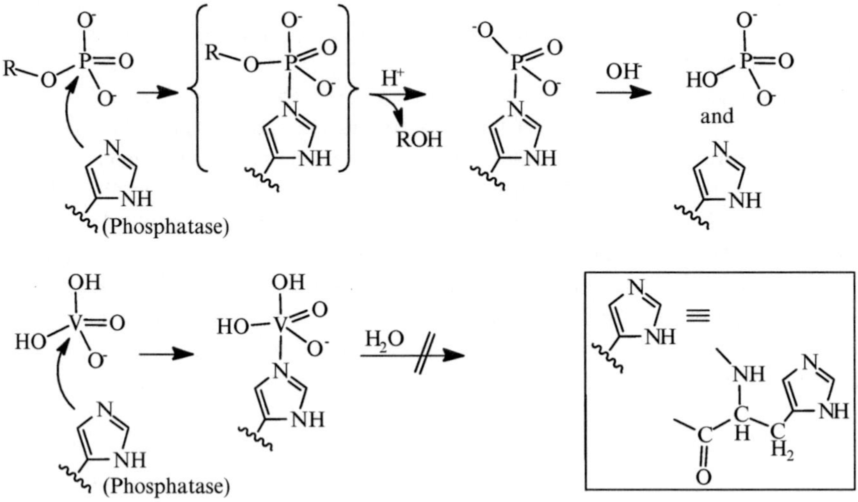

Fig. 11.4 Hydrolysis of the phospho-ester bond catalyzed by purple acid phosphatase via a pentavalent transition state (*top*), and inhibition of the phosphatase by vanadate (*bottom left*) (Rehder 2010; modified). For the active center see also **2** in Fig. 11.3, Sect. 11.3

The active species likely is vanadate $H_2VO_4^-$, a key "end-products" in physiological turn-over of vanadium compounds (vide infra). The mode of action possibly is by inhibition of a protein tyrosine phosphatase at the cytosolic site of the cellular insulin receptor and/or the activation of a tyrosine kinase in the signaling path (Sakurai et al. 2006). So far, BEOV has been the only vanadium compound to be subjected to clinical tests (phase II). The tests have, however, been abandoned due to renal problems with some of the probands.

Other beneficial modes of action of vanadate compounds, as tested ex vivo (with cell cultures) and, in part, in vivo (with animals) include the treatment of certain cancer forms (such as leukemia, Ehrlich ascites tumors, carcinomas of the lungs, prostate, testes, ovaria, and liver), amoebiasis (Maurya et al. 2006), tuberculosis, HIV, and herpes. VO^{2+} complexed to sugars such as the disaccharide trehalose promotes both, glucose consumption in osteoblasts and inhibition of the proliferation of tumoral osteoblasts (Barrio et al. 2003; Etcheverry et al. 2009).

11.5 Uptake, Speciation, Excretion, and Detoxification

The main pathways for uptake, speciation, distribution, and excretion of vanadium compounds are illustrated in Fig. 11.5. Dietary vanadium may be vanadate, vanadyl (VO^{2+}), e.g., in the form of $VOSO_4$, or vanadium compounds with organic ligands (symbolized {V} in Fig. 11.5), such as $VO(acac)_2$ and BEOV. Inhaled vanadium species, mainly ingested by the pulmonary tissue, commonly are particulate vanadium oxides (V_2O_5, V_2O_4, V_2O_3) or essentially oxidic vanadium minerals. Basic

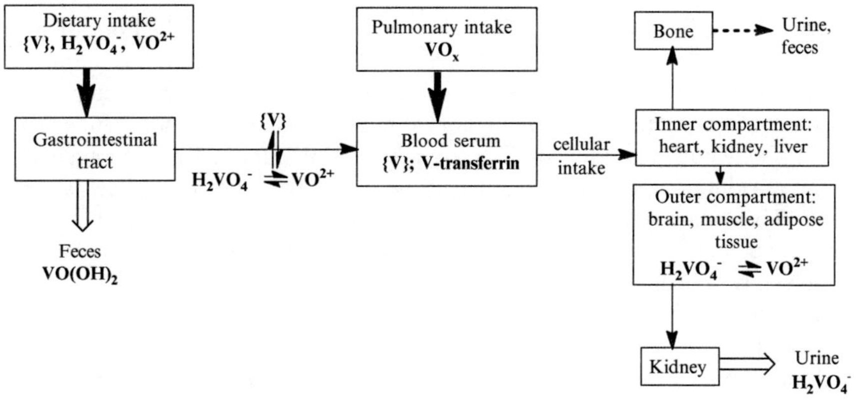

Fig. 11.5 Pathways of vanadium compounds taken in via nutrition or inhalation. {V} stands for any vanadium compound other than vanadate and bare vanadyl, VO_x for any vanadium oxide and particulate vanadium minerals

dietary vanadium species will be oxidatively converted to vanadate(V) in the oral cavity, transformed to decavanadate or even VO_2^+ (Sect. 11.1) in the acidic medium of the stomach, and reconverted to vanadate and vanadyl under the slightly alkaline and reducing conditions in the small intestines. Vanadium *complexes* may or may not survive the medial conditions of the gastrointestinal tract. In case they are unstable under the conditions prevailing there (such as the low pH in the stomach), decomposition and generation of vanadate/vanadyl takes place. Once transferred into the blood, most of the vanadium is complexed by transferrin (Tf) and apoTf. The complexation constant for the binding of VO^{2+} to apoTf, as defined by (11.12), is $10^{-14.7}$ (Kiss et al. 2006), i.e., only those complex {V} species survive the transport through the blood which are thermodynamically more stable than VO^{2+}-Tf, or whose degradation is kinetically hampered. VO^{2+}-Tf can be taken up by the cells via endocytosis; vanadate may also directly enter the cells through phosphate channels. Once arrived within the tissue cells, even stable complexes will be converted to basic inorganic vanadium compounds. Given the commonly reducing conditions in the cytosol, provided by reductants such as NAD(P)H, $FADH_2$, glutathione, ascorbate, and catecholamines, the predominant cytosolic species likely is VO^{2+}, stabilized to some extent by coordination to cytosolic constituents capable of ligating vanadium (and here again, glutathione and ascorbate come in; Rehder et al. 2002). Where VO^{2+} becomes involved in the production of ROS (11.1), or where ROS are formed from other sources, vanadate(V) $H_2VO_4^-$ will form.

$$VO^{2+} + \text{apoTf} \leftrightarrows VO^{2+}\text{–Tf} \qquad (11.12)$$

The retention time of vanadium in tissue is up to 30 min (Yasui et al. 2002). Part of the vanadium is, however, stored in the bones, where it's half-life is 4–5 days (Setyawati et al. 1998): VO^{2+} is absorbed to the bone surface, while $H_2VO_4^-$ is incorporated in the hydroxyapatite lattice, where it can replace phosphate

(Rehder 2008a; Etcheverry et al. 2009). The final excretion of resorbed vanadium is essentially via the kidneys, i.e., most of the vanadium is recovered in the urine.

Vanadium concentrations in serum after intravenous injection to humans in the form of vanadate in a 20% albumin infusion (containing a total amount of 47.6 μg V) declined rapidly within a few hours (<30% after 24 h). Subsequently, serum vanadium concentrations dropped more slowly, approaching zero after a month (Heinemann et al. 2003). In general, vanadium's toxicological and pharmacological potential very much depends on the mode of vanadium intake/application (intravenous, subcutaneous, oral, inhalation, absorption through mucosae), individual response, and modulating factors such as age, sex, exercise, stress, and the nutritional state (Thompson et al. 1998).

While global anthropogenic vanadium emission is not likely to constitute a significant health risk, local exposure of workers in industrial enterprises processing vanadium ores and compounds can result in severe health problems caused by vanadium toxication (see Sect. 11.4). Also, local increase of vanadium contents in soil, originating from the deposition of vanadium oxides in the course of industrial activities or excessive local burning of fossil fuels, may cause a potential health problem for the population in the respective area. Vanadium may further be mobilized from its deposits, and thus arrive in surface and ground water, eventually ending up in drinking water. Mobilization can come about by solubilization of oxidic vanadium, i.e., the formation of water-soluble vanadate, in particular in acidic soils, and by complexation and thus solubilization of vanadyl VO^{2+} in non-oxic environments by siderophores, humic acids, and other organic constituents with suitable ligand properties. Even at vanadium levels <15 μg L^{-1}, the notification level set for drinking water by the US Environmental Protection Agency, accumulation of vanadium up to 2% by weight in the corrosion deposits in lead drinking water pipes can occur – a potential reservoir for human exposure by municipal water systems, if this vanadium becomes re-mobilized by alterations in the drinking water characteristics (Gerke et al. 2009).

The formation of vanadinite, $PbCl_2{\cdot}3Pb_3[V^VO_4]_2$, or $Pb_5[VO_4]_3Cl$ for short, and the re-mobilization of vanadate can be formulated as depicted in the equilibrium (11.13), the Pb^{2+} ions being delivered via, e.g., divalent lead (hydroxy)carbonates and lead phosphate in the lead pipe. Equation (11.13) also demonstrates that remobilization of vanadate from insoluble vanadinite takes place as the medium becomes sufficiently acidic. At higher phosphate concentrations, chloropyromorphite $Pb_5[PO_4]_3Cl$ is more stable than vanadinite (Gerke et al. 2009); excess phosphate thus can destabilize vanadinite. Hence, while the formation of vanadinite deposits in the pipe scales is a means of *de*toxification, a decrease of pH or the presence of excess phosphate (such as provided by orthophosphate-based corrosion inhibitors) can result in *re*-toxification of drinking water (11.14).

$$5Pb^{2+} + 3H_2VO_4{}^- + Cl^- \leftrightarrows Pb_5[VO_4]_3Cl\downarrow + 6H^+ \quad (11.13)$$

$$Pb_5[VO_4]_3Cl + 3HPO_4{}^{2-} + 3H^+ \rightarrow Pb_5[PO_4]_3Cl\downarrow + 3H_2VO_4{}^- \quad (11.14)$$

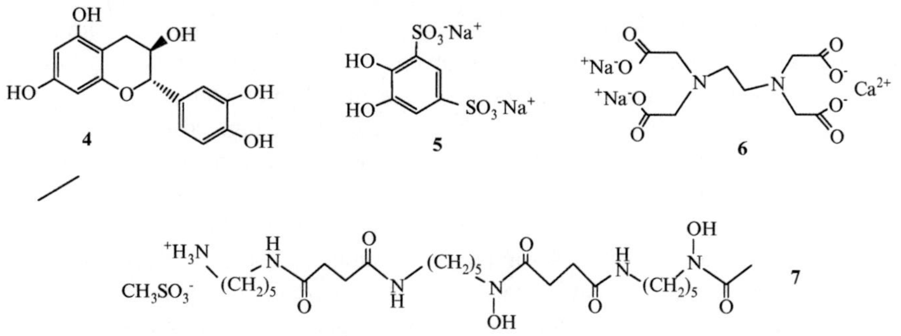

Fig. 11.6 Compounds which have been employed in animal studies to successfully mask vanadium: (−)Catechin (**4**), the disodium salt of Tiron (**5**), the Ca,Na_2 salt of EDTA (**6**) and deferoxamine mesylate (**7**)

Most cases of vanadium toxication will recover on removal from exposure and *symptomatic* treatment. In case of vanadium ingestion, application of ascorbic acid, followed by 2,3-mercapto-1-propanol in a later stage has been recommended (International Programme on Chemical Safety, Health and Safety Guide, no. 42, 1990). Both agents are likely to effect reduction of vanadium(V) to vanadium(IV), which transforms into insoluble and hence essentially harmless vanadyl hydroxides in the small intestines. By extrapolation from animal studies (mice, rats, rabbits, calves), treatment of toxic effects related to vanadium might be achieved by drinking green tea (Soussi et al. 2009), or by treatment with chelating agents such as ethylenediaminetetraacetate (EDTA) (Domingo et al. 1986; Gummow et al. 2006), deferoxamine mesylate (Domingo et al. 1986), or 4,5-dihydroxy-1,3-benzene disulphonate (Tiron) (Shrivastava et al. 2007); Fig. 11.6. Green tea from *Camellia sinensis* contains oligophenols such as catechins (**4** in Fig. 11.6), which can act as antioxidants and as chelators for metal ions such as VO^{2+}, VO^{3+}, and VO_2^+. Similarly, Tiron (**5**) combines these two properties, while EDTA (**6**) and deferoxamine (**7** in Fig. 11.6) are strong chelating agents, which are also used in masking other metal ions.

11.6 Conclusion

The categorization of vanadium, by the WHO and national health organizations, as mutagenic, teratogenic, and potentially cancerogenic affords special awareness in handling this element, and precautions when exposed to it. In the light of the potential hazards, the approval, in North America, of vanadyl sulfate as a food additive ("vanadyl fuel") consumed by athletes for a putative increase of the muscle mass appears to be an unorthodox issue. On the other hand, since less than 1% of bare vanadyl VO^{2+} is absorbed, its adverse effect should be minimal.

On a general basis, exposure to vanadium present in the environment (drinking water, food, aerial dust) is a negligible problem. Acute vanadium poisoning has so far only been observed with workers directly exposed to inhalation of vanadium oxide at the working place, and with animals injected or fed high doses of vanadium compounds. As far as workplace exposure is concerned, a maximum allowable concentration (MAC value) of 0.05 g V m^{-3} has been assessed. The increasing use of vanadium compounds in catalysis (e.g., V_2O_4 in the production of sulfuric acid, vanadate esters and ester chlorides in polymerization catalysis), in silver vanadium oxide batteries, and in vanadium steels, may henceforward increase the contamination, by vanadium, of water resources and the atmosphere. Oncoming developments for industrial applications of vanadium include vanadiumoxide-based nanotubes, -rods and -wires, metal-organic frameworks (MOFs), composite vanadates/silicates, and large polyoxidometalates (POMs). Another source for potentially increasing vanadium pollution and hence increasing health hazards is the burning of petrol and other fossil fuels, which can contain high amounts of porphyrinogenic vanadium compounds. Finally, remobilization of vanadate from vanadinite accumulating in the scales of lead water pipes, is a potential problem. For the decontamination of soil and wet areas containing an overload of vanadium (V), bacteria which reduce V^V (V_2O_5, vanadate) to insoluble vanadyl hydroxide are an option. *Geobacter metallireducens* and *Shewanella oneidensis* are promising candidates.

As is common with elements which, at higher doses and under specific conditioning are toxic, beneficial effects, e.g., the treatment of cancer, amoebiasis, and diabetes mellitus may come in. Bis(ethylmaltolato)oxidovanadium(IV), BEOV, has been a bearer of hope in this respect for the treatment of diabetes for a couple of years. This potential insulin-enhancing drug has faced a draw-back on occasion of clinical phase II tests – which fact does not imply that similar vanadium compounds, or novel developments based on vanadium will be more successful and hence beneficial.

References

Andersson I, Gorzsás A, Kerezsi C, Tóth I, Pettersson L (2005) Speciation of the aqueous $H^+/H_2VO_4^-/H_2O_2$/phosphate system. Dalton Trans 3658–3666

Antipov AN, Lyalikova NN, L'vov NP (2000) Vanadium-binding protein excreted by vanadate-reducing bacteria. IUBMB Life 49:137–141

Barceloux DC (1999) Vanadium. J Toxicol Clin Toxicol 37:265–278

Barrio DA, Williams PAM, Cortizo AM, Etcheverry SB (2003) Synthesis of a new vanadyl(IV) complex with trehalose (TrVO): insulin-mimetic activities in osteoblast-like cells in culture. J Biol Inorg Chem 8:459–468

Bellenger JP, Arnaud-Neu F, Asfari Z, Myeni SCB, Stiefel EI, Kraepiel AML (2007) Complexation of oxoanions and cationic metals by the biscatecholate siderophore azotochelin. J Biol Inorg Chem 12:367–376

Berry RE, Armstrong EM, Beddoes RL, Collison D, Ertok SN, Halliwell M, Garner CD (1999) The structural characterization of amavadin. Angew Chem Int Ed Engl 38:795–798

Cantley LC Jr, Josephson L, Warner R, Yanagisawa M, Lechene C, Guidotti G (1977) Vanadate is a potent (sodium, potassium ion)-dependent ATPase inhibitor in ATP derived from muscle. J Biol Chem 252:7421–7423

Carpentier W, De Smet L, Van Beeumen J, Brigé A (2005) Respiration and growth of *Shewanella oneidensis* MR-1 using vanadate as the sole electron acceptor. J Bacteriol 187:3293–3301

Cohen MD (1996) Vanadium and its immunotoxicology. Toxicol Ecotoxicol News 3:132–135

Crans D, Smee JJ, Gaidamauskas E, Yang L (2004) The chemistry of vanadium and the biological activities exerted by vanadium compounds. Chem Rev 104:849–902

Domingo JL, Llobet JM, Tomas JM, Corbella J (1986) Influence of chelating agents on the toxicity, distribution and excretion of vanadium in mice. J Appl Toxicol 6:337–341

Eady RR (2003) Current status of structure function relationships of vanadium nitrogenase. Coord Chem Rev 237:23–30

Ehrlich VA, Nersesyan AK, Hoelzl C, Ferk F, Bichler J, Valic E, Schaffer A, Schulte-Herrmann R, Fenech M, Wagner K-H, Knasmüller S (2008) Inhalative exposure to vanadium pentoxide causes DNA damage in workers: results of a multiple end point study. Environ Health Perspect 116:1689–1693

Etcheverry SB, Ferrer EG, Gonzales-Baró AC, Parajón-Costa BS, Williams PAM (2009) Vanadis' charms: from the mythology to the bioinorganic chemistry. J Argent Chem Soc 97:127–150

Fautch JM, Fanwick PE, Wilker JJ (2009) Oxidovanadium complexes for the consumption of alkylating toxins. Eur J Inorg Chem 33–37

Gerke TL, Scheckel KG, Schock MR (2009) Identification and distribution of vanadinite ($Pb_5(V^{5+}O_4)_3Cl$) in lead pipe corrosion by-products. Environ Sci Technol 43:4412–4418

Giuli G, Pratesi G, Eeckhout SG, Paris E (2006) Divalent vanadium in Vigarano CV3 meteorite chondrules. Proceedings of the 69th annual meeting of the meteoritical society meeting, Zurich, Switzerland, p 5148, 6–11 Aug 2006

Gummow B, Botha CJ, Williams MC (2006) Chronic vanadium poisoning in calves and its treatment with calcium disodium ethylenediaminetetraacetate. Vet Res Commun 30:807–822

Hamilton EE, Fanwick PE, Wilker JJ (2006) Alkylation of inorganic oxo compounds, and insight on preventing DNA damage. J Am Chem Soc 128:3388–3395

Heinemann G, Fichtl B, Vogt W (2003) Pharmacokinetiks of vanadium in humans after intravenous administration of vanadium containing albumin solution. J Clin Pharmacol 55:241–245

Kiss T, Jakusch S, Bouhsina S, Sakurai H, Enyedy A (2006) Binding constant of $V^{IV}O$ to transferrin. Eur J Inorg Chem 3607–3613

Lee CC, Hu Y, Ribbe MW (2010) Vanadium nitrogenase reduces CO. Science 329:642

Léonard A, Gerber GB (1994) Mutagenicity, carcinogenicity and teratogenicity of vanadium compounds. Mutat Res 317:81–88

Leuschner J, Haschke H, Sturm G (1994) New investigations on acute toxicities of vanadium oxides. Monatsh Chem 125:623–646

Maurya MR, Kumar A, Azam A, Bader C, Rehder D (2006) Dioxo- and oxovanadium(V) complexes of thiohydrazone ONS donor ligands: synthesis, characterization, reactivity, and antiamoebic activity. Inorg Chem 45:1260–1269

McLauchlan CC, Hooker JD, Jones MA, Dymon Z, Backhus EA, Greiner BA, Dorner NA, Youkhana MA, Manus LM (2010) Inhibition of acid, alkaline, and tyrosine (PTP1B) phosphatases by novel vanadium complexes. J Inorg Biochem 104:274–281

Nriagu JO, Pirrone N (1998) Emission of vanadium into the atmosphere. In: Nriagu JO (ed) Vanadium in the environment, part 1. Wiley, New York

Ortiz-Bernad I, Anderson RT, Vrionis HA, Loveley DR (2004) Vanadium respiration by *Geobacter metallireducens*: novel strategy for in situ removal of vanadium from groundwater. Microbiology 70:3091–3095

Rehder D (2008a) Bioinorganic vanadium chemistry. Wiley, Chichester

Rehder D (2008b) Is vanadium a more versatile target in the activity of primordial life forms than hitherto anticipated? Org Biomol Chem 6:957–964

Rehder D (2010) Leben ohne Vanadium? Bioanorganische Chemie des Vanadiums. Chem Unserer Zeit 44:322–331

Rehder D, Costa Pessoa J, Geraldes CFGC, Castro MMCA, Kabanos T, Kiss T, Meier B, Micera G, Pettersson L, Rangel M, Salifoglou A, Turel I, Wang D (2002) *In vitro* study of the insulin-mimetic behaviour of vanadium(IV, V) co-ordination compounds. J Biol Inorg Chem 7:384–396

Rondini EA, Walters DM, Bauer AK (2010) Vanadium pentoxide induces pulmonary inflammation and tumor promotion in a strain-dependent manner. Part Fiber Toxicol 7:9

Sakurai H (1994) Vanadium distribution in rats and DNA cleavage by vanadyl complex: implications for vanadium toxicity and biological effects. Environ Health Perspect 102:35–36

Sakurai H, Kato A, Yoshikawa Y (2006) Chemistry and biochemistry of insulin-mimetic vanadium and zinc complexes. Trial for treatment of diabetes mellitus. Bull Chem Soc Jpn 79:1645–1664

Schmidt H, Andersson I, Rehder D, Pettersson L (2001) A potentiometric ^{51}V NMR study of the aqueous $H^+/H_2VO_4^-/H_2O_2$/L-α-alanyl-L-histidine system. Chem Eur J 7:251–257

Scior T, Guevara-García A, Bernard P, Do Q-T, Domeyer D, Laufer S (2005) Are vanadium compounds drugable? Structure and effects of antidiabetic vanadium compounds: a critical review. Mini Rev Med Chem 5:995–1008

Setyawati IA, Thompson KH, Yuen VG, Sun Y, Battel M, Lyster DM, Vo C, Ruth TJ, Zeissler S, McNeill JH, Orvig C (1998) Kinetic analysis and comparison of uptake, distribution and excretion of ^{48}V labeled compounds in rats. J Appl Physiol 84:569–575

Shrivastava S, Jadon A, Shukla S, Mathur R (2007) Chelation therapy and vanadium: effect on reproductive organs in rats. Ind J Exp Biol 45:515–523

Soussi A, Murat JC, Gaubin Y, Croute F, Soleilhavoup JP, El Feki A (2009) Green tea drinking reduces the effects of vanadium poisoning in rat kidney. Food Sci Technol Res 15:413–422

Stankiewicz PJ, Tracey AS (1995) Stimulation of enzyme activity by oxovanadium complexes. In: Sigel H, Sigel A (eds) Vanadium and its role in life, vol 31, Metal ions in biological systems. Marcel Dekker, New York

Steens N, Ramadam AM, Parac-Vogt TN (2009) When structural and electronic analogy lead to reactivity: the unprecedented phosphodiesterase activity of vanadates. Chem Commun 965–967

Tanaka N, Dumay V, Liao Q, Lange AJ, Wever R (2002) Bromoperoxidase activity of vanadate-substituted acid phosphatases from *Shigella flexneri* and *Salmonella enterica ser typhimurium*. Eur J Biochem 269:2162–2167

Thompson KH, Battell M, McNeill JH (1998) Toxicology of vanadium in mammals. In: Nriagu JO (ed) Vanadium in the environment, part 2. Wiley, New York

Tracey AS, Willsky GR, Takeuchi ES (2007) Vanadium, chemistry, pharmacology and practical applications. CRC, Boca Raton

Ueki T, Kawakami N, Toshishige M, Matsuo K, Gekko K, Michibata H (2009) Characterization of vanadium-binding sites of the vanadium-binding protein vanabin 2 by site-directed mutagenesis. Biochim Biophys Acta 1790:1327–1333

Wang Q, Liu T-T, Fu Y, Wang K, Yang X-G (2010) Vanadium compounds discriminate hepatoma and normal hepatic cells by differential regulation of reactive oxygen species. J Biol Inorg Chem 15:1087–1097

Wenning R, Kirsch N (1988) Vanadium. In: Seiter HG, Sigel H, Sigel A (eds) Handbook of inorganic compounds. Marcel Dekker, New York

Yasui H, Tamura A, Takino T, Sakurai H (2002) Structure-dependent metallokinetics of antidiabetic vanadyl-picolinate complexes in rats: studies on solution structure, insolinomimetic activity and metallokinetics. J Inorg Biochem 91:327–338

Chapter 12
Microbial Remediation of Arsenic Contaminated Soil

Tapan Jyoti Purakayastha

12.1 Introduction

Arsenic (As) contamination in terrestrial and aquatic ecosystems is a very sensitive environmental issue due to its adverse impact on human health. It enters into the terrestrial and aquatic ecosystems through a combination of natural processes such as weathering reactions, biological activity, and volcanic emissions, as well as a result of anthropogenic activities. Excessive use of As-based pesticides and indiscriminate disposal of domestic (sewage) and industrial (timber, tannery, paints, electroplating, etc.) wastes, as well as mining activities, have resulted in widespread As contamination of soils and waterways. Consequently, thousands of As-contaminated sites have been reported around the world (Fig. 12.1, USEPA 1997; ETCS 1998; Smith et al. 1998; Eisler 2004; Lièvremont et al. 2009). The problem of As contamination in groundwater of West Bengal (India) and Bangladesh has been reported due to its serious health hazard. The most severely contaminated areas include West Bengal, India (Das et al. 1996; Chakraborti et al. 2002) and Bangladesh (Nickson et al. 1998; Ohno et al. 2005). The first reported patient with arsenical dermatosis in West Bengal was diagnosed in 1983 (Chakraborty and Saha 1987). Patients with skin disease caused by arsenic have also been observed in Bangladesh (Karim 2000). Dhar et al. (1997) estimated that more than 50 million people in Bangladesh were at risk from arsenic, while Hossain (2006) estimated that 85 million were at risk. There has been a rapidly growing global concern for arsenic contamination in drinking water (WHO 2001). Out of 20 countries in different parts of the world where groundwater arsenic contamination and human suffering therefrom has been reported so far (WHO 2001), the magnitude is considered to be the highest in Bangladesh, followed by West

T.J. Purakayastha (✉)
Division of Soil Science and Agricultural Chemistry, Indian Agricultural Research Institute, New Delhi 110012, India
e-mail: tpurakayastha@gmail.com

I. Sherameti and A. Varma (eds.), *Detoxification of Heavy Metals*, Soil Biology 30,
DOI 10.1007/978-3-642-21408-0_12,

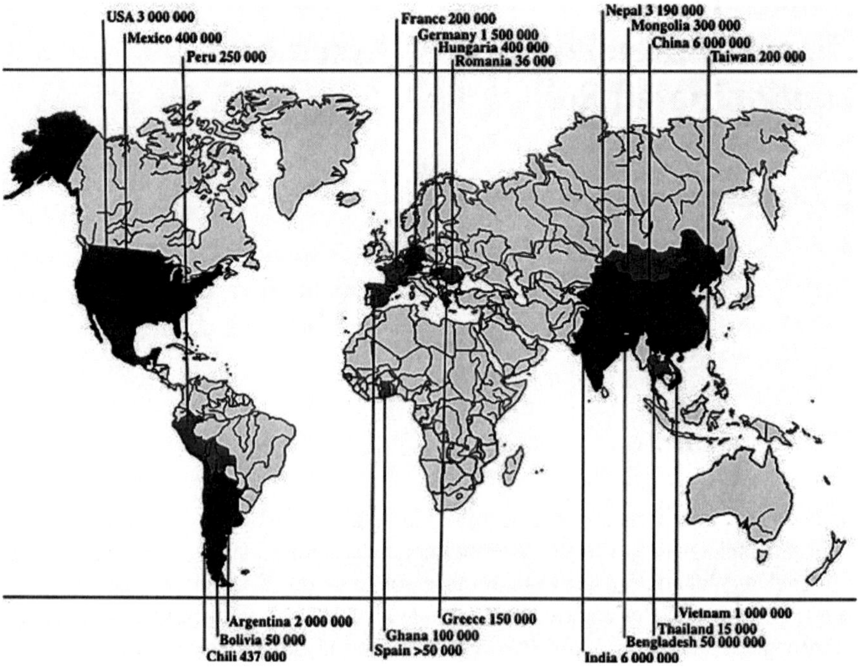

Fig. 12.1 World map of populations at risk, based on the data currently available in the literature (according to Smedley and Kinniburgh (2002). The figures give the number of people whose daily water consumption includes arsenic levels above 10 mg L^{-1}. The shades of *gray* indicate the number of persons contaminated, working from the palest (the lowest numbers) to the *darkest shade* (the highest number) (from Lièvremont et al. 2009)

Bengal, India (Rahman et al. 2001; Chowdhury et al. 2000; Sanyal and Dhillon 2005). In West Bengal, summer rice is grown extensively with As contaminated groundwater in this belt. Therefore, there is a potential risk of As contamination in rice crop leading to food chain contamination. Continuous use of groundwater for crop irrigations in the arsenic belt of West Bengal significantly increased the As level in soils (Sanyal and Dhillon 2005). The input of As to soil may prove detrimental to plant through its uptake to the toxic limit, thereby facilitating its entry into the food chain.

With greater public awareness of As poisoning in animal and human nutrition, there has been growing interest in developing guidelines and remediation technologies for mitigating As-contaminated ecosystems. A range of technologies, including chemical immobilization and bioremediation, has been applied with varying levels of success either to completely remove As from the system or to reduce its biotoxicity. Bioremediation with special reference to microbial approach, an emerging technology that uses microbes to remove or stabilize contaminants, may offer a low cost and ecologically viable means for the mitigation of As toxicity in the environment.

12.2 Sources of Arsenic in Soil

Arsenic occurs naturally as an element, ranks as the 20th most occurring trace element in the earth's crust (NRC 1977) and is widely distributed in the environment. The ultimate source of arsenic on the Earth's surface is igneous activity (Nriagu 1994). Arsenic is widely spread in the upper crust of the Earth, although mainly at very low concentrations, with arsenic concentrations in soil ranging from 0.1 to more than 1,000 ppm (mg kg^{-1}). In atmospheric dust, the range is 503,400 ppm. In seawater, the average arsenic level may be 2.6 ppb and in fresh water about 0.4 ppb. Arsenic at significant levels is all around us (Mukhopadhyay et al. 2002). The global arsenic geocycle elucidates how arsenic enters into the soil, sediment, water, and food chain of living organisms (Fig. 12.2). The major source of As contamination is from naturally existing minerals; however, anthropogenic activities have also contributed extensively (Nordstrom 2002). A range of As compounds, both organic and inorganic, are introduced into the environment through geological (geogenic) and anthropogenic (human activities) sources. Small amounts of As also enter the soil and water through various biological sources (biogenic) that are rich in As. Although the anthropogenic source of As contamination is increasingly becoming important, it should be pointed out that the recent episode of extensive As contamination of groundwaters in Bangladesh and the Indian state of West Bengal is of geological origin, transported by rivers from sedimentary rocks in the Himalayas over tens of thousands of years, rather than anthropogenic.

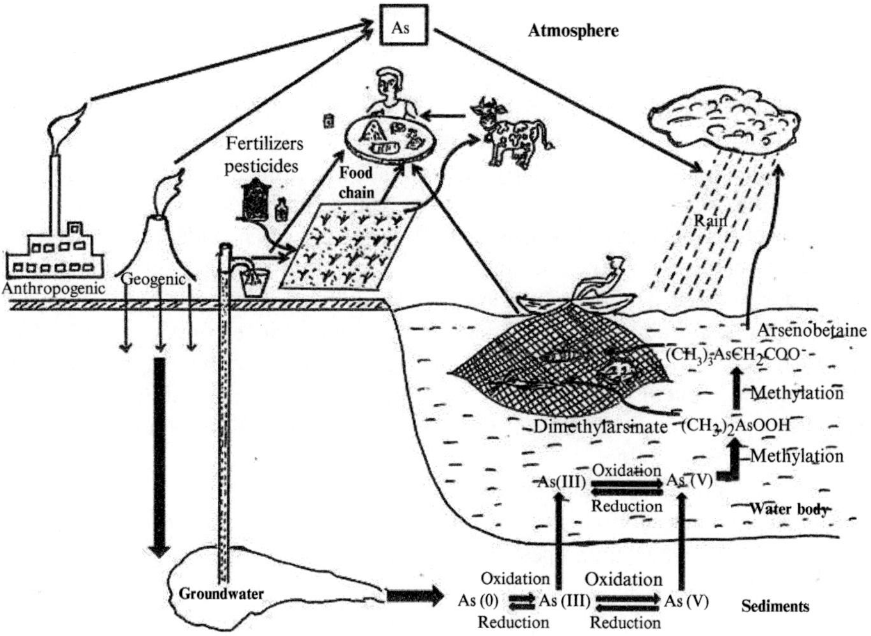

Fig. 12.2 Arsenic geocycle and food chain contamination

Arsenic exists in several oxidation states (−3, 0, +3, and +5), enabling it to mobilize under various environmental conditions and hinders many remediation technologies from efficiently removing it from water. Under oxidizing conditions, As(V) is the dominant form at lower pH while As(III) becomes dominant at higher pH. However, the uncharged form of As(III) [$As(OH)_3$] becomes dominant under reducing environments, which is more toxic and difficult to remove (Smedley and Kinniburgh 2002). Its association with some non-weathering-resistant mineral deposits (e.g., sulfide minerals) has contributed to its release in large amounts into the environment (Murdoch and Clair 1986).

Arsenic is also being introduced into the environment through various anthropogenic activities. These sources release As compounds that differ greatly in chemical nature (speciation) and bioavailability. Major sources of As discharged onto land originate from commercial wastes (40%), coal ash (22%), mining industry (16%), and the atmospheric fallout from the steel industry (13%) (Eisler 2004; Nriagu and Pacyna 1988). Arsenic trioxide (As_2O_3) is used extensively in the manufacturing of ceramic and glass, electronics, pigments and antifouling agents, cosmetics, fireworks, and Cu-based alloys (Leonard 1991). Arsenic is also widely used for wood preservation in conjunction with Cu and chromium (Cr), i.e., copper-chromium-arsenate (CCA). The use of sodium arsenite ($NaAsO_2$) to control aquatic weeds has contaminated small fish ponds and lakes in several parts of the United States with As (Adriano 2001). Arsenic contamination in soil was also reported due to the arsenical pesticides used in sheep and cattle dips to control ticks, fleas, and lice (McBride et al. 1998; McLaren et al. 1998). A study of 11 dip sites in New South Wales indicated considerable surface soil (0–10 cm) contamination with As (37–3,542 mg kg^{-1}) and significant movement of As (57–2,282 mg kg^{-1}) down the soil profile at 20–40 cm depth (McLaren et al. 1998). Continuous application of phosphatic fertilizers that contain trace levels of As also results in As contamination of soil (Peng et al. 2011), thereby reaching the food chain through plant uptake (McLaughlin et al. 1996). Similarly, in New Zealand, timber treatment effluent is considered to be the major source of As contamination in aquatic and terrestrial environments (Bolan and Thiyagarajan 2001). Because As is widely distributed in the sulfide ores of Pb, Zn, Au, and Cu, it is released during their mining and smelting processes. The flue gases and particulate from smelters can contaminate nearby ecosystems downwind from the operation with a range of toxic metal(loid)s, including As (Adriano 2001). Coal combustion not only releases gaseous As into the atmosphere, but also generates fly and bottom ash containing varied amounts of As. Disposal of these materials often leads to As contamination of soil and water (Beretka and Nelson 1994).

Arsenic is present in many pesticides, herbicides, and fertilizers. Arsenic may accumulate in agricultural soils due to the agricultural practices such as the applications of As-containing pesticides and herbicides, pig manure, and phosphorous fertilizers, and it has raised more concerns about the risk of As to the environment and human health (Chirenje et al. 2003; Brouwere et al. 2004; Li and Chen 2005). Industries that manufacture As-containing pesticides and herbicides release As-laden liquid and solid wastes that, upon disposal, are likely to contaminate soil and water bodies. For example, indiscriminate discharge of industrial effluents from the manufacturing of Paris Green pesticide $[Cu(CH_3COO)_2{\cdot}3Cu(AsO_2)_2]$ resulted in the contamination of soil and groundwater in residential area of Calcutta, India (Chatterjee and Mukherjee 1999). The use of horticultural pesticides, lead arsenate (PbAsO4), calcium arsenate ($CaAsO_4$), magnesium arsenate ($MgAsO_4$), zinc arsenate ($ZnAsO_4$), zinc arsenite $[Zn(AsO_2)_2]$, and Paris Green in orchards has contributed to soil As contamination in many parts of the world (Merry et al. 1983; Peryea and Creger 1994). Soil contamination due to the use of organoarsenical herbicides such as monosodium methanearsonate (MSMA) and disodium methanearsonate (DSMA) was also reported (Merry et al. 1983; Peryea and Creger 1994). The accumulated As in agricultural soils can distribute among different soil components, such as organic matter, iron (Fe) and manganese (Mn) oxides, carbonates and sulfides, and such distribution could affect its mobility, bioavailability, and toxicity (Cummings et al. 1999; Islam et al. 2004; deLemos et al. 2006). The distribution and redistribution process of As in soils can be influenced by microbial activities, because microbes could mediate the transformation of As and As adsorbents (Bentley and Chasteen 2002; Mukhopadhyay et al. 2002; Oremland and Stolz 2003, 2005; Islam et al. 2004). It is reported that microbially mediated As release to the groundwater for drinking has threatened the health of millions of people in Bangladesh, West Bengal, and some regions of China (Smith et al. 2000; Shen et al. 2005).

The adverse effects of As in groundwater used for irrigation water on crops and aquatic ecosystems is also of major concern. In addition to potential human health impacts caused by ingestion of food containing As, the potential for reduced crop yield due to its buildup in the soil is an active area of research. The fate of As in agricultural soils is often less well studied compared to groundwater, and in general has been studied in the context of As uptake by different plants (Huq et al. 2001; Das et al. 2004; Al Rmalli et al. 2005; Correll et al. 2006). Crop quality and the effect of As on crop quality and yield is becoming a major worldwide concern, particularly for rice which forms the staple for many South Asian countries where groundwater is widely used for irrigation (Meharg and Rahman 2003). In a recent study it was reported that irrigation has increased in Bangladesh since 1970, while since 1980, the area under groundwater irrigation for the cultivation of *Boro* (winter) rice has increased by almost an order of magnitude (Harvey et al. 2005). Based on available information on the distribution of As concentration in groundwater (BGS and DPHE British Geological Survey and Department of Public Health Engineering 2001) and the area under shallow tubewell irrigation (BADC 2005), Saha (2006) estimated that approximately 1,000 Mg of As is cycled with irrigation water during the dry season of each year. Rice yield has been reported to decrease by 10% at a concentration of 25 mg kg^{-1} As in soil (Xiong et al. 1987). Table 12.1 shows how As might accumulate in soil over time at different concentrations in irrigation water, assuming an annual water application of 1,000 mm. A soil irrigated with 1,000 mm of water containing 100 ppb As receives 1 kg ha^{-1} As per annum. The limited evidence at present suggests that the safe limit of soil As for rice lies somewhere in the range 25–50 mg kg^{-1} (Saha and Ali 2007; Duxbury and Panaullah 2007).

Table 12.1 Potential effect of arsenic concentrations in irrigation water on soils with time (from Brammer and Ravenscroft 2009)

Years of irrigation	Arsenic in irrigation water (ppb)				
	50	100	250	500	1,000
	Arsenic added to soil				
1	0.28	0.56	1.4	2.8	5.6
5	1.4	2.8	7	14	28
10	2.4	5.6	14	28	56
20	5.6	11	28	56	110
30	8.4	17	42	84	170
50	14	28	70	140	280

In Table 12.1, the soil arsenic concentration values between 25–50 and >50 mg kg^{-1} As have been shown when, in principle, these soil concentrations might be reached (Brammer and Ravenscroft 2009). Actual soil loading rates will vary with the amount of irrigation water applied, As concentrations in the water, and losses due to volatilization, leaching, and crop removal. Not all the As delivered by tube wells actually reaches the fields irrigated. In most As-affected areas of South and Southeast Asia, groundwater is rich in iron (e.g. DPHE 1999; Gurung et al. 2005; Postma et al. 2007). That iron is oxidized when the water is exposed to the air and is then precipitated as iron hydroxides which adsorb As. Hossain (2005) reported that topsoil arsenic concentrations at the Faridpur site, which had been irrigated for about 20 years, ranged from 61 mg kg^{-1} in the field nearest the wellhead to 11 mg kg^{-1} in a field near the far side of the command area. Williams et al. (2006), who measured As contents of 37 vegetables, pulses, and spices commonly grown in Bangladesh, found levels were highest in radish leaves (0.79 mg kg^{-1}), arum stolons, spinach, and cucumber, and lowest (0.2 mg kg^{-1}) in most fruits, vegetables, and spices. Roychowdhury et al. (2002) also found great differences between 30 crops and food items from 34 As-affected households in West Bengal, India, inter alia reporting a significant difference between potato skins (0.526 mg kg^{-1}) and potato flesh (0.00728 mg kg^{-1}). A greenhouse study by Abedin et al. (2002) revealed reduced yield of a local variety of rice (BR-11) irrigated with water having As concentrations in the range of 0.2–8 mg L^{-1}. The accumulation of As in rice field soils and its introduction into the food chain through uptake by the rice plant is of major concern (Duxbury et al. 2003).

12.3 Forms of Arsenic in Soil

Generally, As concentrations in uncontaminated soils seldom exceed 10 mg kg^{-1}. However, anthropogenic sources of As have elevated the background concentration of As in soils (Adriano 2001). For example, in areas near As mineral deposits, As levels in soils may reach up to 9,300 mg kg^{-1} (Ashley and Lottermoser 1999).

Depending on the nature of the geogenic and anthropogenic sources, As concentration in soils can range from <1 to 250,000 mg kg^{-1}. However, there is a large fluctuation among countries due to variation in soil parent material, for example, calcareous soils can be expected to have higher levels of As than noncalcareous soils (Ashley and Lottermoser 1999). Arsenic forms solid precipitates with Fe, aluminum (Al), calcium (Ca), magnesium (Mg), and Ni. A number of studies involving solid-phase speciation have shown that As is prevalent mostly in the oxalate fractions associated with amorphous and crystalline Fe and Al oxides, indicating the strong affinity of As for these soil components (Wenzel et al. 2001).

The common valence states of arsenic in nature include −3, 0, +3, and +5 (Leonard, 1991, Jain and Ali 2000). In soils, the most often encountered arsenic forms are inorganic As(III) (arsenite) and As(V) (arsenate) (Cullen and Reimer 1989; Balasoiu et al. 2001). Methylated species, monomethyl arsenic acid (MMAA), dimethyl arsenic acid (DMAA), and trimethyl arsine oxide (TMAO), dominate in biomass, but have also been detected in soils (Leonard 1991). In addition, As(V) and As(III) can be volatilized to arsine (AsH_3), MMAA to monomethyl arsine (CH_3AsH_2, MMA), DMAA to dimethylarsine [$(CH_3)_2AsH$, DMA], and TMAO to trimethyl arsine [$(CH_3)_3As$, TMA] (Cullen and Reimer 1989). The evolution of arsines is thus greatly dependent on the form of arsenic in soil and most often arsines are formed from methylated species (Gao and Burau 1997). The trivalent compounds are generally more toxic than the pentavalent compounds (Cervantes et al. 1994). The most toxic of them all is arsine gas (AsH_3; Leonard 1991). Organic arsenical compounds exist but these are generally nontoxic (Gochfeld 1995).

In solution at neutral pH, arsenic acid exists as the arsenate oxyanion (Rosen 2002). The p*K*a of arsenous acid is 9.2, so that, at neutral pH, it would be primarily present in solution as neutral $As(OH)_3$. As described later, this difference in p*K*a is relevant for the type of transport system that catalyzes uptake of the pentavalent and trivalent forms of arsenic. Compared to As(V), As(III) demonstrates greater mobility. The difference in mobility results from the high affinity of As(V) for Al, Fe, and Mn oxides (Cullen and Reimer 1989). The other aspect of arsenic chemistry relevant to biological activity is reactivity of As(III) as a soft metal ion, forming strong bonds with functional groups such as the thiolates of cysteine residues and the imidazolium nitrogens of histidine residues. The relative proportions of these oxidation states in a given environment depend on the biological processes involved as well as on the local physicochemical conditions, including the redox potential (Eh) and the hydrogen potential (pH), which are important factors. Since the p*K*a contents of arsenate (H_3AsO_4) are $pKa1 = 2.19$, $pKa2 = 6.94$ and $pKa3 = 11.5$, the H_2AsO_4 form predominates in oxidative media with pH levels below 6.9, whereas the $HAsO4^{2-}$ form predominates at higher pH levels. In the case of arsenite, the lowest p*K*a levels recorded are equal to 9.22. In most natural waters with pH levels below 9.2 as well as in slightly reductive environments, $As(OH)_3$ is the main form present (Inskeep et al. 2002). The solubility of arsenic and the resulting bioavailability of this element are closely related to its speciation. Several studies have shown that the reduction of arsenate into arsenite results in the solubilization of this element. However, arsenate may be sequestered after being co-precipitated with ferric iron (Foster 2003) or sulfur (O'Day et al. 2004) or adsorbed by clay, calcite, organic matter, or hydroxides, in particular ferric oxyhydroxides (Morin et al. 2003).

12.4 Arsenic Toxicity of Food Chain

Arsenic can enter through the food chain through drinking water or the food crops raised in arsenic contaminated soil (Fig. 12.2). There are evidences of elevated arsenic levels in the rice grain in regions of West Bengal and Bangladesh where rice fields are irrigated with arsenic contaminated waters (Rahman et al. 2009; Duxbury et al. 2003; Williams et al. 2005; Islam et al. 2004; Khan et al. 2009). Apart from rice crop, elevated levels of arsenic contamination in vegetables was also reported from Bangladesh (Williams et al. 2006). Global normal range of 0.08–0.2 mg As kg^{-1} has been suggested for rice (Zavala and Duxbury 2008) but values may reach as high as 1.8 mg As kg^{-1} have been found in Bangladesh rice (Meharg and Rahman 2003). The most notable case was observed in India and Bangladesh where over 50 million people were exposed to highly contaminated water or food (Hossain 2006). There have been reports of up to 2 mg kg^{-1} of arsenic accumulated in grains and up to 92 mg kg^{-1} of arsenic in straws (Abedin et al. 2002). Zhong et al. (2011) reported that there are different controls on the unloading of inorganic As and dimethylarsinic acid (DMA) in rice grain; the latter accumulated mainly in the caryopsis before flowering, whereas inorganic As was mainly transported into the caryopsis during grain filling. However, Carey et al. (2010) reported that arsenite was retained in the ovular vascular trace and DMA dispersed throughout the external grain parts and into the endosperm.

Arsenic being an carcinogenic element interferers with the cellular components in living creatures. Therefore, long-term intake of arsenic contaminated water and food would cause severe damage to the various metabolic systems in the living body. Arsenic (e.g., As^{3+}) is a potentially hazardous toxic element that interacts with sulfhydryl groups of proteins and enzymes (to denature the proteins and enzymes within the cells; Gebel 2000; Graeme and Pollack 1998) and reactive oxygen species in the cells, consequently causing cell damage (Ahmed et al. 2006). Arsenic can interfere with essential enzymatic functions and transcriptional events in cells, leading ultimately to "multitude of multisystemic noncancer effects that might ensue" (NRC 1995). For example, oxidative stress induced by trivalent methylated arsenicals inhibits glutathione (GSH) reductase (Styblo et al. 1997) and thioredoxin reductase (Lin et al. 1999) with subsequent impairment of cellular protective mechanism against oxidants. While depletion of cellular GSH sensitizes cells to arsenicals and may also contribute to cell transformation (Shimizu et al. 1998), thioredoxin depletion affects gene expression due to the fact that it modulates DNA binding activity of some transcriptional factors (Powis et al. 2000). Arsenite is known to inhibit more than 200 enzymes in the body (Abernathy et al.

1999) and, because arsenate has a similar structure as phosphate, it can substitute for phosphorus in the body, which can lead to replacement of phosphorus in the bone for many years (Arena and Drew 1986; Ellenhorn and Barceloux 1988). Because arsenate is hydrolyzed easily (in the cell), it prevents subsequent transfer of phosphate to adenosine diphosphate (ADP) to form adenosine triphosphate (ATP; the energy currency of the cell) and thus depletes the cell of its energy (Winship 1984). Arsine, the most toxic of the arsenicals (Buchet and Lauwerys 1983; Leonard 1991), is known to cause hemolysis of red blood cells, leading to hemolytic anemia, which is primarily responsible for the development of oliguria renal failure (Fowler and Weissberg 1974; Fowler 1977). It has been suggested that arsine interaction with sulfhydryl group of proteins and enzymes (Levinsky et al. 1970) may be responsible for inhibition of erythrocyte sodium–potassium pump. It also is known that arsenic decreases DNA repair process (Brochmoller et al. 2000) and, hence, enhances susceptibility to cancer (e.g., skin cancer; Wei et al. 1994) and non-cancer-related diseases (Feng et al. 2001).

Arsenic toxicity could affect a wide variety of organisms, including humans (Cervantes et al. 1994). Chronic arsenic effects in humans have been well documented and reviewed (e.g., Pershagen 1983). Organs most affected are those involved with arsenic in absorption, accumulation, and/or excretion. These organs are the gastrointestinal tract, circulatory system, liver, kidney, skin, tissues very sensitive to arsenic and those tissues secondarily affected (e.g., heart; Squibb and Fowler 1983). Signs of chronic arsenic toxicity include dermal lesions (e.g., hyperpigmentation, hyperkeratosis, desquamation, and loss of hair; Zaloga et al. 1985), peripheral neuropathy, skin cancer, and peripheral vascular disease. These signs have been observed mostly in populations whose drinking water contains arsenic (Tseng 1977; Tseng et al. 1968; Zaldivar 1980; Zaldivar and Ghai 1980; Cebrian et al. 1983; Smith et al. 2000). Among these symptoms, dermal lesions were most dominant, and were also known to occur within a period of about 5 years. The skin is known to localize and store arsenic because of its high keratin content, which contains several sulfhydryl groups to which As^{3+} may bind (Kitchin 2001) and may be the reason for its sensitivity to arsenic toxic effect. A study of Tseng (1977) in the Province of Taiwan (China) established a clear dose–response relationship between arsenic and dermal lesions, Blackfoot disease (BFD; a peripheral vascular disorder) and skin cancer. From several studies (Chen and Wu 1962; Chi and Blackwell 1968; Tseng 1977; Chen et al. 1988; Engel et al. 1994), it has been established that peripheral vascular diseases are associated with arsenic in well water in Taiwan. However, vascular disease has also been reported among German vintners (Grobe 1976), inhabitants of Antofagasta and Chile (Borgono et al. 1977). Skin cancers, including in situ cell carcinoma (or Bowen's disease), invasive cell carcinoma, and multiple basal cell carcinomas, are all known to be associated with chronic arsenic exposure (Shneidman and Belizaire 1986; ATSDR 1990). Chen et al. (1995) observed that hypertension was linked to long-term arsenic ingestion as well as cerebrovascular disease (i.e., cerebral infection). Other effects are hematopoietic depression, anhydremia (due to loss of fluid from blood into tissue and the gastrointestinal tract), liver damage characterized by jaundice, portal cirrhosis and ascites, sensory disturbance and peripheral neuritis, anorexia and loss of weight (Webb 1966). Although the effects of arsenic, as recounted previously, result in several kinds of diseases, it certainly may also impact adversely on the immune system, which may predispose to viral/bacterial infections. Several of such diseases resulting from alterations of the immunologic surveillance may not have been known to be due to arsenic and therefore may not have been attributed to arsenic effects. A probing into this area is therefore appropriate. The toxicity of specific arsenic chemical species is mentioned later (Oremland and Stolz 2005).

Arsenic occurs in four oxidation states: As^{5+}, As^{3+}, As^{0}, and As^{3-}. The two highest oxidation states are the most common in nature, whereas the two lowest are rare.

Arsenate: This oxyanion is an analog of phosphate, and as such it is a potent inhibitor of oxidative phosphorylation, the key reaction of energy metabolism in metazoans, including humans.

Arsenite: The most toxic of arsenic oxyanions. It readily binds to reactive sulfur atoms (SH goups) of many enzymes, including those involved in respiration.

Arsenic trioxide (As_2O_3): The most common form of arsenic used for a variety of agricultural, manufacturing, and medical purposes. It is highly toxic, and being soluble in water, as well as colorless and tasteless, it has proved useful in criminal homicide. During the eighteenth century it gained so much notoriety that it was referred to as "inheritance powder."

Methylated forms of arsenate and arsenite: Compounds, such as methylarsonic acid (MMAV), monomethylarsonous acid (MMAIII), and dimethylarsenic acid (DMAV) are produced by algae and as excretory products of animals. They have varying degrees of toxicity, depending on their chemical form and the oxidation state of the arsenic that they contain. They occur in low concentrations in the environment.

Arsines: Arsenic in the 3^{-} oxidation state, occurring as highly toxic gases, such as H_3As and $(CH_3)_3As$. Very little is known about the natural cycles of these substances, as they occur at very low concentrations in the environment.

Organoarsenic compounds: Naturally occurring substances, such as arsenobetaine, are molecular analogs of osmotic-regulating compounds, such as betaine, where arsenic substitutes for the original nitrogen atom. They commonly occur in several marine animals, including shellfish and elasmobranchs. Their physiological role in these organisms is unknown, but they are benign and are not toxic to animals that eat these organisms, including humans.

Synthetic organoarsenic compounds: Substances, such as roxarsone (4-hydroxy-3-nitrophenylarsonic acid), are used as palliatives included in the feed of mass-raised swine and poultry. They are benign, do not accumulate in these organisms, and are ultimately excreted. However, their subsequent breakdown by bacteria in soils will release As(V) into the environment.

12.5 Microbial Transformation of Arsenic

Depending on the physical–chemical conditions of the environment, some arsenic compounds can be easily solubilized in water and taken up by microorganisms, resulting in high levels of bioavailability. Microorganisms have developed various strategies to counteract arsenic toxicity: firstly, active extrusion of arsenic; secondly, intracellular chelation (in eukaryotes) by various metal-binding peptides including glutathione (GSH), phytochelatins (PCs), and metallothioneins (MTs); thirdly, arsenic transformation to various organic forms which could be potentially less toxic (Fig. 12.3). Understanding the molecular and genetic level of arsenic metabolism will be, therefore, an important knowledge base for developing efficient and selective arsenic bioremediation approaches, which has so far been considered as a cost-effective and environmental friendly way for heavy metal removal.

Microbes play an important role in many reactions that have an influence on the speciation of arsenic. The inorganic forms of arsenic, arsenite (As III) and arsenate (As V), can be oxidized or reduced due to microbial activity (Woolson 1977). Inorganic arsenic species can also be methylated to monomethylarsonic acid (MMAA), dimethylarsinic acid (DMAA), and trimethylarsine oxide (TMAO) (Woolson 1977; Cullen and Reimer 1989; Gadd 1993; Turpeinena et al. 1999), while the other microbes can demethylate organic forms back to inorganic species (Sohrin et al. 1997). In soil, water-soluble arsenic species can also be volatilized by microbes to gaseous arsines (Bachofen et al. 1995; Gao and Burau 1997), which are extremely toxic compounds to mammals (Buchet et al. 1981). As(III) can be volatilized to arsine (AsH_3), MMAA to monomethylarsine (MMA, AsH_2 (CH_3), DMAA to dimethylarsine [(DMA, $AsH(CH_3)_2$], and TMAO to trimethylarsine [(TMA, $As(CH_3)_3$] (Cullen and Reimer 1989). The evolution of arsines is thus greatly dependent on the form of arsenic in soil and most often arsines are formed from methylated species (Gao and Burau 1997).

Speciation of arsenic in soil has become an important issue due to the different toxicity of the inorganic As(V) and As(III) species and gaseous species such as arsine, MMA, DMA, and TMA. The most often detected arsenic species in soils is As(V) (Cullen and Reimer 1989). Alexander (1977) and Woolson (1977) proposed that the major pathway by which arsenic is lost from soil is through a reduction of As(V) to As(III) and a series of methylation reactions of As(III) with trimethylarsine (TMA) as a final product. Arsines may travel in air for long distances or they are oxidized rapidly depending on environmental conditions (Pongratz 1995). Oxidation returns arsenic back to inorganic species and the cycle of arsenic is completed because atmospheric inorganic arsenic is deposited back to soil by rain or by dry deposition (Pongratz 1995).

Turpeinen et al. (2002) studied the transformation of arsenic in soil highly contaminated with arsenic from an abandoned wood impregnating plant in Lammi, Southern Finland. Soil dominating species, arsenate [As(V)], under both aerobic and anaerobic conditions to arsenite [As(III)], monomethylarsonic acid (MMAA), dimethylarsinic acid (DMAA), and to volatile trimethylarsine (TMA) was, however,

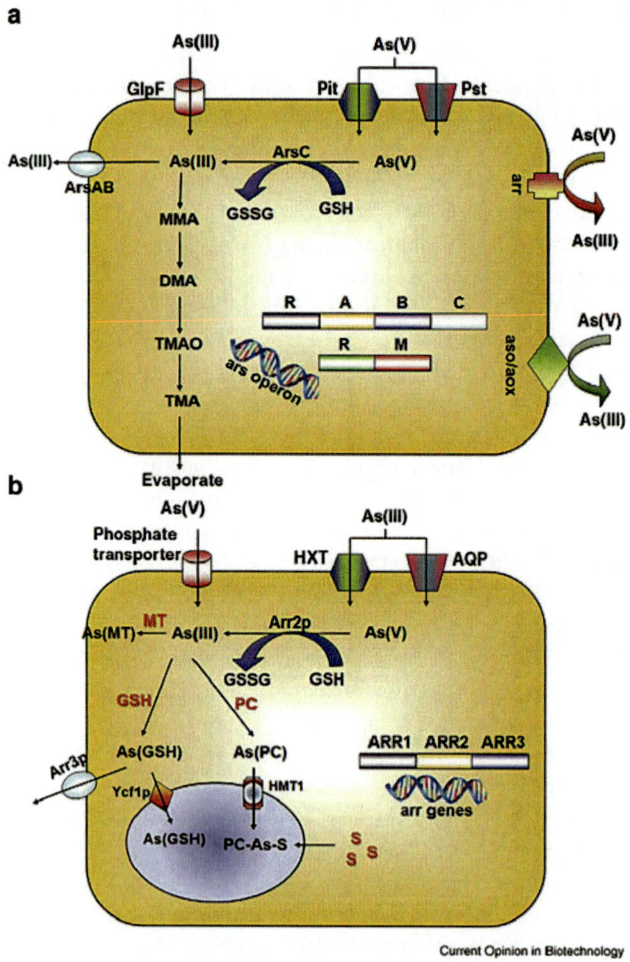

Fig. 12.3 Schematic representations of (**a**) prokaryotes' and (**b**) eukaryotes' processes involved in arsenic metabolism in the environment. In both cases, arsenic enters the cells through transporters. Arsenate is reduced to arsenite by a reductase, which further extrudes out of the cell by a specific membrane pump. In eukaryotes, arsenite can also be detoxified by complexation with Cys-rich peptides such as phytochelatins and storage in the vacuole. In addition, arsenite can serve as an electron donor by oxidation to arsenate. Arsenate can be used as the ultimate electron acceptor during respiration and inorganic arsenic can also be transformed into organic species in a methylation cascade (from Tsai et al. 2009)

less than 0.5%, of which the production of TMA represented 0.02–0.3%. The volatilization process was also verified in the field and in the soil of a dumping area. The "life time" of arsines in air was, however, short and they were rapidly converted back to water soluble species, As(V) and trimethyl arsine oxide (TMAO).

12.6 Microbial Remediation

Although As is highly toxic, microorganisms have evolved various strategies for dealing with it. As(V) can be used as an electron acceptor for anaerobic respiration, or As(III) can be used as an electron donor to support chemoautotrophic fixation of carbon dioxide into cell carbon under aerobic conditions. Microbially mediated methylation–demethylation and oxidation–reduction reactions produce a significant effect on As mobility and toxicity, providing a basis for the As-contaminated soils and groundwater. The major microbial approaches for remediation of arsenic contaminated soils: oxidation–reduction, biomethylation, complexation, solubilization, biosorption, sequestration, and microbially enhanced phytoremediation (Fig. 12.4). Bioaugmentation–biostimulation and genetically engineered microbes have direct applications in the microbial processes described earlier later.

12.6.1 Oxidation

Chemoautotrophs like *Pseudomonas arsenitoxidans* NT-26 that can cause oxidation of As(III) by using O_2, NO_3^-, or Fe(III) as a terminal electron acceptor and CO_2 or HCO_3^- as the carbon source (Santini et al. 2000). This oxidation process provides energy for the microbial growth and development. However, Gihring et al. (2001) found that the ecological role of the As(III) oxidation to As(V) by *Thermus aquaticus* and *Thermus thermophilus* might be the detoxification of As. The oxidation of As(III) to As(V) by bacterial colonies attached to macrophytes occurs immediately when the geothermal fluid is exposed to oxygenated conditions. An estimated half-life was around 20 min for the oxidation of As(III) to As(V) in Hot Creek, Sierra Nevada Mountains of California (Wilkie and Hering 1998). Microbial

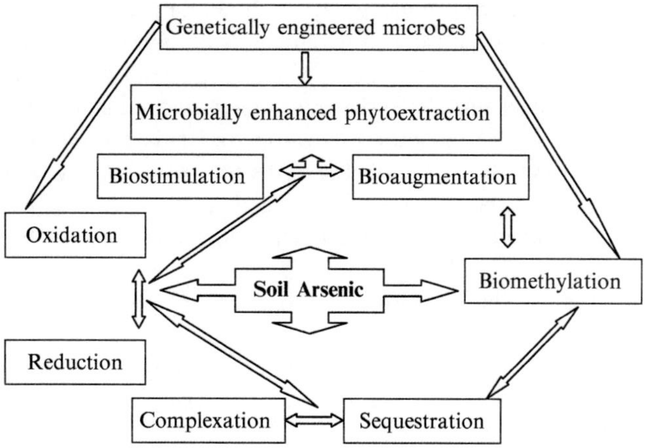

Fig. 12.4 Remediation of arsenic contaminated soils by various microbial approaches

As(III) oxidation was observed in the acid mine waters of the Matsuo sulfur-pyrite mine in Japan (Wakao et al. 1988). Since As(V) is less mobile and toxic than As (III), the bio-oxidation will contribute to As not available by enhancing As immobilization and alleviating As toxicity, especially when the water becomes much less acidic as a result of dilution from inflowing water and neutralization by the alkalinity of the environment.

In *Escherichia coli,* the ArsAB ATPase is a novel system that confers resistance to As(III) and Sb(III). Eukaryotic arsenic resistance transporters include Acr3p and Ycf1p of *Saccharomyces cerevisiae* (Rosen 2002). In both *E. coli* and *S. cerevisiae,* arsenate is brought into cells by phosphate transporters. Arsenate is then reduced to arsenite by the bacterial ArsC or yeast Acr2p enzymes. In both yeast and *E. coli*, glutathione and glutaredoxin serve as the source of reducing potential (Mukhopadhyay et al. 2000; Shi et al. 1999). The trivalent forms of metalloids can be taken up directly by cells using aquaglycerolporins as carriers, GlpF in *E. coli* and Fps1p in yeast. For subsequent detoxification, the trivalent forms of the metalloids are removed from the cytosol by active extrusion. In *E. coli*, arsenite or antimonite is extruded from the cells by the ArsAB pump. In yeast there are two arsenite transporters, Acr3p in the plasma membrane and Ycf1p in the vacuole. Acr3p is a carrier protein that probably uses the membrane potential for energy. Ycf1p is a homologue of the human MRP drug resistance pump and transports As (GS)3 into the vacuole.

12.6.2 Reduction

Anaerobic microorganisms are capable to use As(V) as an electron acceptor for the oxidation of organic matter or H_2 gas, yielding energy to support their growth and activities (Oremland and Stolz 2003). Microorganisms from 16 species of diverse taxonomy, referred to collectively as dissimilatory As(V) reducing prokaryotes (DARPs), have been isolated (Oremland and Stolz 2003). Besides, many microorganisms, including bacteria, archaea, and fungi, exhibit resistance to As(V) toxicity. A common mechanism of resistance involves the reduction of intracellular As(V) to As(III) by As(V) reductases, because As(III) is the substrate of efflux pumps (Oremland and Stolz 2003). The reduction of As(V) to As(III) under anaerobic conditions has been reported to be mediated by a diverse population of anaerobic microorganisms, including methanogens, fermentative bacteria, and sulfate- and iron-reducers (Christensen et al. 2001). The electron donors can be organic matter such as buried peat in sediments, anthropogenic input organic contaminants such as the benzene, toluene, ethylbenzene, xylene (BTEX) group, or organic acids from landfill sites (Wang and Mulligan 2006). Microbially mediated reduction of As(V) can occur even when As is bound to Fe hydroxides (Langner and Inskeep 2000). Zobrist et al. (2000) showed that the As(V) reduction can lead to As mobilization without dissolving the sorbent phase by anaerobically incubating As(V) co-precipitated with Al hydroxide in the presence

Sulforospirillum barnesii. Moreover, the microbial reduction of As(V) to As(III) under aerobic conditions in As-contaminated soils may occur relatively fast, resulting in enhanced As mobilization and transport from contaminated soils to groundwater (Macur et al. 2001). The biologically catalyzed reduction of As(V) to As(III), therefore, may hinder As NA by increasing its mobility.

Six of the seven known species of the epsilon-proteobacterium *Sulfurospirillum* can respire arsenate (*S. barnesii, S. arsenophilum, S. deleyianum, S. multivorans, S. halorespirans,* and *S. carboxydovorans*). Also a new arsenate-respiring *Sulfurospirillum* species, designated strain NP4 was isolated recently from groundwater collected in Maine, USA, containing high As concentrations (MacRae et al. 2007). Bacteria belonging to the genus Geobacter often dominate microbial communities in zones of Fe(III) reduction in freshwater aquifers (Coates et al. 1996; Holmes et al. 2002), and Geobacter-related sequences have also been identified in West Bengal (Islam et al. 2004) and Cambodian sediments (Rowland et al. 2007), as well as in American river floodplain sediments (Saunders et al. 2005) exhibiting high rates of As(III) release and Fe(III) reduction. Phylogenetic analyses using a highly conserved genetic marker (the 16 S rRNA gene) have suggested the involvement of *Sulfurospirillum* and *Geobacter* species in arsenate respiration (Hery et al. 2008). Denaturing gradient gel electrophoresis analysis of 16 S rRNA genes of the total bacterial community revealed that the addition of glucose stimulated uncultivable populations of *Flavobacterium* and *Paenibacillus*. The isolation technique enabled the characterization of 19 arsenic-resistant bacteria, mostly related to the facultative aerobic genera *Bacillus*, *Paenibacillus*, *Staphylococcus,* and to *Rhodococcus* and *Micromonospora* (Corsini et al. 2010). Most of them contained putative arsenate reductase and/or arsenite efflux pump as indicated by the presence of ArsC and/or ArsB genes. Four strains showed the ability to reduce arsenate by an intracellular detoxification mechanism, and one strain was able to oxidize arsenite, indicating that bacteria with the ability to oxidize or reduce arsenic are ubiquitous in soils.

12.6.3 Biomethylation and Demethylation

The methylation of As occurs via alternating reduction of pentavalent As to trivalent As and addition of a methyl group. (Fig. 12.5, Challenger 1945). The conversion of As(V) to small amounts of volatile methylarsines was first described in a pure culture of a methanogen, *Methanobacterium bryantii* (McBride and Wolfe 1971). Recently, several pure cultures of anaerobes, including a methanogen (*Methanobacterium formicicum*), a fermentative bacterium (*Clostridium collagenovorans*), and sulfate-reducing bacteria (*Desulfovibrio vulgaris* and *D. gigas*), were also implicated in the formation of methylarsines (Michalke et al. 2000). As(V) can be converted to monomethylarsine and dimethylarsine by *Achromobacter* sp. and *Enterobacter* sp., and to monomethylarsine, dimethylarsine, and trimethylarsine by *Aeromonas* sp. and *Nocardia* sp. (Cullen 1989). Anaerobic microcosms established from the sediments of microbial methylation

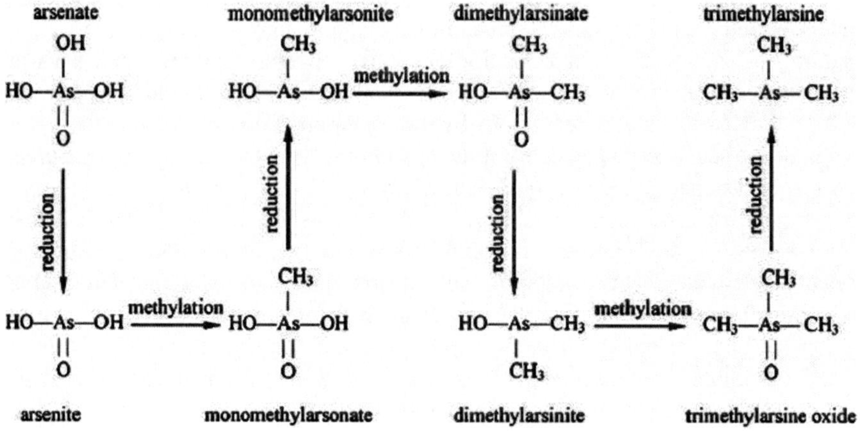

Fig. 12.5 Challenger mechanism for As methylation pathway (from Challenger 1945)

of As to volatile As effectively reduces its toxicity. In soils, the amount of As volatilized by microorganisms is governed by nutrient levels and microbial growth (Sanford and Klein 1988). Further, the addition of microbes with As methylating ability to As polluted soil could be a potential strategy in order to enhance the rate of As volatilization from soils (Sanford and Klein 1988; Huysmans and Frankenberger 1991; Thomas and Rhue 1997). However, Turpeinen et al. (1999) reported that microbes enhanced mobilization of arsenic from soil by 19–24% compared to formaldehyde inhibited controls. Formation of dissolved methylated arsenic species by microbes was low ($<0.1\%$) during the 5-day incubation. Even though methylation may function as a detoxification method, it was of minor importance in the soil tested. Augmentation by As methylating fungi (*Penicillium* sp. and *Ulocladium* sp.) was able to enhance arsine evolution rates in field contaminated soils in Australia (Edvantoro et al. 2004). The amounts of arsine dissipated in augmented long-term contaminated soils and soils spiked with As alone were 3.7 and 8.3 orders of magnitude greater than uninoculated soils, respectively.

Methylation and demethylation may play a significant role in influencing the toxicity and mobility of As in soils and groundwater (Wang and Mulligan 2006). As(III) and As(V) methylation may form volatile species leading to the escape of As from water and soil surfaces by volatilization. However, the contribution of volatilization to As natural attenuation is almost negligible. Though the methylated As species are generally considered less toxic than the inorganic species, the methylation processes do not necessarily contribute to the detoxification mechanism. Recent research has demonstrated that trivalent methylated As species are more effective in destroying DNA. The potency of the DNA damage decreases in the order DMAA(III) > MMAA(III) > [As(III), As(V)] > MMAA(V) > DMAA(V) >trimethylarsine oxide [TMAO(V)].

12.6.4 Complexation and Solubilization

As opposed to metal immobilization, metal mobilization occurs through processes resulting in the production of organic acids, siderophores, or H_2SO_4. Siderophores produced by microbes can mobilize metal by making metal–siderophore complex (Nair et al. 2007). These siderophores, however, have specific purpose to mobilize iron but they are nonspecific for a number of metals (Al, Cd, Co, Cr, Cu, Zn, Mn, Ni, Pb) including arsenic increasing their desorption from soil. Siderophore pyridine-2,6-bis(thiocarboxylic acid) produced by *Pseudomonas stutzeri* KC precipitated As, Cd, Hg, and Pb, conferring resistance to the bacterium (Zawadzka et al. 2007). Similarly, microbially produced biosurfactants including rhamnolipid by *Pseudomonas aeruginosa*, surfactin by *Bacillus subtilis*, and sophorolipid from the yeast *Torulopsis bombicola* have been implicated in metal resistance (Mulligan et al. 2001; Sandrin and Maier 2002). It is important to note that while siderophore and surfactant–metal complexes are less toxic, biosurfactant complexation can increase metal solubility. Consequently, use of these microorganisms in metal remediation has limited application, and has been proposed in soil washing technologies (Hietala and Roane 2009).

12.6.5 Sequestration

Under some conditions, microbial and geological factors work in concert to limit mobility of arsenic in waters and favor sequestration of the contaminant by native sediments. Recent studies have accounted limited arsenic mobility to the anthropogenic introduction of nitrate, which serves as an oxidant for reduced sediments and causes the generation of arsenic-sequestering ferric particles (Senn and Hemond 2004). Gibney and Nu¨sslein (2007) determined the biological and geochemical factors associated with arsenic sequestration in sediments of Upper Mystic Lake (UML, Arlington, MA) that leads to this attenuation. Data indicate that iron, not arsenic, oxidation governs arsenic sequestration in this system, consistent with the results of Senn and Hemond (2004). Sequestration of mobile arsenic occurred immediately upon microbial generation of solid ferric oxides. This occurred for both mobile As(V) and As(III) states, demonstrating that sequestration was arsenic redox-independent. Ferric particles were generated during periods of microbially catalyzed nitrate reduction, while clone libraries demonstrated the presence of populations similar to the genus *Dechloromonas*. Organisms similar to the nitrate-reducing, iron-oxidizing *Dechloromonas* sp. in 16 S rDNA clone libraries were therefore most likely responsible for the generation of arsenic-sequestering Fe(III)-(oxy)hydroxide-containing solids in anoxic sediments collected from Upper Mystic Lake. Paddy rice oxygenates its rhizosphere, resulting in the formation of an iron (Fe) oxyhydroxide plaque (Armstrong 1964). A recent study showed that Fe plaque is composed dominantly of ferrihydrite, goethite, and siderite (Hansel et al. 2001). The Fe hydroxides in soil and solution have a very strong binding affinity for arsenate (Meng et al. 2002; Liu et al. 2004), and a possible capacity to oxidize arsenite to arsenate (Otte et al. 1991). Iron plaque may be a barrier or a buffer to the uptake of As (Liu et al. 2004). The effect of Fe plaque on plant uptake of contaminants or nutrients may depend on the amount of Fe plaque on the roots surface (Otte et al. 1989; Zhang et al. 1998). High concentrations of Fe^{2+} usually exist at the site of ${SO_4}^{2-}$ reduction by sulfate-reducing bacteria, and S^{2-} thus produced immediately reacts with Fe^{2+} to form FeS (Murase and Kimura 1997). FeS is known to take a very long time to reduce ferric iron to form pyrite (FeS_2) (Dent 1986). Fe and Mn oxidizing bacteria, including sheaths (e.g., *Sphaerotilus*), *Leptothrix* group (e.g., *Leptothrix ochracea*), or spirally twisted stalks (e.g., *Gallionella ferruginea*), are able to accelerate Fe(II) oxidation, resulting in As precipitation with respective Fe oxides (DeVitre and Belzile 1990; Katsoyiannis et al. 2004; Mouchet 1992). Leupin and Hug (2005) reported that during the oxidation of Fe(II) to Fe(III), As(III) was partially oxidized in As(V). As can be immobilized through sorption due to the presence of a mixture of Fe oxides, organic material, and microbial biomass following the biological oxidation of Fe. Katsoyiannis and Zouboulis (2004) found that the biotic oxidation of Fe by microorganisms *G. ferruginea* and *Leptothrix ochracea* was effective in removing As from groundwater. As(III) was partially oxidized to As(V), which enabled high As immobilization as the oxidized form was much more strongly sorbed on the biogenic Fe oxides. Thus, the ability of some bacteria belonging to the *Gallionella* and *Leptothrix* genera to oxidize iron and/or manganese has been applied to the development of an innovative technique (Katsoyiannis et al. 2002; Katsoyiannis and Zouboulis 2004; Katsoyiannis and Zouboulis 2006). Similarly, *Acidithiobacillus ferrooxidans* was found to be able to catalyze the oxidation from Fe(II) to Fe(III) and facilitate the formation of schwertmannite under acidic and high ${SO_4}^{2-}$ conditions, which enhances AsNA by immobilizing it through sorption (Fukushi et al. 2003). However, other microbially induced redox reactions may remobilize sorbed As, thus jeopardizing the remediation processes. Cummings et al. (Cummings et al. 1999) reported that the dissimilatory Fe reducing bacterium *Shewanella alga* (strain BrY) enhanced As mobilization from a crystalline ferric arsenate ($FeAsO_4{\cdot}2H_2O$) as well as from sorption sites within whole sediments. This oxidation of sulfide minerals catalyzed by microorganisms, such as *A. ferrooxidans* and *Leptospirillum ferrooxidans*, will release As into surface and groundwaters (Rohwerder et al. 2003).

Hu et al. (2007) reported a significant differences in As uptake into rice between +S and −S treatments. Concentrations of As in rice shoots decreased with increasing rates of S application. The mechanism could be ascribed to sulfur, induced the formation of iron plaque, because concentrations of Fe in iron plaque on quartz sands in the rhizosphere and on the root surface of rice increased with increasing rates of S application. The results suggest that sulfur fertilization may be important for the development approaches to reducing As accumulation in rice.

12.6.6 Biofilm and Biosorption

Biofilms are assemblages of single or multiple populations that are attached to biotic or abiotic surfaces through extracellular polymeric substances (Singh et al. 2006). They are less susceptible to metal toxicity than planktonic cell population (Harrison et al. 2005; Davies et al. 2007) which open new perspectives for biofilm-mediated bioremediation processes (Singh et al. 2006; Harrison et al 2007; Diels et al. 2003; Chang et al. 2006). Haack and Warren (2003) studied microbial biofilms metal dynamics in acid rock drainage providing evidence of the stable accumulation of metals in these biofilms. Thus, microorganisms that secrete polymers are able to immobilize metal compounds by passive sequestration processes. This ability is used to develop bioremediation methods such as biofilm-based reactors (Chang et al. 2006). In this context, bacteria such as *Herminiimonas arsenic oxydans* produce large amounts of exopolymers in the presence of arsenic. Images obtained using transmission electron microscopy (TEM) combined with X-ray energy dispersion spectroscopy (EDS) indicated that this metalloid is sequestered within the exopolysaccharide capsule (Muller et al. 2007). Nevertheless, chemical sensitivity is limited to elemental composition; metal speciation can only be inferred. This finding could be used to detoxify natural waters and wastewaters contaminated with arsenic (Lièvremont et al. 2009).

In order to investigate the mixed adsorption of As(III) on *Halobacillus* sp. Y35 and kaolin, the microbial physiological response and biosorption function were determined. The relative adsorption capacity and adsorption intensity of kaolin for As(III) are higher with strain Y35 than that without strain Y35, demonstrating that it is possible to reduce the toxicity of As(III) to our environment by both using mineral adsorption and biosorption technology. Moreover, this work can provide useful information for As(III) control, transport and removal in geochemical and environmental applications by mixed mineral adsorption and biosorption techniques.

Molecular analysis, based on the 16 S rDNA fragment extracted from the discharged mine water, indicated that *Gallionella* sp. was the predominant microorganism present (Ohnuki et al. 2004). It is known to form stalk-like and sheath-like compounds that were distinguished in the precipitates by transmission electron microscope (TEM) observation. These results demonstrate that As(V) in the discharged mine water is co-precipitated with and/or adsorbed on Fe–S-bearing minerals in the biomat.

12.6.7 Bioaugmentation and Biostimulation

Bioaugmentation is a process in which efficient metal microbial species are added into soil or reactor to remediate the system. Bioaugmentation is the process of introducing metal-immobilizing population or a population which transforms metal into less toxic state into the site. Biostimulation is the process by which a stimulus to the microorganisms that already exist in the site is provided by adding nutrients and other growth substrates, together with electron donors and acceptors. Augmentation by As methylating fungi (*Penicillium* sp. and *Ulocladium* sp.) was able to enhance arsine evolution rates in field contaminated soils and in freshly spiked soils (Edvantoro et al. 2004). The amounts of arsine dissipated in augmented long-term contaminated soils and soils spiked with As alone were 3.7 and 8.3 orders of magnitude greater than uninoculated soils, respectively.

Biosurfactant foam technology can be used to deliver nutrients or microbial populations into the subsurface (Hug et al. 2001; de Koning and Thiesen 2005). As by-products from bacteria and yeast, biosurfactants (e.g., rhamnolipid, surfactin, and sophorolipid) generally are environmentally benign due to their low toxicity and high biodegradability. The use of biosurfactants may increase the availability of Fe(III) and As to the microorganisms due to the decrease in interfacial tension and formation of micelles. An aerobic or anaerobic environment can be created by changing the foaming gas. Aerobic conditions can be created by using air or oxygen as foaming gas. Anaerobic biological processes can proceed with the addition of nitrates, Fe(III) oxides, Mn(IV) oxides, sulfate, and CO_2. Addition of natural organic matter such as humic acids can provide electron acceptors because humic acids can chelate Fe. The presence of humic acids can significantly increase Fe bioavailability (Smedley and Kinniburgh 2002). A preferred pH range can also be achieved with addition of different pH buffers.

12.6.8 Microbially Enhanced Phytoremediation

Hyperaccumulation transfers dissolved As from soils or waters into plant tissues. A combination of hyperaccumulation with phytoremediation can remove or contain As, offering an effective, environmentally nondestructive and cheap remediation method. However, a key issue for an effective phytoremediation process, especially phytoextraction, is to enhance pollutant phytoavailability and to sustain adequate pollutant concentrations in the soil solution for plant uptake (Lombi et al. 2002). In addition, candidate plants for the phytoextraction purpose should tolerate the pollutants and produce large amounts of biomass. Besides the ability of the plants, rhizosphere factors are also important for improving phytoremediation efficiency. The first reported As hyperaccumulator is the Chinese brake fern *Pteris vittata* (Fitz and Wenzel 2002). It can accumulate 12–64 mg As kg^{-1} in its fronds from uncontaminated soils containing 0.5–7.5 mg As kg^{-1}, and up to 22,630 mg As kg^{-1} from a soil amended with 1,500 mg As kg^{-1} (Fitz and Wenzel 2002). Several other fern species, including *Pityrogramma calomelanos*, *Pteris cretica, Pteris longifolia,* and *Pteris umbrosa*, have also been reported to be able to hyperaccumulate As (Han et al. 2003). Other plants, such as *Agrostis tenuis* and *A. stolonifera*, grew on smelter waste in southwest England and accumulated As up to 1% dry weight (Benson et al. 1981). *A. tenuis* accumulated foliage As concentrations between 2,080 and 3,470 mg kg^{-1} (Porter and Peterson 1975).

These plants are tolerant of high concentrations of As(V) but not As(III). An accumulation of As up to 2,000 mg kg^{-1} dry mass in *Lemna gibba* (duckweed) was observed in the tailings water of abandoned uranium mine sites in Saxony, Germany (Mkandawire and Dudel 2005). It has been reported that P-fertilizer, rhizosphere microbes (bacteria and fungi), root exudation, and chelating agents could enhance plant uptake and accumulation of contaminants due to their effects on improving available pollutants for plants (Chen and Cutright 2002; Cao et al. 2003). Besides chemical amendments, biological amendments, such as rhizosphere microbes, have also been investigated for improving metal phytoremediation (Khan 2005; Gohre and Paszkowaski 2006). The possible benefits of rhizosphere microbes on metal phytoremediation may occur in several ways such as stimulation of plant growth by improving plant nutrition (Vogel-Mikus et al. 2006), metal detoxification/resistance by immobilizing metals in the rhizosphere (Wu et al. 2006), and/or enhancement of metal accumulation in the plant (Arriagada et al. 2007).

Mycorrhizal fungi in polluted soils are crucial in maintaining diverse populations of indigenous vegetation and act as a barrier to the uptake of toxic heavy metals by plants (Leyval et al. 1997). Sharples et al. (2000) presented evidence that the ericoid mycorrhizal fungus *Hymenoscyphus ericae* acts as a filter to maintain low arsenic uptake rates by roots of the plant *Calluna vulgaris* when growing in arsenic contaminated soil. Currently, several mycorrhiza-induced phosphate transporters have been identified in different plant species and showed increased expression in arsuscular mycorrhizae (AM) symbiosis. These included StPT3 in *Solanum tuberosum* (potato) (Rausch et al. 2001), LePT3 and LePT4 in *Lycopersicon esculentum* (tomato) (Nagy et al. 2005; Xu et al. 2007), and MtPT4 in *Medicato truncatula* (Javot et al. 2007). Chen et al. (2007) also indicated that both P-starvation and AM colonization could enhance the expression of phosphate transporter genes belonging to members of PHT1 in solanaceous species, i.e., pepper, eggplant, and tobacco. Catarecha et al. (2007) showed that a phosphate transporter PHT1;1 isolated from an As(V)-tolerant mutant of *Arabidopsis thaliana* could enhance As accumulation in wild-type plants. It is envisaged that the significantly enhanced As accumulation in *P. vittata* colonized by the uncontaminated isolates of *G. mosseae* seemed to attribute to the mycorrhiza-induced phosphate transporter. Wang et al. (2002) also suggested that the increased density of phosphate transporters in roots due to P-starvation resulted in enhanced arsenate uptake in *P. vittata*. To fully understand the effects of molecular mechanisms of AM fungi on As accumulation in *P. vittata*, the mycorrhiza-induced phosphate transporters in *P. vittata* should be isolated, identified, and characterized.

In a study of evolved arsenate resistance in cultivars of the grass *Holcus lanatus*, Gonzalez-Chavez et al. (2002) found that colonization by the arbuscular-mycorrhizal fungus *Glomus* suppressed high-affinity arsenate and phosphate transport into the roots. Conversely, mycorrhizal association with the fern *P. vittata* has been reported to stimulate arsenic accumulation by the host (Liu et al. 2005). Under 100 mg As kg^{-1} exposure, percentages of root colonization in HK (Hong Kong) population of *P. vittata* inoculated with the uncontaminated or metal-contaminated isolates of *Glomus mossea* were 54% and 26%, and the corresponding root colonizations in the JCT (Jinchuantang) population were 14% and 8%, respectively (Wu et al. 2009). Regardless of the origin of *G. mossea*, HK population had a significant colonization level than that of JCT population. It was concluded that there were intraspecific differences of AM fungi in their impacts on As accumulation by *P. vittata*.

Jankong et al. (2007) reported that P-fertilizer and rhizosphere bacteria significantly increased biomass and As accumulation in *Pityrogramma calomelanos* grown in soil with arsenic concentration of 269 μg As g^{-1} in Thailand, while rhizofungi reduced significantly As concentration in plants but increased plant biomass. Arsenic content in the roots and the fronds of rhizobacteria-inoculated plants was significantly higher than in both the control and rhizofungi-inoculated plants ($P < 0.05$) (Fig. 12.6). However, there were no significant effects of

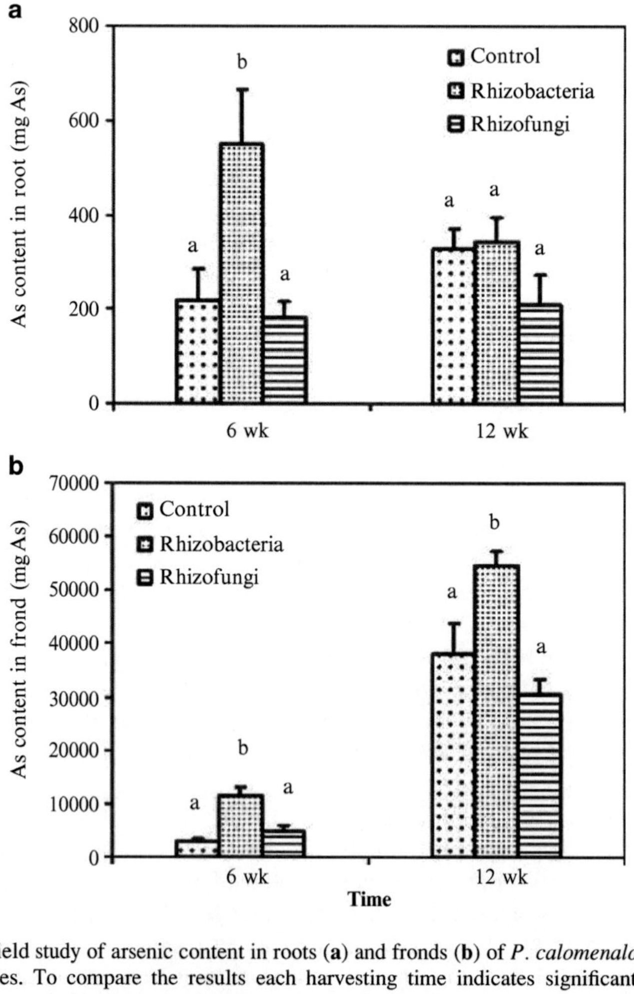

Fig. 12.6 Field study of arsenic content in roots (**a**) and fronds (**b**) of *P. calomenalos* inoculated with microbes. To compare the results each harvesting time indicates significant differences among control, rhizobacteria and rhizofungi inoculation by Duncan's multiple range test at the 5% level, $n = 3$ (from Jankong et al. 2007)

rhizofungi on arsenic content in parts of treated plants compared to the control. In particular, at 12 weeks after planting in the field study, arsenic content in the roots of both rhizobacteria- and rhizofungi-inoculated plants was not significantly different to the control. It was noticeable in the field study that arsenic content in the roots of rhizobacteria-inoculated plants at 12 weeks was lower than at 6 weeks after planting. However, the arsenic content in the fronds of all plants increased markedly from the harvesting time of 6–12 weeks in the field study.

The rhizosphere fungi was reported to exert their effects on phytostabilization. Previous study by their group (Uppanan 2000) revealed that *Alcaligenes* spp. isolated from soil could oxidize arsenite to arsenate which plant could uptake via phosphate transporter as mentioned by Meharg and Macnair (1990). Jankong and Visoottiviseth (2008) investigated the effects of arbuscular mycorrhizal fungi (AMF) (*Glomus mosseae*, *Glomus intraradices*, and *Glomus etunicatum*) on biomass production and arsenic accumulation in *Pityrogramma calomelanos*, *Tagetes erecta*, and *Melastoma malabathricum* were investigated in a soil with higher level of arsenic (243 ± 13 lg As g^{-1}) in Thailand. They reported that As concentrations in petioles and leaves were significantly lower in the AMF inoculated plants, while there were no significant differences for the roots. For *P. calomelanos* and *T. erecta*, AMF reduced only arsenic accumulation in plants but had no significant effect on plant growth. In contrast, AMF improved growth and arsenic accumulation in *M. malabathricum*.

In As phytoremediation, there have also been conflicting results regarding the effects of rhizofungi on As uptake by plants. Leung et al. (2006) reported that the addition of rhizofungi enhanced the uptake and accumulation of As in *P. vittata*. Under the condition of 100 mg As kg^{-1} soil, noncolonized plants accumulated 60.4 mg As kg^{-1} while plants colonized by AMF isolated from an As mine accumulated 88.1 g As kg^{-1} and also enhanced plant growth. On the other hand, Trotta et al. (2006) found that in the same plant species, *P. vittata*, rhizofungi increased plant growth only in the above ground parts but reduced root As concentration without any effect on frond concentration, therefore resulting in a larger As translocation factor. Moreover, in U and As-contaminated soil, Chen et al. (2006) found that rhizofungi depressed growth of *P. vittata* particularly at the early stages and had no effect on As concentration in this plant. These results indicated that the effects of rhizofungi on As uptake is inconsistent even though the same plant species was used. In this work, *P. calomelanos* was used and it was shown that rhizofungi could improve plant biomass but reduce As accumulation. It has been suggested that protection against As accumulation in plants by rhizofungi may occur indirectly by enhancing plant P nutrition and increasing plant growth resulting in a diluting effect on arsenic in the plant, or directly by binding arsenic to the fungal mycelium or roots and immobilization in the rhizosphere (Sharples et al. 2000; Meharg and Hartley-Whitaker 2002). From our previous work (Visoottiviseth and Panviroj 2001), *Penicillium* spp. was the best As-removal fungal species isolated from soil in Ron Phibun district of Thailand.

A glasshouse trial demonstrated that dip-site rhizosphere microbes promoted arsenic accumulation by the grass *Agrostis tenuis* on contaminated dip-site soil without inhibition of growth in Australia and the arsenic content of the shoots was increased by 45% (Chopra et al. 2007). Two plant species growing at the sites, Kikuyu grass (the most abundant plant) and Rainbow fern, exhibited mixed infections of their roots by endomycorrhizal fungi (tentatively identified as *Acaulospora* and *Gigaspora*) and by soil-borne pathogens. Five rhizosphere bacteria were identified to genus level and the effect of arsenic on their growth was determined. The two most prevalent strains differed greatly in their growth sensitivity to arsenate: *Arthrobacter* sp. being the most sensitive while *Ochrobactrum* sp. exhibited exceptional resistance to arsenate. Of the other, less prevalent strains, two were *Bacillus* spp. and the last, *Serratia* sp., were the most resistant to arsenite. These findings show the importance of understanding plant–soil microbe interactions for developing future strategies aimed at a phytoremediation-based approach to removing arsenic from soil at dip sites. Cavalca et al. (2010) reported arsenic resistant bacterial groups predominantly comprised of *Alphaproteobacteria*, *Betaproteobacteria*, and *Gammaproteobacteria* in the rhizosphere of wild thistle *Cirsium arvense* (L.) grown in arsenic contaminated soil in Australia. The arsenic-resistant isolates belonged to 13 genera, the most abundant being those of *Bacillus*, *Achromobacter*, *Brevundimonas*, *Microbacterium*, and *Ochrobactrum*. Most strains possessed the ArsC, ArsB, and ACR3 genes homologous to arsenate reductase and to the two classes of arsenite efflux pumps, respectively, peculiar to the ars operon of the arsenic detoxification system. Some rhizobacteria produced siderophores, indole acetic acid, and 1-amino-cyclopropane-1-carboxylic acid deaminase, thus possessing potential plant growth-promoting traits. Selecting microorganisms that are both metal resistant and able to produce plant growth-promoting compounds could prove useful as inocula in re-vegetation and phytoremediation processes of arsenic contaminated soil.

AMF are widely studied because of their ability to improve plant fitness under stress conditions (Leyval and Joner 2001). Survival studies have shown that AMF often protect plants against high concentrations of nonessential metals, in addition to the improvement of P status, and uptake of some essential elements resulting in greater plant biomass (Smith and Read 2008). Because arsenate (As V) has great similarity in chemical behavior to Pi, it is thought that AMF may also influence As dynamics in soils. In this respect, recently Cozzolino et al. (2010) reported that, with P addition, inoculation with commercial inoculum of AMF (100 AMF infective propagules g^{-1} of product) consisted of a mixture of spores, hyphae, and root fragments colonized by *Glomus intraradices* alleviated the toxicity of excessive As by improving P nutrition without increasing As concentrations in lettuce (*Lactus sativa* L.), emphasizing the role of beneficial microbes and P fertilizer to improve soil fertility in As-contaminated soil.

Azolla is a floating aquatic fern, which is widely distributed in paddy fields, rivers, ponds, and lakes. It can fix nitrogen by its symbiotic partnership with Anabaena which resides in the dorsal cavity of Azolla fronds; for this reason Azolla has been used as a green manure to improve soil fertility and rice production (Wagner 1997). It has been reported that Azolla has a high capacity to accumulate heavy metals such as Cd, Cr, Cu, Ni, and Zn (Sela et al. 1989), and can be used to remove heavy metals from wastewater (Bennicelli et al. 2004; Arora and Saxena 2005; Rakhshaee et al. 2006). Zhang et al. (2008) reported that *Azolla caroliniana* accumulated two times more As than *Azolla filiculoides* owing to a higher influx velocity for arsenate. *A. filiculoides* was more resistant to external arsenate due to a lower uptake. Both strains showed a similar degree of tolerance to internal As. Arsenate and arsenite were the dominant As species in both Azolla strains, with methylated As species accounting for <5% of the total As. *A. filiculoides* had a higher proportion of arsenite than *A. caroliniana*. Both strains effluxed more arsenate than arsenite, and the amount of As efflux was proportional to the amount of As accumulation. They envisaged that the Duckweeds (*Wolffia globosa*), a common macrophyte in paddy and aquatic environments that could reduce the transfer of arsenic from soil and water to rice paddy. Zhang et al. (2009) reported that *W. globosa* was able to accumulate >1,000 mg As kg^{-1} in frond dry weight (DW), and tolerate up to 400 mg As kg^{-1} DW. *W. globosa* decreased arsenate in solution rapidly, but also effluxed arsenite. *Wolffia globosa* is a strong As accumulator and an interesting model plant to study As uptake and metabolism because of the lack of a root-to-frond translocation barrier.

The potential of growing Azolla in paddy fields to reduce As transfer from soil and water to rice should be further evaluated. If this aquatic fern can accumulate As from water substantially, it may provide an effective way to reduce As transfer from soil and irrigation water to paddy rice. Feng et al. (2009) studied the changes in various soil As fractions, i.e., water soluble (F0), exchangeable (F1), bound to carbonates (F2), bound to Fe and Mn oxides (F3), bound to organic matter and sulfides (F4), and residual (F5, mineral matrix). Results showed that most of the As was fixed by mineral matrix (F5, ratios ranging from 46.22 to 96.37%), followed by As bound to Fe and Mn oxides (F3, ratios ranging from 3.14 to 28.18%), and the ratios of the other four fractions (F0, F1, F2, and F4) were mostly less than 10%. With the increase of As addition levels and with the application of manure or chemical NPK fertilizers, As was distributed more in the relatively active fractions (F0, F1, F2, F3, and F4) in the paddy soil mediated by the microbes

Phytostabilization methods using plants can also be applied for long-term remediation of As. This method limits uptake and excludes mobilization of As. One of the major benefits of phytostabilization is that the above-ground vegetative biomass is not contaminated with As, thus reducing the risk of arsenic transfer through food chains (Madejon et al. 2002). Woody species have also been investigated with respect to phytostabilization (French et al. 2006; Vazquez et al. 2006). More recently, four Eucalyptus species were used for phytostabilization of As in gold mine tailings (King et al. 2008). Additionally, natural attenuation processes including many biological applications may transform As to less toxic species and this topic has been recently reviewed (Wang and Mulligan 2006).

12.6.9 Engineered Microbes for Arsenic Remediation

The use of microbial biotechnology in the field remediation of metal contaminated soils with special reference to arsenic is a promising approach. The use of engineered microbes as selective biosorbents is an attractive green cure technology for the low cost and efficient removal of arsenic (Singh et al. 2008a, b). Although efforts have been reported in engineering microbes for the removal of cadmium or mercury by expressing metal-binding peptides such as human MTs (Pazirandeh et al. 1995; Li et al. 2000) or synthetic peptides (Bae et al. 2000, 2001) the relatively low specificity and affinity of these peptides for arsenic make them ineffective for arsenic remediation. The cad operon responsible for cadmium efflux is well characterized in *Staphylococcus aureus* (Nies and Silver 1989). The mer operon in mercury resistance is well understood in a variety of microorganisms (Nies 1999). The czc operon has been described in a variety of bacteria (Abou-Shanab et al. 2007). Development of an arsenic accumulating microbe should comprise the ability to firstly, modify the naturally existing defense mechanisms and secondly, develop novel or hybrid pathways into one easily manipulated microorganism. One of the earliest examples of engineering arsenic accumulation was demonstrated in plants. The bacterial enzymes ArsC (arsenate reductase) and g-ECS (GSH synthase) were expressed in *Arabidopsis thaliana*, resulting in the accumulation of As (V) as GSH–As complexes (Dhankher et al. 2002). A similar effort was subsequently reported by expressing the yeast YCF1 in *A. thaliana* for enhanced As storage in the vacuole (Song et al. 2003). These reports open up the possibility of engineering metabolisms and pathways for arsenic sequestration. On the basis of these early examples, similar efforts have been demonstrated with engineered microbes. In one case, the PC synthase from *A. thaliana* was expressed in *E. coli* (Sauge-Merle et al. 2003). This engineered strain produced PC when exposed to different forms of arsenic, leading to moderate levels of arsenic accumulation. However, the level of GSH, a key PC precursor, became limiting for higher level of PC production and arsenic accumulation. The PC synthase from *S. pombe* (SpPCS) in *E. coli* resulted in higher As accumulation (Singh et al. 2008b). PC production was further increased by coexpressing a feedback desensitized glutamylcysteine synthetase (GshI), resulting in higher PC levels and As accumulation. The significantly increased PC levels were exploited further by coexpressing an arsenic transporter GlpF, leading to an additional 1.5-fold higher As accumulation. These engineering steps were finally combined in an arsenic efflux deletion *E. coli* strain to achieve the highest reported arsenic accumulation in *E. coli* of 16.8 mmol g^{-1} cells.

Naturally, sulfur reducing bacteria are used for As(V) precipitation by the formation of insoluble sulfide complex with H_2S (Rittle et al. 1995). Metabolic engineering approaches have been utilized for intracellular production of H_2S in bacteria, leading to higher cadmium accumulation (Wang et al. 2000). Recently, a yeast strain coexpressing AtPCS and cysteine desulfhydrase, an aminotransferase that converts cysteine into hydrogen sulfide under aerobic condition, to elevate the accumulation of arsenic by the formation of PC-metal-sulfide complexes was engineered (Tsai et al. 2009).

The use of resting cells as a high-affinity biosorbent for arsenic removal has also been exploited. By expressing AtPCS in *S. cerevisiae*, which naturally has a higher level of GSH, the engineered yeast strain accumulated high levels of arsenic and was effective in removing arsenic in resting cell cultures (Singh et al. 2008b). However, the utility of PC producing cells for biosorption necessitates the use of zinc for PC induction, making it difficult to implement in practice. On the other hand, specific arsenic accumulation was achieved in *E. coli* cells by overexpressing the arsenic-specific regulatory protein ArsR. Resting cells expressing ArsR were effective in removing 50 ppb of As(III) within 1 h (Kostal et al. 2004). The concept of resting cell sorbents has been extended to the use of a naturally occurring As-binding MT (Singh et al. 2004). Singh and coworkers developed an engineered *E. coli* strain expressing the fMT from *F. vesiculosus* (Merrifield et al. 2004) isolated from an arsenic contaminated site. When the arsenite-specific transporter GlpF was co-overexpressed with fMT, the engineered *E. coli* accumulated arsenic at high levels even in the presence of tenfold excess amounts of competing heavy metals (Singh et al. 2008a).

Resting cells were able to completely remove 35 ppb of As(III) within 20 min, making this an attractive low-cost option for arsenic remediation.

New irrational approaches such as directed evolution, genome shuffling, and metagenomic studies can be used for developing new arsenic resistant pathways that are suitable for arsenic remediation (Dai and Copley 2004). This was demonstrated by the modification of an arsenic resistance operon using DNA shuffling (Crameri et al. 1997). Cells expressing the optimized operon grew in 0.5 M arsenate, a 40-fold increase in resistance. Along the same line, Chauhan and coworkers constructed a metagenomic library from an industrial effluent treatment plant sludge, and identified a novel As(V) resistance gene (arsN) encoding a protein similar to acetyltransferases. Overexpression of ArsN led to higher arsenic resistance in *E. coli* (Chauhan et al. 2009). These examples highlight the possibility to combine both natural and unnatural pathways for hyperarsenic accumulation.

12.7 Conclusion

Arsenic contamination in soil and ground water is a major global problem and local geochemical cycles have been intensified by either geogenic or irresponsible industrial and mining activities. If these problems are not addressed, these could create disastrous effects on human and animal health as arsenic is carcinogenic in nature. Many microorganisms have already evolved mechanisms to cope with this environmental challenge and these noble organisms could be exploited properly to remediate arsenic contaminated soil and water. The major advantage of microbial remediation is that it is a natural process with huge economic superiority over other methods. The fundamental understanding of the biochemistry and metabolic pathways involved in arsenic resistance are now being gradually translated into strategies for engineering microbes for effective arsenic remediation. Microbially mediated oxidation and reduction reactions may produce less mobile As species and mixed solid phases capable of sorbing As, thus enhancing the immobilization processes. However, the immobilization processes by sorption is reversible and the remobilization of sorbed As may occur when the site biogeochemical conditions change with time. Microbially enhanced phytoextractions hold greater promise to clean up the arsenic contaminated soil because this green cure technology being inexpensive exploits the plant and rhizosphere microorganisms with special reference to arbuscular mycorrhizae. Moreover, the selection of appropriate plant and fungal genotypes could improve phytoremediation technologies. Detailed investigation and long-term continued monitoring are absolutely necessary to ensure that soil ecological factors are optimum for proper functioning of the microbes to alleviate arsenic toxicity. The chemistry of soil and water (i.e., pH and Eh) and predominantly microbial assemblages play a major role in As dynamics. Although bioaccumulation of As in plants and organisms has been reported, its biochemical transformations within the plant and other biota are still largely unknown. Engineering strategies using environmentally benign products may be considered to enhance the remediation rates and efficiency. However, most of these technologies have been tested only at the laboratory and pilot scale levels. Large-scale application of such technologies requires trained personnel for the operation of equipment to treat soils and waters. Although the initial reports are promising, substantial improvements are necessary to move these approaches from the bench to practice. In this respect, new tools in synthetic biology will certainly enable us to increase our efforts toward this end.

References

Abedin MJ, Cresser MS, Meharg A, Feldmann J, Cotter-Howells J (2002) Arsenic accumulation and metabolism in rice (*Oryza sativa* L.). Environ Sci Technol 36:962–968

Abernathy CO, Lui YP, Longfellow D, Aposhian HV, Beck B, Fowler B et al (1999) Arsenic health effects, mechanisms of actions and research issues. Environ Health Perspect 107:593–597

Abou-Shanab RAI, van Berkum P, Angle JS (2007) Heavy metal resistance and genotypic analysis of metal resistance genes in gram-positive and gram negative bacteria present in Ni-rich serpentine soil and the rhizosphere of Alyssum murale. Chemosphere 68:360–367

Adriano DC (2001) Trace elements in terrestrial environments: biogeochemistry, bioavailability and risks of metals, 2nd edn. Springer, New York

Ahmad S, Kitchin KT, Cullen WR (2000) Arsenic species that cause release of iron from ferritin and generation of activated oxygen. Arch Biochem Biophys 382:195–202

Ahmed KM, Huq SMI, Naidu R (2006) Extent and severity of arsenic poisoning in Bangladesh. In: Naidu R, Smit E, Owens G, Bhattacharya P, Nadebaum P (eds) Managing arsenic in the environment: from soil to human health. CSIRO Publishing, Collingwood, pp 525–554

Alexander M (1977) Introduction to soil microbiology, 2nd edn. New York, John Wiley and Sons, Inc

Al Rmalli SW, Haris PI, Harrington CF, Ayub M (2005) A survey of arsenic in foodstuffs on sale in the United Kingdom and imported from Bangladesh. Sci Total Environ 337:23–30

Arena JM, Drew RH (1986) Poisoning: toxicology, symptoms, treatments, 5th edn. Charles C. Thomas, Springfield, IL7

Armstrong W (1964) Oxygen diffusion from the roots of some British bog plants. Nature 204:801–802

Arora A, Saxena S (2005) Cultivation of Azolla microphylla biomass on secondary treated Delhi municipal effluents. Biomass Bioenerg 29:60–64

Arriagada CA, Herrera MA, Ocampo JA (2007) Beneficial effect of saprobe and arbuscular mycorrhizal fungi on growth of *Eucalyptus globules* co-cultured with *Glycine max* in soil contaminated with heavy metals. J Environ Manage 84:93–99

Ashley PM, Lottermoser BG (1999) Arsenic contamination at the Mole River Mine, northern New South Wales. Aust J Earth Sci 46:861–874

ATSDR (1990) ATSDR case studies in environmental medicine. Agency for Toxic Substances and Disease Registry, Atlanta GA, USA

Bachofen R, Birch L, Buchs FP, Flynn I, Gaudenz J, Tahedl H et al (1995) Volatilization of arsenic compounds by microorganisms. In: Hinche RE, Means JL, Burris DR (eds) Bioremediation of inorganics. Battelle Press, Columbus, pp 103–108

BADC (2005) Survey report on irrigation equipment and irrigated area in Boro/2004 season. Bangladesh Agricultural Development Corporation, Dhaka

Bae W, Chen W, Mulchandani A, Mehra RK (2000) Enhanced bioaccumulation of heavy metals by bacterial cells displaying synthetic phytochelatins. Biotechnol Bioeng 70:518–524

Bae W, Mehra RK, Mulchandani A, Chen W (2001) Genetic engineering of *Escherichia coli* for enhanced uptake and bioaccumulation of mercury. Appl Environ Microbiol 67:5335–5338

Bennicelli R, Stezpniewska Z, Banach A, Szajnocha K, Ostrowski J (2004) The ability of Azolla caroliniana to remove heavy metals (Hg(II), Cr(III), Cr(VI)) from municipal waste water. Chemosphere 55:141–146

Benson LM, Porter EK, Peterson PJ (1981) Arsenic accumulation, tolerance, and genotypic variation in plants on arsenical mine wastes in South-West England. J Plant Nutr 3:655–666

Bentley R, Chasteen TG (2002) Microbial methylation of metalloids: arsenic, antimony, and bismuth. Microbiol Mol Biol Rev 66:250–271

Beretka J, Nelson P (1994) The current state of utilization of fly ash in Australia. In: Ash: a valuable resource, vol. 1. South African Coal Ash Association, pp 51–63

BGS and DPHE (British Geological Survey and Department of Public Health Engineering) (2001) Arsenic contamination of groundwater in Bangladesh. Technical report WC/00/19. British Geological Survey, Keyworth, UK

Bolan NS, Thiyagarajan S (2001) Retention and plant availability of chromium in soils as affected by lime and organic amendments. Aust J Soil Res 39:1091–1103

Borgono JM, Vincent P, Venturino H, Infante A (1977) Arsenic in drinking water of the city of Antofagasta: epidemiological and clinical study before and after the installation of a treatment plant. Environ Health Perspect 19:103–105

Brammer H, Ravenscroft P (2009) Arsenic in groundwater: a threat to sustainable agriculture in South and South-east Asia. Environ Int 35:647–654

Brochmoller J, Cascorbi I, Henning S, Meisel C, Roots I (2000) Molecular genetics of cancer susceptibility. Pharmacology 61:212–227

Brouwere K, Smolders E, Merckx R (2004) Soil properties affecting solid-liquid distribution of As (V) in soils. Eur J Soil Sci 55:165–173

Buchet JP, Lauwerys R (1983) Evaluation of exposure to inorganic arsenic in man. In: Fachetti S (ed) Analytical techniques for heavy metals in biological fluids. Elsevier, Amsterdam, pp 75–90

Buchet JP, Lauwerys R, Roels H (1981) Comparison of the urinary excretion of arsenic metabolites after a single oral dose of sodium arsenite, monomethylarsonate or dimethylarsinate in man. Int Arch Occup Environ Health 48:71–79

Cao X, Ma LQ, Shiralipour A (2003) Effects of compost and phosphate amendments on arsenic mobility in soils and arsenic uptake by the hyperaccumulator, *Pteris vittata* L. Environ Pollut 126:157–167

Carey AM, Scheckel KG, Lombi E, Newville M, Choi YS, Norton GJ, Charnock JM, Feldmann J, Price AH, Meharg AA (2010) Grain unloading of arsenic species in rice. Plant Physiol 152:309–319

Catarecha P, Segura MD, Franco-Zorrilla JM, Garcia-Ponce B, Lanza M, Solano R, Paz-Ares J, Leyva A (2007) A mutant of the Arabidopsis phosphate transporter PHT1; 1 displays enhanced arsenic accumulation. Plant Cell 19:1123–1133

Cavalca L, Zanchi R, Corsini A, Colombo M, Romagnoli C, Canzi E, Andreoni V (2010) Arsenic-resistant bacteria associated with roots of the wild *Cirsium arvense* (L.) plant from an arsenic polluted soil, and screening of potential plant growth-promoting characteristics. Syst Appl Microbiol 33:154–164

Cebrian ME, Albores A, Aguilar M, Blakely E (1983) Chronic arsenic poisoning in the North of Mexico. Hum Toxicol 2:121–133

Cervantes C, Ji G, Ramirez JL, Silver S (1994) Resistance to arsenic compounds in microorganisms. FEMS Microbiol Rev 15:355–367

Chakraborti D, Rahman MM, Paul K, Chowdhury UK, Sengupta MK, Lodh D et al (2002) Arsenic calamity in the Indian subcontinent-what lessons have been learned? Talanta 58:3–22

Chakraborty AK, Saha KC (1987) Arsenical dermatosis from tubewell water in West Bengal. Indian J Med Res 85:326–334

Challenger F (1945) Biological methylation. Chem Rev 36:315–362

Chang WC, Hsu GS, Chiang SM, Su MC (2006) Heavy metal removal from aqueous solution by wasted biomass from a combined AS-biofilm process. Bioresour Technol 97:1503–1508

Chatterjee A, Mukherjee A (1999) Hydrogeological investigation of ground water arsenic contamination in South Calcutta. Sci Total Environ 225:249–262

Chauhan NS, Ranjan R, Purohit HJ, Kalia VC, Sharma R (2009) Identification of genes conferring arsenic resistance to *Escherichia coli* from an effluent treatment plant sludge metagenomic library. FEMS Microbiol Ecol 67:130–139

Chen H, Cutright T (2002) The interactive effects of chelator, fertilizer, and rhizobacteria for enhancing phytoremediation of heavy metal contaminated soil. J Soil Sediment 2:203–210

Chen KP, Wu HY (1962) Epidemiologic studies on Blackfoot disease: II. A study of source of drinking water in relation to the disease. J Formos Med Assoc 61:611–618

Chen CJ, Kuo TL, Wu MM (1988) Arsenic and cancers. Lancet 1:414

Chen CJ, Hsueh YM, Lai MS, Shyu MP, Chen SY, Wu MM et al (1995) Increased prevalence of hypertension and long-term arsenic exposure. Hypertension 25:53–60

Chen BD, Zhu YG, Smith FA (2006) Effects of abuscular mycorrhizal inoculation on uranium and arsenic accumulation by Chinese brake fern (*Pteris vittata* L.) from a uranium mining-impacted soil. Chemosphere 62:1464–1473

Chen AQ, Hu J, Sun SB, Xu GH (2007) Conservation and divergence of both phosphate- and mycorrhiza-regulated physiological responses and expression patterns of phosphate transporters in solanaceous species. New Phytol 173:817–831

Chi IC, Blackwell RQ (1968) A controlled retrospective study of Blackfoot disease, an endemic peripheral gangrene disease in Taiwan. Am J Epidemiol 88:7–24

Chirenje T, Ma LQ, Szulczewski M, Littell R, Portier KM, Zillioux E (2003) Arsenic distribution in Florida urban soils: comparison between Gainesville and Miami. J Environ Qual 32:109–119

Chopra BK, Bhat S, Mikheenko IP, Xu Z, Yang Y, Luo X, Chen H, van Zwietene L, McC Lilley R, Zhang R (2007) The characteristics of rhizosphere microbes associated with plants in arsenic-contaminated soils from cattle dip sites. Sci Total Environ 378:331–342

Chowdhury UK, Biswas BK, Chowdhury TR, Samanta G, Mandal BK, Basu GK, Chanda CR, Lodh D, Saha KC, Mukherjee SC, Roy S, Kabir S, Ouamruzzaman Q, Chakraborty D (2000) Groundwater arsenic contamination in Bangladesh and West Bengal, India. Environ Health Perspect 108:393–397

Christensen TH, Kjeldsen P, Bjerg PL, Jensen DL, Christensen JB, Baun A, Albrechtsen HJ, Heron C (2001) Biogeochemistry of landfill leachate plumes. Appl Geochem 16:659–718

Coates JD, Phillips EJP, Lonergan DJ, Jenter H, Lovley DR (1996) Isolation of *Geobacter* species from diverse sedimentary environments. Appl Environ Microbiol 62:1531–1536

Correll R, Huq SMI, Smith E, Owens G, Naidu R (2006) Dietary intake of arsenic from crops. In: Naidu R, Smith E, Owens G, Bhattacharya P, Nadebaum P (eds) Managing arsenic in the environment: from soil to human health. CSIRO Publishing, Melbourne, pp 251–268

Corsini A, Cavalca L, Crippa L, Zaccheo P, Andreoni V (2010) Impact of glucose on microbial community of a soil containing pyrite cinders: role of bacteria in arsenic mobilization under submerged condition. Soil Biol Biochem 42:699–707

Cozzolino V, Pigna M, Di Meo V, Caporale AG, Violante A (2010) Effects of arbuscular mycorrhizal inoculation and phosphorus supply on the growth of *Lactuca sativa* L. and arsenic and phosphorus availability in an arsenic polluted soil under non-sterile conditions. Appl Soil Ecol 45:262–268

Crameri A, Dawes G, Rodriguez E, Silver S, Stemmer WPC (1997) Molecular evolution of an arsenate detoxification pathway DNA shuffling. Nat Biotechnol 15:436

Cullen WR (1989) The metabolism of methylarsine oxide and sulfide. Appl Organometal Chem 3:71–78

Cullen WR, Reimer KJ (1989) Arsenic speciation in the environment. Chem Rev 89:713–764

Cummings DE, Caccavo F, Fendorf S, Rosenzweig RF (1999) Arsenic mobilization by the dissimilatory Fe(III)-reducing bacterium *Shewanella* alga BrY. Environ Sci Technol 33: 723–729

Dai MH, Copley SD (2004) Genome shuffling improves degradation of the anthropogenic pesticide pentachlorophenol by Sphingobium chlorophenolicum ATCC 39723. Appl Environ Microbiol 70:2391–2397

Das D, Samanta G, Mandal BK, Chowdhury TR, Chanda CR, Chowdhury PP et al (1996) Arsenic in groundwater in six districts of West Bengal, India. Environ Geochem Health 18:5–15

Das HK, Mitra AK, Sengupta PK, Hossain A, Islam F, Rabbani GH (2004) Arsenic concentrations in rice, vegetables, and fish in Bangladesh: a preliminary study. Environ Int 30:383–387

Davies JA, Harrison JJ, Marques LL et al (2007) The GacS sensor kinase controls phenotypic reversion of small colony variants isolated from biofilms of *Pseudomonas aeruginosa* PA14. FEMS Microbiol Ecol 59:32–46

Dent D (1986) Acid Sulphate Soils: A Baseline for Research and Development, International Institute for Land Reclamation and Improvement, Wageningen

de Koning J, Thiesen S (2005) Aqua Solaris – an optimized small scale desalination system with 40 litres output per square meter based upon solar-thermal distillation. Desalination 182: 503–509

deLemos JL, Bostick BC, Renshaw CE, Sturup S, Feng XH (2006) Landfill-stimulated iron reduction and arsenic release at the Coakley Superfund Site (Nh). Environ Sci Technol 40:67–73

De Vitre AR, Belzile N, Tessier A (1991) Speciation and adsorption of arsenic on diagenetic iron oxyhrdroxides. Limnol Oceanog 36:1480–1485

Dhankher OP, Li YJ, Rosen BP, Shi J, Salt D, Senecoff JF, Sashti NA, Meagher RB (2002) Engineering tolerance and hyperaccumulation of arsenic in plants by combining arsenate reductase and gamma-glutamylcysteine synthetase expression. Nat Biotechnol 20:1140–1145

Dhar RK, Biswas BK, Samanta G, Mandal BK, Chakraborti D, Roy S et al (1997) Groundwater arsenic calamity in Bangladesh. Curr Sci 73:48–59

Diels L, Spaans PH, Van Roy S et al (2003) Heavy metals removal by sand filters inoculated with metal sorbing and precipitating bacteria. Hydrometallurgy 71:235–241

DPHE/BGS/MML (1999) Groundwater studies for arsenic contamination in Bangladesh. Phase I: rapid investigation phase. BGS/MML Technical report to Department for International Development, UK

Duxbury JM, Panaullah G (2007) Remediation of arsenic for agriculture sustainability, food security and health in Bangladesh. Working paper, Water Service, FAO, Rome

Duxbury JM, Mayer AB, Lauren JG, Hassan N (2003) Food chain aspects of arsenic contamination in Bangladesh: effects on quality and productivity of rice. J Environ Sci Health 38:61–69

Edvantoro BB, Naidu R, Megharaj M, Merrington G, Singleton I (2004) Microbial formation of volatile arsenic in cattle dip site soils contaminated with arsenic and DDT. Appl Soil Ecol 25:207–217

Eisler R (2004) Arsenic hazards to humans, plants, and animals from gold mining. Rev Environ Contam Toxicol 180:133–165

Ellenhorn MJ, Barceloux DG (1988) Arsenic in medical toxicology. Diagnosis and treatment of human poisoning. Elsevier, New York 7, pp 1012–1016

Engel RR, Hopenhayn-Rich C, Receveur O, Smith AH (1994) Vascular effects of chronic arsenic exposure: a review. Epidemiol Rev 16:184–209

European Topic Centre Soil (ETCS) (1998) Topic report: contaminated sites. European Environment Agency, Copenhagen, Denmark

Feng L, Zheng Y-M, He J-Z (2009) Microbes influence the fractionation of arsenic in paddy soils with different fertilization regimes. Sci Total Environ 407:2631–2640

Feng Z, Xia Y, Tian D, Wu K, Schmitt M, Kwok RK, et al. (2001) DNA damage in buccal epithelial cells from individuals chronically exposed to arsenic via drinking water in Inner Mongolia, China. Anticancer Res 21:51–58

Fitz WJ, Wenzel WW (2002) Arsenic transformations in the soil–rhizosphere–plant system: fundamentals and potential application to phytoremediation. J Biotechnol 99:259–278

Foster AL (2003) Spectroscopic investigations of arsenic species in solid phases. In: Welsch AHSKG (ed) Arsenic in groundwater: geochemistry and occurrence. Kluwer Academic Publishers, Boston, pp 27–65

Fowler BA (1977) Toxicology of environmental arsenic. In: Goyer RA, Mehlman MA (eds) Advances in modern toxicology: II. Toxicology of trace elements. Hemisphere Publishing, Washington, DC, pp 79–122

Fowler BA, Weissberg JB (1974) Arsine poisoning. N Engl J Med 291(22):1171–1174

French CJ, Dickson NM, Putwain PD (2006) Woody biomass phytoremediation of contaminated brownfield land. Environ Pollut 141:387–395

Fukushi K, Sasaki M, Sato T, Yanase N, Amando H, Ikeda H (2003) A natural attenuation of arsenic in drainage from an abandoned arsenic dump. Appl Geochem 18:1267–1278

Gadd GM (1993) Microbial formation and transformation of organometallic and organometalloid compounds. FEMS Microbiol Rev 11:297–316

Gao S, Burau RG (1997) Environmental factors affecting rates of arsine evolution from and mineralization of arsenicals in soil. J Environ Qual 26:753–763

Gebel T (2000) Confounding variables in the environmental toxicology of arsenic. Toxicology 144:155–162

Gibney BP, Nüsslein K (2007) Arsenic sequestration by nitrate respiring microbial communities in urban lake sediments. Chemosphere 70:329–336

Gihring TM, Druschel GK, McCleskey RB, Hamers RJ, Banfield JF (2001) Rapid arsenite oxidation by *Thermus aquaticus* and *Thermus thermophilus*: field and laboratory investigations. Environ Sci Technol 35:3857–3862

Gochfeld M (1995) Chemical agents. In: Brooks S, Gochfeld M, Herzstein J et al (eds) Environmental medicine. Mosby, St. Louis, pp 592–614

Gohre V, Paszkowaski U (2006) Contribution of the arbuscular mycorrhizal symbiosis to heavy metal phytoremediation. Planta 223:1115–1122

Gonzalez-Chavez, Harris PJ, Dodd J, Meharg AA (2004) Arbuscular mycorrhizal fungi confer enhanced arsenate resistance on *Holcus lanatus*. New Phytologist 155:163-171

Graeme HM, Pollack JVC (1998) Selected topics: toxicology: part I Arsenic and mercury. J Emerg Med 16:45–56

Grobe JW (1976) Perephere Durchblutungsstorungen und Akrocyanose bei arsengeschadigten Moselwintzern [Peripheral circulatory disorders and acrocyanosis in Moselle valley vineyard workers with arsenic poisoning]. Berufsdermatosen 24:78–84

Gurung JK, Ishiga H, Khadka MS (2005) Geological and geochemical examination of arsenic contamination in groundwater in the Holocene Terai Basin, Nepal. Environ Geol 49:98–113

Haack EA, Warren LA (2003) Biofilm hydrous manganese oxyhydroxides and metal dynamics in acid rock drainage. Environ Sci Technol 37:4138–4147

Han FX, Su Y, Monts DL, Plodinec MJ, Banin A, Triplett GE (2003) Assessment of global industrial-age anthropogenic arsenic contamination. Naturwissenschaften 90:395–401

Hansel CM, Fendorf S, Sutton S, Newville M (2001) Characterization of Fe plaque and associated metals on the roots of mine-waste impacted aquatic plants. Environ Sci Technnol 35:3863–3868

Harrison JJ, Ceri H, Roper NJ, Badry EA, Sproule KM, Turner RJ (2005) Persister cells mediate tolerance to metal oxyanions in *Escherichia coli*. Microbiology 151:3181–3195

Harrison JJ, Ceri H, Turner RJ (2007) Multimetal resistance and tolerance in microbial biofilms. Nat Rev Microbiol 5:928–938

Harvey CF, Swartz CH, Badruzzaman ABM, Keon-Blute N, YuW AMA et al (2005) Groundwater arsenic contamination on the Ganges delta: biogeochemistry, hydrology, human perturbations, and human suffering on a large scale. C R Geosci 337:285–296

Hery M, Gault AG, Rowland HAL, Lear G, Polya DA, Lloyd JR (2008) Molecular and cultivation-dependent analysis of metal-reducing bacteria implicated in arsenic mobilisation in South-East Asian aquifers. Appl Geochem 23(11):3215–3223, Oxford: Elsevier Ltd

Hietala KA, Roane TM (2009) Microbial metal remediation. In: Singh A, Kuhad RC, Ward OP (eds) Soil Biology: Advances in Applied Bioremediation. Springer-Verlag, pp 201–220

Holmes DE, Finneran KT, O'Neil RA, Lovley DR (2002) Enrichment of members of the family *Geobacteraceae* associated with stimulation of dissimilatory metal reduction in uranium-contaminated aquifer sediments. Appl Environ Microbiol 68:2300–2306

Hossain MB (2005) Arsenic distribution in soil and water of a STW command area. CIMMYT/ USGS Symposium on the Behaviour of Arsenic in Aquifers, Soils and Plants:Implications for Management, Dhaka, Jan. 16–18, 2005. Centro Internacional de Mejoramiento de Maíz y Trigo and the U.S. Geological Survey

Hossain MF (2006) Arsenic contamination in Bangladesh – an overview. Agric Ecosyst Environ 113:1–16

Hu ZY, Zhu YG, Li M, Zhang LG, Cao ZH, Smith FA (2007) Sulfur (S)-induced enhancement of iron plaque formation in the rhizosphere reduces arsenic accumulation in rice (*Oryza sativa* L.) seedlings. Environ Pollut 147:387–393

Hug SJ, Canonica L, Wegelin M, Gechter D, Von Gunten U (2001) Solar oxidation and removal of arsenic at circumneutral pH in iron containing waters. Environ Sci Technol 35:2114–2121

Huq SMI, Ara QAJ, Islam K, Zaher A, Naidu R (2001) The possible contamination from arsenic through food chain. In: Jacks G, Bhattacharya P, Khan AA (eds) Groundwater arsenic contamination in the Bengal Delta Plain of Bangladesh, KTH special publication, vol 3084. TRITA-AMI Report, Stockholm, Sweden, pp 91–96

Huysmans KD, Frankenberger WT (1991) Evolution of trimethylarsine by a *Penicillium* sp. isolated from agricultural evaporation pond water. Sci Total Environ 105:13–28

Inskeep WP, McDermott TR, Fendorf S (2002) Arsenic (V)/(III) cycling in soils and natural waters: chemical and microbiological processes. In: Frankenberger WT (ed) Environmental chemistry of arsenic. Marcel Dekker, New York, pp 183–215

Islam FS, Gault AG, Boothman C, Polya DA, Charnock JM, Chatterjee D, Lloyd JR (2004) Role of metal-reducing bacteria in arsenic release from Bengal delta sediments. Nature 430:68–71

Jain CK, Ali I (2000) Arsenic: Occurrence, toxicity and speciation techniques. Water Res 34: 4304–4312

Jankong P, Visoottiviseth P (2008) Effects of arbuscular mycorrhizal inoculation on plants growing on arsenic contaminated soil. Chemosphere 72:1092–1097

Jankong P, Visoottiviseth P, Khokiattiwong S (2007) Enhanced phytoremediation of arsenic contaminated land. Chemosphere 68:1906–1912

Javot H, Penmetsa RV, Terzaghi N, Cook DR, Harrison MJ (2007) A *Medicago truncatula* phosphate transporter indispensable for the arbuscular mycorrhizal symbiosis. Proc Natl Acad Sci USA 104:1720–1725

Karim M (2000) Arsenic in groundwater and health problems in Bangladesh. Water Res 34: 304–310

Katsoyiannis A, Zouboulis AI (2004) Application of biological processes for the removal of arsenic from groundwaters. Water Res 38:17–26

Katsoyiannis IA, Zouboulis AI (2006) Use of iron- and manganese-oxidizing bacteria for the combined removal of iron, manganese and arsenic from contaminated groundwater. Water Qual Res J Can 41:117–129

Katsoyiannis IA, Zouboulis H, Althoff H, Bartel H (2002) As(III) removal from ground waters using fixed-bed upflow bioreactors. Chemosphere 47:325–332

Katsoyiannis IA, Zouboulis AI, Jekel M (2004) Kinetics of bacterial As(III) oxidation and subsequent As(V) removal by sorption onto biogenic manganese oxides during groundwater treatment. Ind Eng Chem Res 43:486–493

Khan AG (2005) Role of soil microbes in the rhizospheres of plants growing on trace metal contaminated soils in phytoremediation. J Trace Elem Med Bio 18:355–364

Khan NI, Owens G, Bruce D, Naidu R (2009) An effective dietary survey framework for the assessment of total dietary arsenic intake in Bangladesh: Part-A—FFQ design. Environ Geochem Health 31:207–220

King DJ, Doronila AI, Feenstra C, Baker AJM, Woodrow IE (2008) Phytostabilisation of arsenical gold mine tailings using four Eucalyptus species: growth, arsenic uptake and availability after five years. Sci Total Environ 406:35–42

Kitchin KT (2001) Recent advances in arsenic carcinogenesis: modes of action, animal model systems, and methylated arsenic metabolites. Toxicol Appl Pharmacol 172:249–261

Kostal J, Yang R, Wu CH, Mulchandani A, Chen W (2004) Enhanced arsenic accumulation in engineered bacterial cells expressing ArsR. Appl Environ Microbiol 70:4582–4587

Langner HW, Inskeep WP (2000) Microbial reduction of arsenate in the presence of ferrihydrite. Environ Sci Technol 34:3131–3136

Leonard A (1991) Arsenic. In: Merian E (ed) Metals and their compounds in the environments: occurrence, analysis, and biological relevance, 2nd edn. VCH, Weinheim, pp 751–773

Leung HM, Ye ZH, Wong MH (2006) Interactions of mycorrhizal fungi with *Pteris vittata* (As hyperaccumulator) in As-contaminated soils. Environ Pollut 139:1–8

Leupin OX, Hug SJ (2005) Oxidation and removal of arsenic (III) from aerated groundwater by filtration through sand and zero-valent iron. Water Res 39:1729–1740

Levinsky WJ, Smalley RV, Hillyer PN, Shindler RL (1970) Arsine hemolysis. Arch Environ Health 20:436–440

Leyval C, Joner E (2001) Bioavailability of heavy metals in the mycorrhizosphere. In: Gobran GR, Wenzel WW, Lombi E (eds) Trace elements in the rhizosphere. CRC Press, New York, pp 165–185

Leyval C, Turnan K, Haselwandter K (1997) Effect of heavy metal pollution on mycorrhizal colonization and function: physiological, ecological and applied aspects. Mycorrhiza 7:139–153

Li Y-x, Chen T-b (2005) Concentrations of addictive arsenic in Beijing pig feeds and the residues in pig manure. Resour Conserv Recy 45:356–367

Li Y, Cockburn W, Kilpatrick J, Whitelam GC (2000) Cytoplasmic expression of a soluble synthetic mammalian metallothioneinalpha domain in *Escherichia coli* – enhanced tolerance and accumulation of cadmium. Mol Biotechnol 16:211–219

Lièvremont D, Philippe BN, Marie-Claire L (2009) Arsenic in contaminated waters: biogeochemical cycle, microbial metabolism and biotreatment processes. Biochimie 91:1229–1237

Lin S, Cullen WR, Thomas DJ (1999) Methylarsenicals and arsinothiols are potent inhibitors of mouse liver thioredoxin reductase. Chem Res Toxicol 12:924–930

Liu ZJ, Boles E, Rosen BP (2004) Arsenic trioxide uptake by hexose permeases in *Saccharomyces cerevisiae*. J Biol Chem 279:17312–17318

Liu Y, Zhu YG, Chen BD, Christie P, Li XL (2005) Influence of the arbuscular mycorrhizal fungus *Glomus mosseae* on uptake of arsenate by the As hyperaccumulator fern *Pteris vittata* L. Mycorrhiza 15:187–192

Lombi E, Zhao FJ, Fuhrmann M, Ma LQ, McGrath SP (2002) Arsenic distribution and speciation in the fronds of the hyperaccumulator *Pteris vittata*. New Phytol 156:195–203

MacRae JD, Lavine IN, McCaffery KA, Ricupero K (2007) Isolation and characterization of NP4, an arsenate-reducing *Sulfurospirillum* from groundwater in Northport, Maine. J Environ Eng 131: 81–88

Macur PE, Wheeler JT, McDermott TR, Inskeep WP (2001) Microbial population associated with the reduction and enhanced mobilization of arsenic in mine tailings. Environ Sci Technol 35:3676–3682

Madejon P, Murillo JM, Maranon T, Cabrera F, Lopez R (2002) Bioaccumulation of As, Cd, Cu, Fe and Pb in wild grasses affected by the Aznalcollar mine spill (SW Spain). Sci Total Environ 290:105–120

McBride BC, Wolfe RS (1971) Biosynthesis of dimethylasrine by a methanobacterium. Biochemistry 10:4312–4317

McBride GL, Holland P, Wilson K (1998) Investigation of contaminated sheep dipping sites in the Waikato. In: Proceedings of waste management institute New Zealand conference. Waste Management Institute, New Zealand Inc., Rotorua, pp 129–137

McLaren RG, Naidu R, Smith J, Tiller KG (1998) Fractionation and distribution of arsenic in soils contaminated by cattle dip. J Environ Qual 27:348–354

McLaughlin MJ, Tiller KG, Naidu R, Stevens DP (1996) The behaviour and environmental impact of contaminants in fertilizers. Aust J Soil Res 34:1–54

Meharg AA, Hartley-Whitaker J (2002) Arsenic uptake and metabolism in arsenic resistant and nonresistant plant species. New Phytol 154:29–43

Meharg AA, Macnair MR (1990) An altered phosphate uptake system in arsenate-tolerant *Holcus lanatus* L. New Phytol 116:29–35

Meng MX, Korfiatis GP, Bang S, Bang KW (2002) Combined effects of anions on arsenic removal by iron hydroxides. Toxicol Letters 133:103–111

Merrifield ME, Ngu T, Stillman MJ (2004) Arsenic binding to Fucus vesiculosus metallothionein. Biochem Biophys Res Commun 324:127–132

Merry RH, Tiller KG, Alston AM (1983) Accumulation of copper, lead, and arsenic in some Australian orchard soils. Aust J Soil Res 21:549–561

Michalke K, Wickenheiser EB, Mehring M, Hirner AV, Hensel R (2000) Production of volatile derivatives of metal(loid)s by microflora involved in anaerobic digestion of sewage sludge. Appl Environ Microbiol 66:2791–2796

Mkandawire M, Dudel EG (2005) Accumulation of arsenic in Lemna gibba L. (duckweed) in tailing waters of two abandoned uranium mining sites in Saxony, Germany. Sci Total Environ 336:81–89

Morin G, Juillot F, Casiot C et al (2003) Bacterial formation of tooeleite and mixed arsenic(III) or arsenic(V) - iron(III) gels in the Carnoule's acid mine drainage, France. A XANES, XRD, and SEM study. Environ Sci Technol 37:1705–1712

Mouchet P (1992) From conventional to biological removal of iron and manganese in France. J Am Water Works Ass 84:158–166

Mukhopadhyay R, Shi J, Rosen BP (2000) Purification and characterization of Acr2p, *the Saccharomyces cerevisiae* arsenate reductase. J Biol Chem 275:21149–21157

Mukhopadhyay R, Rosen BP, Phung LT, Silver S (2002) Microbial arsenic: from geocycles to genes and enzymes. FEMS Microbiol Rev 26:311

Muller D, Me´digue C, Koechler S et al (2007) A tale of two oxidation states: bacterial colonization of arsenic-rich environments. PLoS Genet 3:518–530

Mulligan CN, Yong RN, Gibbs BF (2001) Heavy metal removal from sediments by biosurfactants. J Hazard Mater 85:111–125

Murase J, Kimura M (1997) Anaerobic reoxidation of Mn^{2+}, Fe^{2+}, S^0 and S^{2-} in submerged paddy soils. Biol Fertil Soils 25:302–306

Murdoch A, Clair TA (1986) Transport of arsenic and mercury from gold mining activities through an aquatic system. Sci Total Environ 57:205–216

Nagy R, Karandashov V, Chague V, Kalinkevich K, Tamasloukht M, Xu G, Jakobsen I, Levy AA, Amrhein N, Bucher M (2005) The characterization of novel mycorrhizaspecific phosphate transporters from *Lycopersicon esculentum* and *Solanum tuberosum* uncovers functional redundancy in symbiotic phosphate transporter in solanaceous species. Plant J 42:236–250

Nair A, Juwarkar AA, Singh SK (2007) Production and characterization of siderophores and its application in arsenic removal from contaminated soil. Water Air Soil Pollut 180: 199–212

Nickson R, McArthur J, Burgess W, Ahmed KM, Ravenscroft P, Rahman M (1998) Arsenic poisoning of Bangladesh groundwater. Nature 395:338

Nies DH (1999) Microbial heavy-metal resistance. Appl Microbiol Biotechnol 51:730–750

Nies DH, Silver S (1989) Plasmid-determined inducible efflux is responsible for resistance to cadmium, zinc and cobalt in Alcaligenes eutrophus. J Bacteriol 171:896–900

Nordstrom DK (2002) Public health – worldwide occurrences of arsenic in ground water. Science 296:2143–2145

NRC (1995) Arsenic in drinking water. National Research Council. National Academy Press, Washington,DC

NRC (1977) (National Research Council). Medical and biological effects of environmental pollutants—arsenic. Washington, DC7 National Academy of Sciences

Nriagu JO (1994) Arsenic in the environment: human health and ecosystem effects. Advances in environmental science and technology, vol 27. Wiley, New York

Nriagu JO, Pacyna JM (1988) Quantitative assessment of worldwide contamination of air, water, and soils by trace metals. Nature 333:134–139

O'Day PA, Vlassopoulos D, Root R, Rivera N (2004) The influence of sulfur and iron on dissolved arsenic concentrations in the shallow subsurface under changing redox conditions. Proc Natl Acad Sci USA 101:13703–13708

Ohno K, Furukawa A, Hayashi K, Kamei T, Magara Y (2005) Arsenic contamination of groundwater in Nawabganj, Bangladesh, focusing on the relationship with other metals and ions. Water Sci Technol 52:87–94

Ohnuki T, Kozai N, Samadfam M, Yasuda R, Yamamoto S, Narumi K, Naramoto H, Murakami T (2004) The formation of autunite ($Ca(UO_2)_2(PO_4)2SnH_2O$) within the leached layer of dissolving apatite: incorporation mechanism of uranium by apatite. Chem Geol 211:1–14

Oremland RS, Stolz JF (2003) The ecology of arsenic. Science 300:939–944

Oremland RS, Stolz JF (2005) Arsenic, microbes and contaminated aquifers. Trends Microbiol 13:45–49

Otte ML, Rozema J, Koster L, Haarsma MS, Broekman RA (1989) Iron plaque on roots of Aster tripolium L.: interaction with zinc uptake. New Phytol 111:309–317

Otte ML, Dekkers MJ, Rozema J, Broekman RA (1991) Uptake of arsenic by Aster tripolium in relation to rhizosphere oxidation. Can J Bot 69:2670–2677

Pazirandeh M, Chrisey LA, Mauro JM, Campbell JR, Gaber BP (1995) Expression of the *Neurospora crassa* metallothionein gene in *Escherichia coli* and its effect on heavy-metal uptake. Appl Microbiol Biotechnol 43:1112–1117

Peng X, Wei MX, Ping Y, Zhi Min C, Wang XQ (2011) Eighty-year sedimentary record of heavy metal inputs in the intertidal sediments from the Nanliu River estuary, Beibu Gulf of South China Sea. Environ Pollut 159:92–99

Pershagen G (1983) The epidemiology of human arsenic exposure. In: Fowler BA (ed) Biological and environmental effects of arsenic. Elsevier Science Publishers, Amsterdam, pp 199–232

Peryea FJ, Creger TL (1994) Vertical distribution of lead and arsenic in soils contaminated with lead arsenate pesticide residues. Water Air Soil Pollut 78:297–306

Pongratz W, Endler PC, Poitevin B, Kartnig T (1995) Effect of extremely diluted plant hormone on cell culture. Proc AAAS Ann Meeting, Atlanta

Porter EK, Peterson PJ (1975) Arsenic accumulation by plants on mine waste (United Kingdom). Sci Total Environ 4:365–371

Postma P, Larsen F, Hue NTM, Duc MT, Viet PH, Nhan PQ et al (2007) Arsenic in groundwater of the Red River floodplain, Vietnam: controlling geochemical processes and reactive transport modeling. Geochim Cosmochim Acta 71:5054–5071

Powis G, Mustacich D, Coon A (2000) The role of the redox protein thioredoxin in cell growth and cancer. Free Radic Biol Med 29:312–322

Rahman MM, Chowdhury UK, Mukherjee SC, Mondal BK, Paul K, Lodh D, Biswas BK, Chanda CR, Basu GK, Saha KC, Roy S, Das R, Palit SK, Quamruzzaman Q, Dipankar C (2001) Chronic arsenic toxicity in Bangladesh and West Bengal, India – a review and commentary. J Toxicol Clin Toxicol 39:683–700

Rahman MM, Rahman F, Sansom L, Naidu R, Schmidt O (2009) Arsenic interactions with lipid particles containing iron. Environ Geochem Health 31:201–206

Rakhshaee R, Khosravi M, Ganji MT (2006) Kinetic modeling and thermodynamic study to remove Pb(II), Cd(II), Ni(II) and Zn(II) from aqueous solution using dead and living Azolla filiculoides. J Hazard Mater 134:120–129

Rausch C, Daram P, Brunner S, Jansa J, Laloi M, Leggewie G, Amrhein N, Bucher M (2001) A phosphate transporter expressed in arbuscule-containing cells in potato. Nature 414:462–466

Rittle KA, Drever JI, Colberg PJS (1995) Precipitation of arsenic during bacterial sulfate reduction. Geomicrobiol J 13:1–11

Rohwerder T, Gehrke T, Kinzler K, Sand W (2003) Bioleaching review part A: progress in bioleaching: fundamental and mechanisms of bacterial metal sulfide oxidation. Appl Microbiol Biotechnol 63:239–248

Rosen PB (2002) Biochemistry of arsenic detoxification. FEBS Lett 529:86–92

Rowland HAL, Pederick RL, Polya DA, Pancost RD, van Dongen BE, Gault AG, Vaughan DJ, Bryant C, Anderson B, Lloyd JR (2007) The control of organic matter on microbially mediated iron reduction and arsenic release in shallow alluvial aquifers, Cambodia. Geobiology 5:281–292

Roychowdhury T, Uchino T, Tokunaga H, Ando M (2002) Arsenic and other heavy metals in soils from an arsenic-affected area of West Bengal, India. Chemosphere 49:605–618

Saha GC (2006) Accumulation of arsenic in agricultural soil and selected crops. PhD thesis, Department of Civil Engineering, Bangladesh University of Engineering and Technology (BUET), Dhaka, Bangladesh

Saha GC, Ali MA (2007) Dynamics of arsenic in agricultural soils irrigated with arsenic contaminated groundwater in Bangladesh. Sci Total Environ 379:180–189

Sandrin TR, Maier RM (2002) Effect of pH on cadmium toxicity, speciation, and accumulation during naphthalene biodegradation. Environ Toxicol Chem 21:2075–2079

Sanford RA, Klein DA (1988) Environmental bioremediation for organometallic compounds: microbial growth and arsenic volatilisation from soil and retorted shale. Appl Organomet Chem 2:159–169

Santini JM, Sly LI, Schnagl RD, Macy JM (2000) A new chemolithoautotrophic arsenite-oxidizing bacterium isolated from a gold mine: phylogenetic, physiological, and preliminary biochemical studies. Appl Environ Microbiol 66:92–97

Sanyal SK, Dhillon KS (2005) Arsenic and selenium dynamics in water-soil-plant system: a threat to environmental quality. Indian Society of Soil Science. Proceedings of the International Conference on Soil Water and Environmental Quality, New Delhi, pp 293–263

Sauge-Merle S, Cuine S, Carrier P, Lecomte-Pradines C, Luu DT, Peltier G (2003) Enhanced toxic metal accumulation in engineered bacterial cells expressing *Arabidopsis thaliana* phytochelatin synthase. Appl Environ Microbiol 69:490–494

Saunders JA, Lee MK, Uddin A, Mohammed S, Wilkin RT, Fayek M, Korte NE (2005) Natural arsenic contamination of Hoocene alluvial aquifers by linked tectonic, weathering, and microbial processes. Geosyst 6:1–7

Sela M, Garty J, Tel-or E (1989) The accumulation and the effect of heavy metals on the water fern Azolla filiculoides. New Phytol 112:7–12

Senn DB, Hemond HF (2004) Particulate arsenic and iron during anoxia in a eutrophic, urban lake. EnvironToxicol Chem 23:1610–1616

Sharples JM, Meharg AA, Chambers SM, Cairney JWG (2000) Symbiotic solution to arsenic contamination. Nature 404:951–952
Shen YF, Sun DJ, Zhao XH, Yu GQ (2005) Screening report in areas of endemic arsenism and high content of arsenic in China. Chin J Endemiol (in Chinese) 24:172–175
Shi J, Vlamis-Gardikas A, Aslund F, Holmgren A, Rosen BP (1999) Reactivity of glutaredoxins 1, 2 and 3 from *Escherichia coli* shows that glutaredoxin 2 is the primary hydrogen donor to ArsC-catalyzed arsenate reduction. J Biol Chem 274:36039–36042
Shimizu M, Hochadel JF, Fulmer BA, Waalkes MP (1998) Effects of glutathione depletion and metallothionine gene expression on arsenic-induced cytotoxicity and c-myc expression in vitro. Toxicol Sci 45:204–211
Shneidman D, Belizaire R (1986) Arsenic exposure followed by the development of dermatofibrosarcoma protuberans. Cancer 58:1585–1587
Singh SK, Hawkins C, Clarke ID, Squire JA, Bayani J, Hide T et al (2004) Identification of human brain tumour initiating cells. Nature 432:396–401
Singh R, Paul D, Jain RK (2006) Biofilms: implications in bioremediation. Trends Microbiol 14:389–397
Singh S, Lee W, DaSilva NA, Mulchandani A, Chen W (2008a) Enhanced arsenic accumulation by engineered yeast cells expressing *Arabidopsis thaliana* phytochelatin synthase. Biotechnol Bioeng 99:333–340
Singh S, Mulchandani A, Chen W (2008b) Highly selective and rapid arsenic removal by metabolically engineered *Escherichia coli* cells expressing Fucus vesiculosus metallothionein. Appl Environ Microbiol 74:2924–2927
Smedley PL, Kinniburgh DG (2002) A review of the source, behaviour and distribution of arsenic in natural waters. Appl Geochem 17:517–568
Smith AH, Lingas OE, Rahman M (2000) Contamination of drinking water by arsenic in Bangladesh: A public health emergency. Bull World Health Organ 78:1093–1103
Smith SE, Read DJ (2008) Mycorrhizal symbiosis. Academic Press, New York, London, Burlington, San Diego
Smith E, Naidu R, Alston AM (1998) Arsenic in the soil environment. A review. Adv Agron 64:149–195
Sohrin Y, Matsui M, Kawashima M, Hojo M, Hasegawa H (1997) Arsenic biochemistry affected by eutrophication in Lake Biwa. Jpn Environ Sci Technol 31:2712–2720
Song WY, Sohn EJ, Martinoia E, Lee YJ, Yang YY, Jasinski M, Forestier C, Hwang I, Lee Y (2003) Engineering tolerance and accumulation of lead and cadmium in transgenic plants. Nat Biotechnol 21:914–919
Squibb KS, Fowler BA (1983) The toxicity of arsenic and its compounds. In: Fowler BA (ed) Biological and environmental effects of arsenic. Elsevier Science Publishers, Amsterdam, pp 233–269
Styblo M, Serves SV, Cullen WR, Thomas DJ (1997) Comparative inhibition of yeast glutathione reductase by arsenicals and arsenothiols. Chem Res Toxicol 10:27–33
Thomas JE, Rhue RD (1997) Volatilization of arsenic in contaminated cattle dipping vat soil. Bull Environ Contam Toxicol 59:882–887
Trotta A, Falaschi P, Cornara L, Minganti V, Fusconi A, Drava G, Berta G (2006) Arbuscular mycorrhizae increase the arsenic translocation factor in the As hyperaccumulating fern *Pteris vittata* L. Chemosphere 65:74–81
Tsai SL, Singh S, Chen W (2009) Arsenic metabolism by microbes in nature and the impact on arsenic remediation. Curr Opin Biotechnol 20:659–667
Tseng WP (1977) Effects and dose–response relationships of skin cancer and Blackfoot disease with arsenic. Environ Health Perspect 19:109–119
Tseng WP, Chu HM, How SW, Fong JM, Lin CS, Yeh S (1968) Prevalence of skin cancer in an endemic area of chronic arsenicism in Taiwan. J Natl Cancer Inst 40:453–463
Turpeinen R, Pantsar-Kallio M, Kairesalo T (2002) Role of microbes in controlling the speciation of arsenic and production of arsines in contaminated soils. Sci Total Environ 285:133–145

Turpeinen U, Pantsar-Kallioa M, Häggblomb M, Kairesaloa T (1999) Influence of microbes on the mobilization, toxicity and biomethylation of arsenic in soil. The Sci Tot Environ 236: 173–180

Uppanan P (2000) Screening and characterization of bacteria capable of biotransformation of toxic arsenic compound in soil, M.Sc. Thesis, Mahidol University, Thailand

USEPA (United States Environmental Protection Agency) (1997) Recent development for in-situ treatment of metal contaminated soils. Office of Solid Waste and Emergency Response, EPA-542-R-97-004, pp 8

Vazquez S, Agha R, Granado A, Sarro MJ, Estrban E, Penalosa JM (2006) Use of white lupin plant for phytostabilizzation of Cd and As polluted acid soil. Water Air Soil Pollut 177:349–355

Visoottiviseth P, Panviroj N (2001) Selection of fungi capable of removing toxic arsenic compounds from liquid medium. Sci Asia 27:83–92

Vogel-Mikus K, Pongrac P, Kump P, Necemer M, Regvar M (2006) Colonisation of a Zn, Cd and Pb hyperaccumulator *Thlaspi praecox* Wulfen with indigenous arbuscular mycorrhizal fungal mixture induces changes in heavy metal and nutrient uptake. Environ Pollut 139:362–371

Wagner GM (1997) Azolla: a review of its biology and utilization. Bot Rev 63:1–26

Wakao N, Koyatsu H, Komai Y, Shimokawara H, Sakurai Y, Shiota H (1988) Microbial oxidation of arsenite and occurrence of arsenite-oxidizing bacteria in acid mine water from a sulfur-pyrite mine. Geomicrobiol J 6:11–24

Wang JR, Zhao FJ, Meharg AA, Raab A, Feldmann J, McGrath SP (2002) Mechanisms of arsenic hyperaccumulation in *Pteris vittata*. Uptake kinetics, interactions with phosphate, and arsenic speciation. Plant Physiol 130:1552–1561

Wang S, Mulligan CN (2006) Effect of natural organic matter on arsenic release from soil and sediments into groundwater. Environ Geochem Health 28:197–214

Wang CL, Maratukulam PD, Lum AM, Clark DS, Keasling JD (2000) Metabolic engineering of an aerobic sulfate reduction pathway and its application to precipitation of cadmium on the cell surface. Appl Environ Microbiol 66:4497–4502

Webb JL (1966) Enzymes and metabolic inhibitors vol 3. New York, 7 Academic Press

Wei Q, Matanoski GM, Farmer ER, Hedayati MA, Grossman L (1994) DNA repair and susceptibility to basal cell carcinoma: a case-control study. Am J Epidemiol 140:598–607

Wenzel WW, Kirchbaumer N, Prohaska T, Stingeder G, Lombi E, Adriano DC (2001) Arsenic fractionation in soils using an improved sequential extraction procedure. Anal Chim Acta 436:309–323

WHO (2001) Environmental health criteria 224, arsenic and arsenic compounds. Inter-organization programme for the sound management of chemicals, Geneva

Wilkie JA, Hering JG (1998) Rapid oxidation of geothermal arsenic(III) in stream waters of the eastern Sierra Nevada. Environ Sci Technol 32:657–662

Williams PN, Islam MR, Hussain SA, Meharg AA. Arsenic absorption by rice. CIMMYT/ USGS (2005) Symposium on the behaviour of arsenic in aquifers, soils and plants: implications for management, Dhaka, Jan (16–18). Centro Internacional de Mejoramiento de Maíz y Trigo and the U.S. Geological Survey

Williams PN, Islam MR, Adomako EE, Raab A, Hossain SA, Zhu YG et al. (2006) Increase in rice grain arsenic for regions of Bangladesh irrigating paddies with elevated arsenic in groundwaters. Environ Sci Technol 40:4903–4908

Winship KA (1984) Toxicity of inorganic arsenic salts. Adv Drug React Acute Poisoining Rev 3:129–160

Woolson EA (1977) Generation of alkylarsines from soil. Weed Sci 25:412–416

Wu FY, Ye ZH, Wonga MH (2009) Intraspecific differences of arbuscular mycorrhizal fungi in their impacts on arsenic accumulation by *Pteris vittata* L. Chemosphere 76:1258–1264

Wu SC, Cheung KC, Luo YM, Wong MH (2006) Effects of inoculation of plant growth-promoting rhizobacteria on metal uptake by *Brassica juncea*. Environ Pollut 140:124–135

Xiong XZ, Li PJ, Wang YS, Ten H, Wang LP, Song SH (1987) Environmental capacity of arsenic in soil and mathematical model. Huanjing Kexue 8:8–14

Xu GH, Chague V, Melamed-Bessudo C, Kapulnik Y, Jain A, Raghothama KG, Levy AA, Silber A (2007) Functional characterization of LePT4: a phosphate transporter in tomato with mycorrhiza-enhanced expression. J Exp Bot 58:2491–2501

Zaldivar R (1980) A morbid condition involving cardiovascular, brochopulmonary, digestive and neural lesions in children and young adults after dietary arsenic exposure. Zentralblatt fur Bacteriologie 1 Abt Originale B: Hygiene, Krankenhaushygiene, Betriebshygiene Praventive Medizin 170:44–56

Zaldivar R, Ghai GL (1980) Clinical epidemiological studies on endemic chronic arsenic poisoning in children and adults, including observations on children with high- and low-intake of dietary arsenic. Zentralblatt fur Bacteriologie 1 Abt Originale B: Hygiene, Krankenhaushygiene, Betriebshygiene. Praventive Medizin 170:409–421

Zaloga GP, Deal J, Spurling T, Richter J, Chernow B (1985) Case report: unusual manifestations of arsenic intoxication. Am J Med Sci 289:210–214

Zavala YJ, Duxbury JM (2008) Arsenic in rice: I. Estimating normal levels of total arsenic in rice grain. Environ Sci Technol 42:3856–3860

Zawadzka AM, Crawford RL, Paszczynski AJ (2007) Pyridine-2,6 bis(thiocarboxylic acid) produced by Pseudomonal stutzeri KC reduces chromium (VI) and precipitates mercury, cadmium, lead and arsenic. Biometals 20:145–158

Zhang XK, Zhang FS, Mao DR (1998) Effect of iron plaque outside roots on nutrient uptake by rice (*Oryza sativa* L.): zinc uptake by Fe-deficient rice. Plant Soil 202:33–39

Zhang X, Lin Ai-Jun, Zhao F-J, Guo-Zhong Xu, Duan G-L, Zhu Y-G (2008) Arsenic accumulation by the aquatic fern Azolla: comparison of arsenate uptake, speciation and efflux by *A. caroliniana* and *A. filiculoides*. Environ Pollut 156:1149–1155

Zhang X, Zhao FJ, Huang Q, Williams PN, Sun GX, Zhu YG (2009) Arsenic uptake and speciation in the rootless duckweed *Wolffia globosa*. New Phytol 182:421–428

Zhong ZM, Cai Chao Hu, Ying Sun GuoXin, Williams PN, HaoJie C, Gang L, Jie ZF, Guan ZY (2011) Spatial distribution of arsenic and temporal variation of its concentration in rice. New Phytol 189:200–209

Zobrist J, Dowdle PR, Davis JA, Oremland RS (2000) Mobilization of arsenite by dissimilatory reduction of arsenate. Environ Sci Technol 34:4747–4753

Chapter 13
Fate of Cadmium in Calcareous Soils under Salinity Conditions

Ali Khanmirzaei

13.1 Introduction

Soils containing free Ca or Mg carbonates are termed *calcareous* and usually have a pH above 7 (McBride 1994). The alkalinity in Ca^{2+}- and Mg^{2+}-rich soils is precipitated in the form of calcite ($CaCO_3$), dolomite (Ca, Mg) $(CO_3)_2$, magnesium calcite ($Ca_{1-x}Mg_xCO_3$, $x < 0.2$), and aragonite ($CaCO_3$, polymorph). Calcite and aragonite are two common calcium carbonate minerals in which calcite is in more stable phase, but aragonite is often in the phase deposited biologically, and the conversion to calcite occurs slowly. Generally, the solubility of carbonates in soil is the function of $[H^+]$ and of the partial pressure of CO_2. Assuming the constant CO_2 partial pressure of $10^{-3.5}$ atmosphere, the soils containing carbonates have pH values near 8.25. The presence of $CaCO_3$ in the calcareous soils can greatly increase the capacity of a carbonate solution to resist changes in the soil pH (Butler 1982) and from the principle of solubility products, these carbonates must be completely dissolved out of the soil before the pH can drop. Under these alkalinity conditions, micronutrients such as iron, manganese, and zinc are rendered unavailable to the plants growing in these soils as well as other metal elements. Indeed, the solubility of these elements is regulated by formation of their solid carbonate forms in calcareous soils. The prevalent example for nutrient metal deficiency in these soils is lime-induced chlorosis. The equilibrium concentration of Fe^{3+} in calcareous soil solution at pH 8.3 is 10^{-19} mM (Julian et al. 1983), which gives noticeable iron deficiency in plants not adapted to these conditions. Calcareous soils are distributed in arid and semiarid regions where soil salinization may occur as a consequence of exceeded evaporation.

A. Khanmirzaei (✉)
Department of Soil Science, Karaj Branch, Islamic Azad University, Karaj, Iran
e-mail: alikhanmirzai@yahoo.com; ali.khanmirzaei@kiau.ac.ir

I. Sherameti and A. Varma (eds.), *Detoxification of Heavy Metals*, Soil Biology 30,
DOI 10.1007/978-3-642-21408-0_13,

13.2 Soil Salinity

Soil salinization has emerged as an environmental crisis of global proportions and threatening the quality and sustainability of arable lands. Around 25% of all lands are salt-affected to some degree mostly in arid and semiarid regions. Salinity affects physico-chemical and mineralogical properties, and also microbial communities and activities of the soil (Zahran 1997; Sardinha et al. 2003; Rietz and Haynes 2003). Salinity has a deleterious impact on crop production via osmotic effects and imbalance ion composition which results in some plant nutrient toxicity. Also results of experimental studies have demonstrated an increasing heavy metal mobilization with increasing salinities. The major cations of concern in saline soils and waters are Na^+, Ca^{2+}, Mg^{2+}, and K^+, and the primary anions are Cl^-, SO_4^{2-}, HCO_3^-, CO_3^{2-}, and NO_3^-. Among these ions, chloride has been known as an anion which forms stable complexes with heavy metals especially the case for cadmium (Cd) and can determine Cd availability (Li et al. 1994; McLaughlin et al. 1994, 1997b; Smith and Martell 1981). In the next section, we will focus on the fate of Cd in calcareous soils.

13.3 Fate of Cadmium in Calcareous Soils

Cadmium is a transition element in periodic table of elements with an atomic weight of 112.41 and a melting point of 321°C. This metal is malleable and combines with other heavy metals to form alloys. Cd is the byproduct of the zinc, lead and copper smelting plants and mainly found as impurity in smithsonite ($ZnCO_3$). Zn and Cd have similar electric configuration, but former is an essential element and later is nonessential, even very toxic, to nearly all living organisms (Lindsay 1979). Studies indicate that Cd inhibits seed germination and root elongation of plants (Fargasova 1994). Cd has deleterious impact on photosynthetic pigments as well as inhibition of cellular functions such as photophosphorylation, ATP synthesis, mitochondrial NADH oxidation, and electron-transport system. Adverse renal, kidney, liver, and lung effects are more commonly seen with exposure to Cd by human (Nordberg 1996).

13.3.1 Cadmium in Soil

Soils formed on igneous, metamorphic, and sedimentary rocks averagely contain 0.1–0.3, 0.1–1, and 0.3–11 mg kg^{-1} Cd, respectively (Alloway 1995). Cd tends to accumulate into the soil environment by burning of fossil fuels, mining and smelting of metalliferous ores, metallurgical industries, municipal wastes, fertilizers, pesticides, and sewage (Alloway 1990). The bioavailability of Cd in soil is controlled by the total concentration of Cd as well as the soil factors such as pH, organic matter and clay content (McBride et al. 1997; Yong 2001; Kabata-Pendias and Pendias 2001), soil salinity (McLaughlin et al. 1994), concentration of chloride (Weggler et al. 2004) and carbonate (Renella et al. 2004).

Two main factors which determine the solubility of heavy metals are pH and redox conditions (Sims and Patrick 1978). High mobility of heavy metals usually occurred in acid and reduction conditions, while the effect of pH is well pronounced. Indeed, not only the activity of heavy metals directly and indirectly depends on pH, but also pH controls the adsorption sites. At pH values above 7, cadmium is retained by replacement of calcium and magnesium on clay and organic surfaces making it inaccessible by plants. Precipitation of Cd largely occurred in calcareous soils in the form of CdO, $CdCO_3$, and $Cd_3(PO_4)_2$. Cd can be effectively retained by calcium carbonate surfaces at very low activities of solution Cd^{2+} (McBride 1980; Papadopoulos and Rowell 1988), and the adsorption reactions of Cd on calcite surfaces seemed to be very selective for Cd when Ca was a competing cation (Papadopoulos and Rowell 1988). Maftoun et al.(2004) studied the adsorption isotherms of Cd in 20 highly calcareous soils of Southern Iran, with CCE (calcium carbonate equivalent) being from 25.9 to 63.4%, and proved that the Cd sorption data well fits to two-surface Langmuir adsorption isotherm. They found that CCE is the main factor controlling the adsorption capacity of Cd in these soils as the following:

$$b_2 = -71.9 + 150\,\mathrm{CCE} \quad R^2 = 0.6 \quad p \leq 0.001$$

$$b_2 = -325 + 110\,\mathrm{CCE} + 70.7\mathrm{CEC} \quad R^2 = 0.67\; p \leq 0.001$$

where b (mg kg^{-1}) is the maximum adsorption capacity and CEC (me/100 g soil) is the cation exchange capacity of soil. Similar results have been obtained by Jalali and Moharrami (2007) in ten calcareous soils of Hamadan, Iran. They found a significant correlation of Langmuir b parameter with CCE ($r^2 = 0.69$; $p \leq 0.05$) and CEC ($r^2 = 0.61$; $p \leq 0.05$) in the case of Cd adsorption in these calcareous soils. In this viewpoint, calcareous soils have a natural ability to reduce Cd toxicity via chemiosorption or precipitation of Cd by soil components.

13.3.2 Cadmium Fractionation in Calcareous Soils

Sequential extraction procedures have been widely used for examining the physico-chemical forms of heavy metals, which has been described as "fractionation," and are important tools for investigating the mobility and environmental ecotoxicity of these elements in soils. Recently for the assessment of the contamination levels of heavy metals the total concentration is considered, although it provides no insight into metal mobility and bioavailability. The bioavailability of metals in soil is affected by various physico-chemical forms of metals which are associated with soil constituents (Shuman 1991). In a sequential extraction procedure, a sample is treated with a series of progressively harsher reagents to dissolve increasingly refractory forms. Ideally, the reagents are chosen to selectively attack a specific soil compartment with minimal dissolution of nontargeted phases (Ahnstrom and Parker 1999).

Heavy metals including cadmium were partitioned into five operationally defined fractions: (1) Water soluble + Exchangeable fraction (EXCH); (2) Sorbed-Carbonate bound fraction (CARB); (3) Oxidizable or organically bound fraction (OM); (4) Reducible or associated with Mn and Fe oxides fraction (MNFEO); and (5) Residual fraction (RES). Water soluble and exchangeable forms of metals are the most mobile and bioavailable forms, so that closely related to plant uptake (Shuman 1979), whereas other metal fractions are considered immobile and tightly bound and may not be expected to be released under natural conditions. Renella et al. (2004) found that in a calcareous soil under various management, most of the native Cd of the soil was only slightly available, being mostly bound to CARB and RES Cd.

Han and Banin (1999) studied the transformation of added soluble Cd in a loessial calcareous soil incubated at the field capacity regime under controlled conditions. The soil had a loamy texture, intermediate carbonate content (23.2%) and intermediate CEC, and $Cd(NO_3)_2$ salt was added in the rates of 0, 0.09, 0.54 and 0.9 mg Cd kg^{-1} soil. Then after Cd fractions were determined for 1 h, 1 day, 1, 3, 6, 12, 24, and 48 weeks of incubation time, respectively. They reported that within the first 1 h from the start of incubation time, under the field capacity regime, added soluble Cd was transformed from the Exchangeable + Soluble fraction mainly into the carbonate fraction (60–75%). Jalali and Khanlari (2008) have found that the proportion of Cd associated with the most weakly bound fraction (EXCH) tended to decrease, with corresponding increases in the other five more strongly binding fractions during the incubation. In this study, Cd was added to five calcareous soils at the rate of 8 mg kg^{-1} of Cd as chloride. The samples were incubated for 3 h and 1, 3, 7, 14, 21, and 28 days at 25 C and constant moisture. After incubation, Cd in amended and control soils was fractionated by the sequential extraction procedure. Most of Cd added to soils appeared in the EXCH fraction after 3 h which decreased markedly within the first 3–24 h following Cd addition.

Rajaie et al. (2006) studied the effect of incubation time, soil texture, and application of enriched compost on chemical forms of Cd. In this study, a clay loam calcareous soil [Fine, mixed (calcareous), mesic Typic Calcixerepts] was converted to sandy loam by adding acid-washed pure quartz sand and both the original clay loam and the produced sandy loam were treated with 30 g kg^{-1} of municipal waste compost. The compost had been enriched with different amounts of $CdSO_4$ to obtain Cd concentrations ranging from 5 to 60 mg Cd kg^{-1} in treated soils. After 0, 1, 2, 4, 8, and 16 weeks of incubation, a sequential extraction scheme was used to fractionate Cd of incubated samples into soluble + exchangeable, carbonate-bound, organically bound, Mn-oxide-bound, amorphous Fe-oxide-bound, crystalline Fe-oxide-bound, and residual forms. They found that Cd mostly was converted to soluble + exchangeable, carbonate-bound, and organically bound Cd forms (82–88%) and those for carbonate fraction were dominant and showed the highest capacity for retention. The proportion of soluble + exchangeable fraction in the sandy soil was higher than those for the clay loam after the incubation period suggesting the role of the fine fractions on immobilizing added metals. The study showed that more than 80% of applied Cd immediately converted to carbonate and organic forms and did not change during the remained incubation time. Although the presence of a large amount of cadmium in the first three steps of extraction procedure shows a relatively high Cd bioavailability in soils under study, it can be concluded that the calcareous nature of these soils plays a key role in cadmium retention because a major portion of the soluble Cd entered carbonate fraction immediately after addition to the soils.

13.3.3 Salinity and Cadmium Fractionation

Very limited information is available on the effect of salinity on the bioavailability and fractionation of cadmium in calcareous soils. In an experiment, Abbaspour et al. (2007) evaluated the effects of salinity on distribution of added Cd among soil fractions in a calcareous soil from Central Iran. In this work, soil was treated with 20 mg Cd kg^{-1} as $Cd(NO_3)_2{\cdot}4H_2O$ and 50 mmol kg^{-1} of NaCl, and then incubated at 60% water-holding capacity (60% WHC) and constant temperature (25°C) for 12 weeks. Various fractions of Cd were extracted from the soil after 2 and 12 weeks of incubation using a sequential extraction technique. Results showed that the overall distribution pattern of the unspiked Cd in the calcareous soil was RES > SS > OM > EXC > MnO > AFeO. This may show the positive effect of $CaCO_3$ on the specifically sorbed (SS) Cd. The applied Cd was mainly associated with the SS and MnO fractions, nonetheless addition of NaCl increased the EXC Cd fraction in unspiked and spiked soils, when the time passed a decrease of Cd in the SS fraction and a partial increase of Cd in the MnO, AFeO, RES and OM fractions were observed in the calcareous soil. The study showed slowly transformation and partitionation of Cd applied to the soil in soluble form among the solid-phase component of the soil. EXC forms of Cd are the most mobile and bioavailable form, so that closely related to plant uptake (Shuman 1979).

13.3.4 Salinity and Cadmium Speciation

Cd^{2+} is the predominant ionic species in soil solution, while it can form ionic complexes such as $CdCl^{+}$, $CdOH^{+}$, $CdHCO_3^{+}$, $CdCl_3^{-}$, $CdCl_4^{2-}$, $Cd(OH)_3^{-}$ and $Cd(OH)_4^{2-}$ as well as complexes with organic substances. Studies have shown that chloride-induced salinity, mainly NaCl, increased Cd concentration in cultivated plants (McLaughlin et al. 1994; Weggler-Beaton et al. 2000; Mühling and Läuchli

2003). In a large-scale study, on wheat fields of northeastern North Dakota, Norvell et al. (2000) found that accumulation of Cd in grain was strongly and positively associated with soil salinity. In a greenhouse study, Khoshgoftar et al.(2004) applied different levels of NaCl (0, 60, 120, and 180 mM) and 120 mM $NaNO_3$, via irrigation water, on a Cd-contaminated soil under cultivation of different genotypes of wheat. They found that increasing the NaCl concentration of irrigation water was associated with a significant ($p < 0.05$) increase in Cd concentrations in shoots, whereas application of $NaNO_3$ had no significant effect on Cd concentrations in shoots. They implied that three probable mechanisms are responsible to elevated Cd solution concentration including competition of Na with other cations for sorption sites, complex of cations with added Cl^-, and increased ionic strength in soil solution. Khoshgoftar et al. (2006) found that the major Cd species present in MINTEQA2-calculated soil solution, treated with NaCl-containing irrigation water, were free Cd^{2+} ion, $CdCl^{1+}$, $CdSO_4{}^0$, and $CdHCO^{3+}$ species. McLaughlin et al. (1997b) reported that tuber Cd concentrations of potato (*Solanun tuberosum* L.) were related to activities of the chloro-complexes rather than to Cd^{2+} activities in saline/sodic soil solutions which occurs under high chloride concentration in soil solution. Weggler-Beaton et al. (2003) found that with increases in Cl concentrations in soil solution, Cd–chloro complexes became the dominant species in solution. They found that shoot Cd concentration was most closely correlated with the $CdCl^+$ activity in solution while the activity of free Cd^{2+} was only weakly correlated. The effect of Cl on Cd uptake could be explained by the fact that $CdCl_n{}^{2-n}$ complexes in soil solution are also available for plant uptake (Smolders et al. 1997).

Ghallab and Usman (2007) conducted a greenhouse experiment with two levels of Cd (0.5 and 10 mg Cd kg^{-1}, in the form of $CdCl_2$), and five salinity levels of irrigation water (0, 8.6, 17.1, 34.2, and 68.4 mM NaCl) to determine the effect of NaCl-induced salinity on the solution speciation and availability of Cd in two, clay loam and sandy, calcareous soils. In this study, a chemical equilibrium modeling system (MINEQL+) was applied to predict solution speciation of Cd, which predicted that Cd was present mainly as free Cd^{2+} ions followed by $CdCl^+$ and $CdSO_4{}^0$ in the soils irrigated with deionized water. However, Cd species in the soil solution were significantly altered by increasing chloride concentration, with Cd–chloro complexes becoming the dominant Cd species in the soil solution. They showed that added NaCl resulted in a large decrease in the ratios of Cd^2/Cd_T and $CdSO_4/Cd_T$, and increase in the ratios of $CdCl_2{}^{2-n}/Cd_T$. For example, under the highest level of Cd (10 mg Cd kg^{-1}) and NaCl (68.4 mM), the predicted percentage of free Cd^{2+} decreased from 71.3 to 19.8% and from 50.4 to 17.8%, respectively, in solutions of clay loam and sandy calcareous soils. In addition, the predicted percentage of $CdSO_4$ decreased from 10.3 to 1.1% in clay loam soil and from 16.2 to 1.3% in sandy calcareous soil. In contrast, the predicted percentage of $CdCl_2{}^{2-n}$ increased from 18.4 to 78.2% in clay loam soil and from 33.0 to 80.8% in sandy calcareous soil. It was observed that $CdCl_2{}^{2-n}$ species were present mainly as $CdCl^+$ and, with increasing chloride concentrations in soil solution, followed by the uncharged $CdCl_2^0$. But, the proportion of the negatively charged $CdCl_3^-$ was very small (<1%), even at very high Cl concentrations.

13.4 Cadmium Detoxification

Calcareous soils represent potential adsorptive surfaces for heavy metals by means of carbonate and phosphate components. The presence of carbonate minerals in soils directly affects the mobility and reactivity of heavy metals through the surface interaction and indirectly through their affects on soil pH (McBride 1980). Phosphatic components are the important reaction products of added phosphorus with calcium carbonate as well as in calcareous soils, depending upon Ca:P ratio (Griffin and Jurinak 1973; Chand et al. 1991).

Thakur et al. (2006) conducted a laboratory investigation to determine the mechanism of cadmium sorption by pure calcium carbonate and to examine whether reaction products of phosphate with calcium carbonate serve as a sink for sorption of toxic heavy metal cations like cadmium. Calcium carbonates were treated with orthophosphoric acid in Ca:P ratio of 5:3, 3:2, 4:3 and 1:1, representing the Ca:P ratios of carbonate apatite, tricalcium phosphate, octacalcium phosphate and dicalcium phosphate and then equilibrated for 2 months. They found that cadmium was effectively retained on $CaCO_3$ by the mechanism of chemisorption at lower Cd^{2+} concentrations as the pH of the equilibrium system remained constant (8.6) up to initial Cd^{2+} concentrations of 10^{-4} mol, coinciding to 100% sorption of Cd^{2+} from the solution. At higher concentrations, precipitation of $CdCO_3$ on $CaCO_3$ surface or as a separate phase predominated. Enrichment of $CaCO_3$ with P reduced the Cd-sorption. The chemisorption of Cd probably involved the exchange of Ca^{2+} by Cd^{2+} from $CaCO_3$ surface.

Brown et al. (2004) showed that soil amendments containing phosphorus reduce Cd availability in soils near a former Zn and Pb smelter in Joplin, Missouri, USA. Soil collected from the field was amended in the laboratory with phosphorus added in different ways, including as 1% P-H_3PO_4, a high-Fe by-product + P-triple superphosphate (TSP) (2.5% Fe + 1% P-TSP), 1% P-TSP, 3.2% P-TSP, 1% P-phosphate rock, sludge compost at 10% + 0.32% P-TSP, and sludge compost at 10% + 1% P-TSP. After treatment of the plots, $Ca(OH)_2$ (71% purity) was applied evenly and rototilled into each plot to bring the pH to 7.0. The amount of lime required ranged from 200 ton ha^{-1} (3.2% P-TSP) to 50 ton ha^{-1} (10% compost + 0.32% P-TSP) and no lime for the compost-alone treatment. The indicator plant was tall fescue (*Festuca arundinaceae*), which grew in the field. They found that P-TSP and 1% P-H_3PO_4 were the most effective treatments for reducing plant concentrations of Pb, Zn, and Cd. In this study, they applied high amounts of Ca $(OH)_2$ to maintain soil pH about 7.0 and this would be an important factor controlling solubility of heavy metals and the formation of carbonate components for these metals should be addressed.

13.5 Remediation Techniques

Soil remediation is one of the permanent alternatives to remove metal contaminants from soils. The remediation of metal-contaminated soils involves physical, chemical, and biological techniques. Physical techniques are based on approaches generally applied in mining and the mineral processing industry to extract the desired metal-bearing particles from mineral ores (Dermont et al. 2008). These approaches involve mechanical screening, hydrodynamic classification, gravity concentration, froth flotation, magnetic separation, electrostatic separation, and attrition scrubbing (Dermont et al. 2008). Chemical techniques are based on application of leaching solutions containing chemical reagents to enhance the solubility of metals and to transfer the metals from the soils into extractant solution. Depending on the metal type, degree of contamination, and soil characteristics different chemical reagent including acids, salts and high-concentration chloride solutions, chelating agents, surfactants, and reducing or oxidizing agents can be used in chemical extractions. The metal partitioning in the soil fractions also has a key role in reagent selection. For example, the carbonate-bound fraction is the predominant metal fraction in calcareous soils so that application of a chelating agent such as ethylenediaminetetraacetic acid (EDTA) causes co-dissolution of $CaCO_3$ and Ca^{2+} interferences in target heavy metal complexation by EDTA in high Ca^{2+} solution (Di Palma and Ferrantelli 2005; Papassiopi et al. 1999). Papassiopi et al. (1999) reported that 90% of applied EDTA was consumed to dissolve $CaCO_3$ and less than 10% of remained EDTA was utilized to complex heavy metals in a leaching experiment of a calcareous soil.

13.6 Phytoremediation

Most of the physical and chemical approaches are economically prohibitive due to their modern engineering-type techniques. On the other hand, more ecologically friendly approaches such as phytoremediation had been suggested widely. The use of green plants for reclamation of contaminated land can be a low-cost and ecologically sustainable alternative to other techniques. The effectiveness of phytoremediation of heavy metal-polluted soil depends on (1) the soil properties, (2) the metal content, (3) climatic factors, and (4) the plant. Among the soil properties, pH is closely related to heavy metal bonding in soils (Sauvé et al. 2000) with weak or negligible bonding at acidic pH (Jensen et al. 2000) and strong bonding at neutral to alkaline pH (Sukreeyapongse et al. 2002). Two characteristics of the selected plant to be considered in phytoremediation approaches include their ability to accumulate large concentrations of heavy metals in their tissue which is the so-called "hyper accumulator" and the high biomass producers that accumulate lower metal concentrations in their tissue, but compensates for this by a large production of biomass. Jensen et al. (2009) conducted a field and growth chamber experiment to assess the suitability of willow (*Salix viminalis*) for remediation of two, strongly and moderately, polluted calcareous soils. They stated that although the concentration of Cd in willow leaves grown on strongly polluted calcareous soils was high (80 mg kg^{-1}), it is unsuited on this strongly polluted soil because of poor growth. For a tree species to be suitable for phytostabilization, the root system should be able to both retain and tolerate high concentrations of available metals (Domínguez et al. 2009). Domínguez et al. (2009) investigated the performance of Holm oak (*Quercus ilex* subsp. ballota) seedlings in Cd stabilization in another greenhouse study. They found that seedlings had a high Cd retention capacity in fine roots (up to 7 g kg^{-1}) and low rates of Cd translocation to the leaves (transfer coefficients below 0.03). Also the chlorophyll fluorescence measurements (an indicator of plant stress) only differed slightly from the control treatment and showed that this species has a relatively high tolerance to Cd; therefore, may be useful for the phytostabilization of soils contaminated by Cd. *Thlaspi caerulescens* is one of the hyper accumulators known to accumulate Cd (McGrath 1998; Wang et al. 2006). These species are able to uptake and translocate high amounts of Cd to above-ground parts as a defense mechanism minimizing metal phytotoxicity. Nevertheless, in a critical review Kirkham (2006) stated that plants are limited in the amount of Cd that they can accumulate, even when chelates are used to solubilize the Cd for uptake. Consequently, phytoremediation will be of limited use in removing Cd from the soil.

13.7 Conclusion

This chapter has shown that calcareous soils have a natural ability to retain Cd in insoluble forms with limited treatment on living organism compared to neutral and acidic soils. Nevertheless, under the salinity condition the presence of chloride anions are known to reduce soil sorption of Cd, and an increase in Cl concentration in the soil or soil solution has been shown to increase Cd concentration in plants. Three probable mechanisms are responsible for elevated Cd solution concentration: competition of Na with other cations for sorption sites, complex of cations with added Cl^-, and increased ionic strength in soil solution.

References

Abbaspour A, Kalbasi M, Hajrasuliha S, Golchin A (2007) Effects of plant residue and salinity on fractions of cadmium and lead in three soils. Soil Sediment Contam Int J 16:539–555

Ahnstrom ZS, Parker DR (1999) Development and assessment of a sequential extraction procedure for the fractionation of soil cadmium. Soil Sci Soc Am J 63:1650–1658

Alloway BJ (1990) Heavy metals in soils. Blackie and Sons, UK

Alloway BJ (1995) Heavy metals in soils. Blackie Academic and Professional, London, England

Brown S, Chaney R, Hallfrisch J, Ryan JA, Berti WR (2004) In situ soil treatments to reduce the phyto- and bioavailability of lead, zinc, and cadmium. J Environ Qual 33:522–531
Butler JN (1982) Carbon dioxide equilibria and their applications. Harvard university, USA
Chand T, Tomar NK, Singh JP (1991) Effect of soil properties on the forms inorganic phosphorus in alkaline calcareous soils of different agroclimatic zones. Arid Soil Res Rehabil 5:199–210
Dermont G, Bergeron M, Mercier G, Richer-Laflèche M (2008) Soil washing for metal removal: a review of physical/chemical technologies and field applications. J Hazard Mater 152:1–31
Di Palma L, Ferrantelli P (2005) Copper leaching from a sandy soil: mechanism and parameters affecting EDTA extraction. J Hazard Mater 122:85–90
Domínguez MT, Madrid F, Marañón T, Murillo JM (2009) Cadmium availability in soil and retention in oak roots: potential for phytostabilization. Chemosphere 76:480–486
Fargasova A (1994) Effect of Pb, Cd, Hg, As, and Cr on germination and root growth of *Sinapis alba* seeds. Bull Environ Contam Toxicol 52:452–462
Ghallab A, Usman ARA (2007) Effect of sodium chloride-induced salinity on phyto- availability and speciation of Cd in soil solution. Water Air Soil Pollut 185:43–51
Griffin RA, Jurinak JJ (1973) The interaction of phosphate with calcite. Soil Sci Soc Am J 37:847–850
Han FX, Banin A (1999) Long-term transformation and redistribution of potentially toxic heavy metals in arid-zone soils: II. Incubation at the field capacity moisture content. Water Air Soil Pollut 114:221–250
Jalali M, Khanlari ZV (2008) Cadmium availability in calcareous soils of agricultural lands in Hamadan, Western Iran. Soil Sediment Contam Int J 17(3):256–268
Jalali M, Moharrami S (2007) Competitive adsorption of trace elements in calcareous soils of western Iran. Geoderma 140:156–163
Jensen DL, Holm PE, Christensen TH (2000) Soil and groundwater contamination with heavy metals at two scrap iron and metal recycling facilities. Waste Manag Res 18:52–63
Jensen JK, Holm PE, Nejrup J, Larsen MB, Borggaard OK (2009) The potential of willow for remediation of heavy metal polluted calcareous urban soils. Environ Pollut 157:931–937
Julian G, Cameron HJ, Olsen RA (1983) Role of chelation by ortho dihydroxy phenols in iron absorption by plant roots. J Plant Nutr 6:163–175
Kabata-Pendias A, Pendias H (2001) Trace elements in soils and plants, 2nd edn. CRC, Boca Raton, FL
Khoshgoftar AH, Shariatmadari H, Karimian N, Kalbasi M, Van der Zee SEATM, Parker DR (2004) Salinity and zinc application effects on phytoavailability of cadmium and zinc. Soil Sci Soc Am J 68:1885–1889
Khoshgoftar AH, Shariatmadari H, Karimian N, Kalbasi M, van der Zee SEATM (2006) Cadmium and zinc in saline soil solutions and their concentrations in wheat. Soil Sci Soc Am J 70:582–589
Kirkham MB (2006) Cadmium in plants on polluted soils: effects of soil factors, hyperaccumulation, and amendments. Geoderma 137:19–32
Li Y, Chanet RL, Schneiter AA (1994) Effect of soil chloride level on cadmium concentration in sunflower kernels. Plant Soil 167:275–280
Lindsay WL (1979) Chemical equilibria in soils. Wiley, New York, USA
Maftoun M, Rassooli F, Alinejad Z, Karimian N (2004) Cadmium sorption behavior in some highly calcareous soils of Iran. Commun Soil Sci Plant Anal 35:1271–1282
McBride MB (1980) Chemisorption of Cd^{2+} on calcite surfaces. Soil Sci Soc Am J 44:26–28
McBride MB (1994) Environmental chemistry of soils. Oxford University Press, New York
McBride MB, Richards BK, Steenhuis T, Russo JJ, Sauve S (1997) Mobility and solubility of toxic metals and nutrients in soil fifteen years after sludge application. Soil Sci 162:487–500
McGrath SP (1998) Phytoextraction for soil remediation. In: Brooks RR (ed) Plants that hyperaccumulate heavy metals. CAB International, Wallingford, Oxon, United Kingdom, pp 261–287

McLaughlin MJ, Palmer LT, Tiller KG, Beech TA, Smart MK (1994) Increased soil salinity causes elevated cadmium concentrations in field grown potato tubers. J Environ Qual 23:1013–1018

McLaughlin MJ, Maier NA, Rayment GE, Sparrow LABG, McKay A, Milham P, Merry RH, Smart MK (1997a) Cadmium in Australian potato tubers and soils. J Environ Qual 26:1644–1649

McLaughlin MJ, Tiller KG, Smart MK (1997b) Speciation of cadmium in soil solution of saline/sodic soils and relationship with cadmium concentrations in potato tubers (*Solanum tuberosum* L.). Aust J Soil Res 35:183–198

Mühling KH, Läuchli A (2003) Interaction of NaCl and Cd stress on compartmentation pattern of cations, antioxidant enzymes and protein in leaves of two wheat genotypes differing in salt tolerance. Plant Soil 253:219–231

Nordberg GF (1996) Current issues in low-dose cadmium toxicology: nephrotoxicity and carcinogenicity. Environ Sci 4:133–150

Norvell WA, Wu J, Hopkins DG, Welch RM (2000) Association of cadmium in durum wheat grain with soil chloride and chelate extractable soil cadmium. Soil Sci Soc Am J 64:2162–2168

Papadopoulos P, Rowell DL (1988) The reaction of cadmium with calcium carbonate surfaces. J Soil Sci 39:23–36

Papassiopi N, Tambouris S, Kontopoulos A (1999) Removal of heavy metals from calcareous contaminated soils by EDTA leaching. Water Air Soil Pollut 109:1–15

Rajaie M, Karimian N, Maftoun M, Yasrebi J, Assad MT (2006) Chemical forms of cadmium in two calcareous soil textural classes as affected by application of cadmium-enriched compost and incubation time. Geoderma 136:533–541

Renella G, Adamo P, Bianco MR, Landi L, Violante P, Nannipieri P (2004) Availability and speciation of cadmium added to a calcareous soil under various management. Eur J Soil Sci 55:123–133

Rietz DN, Haynes RJ (2003) Effects of irrigation-induced salinity and sodicity on soil microbial activity. Soil Biol Biochem 35:845–854

Sardinha M, Müller T, Schmeisky H, Joergensen RG (2003) Microbial performance in soils along a salinity gradient under acidic conditions. Appl Soil Ecol 23(3):237–244

Sauvé S, Hendershot W, Allen HE (2000) Solid-solution partitioning of metals in contaminated soils. Dependence on pH, total metal burden, and organic matter. Environ Sci Technol 34:1125–1131

Shuman LM (1979) Zinc, manganese and copper in soil fractions. Soil Sci 127:10–17

Shuman LM (1991) Chemical forms of micronutrient in soils. In: Morvedt JJ et al (eds) Micronutrients in agriculture, vol 2, SSSA Book Series No. 4. Soil Science Society of America, Madison, WI, pp 113–144

Sims JL, Patrick JL (1978) The distribution of micronutrient cations in soil under conditions of varying redox potential and pH. Soil Sci Soc Am J 42:258

Smith RM, Martell AE (1981) Critical stability constants, vol 4. Plenum Press, New York

Smolders E, Lambrechts RM, McLaughlin MJ, Tiller KG (1997) Effect of soil solution chloride on Cd availability to Swiss chard. J Environ Qual 27:426–431

Sukreeyapongse O, Holm PE, Strobel BW, Panichsakpatana S, Magid J, Hansen HCB (2002) pH-dependent release of cadmium, copper, and lead from natural and sludge-amended soils. J Environ Qual 31:1901–1909

Thakur SK, Tomar NK, Pandeya SB (2006) Influence of phosphate on cadmium sorption by calcium carbonate. Geoderma 130:240–249

Wang AS, Angle JS, Chaney RL, Delorme TA, Reeves RD (2006) Soil pH effects on uptake of Cd and Zn by *Thlaspi caerulescens*. Plant Soil 281:325–337

Weggler K, McLaughlin MJ, Graham RD (2004) Effect of chloride in soil solution on the plant availability of biosolid-borne cadmium. J Environ Qual 33:496–504

Weggler-Beaton K, McLaughlin MJ, Graham RD (2000) Salinity increases cadmium uptake by wheat and Swiss chard from soil amended with biosolids. Aust J Soil Res 38:37–45

Weggler-Beaton K, Graham RD, McLaughlin MJ (2003) The influence of low rates of air-dried biosolids on yield and phosphorus and zinc nutrition of wheat (*T. durum*) and barley (*H. vulgare*). Aust J Soil Res 41:293–308

Yong R (2001) Geoenvironmental engineering: contaminated soils, pollutant fate and mitigation. CRC Press, Boca Raton, FL

Zahran HH (1997) Diversity, adaptation and activity of the bacterial flora in saline environments. Biol Fertil Soils 25:211–223

Chapter 14
Organellar Proteomics: A High-Throughput Approach for better Understanding of Heavy Metal Accumulation and Detoxification in Plants

Nagib Ahsan, Byung-Hyun Lee, and Setsuko Komatsu

14.1 Introduction

Heavy metals are metallic elements with densities exceeding 5 g/cm^3. A number of heavy metals, such as copper (Cu), zinc (Zn), and iron (Fe), serve as essential micronutrients for normal plant growth and development when they are present at low concentrations. However, the same metals can also be considered as toxicants when present in excessive amounts due to their potential to inhibit many physiological and biochemical processes (Hartley-Whitaker et al. 2001; Ahsan et al. 2007a; Bona et al. 2007).

A number of nonessential heavy metals and metalloids are potentially toxic to plants even at low concentrations, including aluminum (Al), cadmium (Cd) (Sarry et al. 2006; Ahsan et al. 2007b; Kieffer et al. 2008), mercury (Hg), lead (Pb), arsenic (As) (Ahsan et al. 2008, 2010), chromium (Cr) (Labra et al. 2006), and cesium (Cs) (Le Lay et al. 2006). Some heavy metals are close chemically homologous to essential elements for plant growth and therefore can easily enter or be taken up by roots and transported to other aerial parts of plants by the same pathways as essential nutrients or by other transport mechanisms (Sanita di Toppi and Gabbrielli

N. Ahsan (✉)
National Institute of Crop Science, 2-1-18 Kannondai, Tsukuba 305-8518, Japan

Present address: Department of Biochemistry and Interdisciplinary Plant Group, University of Missouri-Columbia, Columbia, MO 65211, USA
e-mail: nagib_ahsan@yahoo.com

B.-H. Lee
Division of Applied Life Sciences (BK21 Program), IALS, PMBBRC, College of Agriculture, Gyeongsang National University, Jinju 660-701, Republic of Korea

S. Komatsu (✉)
National Institute of Crop Science, 2-1-18 Kannondai, Tsukuba 305-8518, Japan
e-mail: skomatsu@affrc.go.jp

I. Sherameti and A. Varma (eds.), *Detoxification of Heavy Metals*, Soil Biology 30,
DOI 10.1007/978-3-642-21408-0_14,

1999; Meharg and Hartley-Whitaker 2002; Ma et al. 2008). Generally, however, the uptake of heavy metals by plant cells results in severe disruption of various physiological and biochemical processeses by the interaction of metals with sulfhydryl groups in proteins, which leads to an inhibition of protein activity, disruption of protein structure, or displacement of essential elements. The uptake of heavy metals thus results in deficiency effects, restriction of plant growth, and ultimately, cell death (Van Assche and Clijsters 1990).

Plants employ a range of cellular mechanisms to detoxify heavy metal-induced toxicities in cells. Heavy metal tolerance or accumulation and detoxification mechanisms in plants apparently involve pathways in several cellular organelles, including the cell wall, plasma membrane, cytosol, and vacuolar compartments (Hall 2002). Earlier studies suggest that heavy metal tolerance or accumulation and detoxification in plants depend on a series of cell-specific pathways in which metals enter cells through their potential to bind cell walls, enter the cytosol by the efflux pumping activity of the plasma membrane, and are then transported and deposited into vacuoles by pathways involving various amino acids, peptides, and organic acids. Several genes and proteins are likely involved in these detoxification pathways. Identifying the functional genes or proteins involved in organ-specific heavy metal response pathways is critical for better understanding of the molecular mechanisms underlying accumulation and detoxification of heavy metals in plant cells. This can be addressed by organ-specific transcriptome and/or proteome analyses. Proteomics, a high-throughput technology, is now frequently used to develop a more comprehensive picture of how organisms respond at the protein level to stress conditions and the possible interactions and regulation of the complex molecular networks that respond to stressors (Sarry et al. 2006; Ahsan et al. 2008; Kieffer et al. 2008).

Although studies of gene expression at the mRNA level have enhanced understanding of plant responses to heavy metals (Weber et al. 2006; Norton et al. 2008), transcriptional analysis has a number of limitations, including poor correlation between changes in mRNA expression and expression of corresponding proteins (Ross et al. 2004; Gygi et al. 1999). Moreover, protein expression is regulated not only at the transcriptional level, but also at the translational and posttranslational levels. Therefore, obtaining information at the translational and posttranslational levels using proteomics approaches can provide deeper insights into the responses and functional interactions of mature proteins compared to genome-based predictions. High-throughput proteomics is an extremely efficient tool for studying proteins at the translational and posttranslational levels and is therefore one of the best platforms for studying complex biological protein networks under normal and stressful conditions.

This chapter provides an overview of current organellar proteomics applications in studies of heavy metal toxicity and detoxification in plants. We address some points that should be considered in organelle proteomic analysis of responses to heavy metal stress in plants that will provide a deeper understanding of heavy metal detoxification and biotransformation processes in plant cells. Although this chapter does not cover all of the heavy metal detoxification mechanisms that may operate in each tissue at the subcellular level, it does illuminate the general responses of each organ at the subcellular level.

14.2 Proteomic Studies in Response to Heavy Metal Toxicity

Proteomics is a high-throughput analytical technique that has been applied to the study of a number of biological processes, such as protein identification, determination of the protein expression profile during normal and stressful conditions, analysis of posttranslational modifications (PTMs), and studies of protein–protein interactions. The major advantage of proteomics is that it focuses on the functional translated portion of the genome. Thus, the use of proteomics is expanding rapidly in the field of heavy metal stress biology. Ahsan et al. (2009) recently reviewed developments in proteomic studies of plant responses to heavy metal stress, indicating that proteomic studies of metal stress have been far less numerous and comprehensive compared with studies on other environmental stresses in plants (Jorrín et al. 2007).

Currently, many studies seeking to identify heavy metal toxicity-induced differentially expressed proteins or heavy metal stress-responsive proteins in plants rely on a classical gel-based proteomic approach. This approach typically involves two-dimensional polyacrylamide gel electrophoresis (2-DE) coupled with Edman sequencing, Matrix-Assisted Laser Desorption Ionization Time-of-Flight (MALDI-TOF) mass spectrometry (MS), or nanoscale liquid chromatography (nano-LC) electrospray ionization (ESI)-MS/MS analysis (Ahsan et al. 2009).

An improvement to the gel-based proteomics approach allows several samples to be separated simultaneously and visualized in one gel by labeling proteins with various fluorescent reagents before 2-DE. This approach, known as two-dimensional difference gel electrophoresis (2D-DIGE), has been used to successfully identify heavy metal-induced differentially regulated proteins (Kieffer et al. 2008). The method considerably reduces gel-to-gel variation and improves the power of quantitative analyses through the use of a common standard labeled by one of three dyes (e.g., Cy2, Cy3, or Cy5) and used on each gel of an experiment. This technique can reveal the presence of several hundred proteins in a highly reproducible manner and remain unchallenged as the most efficient method for analyzing complex protein mixtures.

Blue Native (BN) and denaturing SDS-PAGE gels provide a better separation of high molecular weight protein complexes associated with the proteomes of organelles such as chloroplasts and mitochondria. BN-PAGE allows proteins to be efficiently resolved under native conditions and also permits direct quantitative assessment of differential changes in a given proteome. Two-dimensional BN-SDS-PAGE coupled with nano-LC-ESI-MS/MS has also been used to analyze the heavy metal stress responses of multiprotein complexes in the leaf apoplast proteome and thylakoid membrane (Führs et al. 2008; Fagioni et al. 2009).

Protein–metabolite interaction analysis is an efficient way to demonstrate heavy metal sequestration and the interaction of metals with proteins and metabolites. Pull-down assays using various affinity columns, such as glutathione (GSH)-sepharose beads, could be a useful technique for determining the response of GSH-binding proteins (Smith et al. 2004). While GSH itself is a metal-binding metabolite, it is also a key precursor of the well-known metal-binding metabolite phytochelatin. Using this technique coupled with MS analyses, several Cu- and Cd-responsive glutathione-*S*-transferases (GSTs) have been identified in *Arabidopsis* (Smith et al. 2004; Roth et al. 2006). Immobilized metal-affinity chromatography (IMAC) is another technique for directly isolating metal-binding proteins. To obtain a better understanding of Cu transport, chelation, and sequestration, Kung et al. (2006) screened a total of 35 unique Cu-interacting proteins in *Arabidopsis* roots using Cu-IMAC. Thus, the identification of Cu-binding proteins could help in elucidating the precise relationships between Cu and other metal ions in various biological processes.

Although classical proteomic approaches have been used to reproducibly identify several hundred heavy metal-responsive proteins from plant samples, they are not satisfactory for separating and identifying soluble/insoluble and/or highly hydrophobic core components of multisubunit complexes. A number of recently developed second-generation MS-based proteomic technologies (nongel-based technologies/shotgun analysis), including multidimensional protein identification technology (MudPIT), stable isotope labeling of amino acids in cell culture (SILAC), and isotope tagging for relative and absolute protein quantitation (iTRAQ), can identify almost all types of proteins, including highly acidic, basic, and hydrophobic proteins, as well as protein complexes (Ong et al. 2002; Ross et al. 2004). However, MudPIT has a major limitation in that it is incapable of performing quantitative expression analysis of multiple samples from separate experiments. The SILAC and iTRAQ approaches are based on differential isotope labeling of proteins from cells cultured under two or more experimental conditions, allowing quantitation of the relative amounts of a given protein expressed under each condition. The major advantages of these techniques are that they provide protein identification with a quantitative annotation of protein properties, including dynamics, turnover, and interaction partners. Currently, SILAC and iTRAQ are considered to be among the most promising techniques available for large-scale quantitative proteome analyses in plant biology (Thelen and Peck 2007).

However, advanced proteomic techniques are still limited with regard to their application to the analysis of heavy metal-responsive proteins in plants. Thus far, only a few reports describing quantitative shotgun proteomic approaches for identification of metal toxicity-induced proteins in plants have been published (Patterson et al. 2007; Alvarez et al. 2009; Schneider et al. 2009). Using the iTRAQ approach, Patterson et al. (2007) analyzed boron toxicity-induced proteins in boron tolerant and intolerant barley cultivars differing in two B-tolerance quantitative trait loci. The identification of a large number of proteins in the roots of these plants, including three enzymes involved in siderophore production (Iron Deficiency Sensitive 2 (IDS2), IDS3, and a methylthio-ribose kinase) that increase in abundance in the B-tolerant barley, suggested that iTRAQ has considerable potential for identifying and quantifying the expression of proteins in conjunction with bulked segregant analysis (Patterson et al. 2007).

To unravel the role of vacuolar transporters in Cd-detoxification processes, Schneider et al. (2009) used iTRAQ to identify a number of membrane proteins, including transporters such as the cation/proton exchanger (CAX1), in the barley leaf tonoplast proteome. Comparative proteomic approaches, including both gel-based (2D-DIGE) and gel-free (iTRAQ) techniques, were used to investigate Cd-responsive proteins in *Brassica juncea* roots, and these experiments revealed that only 12% of the differentially expressed proteins were identified by both approaches (Alvarez et al. 2009). As expected, membrane proteins as well as low abundance proteins were primarily identified by the iTRAQ method; however, 2D-DIGE analysis identified many differentially expressed posttranslationally modified proteins, suggesting that each approach has its own strengths and thus should be used in a complementary manner.

14.3 The Role of Functional Analysis of Organellar Proteins in Understanding Heavy Metal Detoxification Mechanisms

The analysis of total soluble protein in response to heavy metal stress provides valuable information about cytosolic soluble proteins related to the toxicity response to particular heavy metals. However, such an approach does not allow for exploration of the mechanisms and proteins associated with heavy metal transport, sequestration, and deposition/detoxification processes. It is therefore necessary to investigate the responses of specific tissues and subcellular compartments that are directly involved in heavy metals translocation, transformation, extrusion, and sequestration within cells, such as the xylem, cell wall, plasma membrane, vacuole, and apoplast (Hall 2002). To understand the function of organelles in heavy metal detoxification processes, it is important to analyze the subcellular proteome with advanced proteomic technologies.

In plants, some organelles, such as the nucleus, mitochondria, and chloroplasts, are relatively easy to obtain in a pure form, whereas other organelles, such as the cell wall and plasma membrane, are very difficult to isolate using existing techniques (Szponarski et al. 2004; Pan et al. 2005). To obtain significant biological results from subcellular proteome analyses, it is vital that a pure proteome is obtained or extracted; therefore, one of the most critical points that need to be considered in relation to plant subcellular proteomic analysis is confirming the purity of the subcellular organs/proteins. The efficiency of plant organelle proteomic studies also depends on the quality of the biological samples that are used. Several high-throughput methods have emerged recently that have overcome the need to produce relatively pure organellar fractions for analysis (Dunkley et al. 2006; Konishi et al. 2005; Komatsu 2007; Eubel et al. 2007; van Wijk et al. 2007; Salvi et al. 2008).

14.3.1 Cell Wall and Microsomal Proteomics

The root cell wall is the first organ to contact heavy metals in the soil. The cell wall is important not only in maintaining the shape, size, and rigidity of cells, but also in providing a first line of defense against heavy metals by secreting peptides or changing the abundance of several cell wall proteins that interact with metals (Isaacson and Rose 2006; Douchiche et al. 2010; Jiang and Liu 2010). Organic acids secreted by root cell walls bind toxic metals and thus reduce their toxicity, while alteration in the cell wall structure through changes in its protein composition inhibits the entry of metals. It has been reported that under Al stress, a number of Al-tolerant plant species secrete organic acids such as citrate, oxalate, and malate from their roots which form sufficiently strong complexes with Al to protect plant roots (Ma et al. 2001; Ryan et al. 2001). Cell wall structure modulation in response to Cd stress has been observed in several studies (Douchiche et al. 2007, 2010). For instance, flax seedlings exposed to Cd exhibited cell wall thickening and significant alternation of pectin structure (Douchiche et al. 2007), and recently it has been shown that under Cd stress flax adapts the structure of cortical tissue walls by modulating the methyl-esterification pattern of homogalacturonan in various domains (Douchiche et al. 2010). Moreover, Xiong et al. (2009) demonstrated that exogenous application of nitric oxide enhanced Cd tolerance in rice by increasing pectin and hemicellulose in root cell walls.

In the case of Pb stress, the metal accumulates in roots and localizes in the insoluble fraction of the cell walls and nuclei (Piechalak et al. 2002; Jiang and Liu 2010). Using electron microscopy and cytochemistry, Jiang and Liu (2010) investigated ultrastructural alterations and the synthesis and distribution of cysteine-rich proteins in the *Allium sativum* root cell wall in response to Pb stress. Their experiments revealed that Pb ions localized and accumulated in the cell wall, suggesting that the cysteine-rich proteins presented at the cell wall interact with the Pb ions to immobilize them. Taken together, physiological and biochemical studies have clearly indicated that root cell wall proteins are the first line of defense against numerous toxic metals. However, biochemical analyses have identified only a portion of the cell wall proteins involved in metal responses. Analyzing the cell wall proteome could thus be a potentially useful strategy for increasing understanding of the precise role of cell wall proteins in plant toxic metal responses. The major problem associated with cell wall proteomics is that of extracting proteins from the cell wall; specific methods that depend on the biochemical properties of cell wall proteins must be developed.

Jamet et al. (2008) recently reviewed the status of research pertaining to the cell wall proteome. In addition to efforts aimed at generating a proteome map, researchers have recently begun to show considerable interest in identifying cell wall proteins that are differentially secreted as part of the cellular response to adverse conditions or biotic stresses (Ndimba et al. 2003; Oh et al. 2005; Chivasa et al. 2005; Wen et al. 2007). The terms "secretome" and "secretomics" have recently emerged in cell wall proteomics to describe the global study of proteins that are secreted by cells at any given time or under certain conditions. However, to date there have been no reports of large-scale analysis of cell wall proteins secreted in response to heavy metal stress. In addition, the relatively small number of heavy metal-responsive cell wall proteins that have been identified suggests that much work remains in the field of cell wall proteomics.

Membrane proteins play important roles in various cellular processes, such as the modulation of diverse signaling pathways. Plant cells possess a number of transporters, including the cation diffusion facilitators (CDFs), the Zrt- and Irt-like proteins (ZIPs), the cation exchangers (CAXs), the copper transporters (COPTs), the heavy metal P-type ATPases (HMAs), the natural resistance-associated macrophage proteins (NRAMPs), and the ATP-binding cassette (ABC) transporters (Cobbett et al. 2003; Hall and Williams 2003; Hanikenne et al. 2005), all of which are directly involved in metal uptake and homeostasis. Up to now, only a few of the heavy metal transporters have been identified, and most of them are encoded by multigene families. For example, in *Arabidopsis thaliana*, a total of 15 ZIP genes, 12 metal tolerance protein (MTP) genes, and 8 HMA genes have been described that are associated with metal transport or similar functions (Cobbett et al. 2003; Delhaize et al. 2003). However, the transport specificities, patterns of expression, or subcellular localizations of many metal transport proteins are still largely unknown.

The plasma membrane is the first living barrier to free inward diffusion of heavy metals into the cell, and it is considered to be the initial site of heavy metal toxicity in plant root cells. Knowledge regarding the mechanisms controlling the entrance and translocation of metals from the root to the shoot is still rudimentary, however. Identifying the genes encoding membrane proteins associated with metal entrance and translocation is key to understanding heavy metal detoxification processes in plant cells. Analysis of the microsomal and/or plasma membrane proteomes, which contain a number of proteins involved in heavy metal detoxification, including receptors, channel proteins, transporters, and membrane-associated signaling proteins, represents one potential analytical strategy for the comprehensive identification of membrane proteins associated with responses to specific heavy metals. However, a number of factors make extracting and purifying plasma membrane proteins a difficult task, including the presence of the rigid cell wall, interactions between the varying sizes of cytoplasmic and extracellular loops, the presence of posttranslational modifications (e.g., glycosylation and phosphorylation), and the diverse physicochemical properties of integral membrane proteins (Komatsu 2008). Despite these limitations, however, several methods that enable successful identification of plasma membrane proteins have been developed recently (Alexandersson et al. 2008; Komatsu 2008; Whiteman et al. 2008; Mitra et al. 2009). These technologies have been used with some modifications to investigate changes in the plasma membrane proteome in development and in response to stress conditions (Cheng et al. 2009; Nilsson et al. 2010; Nouri and Komatsu 2010). These global analyses have helped to elucidate the roles of membrane proteins by determining their localization to specific tissues and organelles and have identified key proteins involved in membrane transport and regulation of the proton pumping activity of plasma membrane ATPases. Although several studies have investigated the response of plasma membrane proteins to stressful conditions, information regarding plasma membrane proteomic analysis to heavy metal stress is quite limited (Lanquar et al. 2007). Therefore, the field is open to plasma membrane proteomics studies that may advance understanding of the precise role membrane proteins play in the transport of metals from the environment to plant cells.

14.3.2 Cytosolic Proteomics

The activities of heavy metals that result in toxicity occur in the cytosol of cells, and thus a number of cytosolic proteins are differentially regulated in response to heavy metal toxicity in plants (Ahsan et al. 2009). Although most cytosolic proteins are not directly involved in heavy metal detoxification processes, several proteomic studies have shown that, together with upregulation of antioxidative and defense-responsive proteins, a variety of signaling pathway proteins [e.g., those involving jasmonic acid (JA), ethylene, or salicylic acid (SA) signaling] are also differentially expressed under metal stress (Ahsan et al. 2009). Metabolites such as GSH, phytochelatins (PCs), organic acids, and flavonoids are generally synthesized in the cytosol. During metal stress, these metabolites bind heavy metals and thus play a vital role in tolerance and detoxification mechanisms (Clemens 2006; Kovácik and Klejdus 2008). The genes and proteins involved in the direct and indirect biosynthetic pathways of these metabolites are also primarily located in the cytosol. Therefore, proteomic analysis of the cytosol is very important for obtaining a better understanding of heavy metal detoxification processes in plant cells.

Upregulation of ethylene biosynthesis-related proteins or their precursors, such as S-adenosyl-L-methionine synthetase (SAMS), has been demonstrated in several proteomic studies of the response to stress from metals such as Cd (Sarry et al. 2006), Al (Yang et al. 2007), and As (Ahsan et al. 2008). The formation of SAM from L-methionine and ATP is catalyzed by SAMS and via trans-sulfuration reactions. SAM can also serve as a precursor of GSH, a versatile and pervasive metal-binding ligand that plays an important role in metal transport, storage, and metabolism (Wang and Ballatori 1998). Moreover, adding SAM to Cd-treated cells protects them against the deleterious effects of this metal, suggesting that SAM itself can play a protective role during heavy metal stress (Noriega et al. 2007).

Jasmonic acid is a well-known growth regulator in plants; however, recent studies have shown that JA levels or the expression of JA biosynthesis-related proteins increase in response to heavy metal exposure (Rodríguez-Serrano et al. 2006; Ahsan et al. 2008). These results suggest that JA is either directly involved in the mechanism of heavy metal toxicity or indirectly regulates the GSH biosynthesis pathway (Xiang and Oliver 1998). In addition, a number of research papers have shown that proteins involved in the biosynthesis of GSH and PCs are directly involved in heavy metal detoxification process and are differentially regulated.

Among detoxification-related proteins, cysteine synthase (CS) and GST are the most common proteins that increase in response to a wide range of heavy metals, including Cd, Cu, Al, As, and Pb (Ahsan et al. 2009). Cysteine synthase is a key enzyme in the cysteine biosynthetic pathway, and cysteine is one of the main precursors in the biosynthetic pathways for GSH and the PCs. Plants have a large number of GST isozymes that are encoded by multigenic families and that are potentially involved in several cellular functions, including heavy metal detoxification. For instance, under Cu stress a total of eight GST proteins (GSTF 2, GSTFs 6–10, GSTU 19, and GSTU 20) are differentially expressed in *Arabidopsis* (Smith et al. 2004), whereas GSTFs 2 and 6–9 are also upregulated in *Arabidopsis* root following Cd treatment (Roth et al. 2006), suggesting that they may be involved in a common response to heavy metal stress in *Arabidopsis*. Moreover, we found that the expression of an omega-domain containing GST was markedly increased by arsenate treatment in rice root cells, whereas its expression was unchanged in response to other heavy metals, such as arsenite, Cu, or Al, suggesting that GST-omega may be involved in As biotransformation and metabolism. Taken together, these results indicate that a number of cytosolic proteins involved in heavy metal detoxification process in plants are still unknown and that these proteins may work either directly or indirectly through several different pathways.

14.3.3 Vacuolar Proteomics

Vacuoles are the mostly occupied (more than 30%) fluid-filled organelles in cells and have a number of diverse functions including the storage and sequestration of toxic compounds. Much is known about the general functions of plant vacuoles; however, knowledge concerning the proteins that are targeted to the vacuoles and the underlying molecular mechanisms involved in vacuole-mediated heavy metal detoxification is still far from comprehensive. Several vacuolar heavy metal transporters, such as HMA, the ABC transporter, HMT, and CAX, transport either free or conjugated heavy metals across biological membranes extruding from the cytosol and thus play an important role in the deposition, sequestration, and storage of excess and/or toxic metals in vacuoles (Martinoia et al. 2007; Schneider et al. 2009).

To understand the complete array of vacuolar functions in response to heavy metal stress, a comprehensive picture of the protein content of plant vacuoles is required. It is therefore necessary to isolate highly enriched vacuolar preparations and subject them to proteomic analyses. Unfortunately, difficulties associated with isolating vacuoles in a state of high purity have impeded the biochemical characterization of vacuolar proteins. As a result, vacuolar and/or tonoplast proteome analyses have been carried out on only a few plant species, including *Arabidopsis* (Jaquinod et al. 2007), rice (Konishi et al. 2005), barley (Endler et al. 2006), and cauliflower (Schmidt et al. 2007). These studies led to the identification of a large number of vacuolar proteins, including 416 proteins from *Arabidopsis* (Jaquinod et al. 2007), 316 proteins from cauliflower (Schmidt et al. 2007), 101 proteins from barley (Endler et al. 2006), and 43 tonoplast proteins from rice (Tanaka et al. 2004). A comparison between recently published analyses of monocot and dicot vacuolar proteomes showed that more than 45% of vacuolar proteins are unique to, or have been identified only in, the monocot system. These results suggest that monocot vacuolar proteins may be different from dicot vacuolar proteins. The considerable differences in the vacuolar proteomes of monocot and dicot species may be due to the specialized functions of vacuolar proteins in the different cell types, or the differences may be due to the synthesis or storage activities of various biomolecules (Endler et al. 2006).

The proportion of monocot and dicot vacuolar proteins that have been identified is very low (Schneider et al. 2009. While earlier global studies identified many novel vacuolar transporters and other proteins, and suggested possible functions for these proteins under various conditions (Endler et al. 2006; Jaquinod et al. 2007; Schmidt et al. 2007), only limited information is available regarding heavy metal-responsive vacuolar proteins. Recently, however, a quantitative proteomic approach was used to investigate Cd-regulated proteins in the vacuoles of barley, and this study identified a number of transporters that are directly involved in Cd-detoxification process, including CAX1, γ-tonoplast intrinsic protein, and MRP3-like proteins (Schneider et al. 2009).

In order to develop a thorough understanding of the mechanisms that regulate cytosolic homeostasis under metal stress, the identity and function of as many vacuolar transporter proteins as possible must be elucidated. Comparative proteomic studies of the vacuoles of metal hyperaccumulating and nonhyperaccumulating plants will help determine the precise role of vacuolar proteins in heavy metal deposition and accumulation. The protocols developed by Jaquinod et al. (2007) and Schmidt et al. (2007) would be well suited for investigations of the vacuolar and tonoplast heavy metal stress response proteomes, respectively.

14.4 Conclusions

Although high-throughput proteomic analyses have led to an enhancement in our understanding of the global protein networks involved in plant responses to various heavy metal stresses, very limited information is available regarding the proteomic changes at the subcellular level during metal-induced stress. Therefore, a considerable amount of research is still needed to understand the particular response of subcellular organs to heavy metals. Based on previous reports it could be suggested that four subcellular organs (cell wall, plasma membrane, cytosol, and vacuoles) play the most important role in heavy metal stress tolerance and detoxification in plants (Fig. 14.1). Moreover, it is quite clear that more than one mechanism may be involved in heavy metal detoxification in plants, and it is possible that there is no single mechanism that can account for the wide range of heavy metal tolerance found in plants (Hall 2002; Milner and Kochian 2008). Although a number of transporters

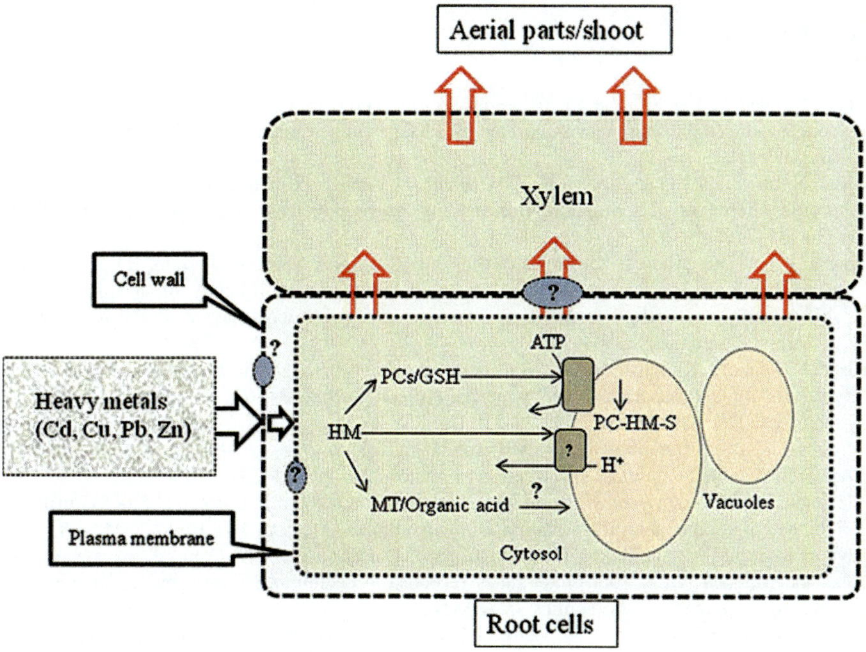

Fig. 14.1 Schematic representation of the possible mechanisms of different subcellular organs for transportation, detoxification, and deposition of heavy metals in plant root cells. Heavy metal transportation, detoxification, and deposition in plant cells are dependent on the direct involvement of a number of organ-specific pathways, such as entering cells via their cell wall binding potential and entering the cytosol via the efflux pumping activity of the plasma membrane. Within the cytosol, heavy metals are transported, detoxified, and deposited through the involvement of several low molecular weight thiols, peptides, and organic acids and thus transported or deposited in the vacuoles

have been identified that are directly involved in transporting various heavy metals from the plasma membrane to vacuoles, they constitute a small proportion of the known metal transporter protein families in plants. In addition, very few reports have been published regarding the interactions of metal transporters with the other proteins or metabolites. Transporter protein–metabolite interactions under heavy metal stress in plants are still poorly understood. Therefore, studies should be undertaken to analyze tissue-specific or subcellular organ-based proteomes of plants in response to heavy metal stress. These studies should identify the complete network of proteins involved with tolerance of plants to detoxification of particular heavy metals. Moreover, a comparative proteomic analysis of the subcellular organs of hyperaccumulating and nonhyperaccumulating plants undergoing metal stress would certainly enhance our understanding of the mechanisms underlying heavy metal tolerance and accumulation in these types of plants.

References

Ahsan N, Lee DG, Lee SH, Kang KY, Lee JJ, Kim PJ, Yoon HS, Kim JS, Lee BH (2007a) Excess copper induced physiological and proteomic changes in germinating rice seeds. Chemosphere 67:1182–1193

Ahsan N, Lee SH, Lee DG, Lee H, Lee SW, Bahk JD, Lee BH (2007b) Physiological and protein profiles alternation of germinating rice seedlings exposed to acute cadmium toxicity. C R Biol 330:735–746

Ahsan N, Lee DG, Alam I, Kim PJ, Lee JJ, Ahn YO, Kwak SS, Lee IJ, Bahk JD, Kang KY, Renaut J, Komatsu S, Lee BH (2008) Comparative proteomic study of arsenic-induced differentially expressed proteins in rice roots reveals glutathione plays a central role during As stress. Proteomics 8:3561–3576

Ahsan N, Renaut J, Komatsu S (2009) Recent developments in the application of proteomics to the analysis of plant responses to heavy metals. Proteomics 9:2602–2621

Ahsan N, Lee DG, Kim KH, Alam I, Lee SH, Lee KW, Lee H, Lee BH (2010) Analysis of arsenic stress-induced differentially expressed proteins in rice leaves by two-dimensional gel electrophoresis coupled with mass spectrometry. Chemosphere 78:224–231

Alexandersson E, Gustavsson N, Bernfur K, Karlsson A, Kjellbom P, Larsson C (2008) Purification and proteomic analysis of plant plasma membranes. Methods Mol Biol 432:161–173

Alvarez S, Berla BM, Sheffield J, Cahoon RE, Jez JM, Hicks LM (2009) Comprehensive analysis of the Brassica juncea root proteome in response to cadmium exposure by complementary proteomic approaches. Proteomics 9:2419–2431

Bona E, Marsano F, Cavaletto M, Berta G (2007) Proteomic characterization of copper stress response in *Cannabis sativa* roots. Proteomics 7:1121–1130

Cheng Y, Qi Y, Zhu Q, Chen X, Wang N, Zhao X, Chen H, Cui X, Xu L, Zhang W (2009) New changes in the plasma-membrane-associated proteome of rice roots under salt stress. Proteomics 9:3100–3114

Chivasa S, Simon WJ, Yu XL, Yalpani N, Slabas AR (2005) Pathogen elicitor-induced changes in the maize extracellular matrix proteome. Proteomics 5:4894–4904

Clemens S (2006) Toxic metal accumulation, responses to exposure and mechanisms of tolerance in plants. Biochimie 88:1707–1719

Cobbett CS, Hussain D, Haydon MJ (2003) Structural and functional relationships between type 1B heavy metal-transporting P-type ATPases in *Arabidopsis*. New Phytol 159:315–321

Delhaize E, Kataoka T, Hebb DM, White RG, Ryan PR (2003) Genes encoding proteins of the cation diffusion facilitator family that confer manganese tolerance. Plant Cell 15:1131–1142

Douchiche O, Rihouey C, Schaumann A, Driouich A, Morvan C (2007) Cadmium-induced alterations of the structural features of pectins in flax hypocotyl. Planta 225:1301–1312

Douchiche O, Driouich A, Morvan C (2010) Spatial regulation of cell-wall structure in response to heavy metal stress: cadmium-induced alteration of the methyl-esterification pattern of homogalacturonans. Ann Bot 105:481–491

Dunkley TP, Hester S, Shadforth IP, Runions J, Weimar T, Hanton SL, Griffin JL, Bessant C, Brandizzi F, Hawes C, Watson RB, Dupree P, Lilley KS (2006) Mapping the *Arabidopsis* organelle proteome. Proc Natl Acad Sci USA 103:6518–6523

Endler A, Meyer S, Schelbert S, Schneider T, Weschke W, Peters SW, Keller F, Baginsky S, Martinoia E, Schmidt UG (2006) Identification of a vacuolar sucrose transporter in barley and Arabidopsis mesophyll cells by a tonoplast proteomic approach. Plant Physiol 141:196–207

Eubel H, Lee CP, Kuo J, Meyer EH, Taylor NL, Millar AH (2007) Free-flow electrophoresis for purification of plant mitochondria by surface charge. Plant J 52:583–594

Fagioni M, D'Amici GM, Timperio AM, Zolla L (2009) Proteomic analysis of multiprotein complexes in the thylakoid membrane upon cadmium treatment. J Proteome Res 8:310–326

Führs H, Hartwig M, Molina LE, Heintz D, Van Dorsselaer A, Braun HP, Horst WJ (2008) Early manganese-toxicity response in *Vigna unguiculata* L.-a proteomic and transcriptomic study. Proteomics 8:149–159

Gygi SP, Rochon Y, Franza BR, Aebersold R (1999) Correlation between protein and mRNA abundance in yeast. Mol Cell Biol 19:1720–1730

Hall JL (2002) Cellular mechanisms for heavy metal detoxification and tolerance. J Exp Bot 53:1–11

Hall JL, Williams LE (2003) Transition metal transporters in plants. J Exp Bot 54:2601–2613

Hanikenne M, Krämer U, Demoulin V, Baurain D (2005) A comparative inventory of metal transporters in the green alga *Chlamydomonas reinhardtii* and the red alga *Cyanidioschizon merolae*. Plant Physiol 137:428–446

Hartley-Whitaker J, Ainsworth G, Meharg AA (2001) Copper and arsenate-induced oxidative stress in *Holcus lanatus* L. clones with differential sensitivity. Plant Cell Environ 24:713–722

Isaacson T, Rose JK (2006) Surveying the plant cell wall proteome or secretome. In: Christine F (ed) Plant proteomics. Ann Plant Rev 28:212–243

Jamet E, Albenne C, Boudart G, Irshad M, Canut H, Pont-Lezica R (2008) Recent advances in plant cell wall proteomics. Proteomics 8:893–908

Jaquinod M, Villiers F, Kieffer-Jaquinod S, Hugouvieux V, Bruley C, Garin J, Bourguignon J (2007) A proteomics dissection of *Arabidopsis thaliana* vacuoles isolated from cell culture. Mol Cell Proteomics 6:394–412

Jiang W, Liu D (2010) Pb-induced cellular defense system in the root meristematic cells of *Allium sativum* L. BMC Plant Biol 10:40

Jorrín JV, Maldonado AM, Castillejo MA (2007) Plant proteome analysis: a 2006 update. Proteomics 7:2947–2962

Kieffer P, Dommes J, Hoffmann L, Hausman JF, Renaut J (2008) Quantitative changes in protein expression of cadmium-exposed poplar plants. Proteomics 8:2514–2530

Komatsu S (2007) Extraction of nuclear proteins. Methods Mol Biol 355:73–77

Komatsu S (2008) Plasma membrane proteome in *Arabidopsis* and rice. Proteomics 8:4137–4145

Konishi H, Maeshima M, Komatsu S (2005) Characterization of vacuolar membrane proteins changed in rice root treated with gibberellin. J Proteome Res 4:1775–1780

Kovácik J, Klejdus B (2008) Dynamics of phenolic acids and lignin accumulation in metal-treated *Matricaria chamomilla* roots. Plant Cell Rep 27:605–615

Kung CC, Huang WN, Huang YC, Yeh KC (2006) Proteomic survey of copper-binding proteins in *Arabidopsis* roots by immobilized metal affinity chromatography and mass spectrometry. Proteomics 6:2746–2758

Labra M, Gianazza E, Waitt R, Eberini I, Sozzi A, Regondi S, Grassi F, Agradi E (2006) *Zea mays* L. protein changes in response to potassium dichromate treatments. Chemosphere 62:1234–1244

Lanquar V, Kuhn L, Lelièvre F, Khafif M, Espagne C, Bruley C, Barbier-Brygoo H, Garin J, Thomine S (2007) ^{15}N-metabolic labeling for comparative plasma membrane proteomics in *Arabidopsis* cells. Proteomics 7:750–754

Le Lay P, Isaure MP, Sarry JE, Kuhn L, Fayard B, Le Bail JL, Bastien O, Garin J, Roby C, Bourguignon J (2006) Metabolomic, proteomic and biophysical analyses of *Arabidopsis thaliana* cells exposed to a caesium stress. Influence of potassium supply. Biochimie 88:1533–1547

Ma JF, Ryan PR, Delhaize E (2001) Aluminium tolerance in plants and the complexing role of organic acids. Trends Plant Sci 6:273–278

Ma JF, Yamaji N, Mitani N, Xu XY, Su YH, McGrath SP, Zhao FJ (2008) Transporters of arsenite in rice and their role in arsenic accumulation in rice grain. Proc Natl Acad Sci USA 105:9931–9935

Martinoia E, Maeshima M, Neuhaus HE (2007) Vacuolar transporters and their essential role in plant metabolism. J Exp Bot 58:83–102

Meharg AA, Hartley-Whitaker J (2002) Arsenic uptake and metabolism in arsenic resistant and nonresistant plant species. New Phytol 154:29–43

Milner MJ, Kochian LV (2008) Investigating heavy-metal hyperaccumulation using *Thlaspi caerulescens* as a model system. Ann Bot 102:3–13

Mitra SK, Walters BT, Clouse SD, Goshe MB (2009) An efficient organic solvent based extraction method for the proteomic analysis of *Arabidopsis* plasma membranes. J Proteome Res 8:2752–2767

Ndimba BK, Chivasa S, Hamilton JM, Simon WJ, Slabas AR (2003) Proteomic analysis of changes in the extracellular matrix of *Arabidopsis* cell suspension cultures induced by fungal elicitors. Proteomics 3:1047–1059

Nilsson R, Bernfur K, Gustavsson N, Bygdell J, Wingsle G, Larsson C (2010) Proteomics of plasma membranes from poplar trees reveals tissue distribution of transporters, receptors, and proteins in cell wall formation. Mol Cell Proteomics 9:368–387

Noriega GO, Balestrasse KB, Batlle A, Tomaro ML (2007) Cadmium induced oxidative stress in soybean plants also by the accumulation of delta-aminolevulinic acid. Biometals 20:841–851

Norton GJ, Lou-Hing DE, Meharg AA, Price AH (2008) Rice-arsenate interactions in hydroponics: whole genome transcriptional analysis. J Exp Bot 59:2267–2276

Nouri MZ, Komatsu S (2010) Comparative analysis of soybean plasma membrane proteins under osmotic stress using gel-based and LC MS/MS-based proteomics approaches. Proteomics 10:1930–1945

Oh IS, Park AR, Bae MS, Kwon SJ, Kim YS, Lee JE, Kang NY, Lee S, Cheong H, Park OK (2005) Secretome analysis reveals an *Arabidopsis* lipase involved in defense against *Alternaria brassicicola*. Plant Cell 17:2832–2847

Ong SE, Blagoev B, Kratchmarova I, Kristensen DB, Steen H, Pandey A, Mann M (2002) Stable isotope labeling by amino acids in cell culture, SILAC, as a simple and accurate approach to expression proteomics. Mol Cell Proteomics 1:376–386

Pan SQ, Caeter CJ, Raikhel NV (2005) Understanding protein trafficking in plant cells through proteomics. Expert Rev Proteomics 2:781–792

Patterson J, Ford K, Cassin A, Natera S, Bacic A (2007) Increased abundance of proteins involved in phytosiderophore production in boron-tolerant barley. Plant Physiol 144:1612–1631

Piechalak A, Tomaszewska B, Baralkiewicz D, Malecka A (2002) Accumulation and detoxification of lead ions in legumes. Phytochemistry 60:153–162

Rodríguez-Serrano M, Romero-Puertas MC, Zabalza A, Corpas FJ, Gómez M, Del Río LA, Sandalio LM (2006) Cadmium effect on oxidative metabolism of pea (*Pisum sativum* L.) roots. Imaging of reactive oxygen species and nitric oxide accumulation in vivo. Plant Cell Environ 8:1532–1544

Ross JK, Bashir S, Giovannoni JJ, Jahn MM, Saravanan RS (2004) Tackling the plant proteome: practical approaches, hurdles and experimental tools. Plant J 39:715–733

Roth U, von Roepenack-Lahaye E, Clemens S (2006) Proteome changes in *Arabidopsis thaliana* roots upon exposure to Cd^{2+}. J Exp Bot 57:4003–4013

Ryan P, Delhaize E, Jones D (2001) Function and mechanism of organic anion exudation from plant roots. Annu Rev Plant Physiol Plant Mol Biol 52:527–560

Salvi D, Rolland N, Joyard J, Ferro M (2008) Purification and proteomic analysis of chloroplasts and their sub-organellar compartments. Methods Mol Biol 432:19–36

Sanita di Toppi L, Gabbrielli R (1999) Response to cadmium in higher plants. Environ Exp Bot 41:105–130

Sarry JE, Kuhn L, Ducruix C, Lafaye A, Junot C, Hugouvieux V, Jourdain A, Bastien O, Fievet JB, Vailhen D, Amekraz B, Moulin C, Ezan E, Garin J, Bourguignon J (2006) The early responses of *Arabidopsis thaliana* cells to cadmium exposure explored by protein and metabolite profiling analyses. Proteomics 6:2180–2198

Schmidt UG, Endler A, Schelbert S, Brunner A, Schnell M, Neuhaus HE, Marty-Mazars D, Marty F, Baginsky S, Martinoia E (2007) Novel tonoplast transporters identified using a proteomic approach with vacuoles isolated from cauliflower buds. Plant Physiol 145:216–229

Schneider T, Schellenberg M, Meyer S, Keller F, Gehrig P, Riedel K, Lee Y, Eberl L, Martinoia E (2009) Quantitative detection of changes in the leaf-mesophyll tonoplast proteome in dependency of a cadmium exposure of barley (*Hordeum vulgare* L.) plants. Proteomics 9:2668–2677

Smith AP, DeRidder BP, Guo WJ, Seeley EH, Regnier FE, Goldsbrough PB (2004) Proteomic analysis of *Arabidopsis* glutathione *s*-transferases from benoxacor- and copper-treated seedlings. J Biol Chem 279:26098–26104

Szponarski W, Sommerer N, Boyer JC, Rossignol M, Gibart R (2004) Large-scale characterization of integral proteins from *Arabidopsis* vacuolar membrane by two-dimensional liquid chromatography. Proteomics 4:397–406

Tanaka N, Fujita M, Hanada H, Murayama S, Uemura M, Kawamura Y, Mitsui T, Mikami S, Tozawa Y, Yoshinaga T, Komatsu S (2004) Proteomics of the rice cell: systematic identification of the protein populations in subcellular compartments. Mol Genet Genomics 271:566–576

Thelen JJ, Peck SC (2007) Quantitative proteomics in plants: choices in abundance. Plant Cell 19:3339–3346

Van Assche F, Clijsters H (1990) A biological test system for the evaluation of the phytotoxicity of metal-contaminated soils. Environ Pollut 66:157–172

van Wijk KJ, Peltier JB, Giacomelli L (2007) Isolation of chloroplast proteins from *Arabidopsis thaliana* for proteome analysis. Methods Mol Biol 355:43–48

Wang W, Ballatori N (1998) Endogenous glutathione conjugates: occurrence and biological functions. Pharmacol Rev 50:335–356

Weber M, Trampczynska A, Clemens S (2006) Comparative transcriptome analysis of toxic metal reponses in *Arabodopsis thaliana* and the Cd^{2+}-hypertolerant facultative metallophyte *Arabidopsis halleri*. Plant Cell Environ 29:950–963

Wen F, VanEtten HD, Tsaprailis G, Hawes MC (2007) Extracellular proteins in pea root tip and border cell exudates. Plant Physiol 143:773–783

Whiteman SA, Nühse TS, Ashford DA, Sanders D, Maathuis FJ (2008) A proteomic and phosphoproteomic analysis of *Oryza sativa* plasma membrane and vacuolar membrane. Plant J 56:146–156

Xiang C, Oliver DJ (1998) Glutathione metabolic genes coordinately respond to heavy metals and jasmonic acid in *Arabidopsis*. Plant Cell 10:1539–1550

Xiong J, An L, Lu H, Zhu C (2009) Exogenous nitric oxide enhances cadmium tolerance of rice by increasing pectin and hemicellulose contents in root cell wall. Planta 230:755–765

Yang Q, Wang Y, Zhang J, Shi W, Qian C, Peng X (2007) Identification of aluminum-responsive proteins in rice roots by a proteomic approach: cysteine synthase as a key player in Al response. Proteomics 7:737–749

Chapter 15
Sulfur Metabolism as a Support System for Plant Heavy Metal Tolerance

Laura A. Hardulak, Mary L. Preuss, and Joseph M. Jez

15.1 Introduction

Sulfur is a critical nutrient for the growth and development of plants, as well as all living organisms, and it plays a central role in plant defense responses against biotic and abiotic stresses. It is a component of the amino acids cysteine and methionine, cofactors, metal clusters, and a diverse range of primary and secondary metabolites, such as glutathione, phytochelatins, and glucosinolates that protect plants from oxidative and environmental stresses (Rausch and Wachter 2005). Sulfur is required for chlorophyll production and the conversion of inorganic nitrogen into protein. As an anionic solute, plants take up sulfur from the soil through sulfate transporters in their roots, which have differential affinities for sulfate (Bick and Leustek 1998). From the atmosphere, sulfur enters leaves by passive diffusion of sulfur dioxide (SO_2) and hydrogen sulfide (H_2S) (Saito 2000). Once internalized, sulfate must be converted into sulfide and then incorporated into various metabolites; it is either accumulated and stored in a vacuole, or incorporated into organic compounds (Bick and Leustek 1998). Sulfur can be assimilated in either the oxidized or the reduced form (Rotte and Leustek 2000). In the oxidized form, the nonreductive pathway directly incorporates sulfate into metabolites such as glucosinolates, hormones, flavonoids, and jasmonates (Varin et al. 1997; Mugford et al. 2010). Metabolically, sulfur metabolism is a core pathway for the synthesis of molecules required for heavy metal tolerance in plants.

L.A. Hardulak • M.L. Preuss (✉)
Department of Biological Sciences, Webster University, 470 East Lockwood Avenue, Webster Groves, MO 63119, USA
e-mail: marypreuss34@webster.edu

J.M. Jez (✉)
Department of Biology, Washington University, One Brookings Drive, Campus Box 1137, St. Louis, MO 63130, USA
e-mail: jjez@biology2.wustl.edu

I. Sherameti and A. Varma (eds.), *Detoxification of Heavy Metals*, Soil Biology 30,
DOI 10.1007/978-3-642-21408-0_15,

15.2 Overview of Sulfur Metabolism

In plants, the reductive sulfate assimilation pathway consists of two phases: the reduction of sulfate to sulfide and the assimilation of sulfide into cysteine (Fig. 15.1). These phases also occur in distinct organelles. Sulfate reduction takes place in the plastids with cysteine synthesis occurring in the plastids as well as in the mitochondria and cytosol (Wu et al. 2010). After uptake from the environment, sulfate is first activated by the enzyme ATP sulfurylase (ATPS) via conversion of ATP to adenosine-5′-phosphosulfate (APS) and inorganic pyrophosphate (Lunn et al. 1990). The high-energy compound APS is then reduced to sulfite and AMP by adenosine 5′-phosphosulfate reductase (APSR), which transfers two electrons to APS to yield sulfite (Setya et al. 1996). Next, sulfite reductase (SiR) converts sulfite to sulfide (Saitoh et al. 2006). In second phase of the assimilatory pathway, sulfide is the source of sulfur for cysteine synthesis. Serine acetyltransferase (SAT) forms O-acetylserine from serine and acetyl-coenzyme A (Saito et al. 1995). The final step of cysteine synthesis, combining O-acetylserine and sulfide, is catalyzed by O-acetylserine sulfhydrylase (OASS; also known as O-acetylserine(thiol)lyase (Bonner et al. 2005). Cysteine can then be used for the synthesis of glutathione and phytochelatin peptides in response to oxidative stresses like heavy metal exposure (Cobbett and Goldsbrough 2002).

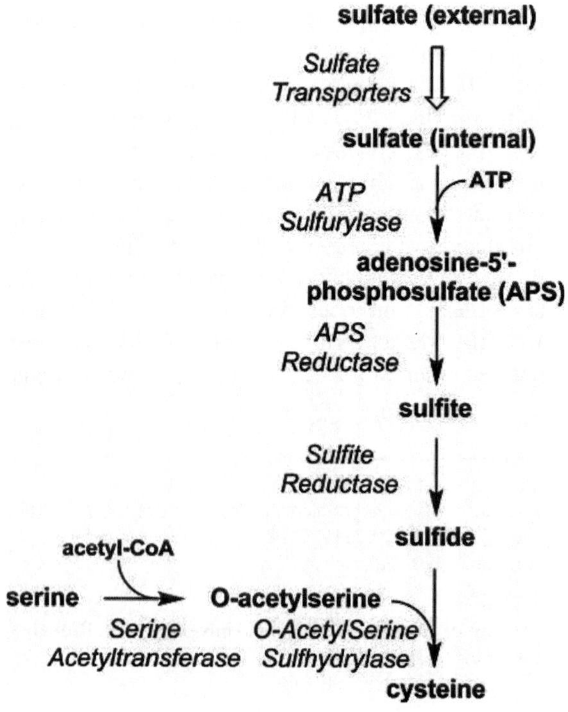

Fig. 15.1 Overview of sulfur metabolism in plants

Toxic heavy metals and metalloids are major environmental pollutants that cause abiotic stresses in plants. Lead and cadmium, for example, decrease the photosynthetic ability and biomass of soybeans (Huang et al. 1974). Flowering has been shown in *Hieracium piloselloides* to be very sensitive to heavy metal contamination of the soil with reductions in leaf life span and delayed reproduction (Ryser and Saunder 2005). Adaptation of the sulfur metabolic circuit is a crucial component of plant survival under stress (Nocito et al. 2007). Plants grown in these conditions concomitantly increase sulfate uptake and increase synthesis of low molecular weight thiols, which bind heavy metals. Cysteine is a precursor of many other cellular compounds, the most widely studied of which is the major cellular redox buffer glutathione (Meister and Anderson 1983). This tripeptide also forms the repeating subunit constituent of phytochelatins, the cysteine-rich peptides that act as heavy-metal chelators in plants and fungi (Grill et al. 1985). Phytochelatin synthesis is part of the cellular heavy metal detoxification network (Cobbett and Goldsbrough 2002). Considering the vast amount of research that has been published on the subject of glutathione and phytochelatins, in this chapter we only consider the reductive sulfur assimilatory pathway prior to the production of glutathione, as it applies to the heavy metal support system of plants.

Sulfur metabolism is profoundly affected by heavy metals. The interplay of a wide range of adaptive responses involving sulfur and sulfur-containing molecules supports plant survival under heavy metal and other oxidative stresses. Exposing sulfur-deficient maize (*Zea mays)* to cadmium stimulated expression of ATPS and OASS (Astolfi et al. 2004). Expression profiling of *Crambe abyssinica* under arsenic stress shows that genes encoding SiR and ATPS were among those highly upregulated in response to arsenic exposure (Paulose et al. 2010). OASS was also upregulated in arsenic-stressed rice (Ahsan et al. 2008). Expression of APSR, SiR, and OASS is elevated in cadmium-treated Indian mustard (*Brassica juncea*) (Minglin et al. 2005; Alvarez et al. 2009). Because glutathione is crucial as an antioxidant and as a precursor to phytochelatins, its cellular concentrations can be depleted under stress. If either sulfur supply or assimilation becomes limiting, depletion of cysteine will limit the biosynthesis and replacement of glutathione. However, if sulfur is not limiting, the challenge is to increase the flux through the sulfur metabolic circuit during heavy metal stresses.

15.3 Sulfur Transporters

The first step in the assimilation of sulfur into a biologically useful molecule for a plant is uptake via sulfate transporters. The processes of uptake and assimilation are energy dependent, driven by a proton gradient across the plasma membranes of root epidermis and root hair cells containing these transporters (Takahashi 2010). Hydrolysis of ATP maintains the proton gradient, as proton-ATPase moves protons from the cytosol into the extracellular space (Lass and Ullrich-Eberius 1984; Hawkesford et al. 1993). Movement of sulfate from the cytoplasm into the apoplast is favored because of the lower apoplastic pH. Reduction from sulfate to sulfide occurs via a series of enzymatic reactions in plastids.

Plant sulfate transporters were first identified in the legume *Stylosanthes hamata*. Their basic structure is similar to predictions based on their orthologs in fungi (Ketter et al. 1991; Smith et al. 1995; Cherest et al. 1997). Multiple sulfate transporters have been identified, mostly in the epidermis and cortex of root systems (Takahashi 2010). The plant sulfate transporter family consists of five groups, based on their amino acid sequences, as observed in *Arabidopsis thaliana* (Nocito et al. 2007). They differ in their affinities for sulfate, rates of transport, and localization in the tissue types.

The locations and capacities of sulfate transporters are greatly influenced by the availability of sulfur in the environment (Takahashi et al. 1997). Low sulfur conditions are associated with abundant expression of high-affinity isoforms. Takahashi et al. (1997) reported the first comprehensive study on changes in gene expression in response to sulfur starvation in Arabidopsis. Transcription of *SULTR1;1* and *SULTR1;2*, the main high-affinity transporters in the roots, accumulated in response to sulfur starvation, but *SULTR1;2* is also abundantly transcribed in conditions of high sulfur. They are predominantly located in root hairs, epidermis, and cortex of roots. Coordinate transcriptional changes for several enzymes, including those involved in cysteine synthesis, occurred as well. Mutant plants defective in *SULTR1;2* are tolerant of selenate, a toxic analog of sulfate (Smith et al. 1995; Cherest et al. 1997). Growth of double mutants defective in both *SULTR1;1* and *SULTR1;2* was stunted under low-sulfur conditions, but adequate sulfur supply permitted growth. This suggests that initial sulfate uptake can be partially substituted by a lower affinity transport system (Barberon et al. 2008; Yoshimoto et al. 2007). Through genetic screening of mutant Arabidopsis plants impaired in their ability to respond to conditions of sulfur starvation, a gene was identified which plays a key role in the demand-driven regulation of sulfate uptake and assimilation (Maruyama-Nakashita et al. 2006). The Sulfur Limitation 1 gene, *SLIM1,* is responsible for transcription of *SULTR1;2*, and is critical for maintaining balance of the entire biochemical circuit of sulfur metabolism in the plant, from uptake all the way through glutathione biosynthesis.

The other three phylogenetic groups of sulfate transporters are less well understood. The group 2 transporters have low affinities to sulfate and the group 3 transporters are relatively uncharacterized, although there is evidence that the SULTR3;5 isoform may form a heterodimer with SULTR2;1 which facilitates the root-to-shoot transport of sulfur in Arabidopsis (Kataoka et al. 2004). Moreover, the group 3 proteins are expressed preferentially in the leaves (Takahashi 2010). In chromium-treated *B. juncea, sulfate* uptake rates were decreased, concomitant with repression of the low-affinity transporter BjST1 (Schiavon et al. 2008), suggesting involvement of sulfate carriers in the transport of chromate. Group 4 transporters are primarily involved in the mobilization of vacuolar sulfate pools. Increased expression of SULTR4;1 and SULTR1:1 in wheat grown on low-sulfur augmented accumulation of selenium and molybdenum in the grain, suggesting efficient mobilization of these metals (Shinmachi et al. 2010). The putative transporters that make up group 5 have short amino acid sequences about which little is known. SULTR5;2 is thought to be involved in molybdenum accumulation in Arabidopsis, but its expression patterns in sulfur-deprived wheat do not fully explain distribution patterns of molybdenum throughout plant tissue (Shinmachi et al. 2010).

Although the basic phylogenetic groups of sulfate transporters and their putative functions have been identified, their exact mechanisms of action and regulation remain elusive. How sulfur is sensed at the molecular level and the subsequent interactions of proteins, metals, and gene expression changes is an area of ongoing research.

15.4 Sulfur Assimilation

15.4.1 ATP Sulfurylase

Once sulfate has been internalized, ATPS catalyzes the entry step into the reductive assimilatory pathway. Because this enzyme catalyzes the first committed step in sulfur assimilation, it has been targeted for the generation of transgenic plants to increase tolerance to cadmium, selenium, and other heavy metals (Khan et al. 2009). It appears to have a rate-limiting role in the accumulation of and tolerance to heavy metals.

Using the heavy metal accumulator *B. juncea*, Heiss et al. (1999) showed that ATPS (and APSR) are induced following cadmium treatment. The increased expression corresponded with elevated demand for thiol metabolism under this stress condition. Although analysis of transgenic *B. juncea* overexpressing ATPS showed that these plants have higher levels of glutathione and total thiols compared to wild-type plants, the accumulation of cadmium, zinc, chromium, copper, lead, and manganese showed no change versus wild type (Bennett et al. 2003). Later studies revealed that overexpression of ATPS in *B. juncea* confers a 2.5-fold increase in selenium tolerance and accumulation to levels observed in the natural selenium hyperaccumulator *Stanleya pinnata* (Van Huysen et al. 2004). Subsequent studies on metal tolerance and accumulation suggest that plants overexpressing ATPS overexpression may be promising for phytoextraction and phytoremediation (Wangeline et al. 2004).

Because photosynthesis is the source of the carbon backbones of amino acids, it might be expected that enhanced photosynthetic potential would also increase pathway flux in order to meet increased demands for cysteine and glutathione under heavy metal stress. In a study by Khan et al. (2009), two varieties of *B. juncea* (Cv. Varuna and RH30) were exposed to sublethal cadmium levels. The activity of ATPS and levels of cysteine and glutathione were increased in both plants, but were much higher in cv. Varuna with RH30 showing symptoms of greater oxidative stress. These results also support the idea that increasing ATPS activity can enhance sulfur metabolism and hence heavy metal tolerance.

15.4.2 APS Reductase

APS Reductase (APSR), responsible for the reduction of APS to sulfite and AMP, is among a group of genes upregulated in response to cadmium stress (Minglin et al. 2005). Transcription of APSR, as well as ATPS, increases in response to depletion of cysteine and glutathione in cadmium-stressed *B. juncea* (Heiss et al. 1999). The coordinate changes resulted in increased cysteine concentrations, especially in the roots, supporting the importance of APSR in regulation of cysteine levels. Although APSR has been shown to be upregulated in response to cadmium in *B. juncea*, no reports describing overexpression of this protein and its effect on heavy metal tolerance and accumulation have been reported to date.

ATPS and APSR catalyze thermodynamically linked steps in sulfur assimilation (Yi et al. 2010). The reaction catalyzed by ATPS is energetically unfavorable in the forward reaction and requires the presence of APSR to maintain metabolic flux through the pathway (Renosto et al. 1993). Therefore, it is possible that the limited success in using ATPS to improve heavy metal tolerance and accumulation results from insufficient APSR to match elevated ATPS levels in transgenic plants. Future efforts could focus on the effect of co-expression of ATPS and APSR to overcome the energetic bottleneck in sulfur assimilation.

15.4.3 Sulfite Reductase

SiR is the last enzyme in the pathway of reduction from sulfate to sulfide. It is constitutively expressed and is not subject to allosteric effectors or posttranslational modification (Nakayama et al. 2000). Although its role in control of flux through the pathway has sometimes been neglected, it has been shown that decreasing its activity creates a bottleneck effect for the entire reductive pathway and downstream metabolites (Khan et al. 2010). In *Arabidopsis thaliana*, SiR is encoded by the only single-copy gene in primary sulfur metabolism. Arabidopsis lines with T-DNA insertions in the promoter region of *SiR* were used to downregulate the expression of the enzyme to different degrees (Khan et al. 2010). Seedlings with only 14% of the *SiR* transcript levels did not survive, which shows that the reductive sulfate pathway is essential for plant life. Seedlings expressing 44% *SiR* transcript levels were severely limited in growth, and they were also sensitive to cadmium, indicating that repression of SiR affects not only sulfur assimilation, but also downstream metabolites in the heavy metal defense network (Khan et al. 2010). No overexpression studies with SiR and analysis of heavy metal tolerance and accumulation have been reported to date.

15.5 Cysteine Synthesis

15.5.1 Serine Acetyltransferase

SAT is the first enzyme in the cysteine synthesis pathway, and is the limiting reaction in the production of this amino acid (Sirko et al. 2004; Yi et al. 2010). SAT plays important roles in heavy metal tolerance and accumulation (Freeman and Salt 2007). Gene expression studies in Arabidopsis show increased expression of SAT isoforms following exposure to cadmium (Howarth et al. 2003). The activity of SAT has also been directly linked to hyperaccumulation of nickel in *Thlaspi goesingense* (Freeman et al. 2004; Freeman and Salt 2007). Comparison of hyperaccumulator and nonaccumulator species of Thlaspi indicates that increased SAT activity correlates with the ability to hyperaccumulate nickel (Freeman et al. 2004). In transgenic Arabidopsis, heterologous expression of *T. goesingense* SAT hyperactivates sulfur assimilation, contributing to elevated shoot concentrations of glutathione, thereby conveying a high level of nickel tolerance (Freeman and Salt 2007). It also contributed to increased tolerance of zinc and cobalt, a slight increase in cadmium resistance, and increased resistance to general oxidative stress (Freeman and Salt 2007). These findings indicate excellent potential for the utilization of mitochondrial SAT for genetic engineering for the application of phytoremediation.

As a limiting enzyme in the production of cysteine, Wawrzynski et al. (2006) explored the strategy of combining overexpression of SAT with enzymes involved in glutathione and phytochelatin synthesis (i.e., glutamate-cysteine ligase and phytochelatin synthase) in transgenic tobacco. Following cadmium treatment, plants expressing the three transgenes all contained higher nonprotein thiol content and increased cadmium accumulation. Although these studies support a key role for sulfur metabolism, they also suggest that other factors, possibly transport systems, limit cadmium accumulation.

15.5.2 O-Acetylserine Sulfhydrylase (O-Acetylserine-Thiolyase)

The final step of the synthesis of cysteine from serine is catalyzed by OASS. Interaction of SAT and OASS forms the decameric cysteine regulatory complex with cellular concentrations of sulfide and *O*-acetylserine affects the reversible association of the complex (Kumaran et al. 2009). Analysis of gene expression in *B. juncea* exposed to cadmium revealed a modest increase in OASS levels, but suggested that coordinated regulation of the genes encoding the enzymes of sulfate uptake and reductive assimilation is required to increase flux through the pathway to meet increased demands due to heavy metal stress (Schäfer et al. 1998). This observation focused efforts to test the role of OASS in heavy metal protection.

Analysis of OASS gene expression in Arabidopsis showed that the cytosolic OASS isoform increased sevenfold in response to cadmium stress (Dominguez-Solís et al. 2001). This corresponded with increased cysteine and glutathione levels required to support the synthesis of phytochelatin peptides. Moreover, plants overexpressing the cytosolic OASS were ninefold more tolerant to cadmium than wild-type plants and are potential phytoremediation tools (Domínguez-Solís et al. 2004). Later studies using a knockout of the cytosolic OASS in Arabidopsis demonstrated that antioxidant activity is essential to maintain redox balance in these plants (López-Martín et al. 2008). Additional studies using OASS to generate transgenic tobacco show enhanced tolerance to cadmium, selenium, and nickel, as well as improved cadmium accumulation in the shoots (Kawashima et al. 2004; Ning et al. 2010). In *Typha latifolia* L. and *Phragmites australis*, activity of OASS increased in response to abscisic acid under stress conditions (Fediuc et al. 2005). The increased OASS activity contributed substantially to replenishing cysteine during conditions of increased demand due to cadmium and sodium chloride stress.

15.6 Alternative Sources of Sulfide Under Stress

As discussed earlier, increased flux through the sulfur assimilation pathway is necessary to maintain glutathione levels for protection against heavy metals; however, there are other ways to help meet the metal-induced demand for increased sulfide. In vitro synthesis of cysteine has been observed, using the enzyme rhodanese to transfer sulfur that has been reduced by thiosulfate reductase to OASS (Louie et al. 2003). This coupling of thiosulfate reductase to the cysteine synthetic pathway by a sulfur transferase was induced in *Schizosaccharomyces pombe* by addition of cadmium. This activity may provide an alternative pathway to provide more sulfide for increased cysteine and glutathione synthesis (Louie et al. 2003). The role of sulfur transferases in cysteine synthesis in plants is an area in which more research is needed, but studies such as this suggest that heavy metal stress may induce this link.

Glucosinolates are an important class of secondary metabolites in plants whose synthesis is linked to sulfate assimilation (Yatusevich et al. 2010). In *Arabidopsis thaliana* and other Brassicaceae, glucosinolates are broken down by the enzyme myrosinase. This degradation produces toxins that deter herbivores. Sulfur released by glucosinolate degradation can be recycled through the biochemical pathways in the roots and leaves. Synthesis of glucosinolates can be induced by biotic stresses, such as heat shock, which is highly genetically correlated to the response of plants to cadmium (Louie et al. 2003). There are two main groups of transcription factors, each regulating the synthesis of one of the two major groups of glucosinolates. The MYB28, MYB76, and MYB29 transcription factors are associated with elevated synthesis of aliphatic glucosinolates and the MYB51, MYB122, and

MYB34 transcription factors affect synthesis of indole glucosinolates. Microarray experiments have predicted that these transcription factors also regulate some of the genes involved in the primary reduction of sulfate. A trans-activation assay provided direct evidence for this link (Yatusevich et al. 2010). Overexpression of the MYB transcription factors also increased mRNA levels of ATPS and APS kinase. APS kinase catalyzes the formation of 3′-phosphoadenosine 5′-phosphosulfate (PAPS), the sulfate donor for glucosinolate biosynthesis.

15.7 Natural Hyperaccumulators

Several plant genera, as well as some microbes and fungi, are natural hyperaccumulators of heavy metals. It is suspected that their enhanced abilities to accumulate, translocate, and detoxify and sequester heavy metal ions evolved in some taxa to protect against disease and insect herbivores, similar to the function of glucosinolates (Salt 2006). The Brassicaceae family contains the largest number of hyperaccumulators with 11 genera (Prasad and Freitas 2003). *Thlaspi* species can accumulate cadmium, nickel, lead, and zinc. Cysteine and other low molecular weight thiols have been implicated in the ability of *Thlaspi caerulescens* to hyperaccumulate cadmium (Hernández-Allica et al. 2006). *B. juncea* (Indian mustard) is known for its ability to accumulate cadmium, chromium, copper, nickel, lead, and zinc. Several aquatic species, such as duckweed (*Lemma minor*) and water hyacinth (*Eichhornia crassipes*) take up metal ions from contaminated water. Sunflower has been shown to remove lead, uranium, cesium, and strontium from hydroponic solution (Prasad and Freitas 2003). Physiological and biochemical comparisons of hyperaccumulators and their nonhyperaccumulating counterparts have discovered genes at key steps in accumulation, translocation, and vacuolar sequestration (Salt 2006). By comparison with *A. thaliana, A. halleri* appears to derive its ability to hyperaccumulate zinc from overexpression of zinc transporters from the ZIP family (Becher et al. 2004). Constitutive overexpression of the CDF-family member *MTP1* contributes to the abilities of several species to compartmentalize cadmium, nickel, and zinc into vacuoles (Küpper et al. 1999, 2001; Krämer et al. 2000; Ma et al. 2005). A group of genes involved in iron balance, including *NAS2*, *NAS3*, *IRT1*, and FRO2 are also overexpressed in hyperaccumulators (Lombi et al. 2001). Many of the important genes have become useful in genetic engineering, in order to exploit these traits for the application of phytoremediation. The problem with using most natural hyperaccumulators for phytoremediation of polluted soils is that they are slow growing and tend to have low biomass. For these reasons there is ongoing research in the incorporation of certain genes into transgenic high-biomass crop plants.

15.8 Conclusion

The uptake, storage, and metabolism of sulfur are highly regulated both locally and at the whole plant level. The sulfate transport system responds to environmental sulfur availability, as well as to cellular levels of cysteine, and glutathione, and other compounds. Nucleophilic properties of the sulfhydryl group convey a unique ability to react with free radicals, active oxygen species, and other substances involved in oxidative stress. Transcriptional regulation of sulfate transporters and enzymes of sulfate reduction and cysteine biosynthesis controls flux through the metabolic pathways. Some plants have evolved remarkable resistances to toxic heavy metals through biochemical responses, most prominently those involving glutathione and phytochelatins. How sulfur and sulfur-containing compounds are initially sensed at a cellular level and provide a signal for the multitude of adaptive responses addresses questions for further research.

Acknowledgements The authors acknowledge funding from the International Center for Advanced Renewable Energy and Sustainability (I-CARES) of Washington University in Saint Louis, a grant from the National Science Foundation (MCB-0824492), and an NSF Research Opportunity Award Supplement to M.L.P and J.M.J.

References

Ahsan N, Lee DG, Alam I, Kim PJ, Lee JJ, Ahn YO, Kwak SS, Lee IJ, Bahk JD, Kang KY, Renaut J, Komatsu S, Lee BH (2008) Comparative proteomic study of arsenic-induced differentially expressed proteins in rice roots reveals glutathione plays a central role during As stress. Proteomics 17:3561–3576

Alvarez S, Berla BM, Sheffield J, Cahoon RE, Jez JM, Hicks LM (2009) Comprehensive analysis of the *Brassica juncea* root proteome in response to cadmium exposure by complementary proteomic approaches. Proteomics 9:2419–2431

Astolfi S, Zuchi S, Passera C (2004) Role of sulphur availability on cadmium-induced changes of nitrogen and sulphur metabolism in maize (*Zea mays* L.) leaves. J Plant Physiol 161:795–802

Barberon M, Berthomieu P, Clairotte M, Shibagaki N, Davidian JC, Gosti F (2008) Unequal functional redundancy between the two *Arabidopsis thaliana* high-affinity sulphate transporters SULTR1;1 and SULTR1;2. New Phytol 180:608–619

Becher M, Talke IN, Krall L, Krämer U (2004) Cross-species microarray transcript profiling reveals high constitutive expression of metal homeostasis genes in shoots of the zinc hyper-accumulator *Arabidopsis halleri*. Plant J 37:251–268

Bennett LE, Burkhead JL, Hale KL, Terry N, Pilon M, Pilon-Smits EA (2003) Analysis of transgenic Indian mustard plants for phytoremediation of metal-contaminated mine tailings. J Environ Qual 32:432–440

Bick JA, Leustek T (1998) Plant sulfur metabolism – the reduction of sulfate to sulfite. Curr Opin Plant Biol 1:240–244

Bonner ER, Cahoon RE, Knapke SM, Jez JM (2005) Molecular basis of cysteine biosynthesis in plants: structural and functional analysis of O-acetylserine sulfhydrylase from *Arabidopsis thaliana*. J Biol Chem 280:38803–38813

Cherest H, Davidian JC, Thomas D, Benes V, Ansorge W, Surdin-Kerjan Y (1997) Molecular characterization of two high affinity sulfate transporters in Saccharomyces cerevisiae. Genetics 145:627–635

Cobbett C, Goldsbrough P (2002) Phytochelatins and metallothioneins: roles in heavy metal detoxification and homeostasis. Annu Rev Plant Biol 53:159–182

Dominguez-Solís JR, Gutierrez-Alcalá G, Vega JM, Romero LC, Gotor C (2001) The cytosolic O-acetylserine(thiol)lyase gene is regulated by heavy metals and can function in cadmium tolerance. J Biol Chem 276:9297–9302

Domínguez-Solís JR, López-Martín MC, Ager FJ, Ynsa MD, Romero LC, Gotor C (2004) Increased cysteine availability is essential for cadmium tolerance and accumulation in Arabidopsis thaliana. Plant Biotechnol J 2:469–476

Fediuc E, Lips SH, Erdei L (2005) O-acetylserine (thiol) lyase activity in Phragmites and Typha plants under cadmium and NaCl stress conditions and the involvement of ABA in the stress response. J Plant Physiol 162:865–872

Freeman JL, Salt DE (2007) The metal tolerance profile of *Thlaspi goesingenseis* mimicked in *Arabidopsis thaliana* heterologously expressing serine acetyltransferase. BMC Plant Biol 7:63

Freeman JL, Persans MW, Nieman K, Albrecht C, Peer W, Pickering IJ, Salt DE (2004) Increased glutathione biosynthesis plays a role in nickel tolerance in Thlaspi nickel hyperaccumulators. Plant Cell 16:2176–2191

Grill E, Winnacker EL, Zenk MH (1985) Phytochelatins: the principal heavy-metal complexing peptides of higher plants. Science 230:674–676

Hawkesford MJ, Davidian JC, Grignon C (1993) Sulphate/proton cotransport in plasma-membrane vesicles isolated from roots of *Brassica napus* L.: increased transport in membranes isolated from sulphur-starved plants. Planta 190:297–304

Heiss S, Schäfer HJ, Haag-Kerwer A, Rausch T (1999) Cloning sulfur assimilation genes of Brassica juncea L.: cadmium differentially effects the expression of a putative low-affinity sulfate transporter and isoforms of ATP sulfurylase and APS reductase. Plant Mol Biol 39:847–857

Hernández-Allica J, Garbisu C, Becerril JM, Barrutia O, García-Plazaola JI, Zhao FJ, Mcgrath SP (2006) Synthesis of low molecular weight thiols in response to Cd exposure in *Thlaspi caerulescens*. Plant Cell Environ 29:1422–1429

Howarth JR, Domínguez-Solís JR, Gutiérrez-Alcalá G, Wray JL, Romero LC, Gotor C (2003) The serine acetyltransferase gene family in *Arabidopsis thaliana* and the regulation of its expression by cadmium. Plant Mol Biol 51:589–598

Huang C, Bazzaz F, Vanderhoef L (1974) The inhibition of soybean metabolism by cadmium and lead. Plant Physiol 54:122–124

Kataoka T, Hayashi N, Yamaya T, Takahashi H (2004) Root-to-shoot transport of sulfate in *Arabidopsis*. Evidence for the role of SULTR3;5 as a component of low-affinity sulfate transport system in the root vasculature. Plant Physiol 136:4198–4204

Kawashima CG, Noji M, Nakamura M, Ogra Y, Suzuki KT, Saito K (2004) Heavy metal tolerance of transgenic tobacco plants over-expressing cysteine synthase. Biotechnol Lett 26:153–157

Ketter JS, Jarai G, Fu YH, Marzluf GA (1991) Nucleotide sequence, messenger RNA stability, and DNA recognition elements of cys-14, the structural gene for sulfate permease II in Neurosporacrassa. Biochemistry 30:1780–1787

Khan NA, Anjum NA, Nazar R, Iqbal N (2009) Increased activity of ATP-sulfurylase and increased contents of cysteine and glutathione reduce high cadmium-induced oxidative stress in mustard cultivar with high photosynthetic potential. Russ J Plant Physiol 56:670–677

Khan MS, Haas FH, Samami AA, Gholami AM, Bauer A, Fellenberg K, Reichelt M, Hänsch R, Mendel RR, Meyer AJ, Wirtz M, Hell R (2010) Sulfite reductase defines a newly discovered bottleneck for assimilatory sulfate reduction and is essential for growth and development in Arabidopsis thaliana. Plant Cell 22:1216–1231

Krämer U, Pickering IJ, Prince RC, Raskin I, Salt DE (2000) Subcellular localization and speciation of nickel in hyperaccumulator and nonaccumulator Thlaspi species. Plant Physiol 122:1343–1353

Kumaran S, Yi H, Krishnan HB, Jez JM (2009) Assembly of the cysteine synthase complex and the regulatory role of protein-protein interactions. J Biol Chem 284:10268–10275

Küpper H, Zhao FJ, McGrath SP (1999) Cellular compartmentation of zinc in leaves of the hyperaccumulator *Thlaspi caerulescens*. Plant Physiol 119:305–312

Küpper H, Lombi E, Zhao FJ, Wieshammer G, McGrath SP (2001) Cellular compartmentation of nickel in the hyperaccumulators Alyssum lesbiacum, Alyssum bertolonii, and Thlaspi goesingense. J Exp Bot 52:2291–2300

Lass B, Ullrich-Eberius CL (1984) Evidence for proton/sulfate cotransport and its kinetics in Lemnagibba G1. Planta 161:53–60

Lombi F, Zhao FJ, McGrath SP, Young SD, Sacchi GA (2001) Physiological evidence for a high-affinity cadmium transporter highly expressed in a *Thlaspi caerulescensecotype*. New Phytol 149:53–60

López-Martín MC, Becana M, Romero LC, Gotor C (2008) Knocking out cytosolic cysteine synthesis compromises the antioxidant capacity of the cytosol to maintain discrete concentrations of hydrogen peroxide in Arabidopsis. Plant Physiol 147:562–572

Louie M, Kondor N, DeWitt JG (2003) Gene expression in cadmium-tolerant Daturainnoxia: detection and characterization of cDNAs induced in response to Cd^{2+}. Plant Mol Biol 52:81–89

Lunn JE, Droux M, Martin J, Douce R (1990) Localization of ATP-sulfurylase and O-acetylserine (thiol) lyase in spinach leaves. Plant Physiol 94:1345–1352

Ma JF, Ueno D, Zhao FJ, McGrath SP (2005) Subcellular localisation of Cd and Zn in the leaves of a Cd-hyperaccumulating ecotype of *Thlaspi caerulescens*. Planta 220:731–736

Maruyama-Nakashita A, Nakamura Y, Tohge T, Saito K, Takahashi H (2006) Arabidopsis SLIM1 is a central transcriptional regulator of plant sulfur response and metabolism. Plant Cell 18:3235–3251

Meister A, Anderson ME (1983) Glutathione. Annu Rev Biochem 52:711–760

Minglin L, Yuxiu Z, Tuanyao C (2005) Identification of genes up-regulated in response to Cd exposure in *Brassica juncea* L. Gene 363:151–158

Mugford SG, Matthewman CA, Hill L, Kopriva S (2010) Adenosine-5′-phosphosulfate kinase is essential for Arabidopsis viability. FEBS Lett 584:119–123

Nakayama M, Akashi T, Hase T (2000) Plant sulfite reductase: molecular structure, catalytic function and interaction with ferredoxin. J Inorg Biochem 82:27–32

Ning H, Zhang C, Yao Y, Yu D (2010) Overexpression of a soybean O-acetylserine (thiol) lyase-encoding gene GmOASTL4 in tobacco increases cysteine levels and enhances tolerance to cadmium stress. Biotechnol Lett 32:557–564

Nocito FF, Lancilli C, Giacomini B, Attilio-Sacchi G (2007) Sulfur metabolism and cadmium stress in higher plants In Plant Stress. Global Science Books, UK

Paulose B, Kandasamy S, Dhankher OP (2010) Expression profiling of Crambe abyssinica under arsenate stress identifies genes and gene networks involved in arsenic metabolism and detoxification. BMC Plant Biol 10:108

Prasad MNV, Freitas H (2003) Metal hyperaccumulation in plants – biodiversity prospecting for phytoremediation technology. J Biotechnol 6:275–321

Rausch T, Wachter A (2005) Sulfur metabolism: a versatile platform for launching defence operations. Trends Plant Sci 10:503–509

Renosto F, Patel HC, Martin RL, Thomassian C, Zimmerman G, Segel IH (1993) ATP sulfurylase from higher plants: kinetic and structural characterization of the chloroplast and cytosol enzymes from spinach leaf. Arch Biochem Biophys 307:272–285

Rotte C, Leustek T (2000) Differential subcellular localization and expression of ATP sulfurylase and 5′-adenylylsulfate reductase during ontogenesis of Arabidopsis leaves indicates that cytosolic and plastid forms of ATP sulfurylase may have specialized functions. Plant Physiol 124:715–724

Ryser P, Saunder WR (2005) Effects of heavy-metal-contaminated soil on growth, phenology, and biomass turnover of Hieracium pioselloides. Environ Pollut 140:152–161

Saito K (2000) Regulation of sulfate transport and synthesis of sulfur-containing amino acids. Curr Opin Plant Biol 3:188–195

Saito K, Yokoyama H, Noji M, Murakoshi I (1995) Molecular cloning and characterization of a plant serine acetyltransferase playing a regulatory role in cysteine biosynthesis from watermelon. J Biol Chem 270:16321–16326

Saitoh T, Ikegami T, Nakayama M, Teshima K, Akutsu H, Hase T (2006) NMR study of the electron transfer complex of plant ferredoxin and sulfite reductase: mapping the interaction sites of ferredoxin. J Biol Chem 281:10482–10488

Salt DE (2006) An extreme plant lifestyle: metal hyperaccumulation. Plant Physiol Online 5:26

Schäfer HJ, Haag-Kerwer A, Rausch T (1998) cDNA cloning and expression analysis of genes encoding GSH synthesis in roots of the heavy-metal accumulator *Brassica juncea* L.: evidence for Cd-induction of a putative mitochondrial gamma-glutamylcysteine synthetase isoform. Plant Mol Biol 37:87–97

Schiavon M, Pilon-Smits EA, Wirtz M, Hell R, Malagoli M (2008) Interactions between chromium and sulfur metabolism in *Brassica juncea*. J Environ Qual 37:1536–1545

Setya A, Murillo M, Leustek T (1996) Sulfate reduction in higher plants: molecular evidence for a novel 5′-adenylylsulfate reductase. Proc Natl Acad Sci USA 93:13383–13388

Shinmachi F, Buchner P, Stroud JL, Parmar S, Zhao FJ, McGrath SP, Hawkesford MJ (2010) Influence of sulfur deficiency on the expression of specific sulfate transporters and the distribution of sulfur, selenium, and molybdenum in wheat. Plant Physiol 153:327–336

Sirko A, Blaszczyk A, Liszewska F (2004) Overproduction of SAT and/or OASTL in transgenic plants: a survey of effects. J Exp Bot 55:1881–1888

Smith FW, Hawkesford MJ, Prosser IM, Clarkson DT (1995) Isolation of a cDNA from Saccharomyces cerevisiae that encodes a high affinity sulphate transporter at the plasma membrane. Mol Gen Genet 247:709–715

Takahashi H (2010) Regulation of sulfate transport and assimilation in plants. Int Rev Cell Mol Biol 281:129–159

Takahashi H, Yamazaki M, Sasakura N, Watanabe A, Leustek T, De Almeida EJ, Engler G, Van Montagu M, Saito K (1997) Regulation of sulfur assimilation in higher plants: a sulfate transporter induced in sulfate-starved roots plays a central in *Arabidopsis thaliana*. Proc Natl Acad Sci USA 94:11102–11107

Van Huysen T, Terry N, Pilon-Smits EA (2004) Exploring the selenium phytoremediation potential of transgenic Indian mustard overexpressing ATP sulfurylase or cystathionine-gamma-synthase. Int J Phytoremediation 6:111–118

Varin L, Marsolais F, Richard M, Rouleau M (1997) Sulfation and sulfotransferases: VI. Biochemistry and molecular biology of plant sulfotransferases. FASEB J 11:517–525

Wangeline AL, Burkhead JL, Hale KL, Lindblom SD, Terry N, Pilon M, Pilon-Smits EA (2004) Overexpression of ATP sulfurylase in Indian mustard: effects on tolerance and accumulation of twelve metals. J Environ Qual 33:54–60

Wawrzynski A, Kopera E, Wawrzynska A, Kaminska J, Bal W, Sirko A (2006) Effects of simultaneous expression of heterologous genes involved in phytochelatin biosynthesis on thiol content and cadmium accumulation in tobacco plants. J Exp Bot 57:2173–2182

Wu Y, Zhao Q, Gao L, Yu X-M, Fang P, Oliver DJ, Xiang CB (2010) Isolation and characterization of low-sulphur-tolerant-mutants of Arabidopsis. J Exp Bot 61:3407–3422

Yatusevich R, Mugford SG, Matthewman C, Gigolashvili T, Frerigmann H, Delaney S, Koprivova A, Flügge UI, Kopriva S (2010) Genes of primary sulfate assimilation are part of the glucosinolate biosynthetic network in *Arabidopsis thaliana*. Plant J 62:1–11

Yi H, Galant A, Ravilious GE, Preuss ML, Jez JM (2010) Sensing sulfur conditions: simple to complex protein regulatory mechanisms in plant thiol metabolism. Mol Plant 3:269–279

Yoshimoto N, Inoue E, Watanabe-Takahashi A, Saito K, Takahashi H (2007) Posttranscriptional regulation of high-affinity sulphate transporters in Arabidopsis by sulphur nutrition. Plant Physiol 145:378–388

Chapter 16
Cd(II)-Activated Synthesis of Phytochelatins

Etsuro Yoshimura

16.1 Introduction

Sequestering metal ions is a strategy adopted by organisms to ameliorate metal toxicity. Although their biological function has not been fully elucidated, phytochelatins (PCs), a family of peptides with the general structure of (γ-Glu-Cys)$_n$-Gly (PC$_n$), where $n \geq 2$ (Fig. 16.1), play such a role (Zenk 1996). Thiol groups coordinate metal ions, particularly soft metal ions such as Cd(II) and Cu(I), thereby reducing metal ion activity. The peptides, which are designated as cadystins and have the same structure as PC_2 and PC_3, were first identified in the fission yeast *Schizosaccharomyces pombe* grown in Cd(II)-containing medium (Kondo et al. 1984). Shortly after the initial identification, the peptides were found to occur in suspension culture cells of the higher plant *Rauvolfia serpentina* exposed to Cd(II) (Grill et al. 1985). Currently, PCs have been identified in higher plants, algae, and some fungi exposed to toxic levels of heavy metal ions (Zenk 1996).

It was suggested early on that the enzyme(s) are involved in PC synthesis, as the peptide possesses γ-glutamyl peptide bonds. PC synthesis in vitro was first demonstrated via reactions catalyzed by the enzyme designated PC synthase (PCS), which was purified from *Silene cucubalus* cell suspension cultures (Grill et al. 1989). The reaction depends on Cd(II) (Loeffler et al. 1989). Thus, adding Cd(II) to a solution containing GSH and PCS initiates PC synthesis, and supplementing the solution with Cd(II) chelators, such as uncomplexed PC$_n$ mixtures or ethylenediaminetetraacetic acid, immediately terminates PC synthesis. These observations led to the PC synthesis mechanism in which the enzyme is activated through binding of Cd(II) to the enzyme, although these observations are insufficient to support this mechanism in the strict sense.

E. Yoshimura (✉)
Department of Applied Biological Chemistry, School of Agricultural and Life Sciences, The University of Tokyo, 1-1-1 Yayoi, Bunkyo, Tokyo 113-8657, Japan
e-mail: ayoshim@mail.ecc.u-tokyo.ac.jp

I. Sherameti and A. Varma (eds.), *Detoxification of Heavy Metals*, Soil Biology 30,
DOI 10.1007/978-3-642-21408-0_16, © Springer-Verlag Berlin Heidelberg 2011

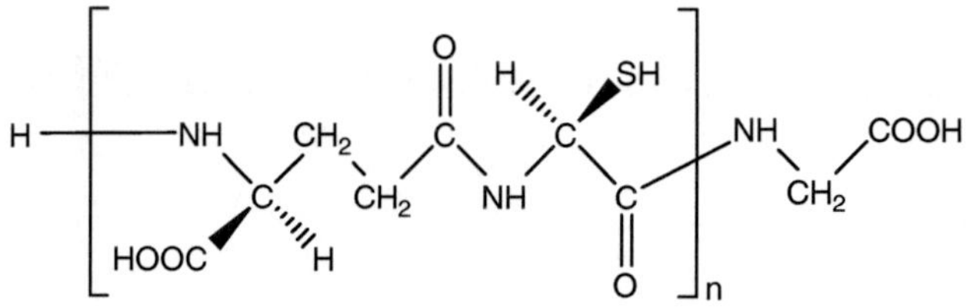

Fig. 16.1 Structure of phytochelatin. A general structure of the peptides is denoted by (γ-Glu-Cys)*n*-Gly with *n* being greater or equal to two, and γ-Glu-Cys indicates a γ-glutamyl linkage, which is a peptide bond formed between the amino group of Cys and the γ-carboxyl group of Glu

It took more than a decade from the first PCS demonstration to identify the genes encoding PCSs, which was performed by three groups independently. Screening Cd (II) tolerant mutant cells from *Saccharomyces cerevisiae* cells carrying wheat (*Triticum aestivum*) or *Arabidopsis thaliana* cDNAs led to the isolation of higher plant PCSs (Clemens et al. 1999; Vatamaniuk et al. 1999). Positional cloning of the *CAD*1 gene, which is responsible for PC-dependent Cd(II) tolerance in *A. thaliana*, also resulted in the isolation of a gene encoding a PCS (Ha et al. 1999). Subsequently, PCS sequence data have accumulated, making it feasible to prepare recombinant PCS enzymes in large quantities with a high degree of purity. Due to the use of recombinant enzymes, considerable progress has been made regarding the synthetic mechanisms of PCSs. In this chapter, the enzymatic mechanisms of action for PCS in relation to metal ion activation are reviewed, focusing on the metal activation enzyme reaction.

16.2 Dipeptidyl Transfer of PCS

16.2.1 Direction of PC Synthesis

The time course of in vitro PC synthesis is shown in Fig. 16.2 (Grill et al. 1989). PC_2 was produced immediately after adding Cd(II) to a solution containing glutathione (γ-Glu-Cys-Gly; GSH) and PCS purified from *S. cucubalus* cell suspension cultures. After a time lag of 15 min, PC_3 began to be formed, followed by PC_4. These observations indicate that Cd(II) is required for PC synthesis, and that PC synthesis proceeds through PC_2 production by a transpeptidation of the γ-Glu-Cys group of GSH to another GSH, following a subsequent peptide elongation by stepwise acquisition of a γ-Glu-Cys unit (*n*) to the peptides. Two mechanisms can be proposed for this peptide elongation, as shown below.

$$\gamma\text{-Gly-Cys-Gly}* + (\gamma\text{-Gly-Cys})_n\text{-Gly} \rightarrow (\gamma\text{-Gly-Cys})_{n+1}\text{-Gly} + \text{Gly}*, \quad (16.1)$$

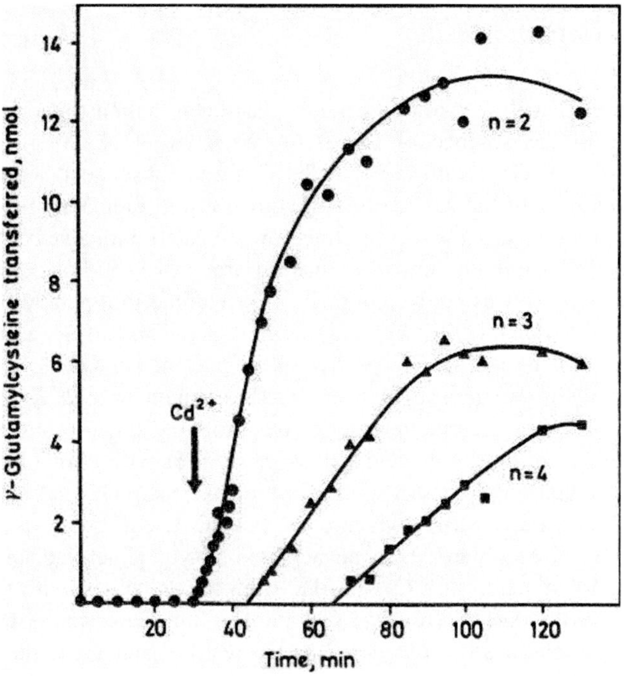

Fig. 16.2 Time course of in vitro phytochelatin synthesis. Adding Cd(II) into a solution containing glutathione (GSH) and phytochelatin synthase (PCS) purified from *Silene cucubalus* cell suspension cultures immediately produced PC_2, followed by PC_3 and PC_4 with lag periods of 15 and 20 min, respectively (taken from Grill et al. 1989, with permission)

$$\gamma\text{-Gly-Cys-Gly}* + (\gamma\text{-Gly-Cys})_n\text{-Gly} \rightarrow (\gamma\text{-Gly-Cys})_{n+1}\text{-Gly}* + \text{Gly}. \quad (16.2)$$

Equation (16.1) indicates peptide elongation in a C to N direction, in which the γ-Gly-Cys moiety of GSH is transferred to form PC_n, thereby releasing the Gly attached to the GSH molecule. Conversely, (16.2) demonstrates an N to C elongation in which the Gly attached on PC_n is replaced by GSH with the release of Gly.

The early observation that GSH and PC_3 were formed when PC_2 was used as a sole substrate implies the dipeptide transfer mechanism shown by (16.1) (Grill et al. 1989). In support of this notion, evidence was afforded using PC_2 and GSH with isotopically labeled Gly as substrates in a PC_3 synthetic reaction catalyzed by AtPCS1 C-terminally tagged with FLAG (AtPCS1-FLAG), together with a parallel experiment in which PC_2 and GSH with isotopically labeled Cys were used (Vatamaniuk et al. 2004). An analysis of radioisotope incorporation into the peptides in a Cd(II)-activated reaction substantiates that PCs are synthesized through the dipeptide transfer reaction (16.1).

16.2.2 Acylation of PCS

Acylation of PCS with a γ-Glu-Cys moiety has been demonstrated using radiolabeled GSH in the absence of Cd(II) (Vatamaniuk et al. 2004). Incubating AtPCS1-FLAG and GSH with radiolabeled Cys and subsequent Sephadex G-50 gel-chromatography of the acid-treated mixture showed radioactivity in the void volume, and protein elution was ascertained by a Western blot analysis using anti-FLAG monoclonal antibody, implying that the protein was radiolabeled. By contrast, no radioactivity was recovered in the void volume in a parallel experiment using AtPCS1-FLAG and GSH with radiolabeled Gly. Thus, PCS was acylated by the γ-Glu-Cys moiety of GSH irrespective of the presence of Cd(II).

Accumulating information about the amino acid sequences of PCSs from a range of organisms has uncovered the features of this protein class. A prominent finding is the resemblance of the amino acid sequences of PCSs with those of papain-type proteases, particularly the conservation of a protease catalytic triad in PCSs (Rea et al. 2004). The catalytic triad, which consists of Cys, His, and Asp (or Asn), is essential to the protease, in which an acylated-enzyme intermediate is formed through the thiolate anion of a Cys residue from the proteases, with the aid of a His residue to withdraw H^+ from the Cys thiol group and of the Asp residue to stabilize the resultant positive charge of the His residue, enhancing the nucleophilicity of the thiol. A logical extension of the conservation of the triad in PCSs will be the formation of a similar acylated-enzyme intermediate during PC synthesis.

A point mutation of AtPCS1-FLAG and heterologous expression of the proteins demonstrated the PCS triad to be practically functional (Romanyuk et al. 2006). Among the yeast cells carrying PCS genes with a Cys residue mutated with a Ser, a C56S mutant alone was unable to grow under increasing concentrations of Cd(II). Furthermore, no apparent Cd(II)-dependent PC synthesis activity was found in the cell supernatant of the C56S transformant alone, despite expression of the protein at a level similar to that of other transformants. These results imply the essentiality of the Cys56 residue, a member of the hypothesized catalytic triad. Similarly, the mutation experiments unequivocally demonstrated the association of His162 and Asp189 during PC synthesis. It was therefore demonstrated that the PCS triad plays a catalytic role with the Cys56 residue as a γ-Glu-Cys acylation site, a reasonable deduction from the catalytic reaction of the proteases.

16.3 Kinetic Analysis of Enzyme Reaction

16.3.1 Assignment of the Substrates in a PC Synthetic Reaction

PC synthesis reaction rates have been determined as a function of GSH concentration at a constant total Cd(II) concentration. However, some complexity arises from the nature of the GSH to form complexes with Cd(II). There are two known types of

GSH complexes, namely, 1:1 and 1:2 Cd(II) GSH complexes, which are denoted as Cd(II)–GSH and Cd(II)–GSH_2, respectively. The respective conditional affinity constants are defined by the following equations:

$$K'_1 = \frac{[\text{Cd(II)}-\text{GSH}]}{[\text{Cd(II)}][\text{GSH}]}, \tag{16.3}$$

$$\beta'_2 = \frac{[\text{Cd(II)}-\text{GSH}_2]}{[\text{Cd(II)}][\text{GSH}]^2}, \tag{16.4}$$

where [Cd(II)–GSH] and [Cd(II)–GSH_2] denote the concentrations of Cd(II)–GSH and Cd(II)–GSH_2, respectively, and [Cd(II)] and [GSH] represent the concentrations of free Cd(II) and GSH, respectively. Therefore, it is apparent that while an increase in GSH concentration means an elevated substrate concentration, it results in a decrease in free Cd(II) concentration concomitantly. To avoid this complexity, an enzyme assay system was designed in which the free Cd(II) level was maintained constantly against a change in GSH concentration. For this purpose, the concentration of total Cd(II) was changed with GSH concentration in such a way that the $[\text{Cd(II)}]_t/[\text{GSH}]_t^{\,2}$ ratio was kept constant, where $[\text{Cd(II)}]_t$ and $[\text{GSH}]_t$ denoted the total concentrations of Cd(II) and GSH, respectively. Thus, it was demonstrated from the conditional stability constants that the majority of Cd(II) added to the assay solution is in the form of Cd(II)–GSH_2 complexes (Ogawa and Yoshimura 2010). In addition, as the GSH concentration bound to Cd(II) is negligible with respect to that of total GSH, it follows that [GSH] can be equated to $[\text{GSH}]_t$. This leads to a free Cd(II) level, which is shown by the following equation:

$$[\text{Cd(II)}] = \frac{[\text{Cd(II)}]_t}{\beta'_2[\text{GSH}]_t^2}. \tag{16.5}$$

Figure 16.3a shows PCS activity as a function of total GSH concentration at a ratio of $[\text{Cd(II)}]_t/[\text{GSH}]_t^{\,2}$ of 0.05 M^{-1} (i.e., a free Cd(II) level of 122 pM). The recombinant enzyme that possesses the *A. thaliana* PCS1 amino acid sequence was used (Ogawa et al 2011). The reaction rate does not follow simple Michaelis–Menten-type kinetics, because a Hanes plot of $[\text{GSH}]_t$/activity versus $[\text{GSH}]_t$ demonstrated nonlinearity under the conditions employed: the plots fell on a linear line at GSH concentrations greater than 25 mM, whereas they turned upward from the line at GSH concentrations less than 25 mM (Fig. 16.3b).

Possible substrates of a PCS-mediated reaction are GSH, Cd(II)–GSH, and Cd (II)–GSH_2. The reaction seems to proceed via a substituted-enzyme mechanism, because the enzyme acylated by a γ-Glu-Cys group has been identified as an intermediate (Vatamaniuk et al. 2004). Using the maximal reaction rate, V_{max}, the enzyme reaction rate, v, for a substituted-enzyme mechanism can be expressed as:

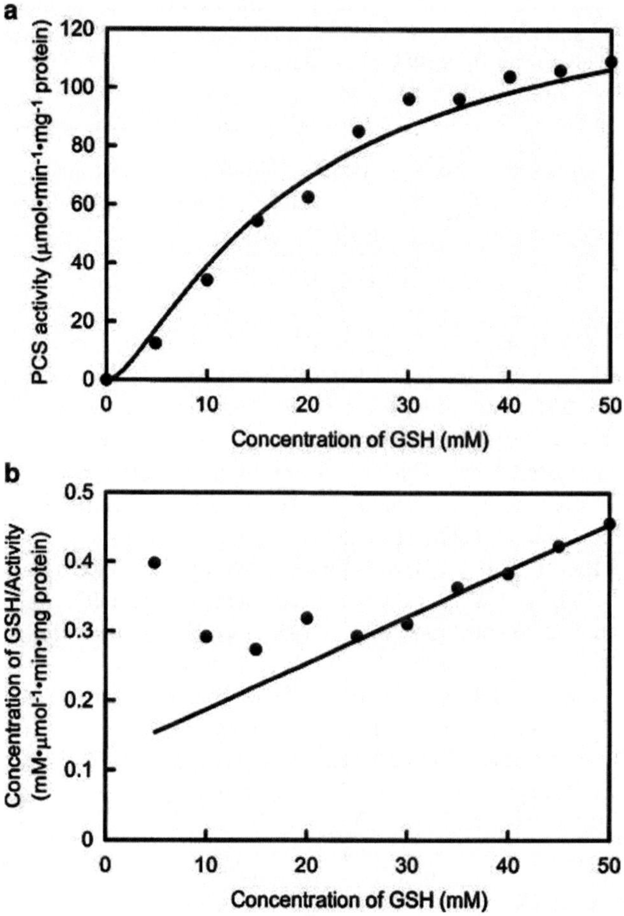

Fig. 16.3 Effect of phytochelatin synthase (PCS) activity on glutathione (GSH) concentration in the presence of a constant free Cd(II) (a), and a Hanes plot of $[GSH]_t$/activity as a function of $[GSH]_t$ (b). The ratio of $[Cd(II)]_t/[GSH]_t^2$ was kept at 0.05 M^{-1} with an increase in $[GSH]_t$, and a free Cd(II) level of 122 pM was attained. The nonlinearity nature of the Hanes plot, together with the finding that GSH reacts with PCS to form a γ-Glu-Cys acylated intermediate in the absence of Cd(II) demonstrated that the GSH and Cd(II)–GSH_2 complex are the substrates in a PCS-mediated PC synthesis reaction. A nonlinear least-square analysis established that the Michaelis constants for GSH and Cd(II)–GSH_2 were 18.0 ± 3.2 mM and 5.14 ± 1.22 μM, respectively. The simulated activity (a *solid line* in **a**) and the determined activity were in agreement (*black circles* in **a**) (Ogawa et al. 2011)

$$v = \frac{V_{\max}[\mathrm{A}][\mathrm{B}]}{K_{\mathrm{mB}}[\mathrm{A}] + K_{\mathrm{mA}}[\mathrm{B}] + [\mathrm{A}][\mathrm{B}]}, \tag{16.6}$$

where A and B denote substrates bound first and second to the enzyme, respectively, and K_{mA} and K_{mB} represent the Michaelis constants for substrates A and B, respectively (Bisswanger 2008). During PCS-mediated PC synthesis, γ-Glu-Cys acylation of the enzyme occurs in a Cd(II)-independent manner, demonstrating that the substrate first bound to the enzyme should be free GSH. If GSH or Cd(II)–GSH act as a second substrate, then the reaction will follow simple Michaelis–Menten-type kinetics. When GSH is the second substrate, the reaction rate can be reduced to

$$v = \frac{V_{\mathrm{max}}[\mathrm{GSH}]_{\mathrm{t}}}{(K_{\mathrm{mA}} + K_{\mathrm{mB}}) + [\mathrm{GSH}]_{\mathrm{t}}}. \tag{16.7}$$

When Cd(II)–GSH is the second substrate, the concentration of the Cd(II)–GSH complex can be shown as:

$$[\mathrm{Cd\text{-}GSH}] = K_1'[\mathrm{Cd(II)}][\mathrm{GSH}]_{\mathrm{t}}. \tag{16.8}$$

Therefore, the rate can be derived by:

$$v = \frac{V_{\mathrm{max}}[\mathrm{GSH}]_{\mathrm{t}}}{\{K_{\mathrm{mA}} + K_{\mathrm{mB}}/(K_1'[\mathrm{Cd(II)}])\} + [\mathrm{GSH}]_{\mathrm{t}}}. \tag{16.9}$$

In contrast, simple Michaelis–Menten-type kinetics likely would not be realized if the Cd(II)–GSH_2 complex were a second substrate. Thus, the concentration of the Cd(II)–GSH_2 complex can be given by the following equation:

$$[\mathrm{Cd\text{-}GSH_2}] = \beta_2'[\mathrm{Cd(II)}][\mathrm{GSH}]_{\mathrm{t}}^2. \tag{16.10}$$

Therefore, the PC synthesis rate can be expressed by:

$$v = \frac{V_{\mathrm{max}}'[\mathrm{GSH}]_{\mathrm{t}2}}{\frac{K_{\mathrm{mB}}}{\beta'_2[\mathrm{Cd(II)}]} + K_{\mathrm{mA}}[\mathrm{GSH}]_{\mathrm{t}} + [\mathrm{GSH}]_{\mathrm{t}}^2}. \tag{16.11}$$

The derivation of this equation leads to the relationship of $[\mathrm{GSH}]_{\mathrm{t}}$/activity vs. $[\mathrm{GHS}]_{\mathrm{t}}$ as follows:

$$\frac{[\mathrm{GSH}]_{\mathrm{t}}}{v} = \frac{1}{V_{\mathrm{max}}'}\left\{[\mathrm{GSH}]_{\mathrm{t}} + K_{\mathrm{mA}} + \frac{K_{\mathrm{mB}}}{\beta_2'[\mathrm{Cd(II)}][\mathrm{GSH}]_{\mathrm{t}}}\right\}. \tag{16.12}$$

This equation is consistent with the Hanes plot of enzyme activity determined at a constant free Cd(II) level (Fig. 16.3b). A nonlinear least-square analysis of the PCS activity data to fit (16.11) yielded $V_{\mathrm{max}}' = 149 \pm 9$, $K_{\mathrm{mA}} = 18.0 \pm 3.2$ mM, and $K_{\mathrm{mB}} = 5.14 \pm 1.22$ μM, which gave a simulated profile significantly consistent with the measured rates. These observations unequivocally demonstrate that PC synthesis proceeds using a GSH molecule and a Cd(II)–GSH_2 complex as substrates that bind first and second to rAtPCS1, as suggested by Vatamaniuk et al. (2000).

16.3.2 Binding of Cd(II) to Activate PCS

Although Cd(II) is essential for PCS-mediated PC synthesis to proceed, the role of Cd(II) in the reaction is equivocal. One of the possible mechanisms is that Cd(II) is solely required to form the substrate Cd(II)–GSH_2 complex in which the enzyme is already in an active form without Cd(II) binding, while another mechanism is that the enzyme needs to bind Cd(II) for activation. The evidence supporting the latter notion was obtained from a PCS assay at a constant total Cd(II) concentration (Fig. 16.4) (Ogawa et al. 2011). When the assay solution contained 10 μM total Cd (II), activity increased and reached a maximum with an increase in GSH concentration up to 15 mM, and then decreased with further increases in GSH concentration (Fig. 16.4, black squares). A similar trend was observed for the enzyme assay system containing 1 or 5 μM Cd(II) (Fig. 16.1, black circles and triangles, respectively). It was also evident that an increase in Cd(II) concentration shifted the maximal activity toward higher GSH concentrations. Thus, activity was maximized at GSH concentrations of approximately 10 and 15 mM in the presence of 5 and

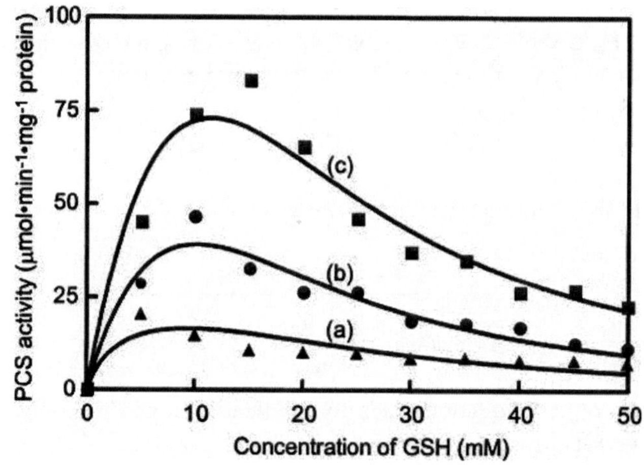

Fig. 16.4 Effect of glutathione (GSH) concentration on phytochelatin synthase (PCS) activity in the presence of total Cd(II) concentrations of 1 μM (*black triangles*), 5 μM (*black circles*), and 10 μM (*black squares*) in the assay solution. Decreased activity at higher GSH concentrations can be attributed to a reduced free Cd(II) level arising from the enhanced formation of Cd(II)–GSH and Cd(II)–GSH_2 complexes, thereby lowering a fraction of active Cd(II)-bound enzyme. A nonlinear least-square analysis using the one-site model (16.15) yielded a V_{max} of 193 ± 17 μmol min^{-1} mg^{-1} protein and a K_{E1} of (28.4 ± 5.1) × 10^9 M^{-1} with an R^2 of 0.926 when the activity data at 5 μM Cd(II) were used and a V_{max} of 302 ± 25 μmol min^{-1} mg^{-1} protein and a K_{E1} of (15.9 ± 2.7) × 10^9 M^{-1} with an R^2 of 0.938 when the activity data at 10 μM Cd(II) were used. Simulated activity (*solid lines* **b** and **c**) was in good agreement with determined activity. In contrast, a nonlinear least-square analysis of the data determined at 1 μM Cd(II) produced a low correlation coefficient (R^2 = 0.681) with a rather inconsistent result between determined activity (*black triangles*) and simulated activity (*solid line* **a**), probably due to the poor activity data set (Ogawa et al. 2011)

10 μM total Cd(II), respectively. At 1 μM total Cd(II), the maximal activity seemed to occur at a concentration less than or equal to 5 mM GSH, although the maximum did not appear unequivocally.

The observation that PCS activity decreased at higher GSH concentrations is consistent with the notion that the enzyme possesses Cd(II)-binding site(s) at which Cd(II) ions bind to activate it. An increase in GSH concentration would decrease free Cd(II) levels by enhancing the formation of Cd(II)–GSH and Cd(II)–GSH_2 complexes, thereby reducing the level of the enzyme bound to Cd(II). Given the binding constant of Cd(II) to rAtPCS1, K_{E1}, the active enzyme concentration can be expressed by the following equation:

$$[\mathrm{Cd(II)\text{-}rAtPCS1}] = K_{\mathrm{E1}}[\mathrm{Cd(II)}][\mathrm{rAtPCS1}]. \tag{16.13}$$

As the free Cd(II) level in an assay solution is expressed by (16.5), a fraction of PCS bound by Cd(II), f_1, is shown by:

$$f_1 = \frac{1}{1 + \frac{\beta'_2}{K_{\mathrm{E1}}} \cdot \frac{[\mathrm{GSH}]_{\mathrm{t}}^2}{[\mathrm{Cd(II)}]_{\mathrm{t}}}}. \tag{16.14}$$

Assuming that only rAtPCS1 bound to Cd(II) has PCS activity, the reaction rate can be derived as a one-site model, as follows:

$$v = \frac{f_1 V_{\max}[\mathrm{GSH}]_{\mathrm{t}}}{K_{\mathrm{mA}} + (1 + \frac{K_{\mathrm{mB}}}{[\mathrm{Cd(II)}]_{\mathrm{t}}})[\mathrm{GSH}]_{\mathrm{t}}}. \tag{16.15}$$

Using the K_{mA} and K_{mB} values obtained, V_{max} and K_{E1} were optimized by a nonlinear least-square analysis using the PCS activity data shown in Fig. 16.4. The V_{max} and K_{E1} values for the activity data sets at total concentrations of 5 and 10 μM were consistent, as revealed by the correlation coefficients, which were close to 0.93. Furthermore, the parameters obtained under both conditions were parallel to each other. In contrast, a low correlation coefficient was obtained for activity at a total Cd(II) concentration of 1 μM ($R^2 = 0.681$), probably due to poor data quality, in which all of the activity plots except that of the origin revealed a decreasing profile. Therefore, omitting the parameters for 1 μM total Cd(II), it can be concluded that a PCS molecule possesses a Cd(II)-binding site with dissociation constants ($1/K_{E1}$) of 35–63 pM to which the ion binds to activate the enzyme.

A possible alternative to this notion may be that the enzyme possesses an inhibitory GSH-binding site(s) at which GSH binds to inhibit PC synthesis. However, this possibility can be ruled out, as PCS activity steadily increased with GSH concentration in the presence of a constant free Cd(II) level (Fig. 16.3). This is in contrast with the reaction rate determined at constant total Cd(II) concentrations in which a significant reduction in rate was found despite the fact that the free GSH concentrations were substantially equal to those at a constant free Cd(II) concentration.

16.3.3 *Inhibitory Second Cd(II) Binding Site of rAtPCS1*

As shown in Fig. 16.5, PCS activity at a constant total Cd(II) concentration still had a maximum for the assay solution containing higher levels of total Cd(II): the activity maximized at 30 and 60 mM GSH for the assay solution containing 50 μM (black squares) and 500 μM Cd(II) (black circles), respectively. However, a sigmoidal increase in activity was observed at GSH concentrations less than 20 mM. A one-site Cd(II)-binding model appeared to be inappropriate to rationalize the activity at higher Cd(II) concentrations, which can be recognized by the poor correlation coefficient ($R^2 = 0.870$ at 50 μM Cd(II) and $R^2 = 0.771$ at 500 μM Cd(II)). In addition, PCS activity determined as a function of the concentration of Cd(II) revealed a maximum (Fig. 16.6). These findings led to the assumption that AtPCS1 possesses an inhibitory second Cd(II)-binding site. Binding of Cd(II) to this site

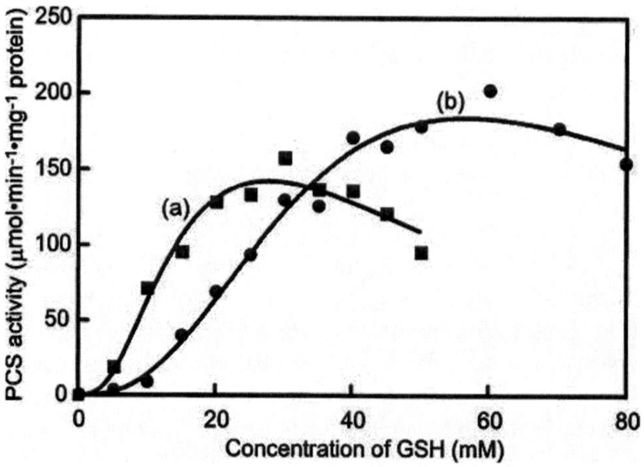

Fig. 16.5 Effect of glutathione (GSH) concentration on phytochelatin synthase (PCS) activity in the presence of total Cd(II) concentrations of 50 μM (*black squares*) and 500 μM (*black circles*) in the assay solution. A sigmoidal increase in the activity is apparent at lower GSH concentrations, implying that the enzyme possesses a second Cd(II)-binding site to which Cd(II) binds to inactivate the enzyme. Although the one-site model resulted in a poor simulation, a nonlinear least-square analysis using the two-site model (16.18) yielded consistent results and showed a V_{max} of 399 ± 62 μmol min^{-1} mg^{-1} protein, a K_{E1} of (12.4 ± 4.6) × 10^9 M^{-1}, and a K_{E2} of (0.961 ± 0.442) × 10^9 M^{-1} with an R^2 of 0.971 when activity data determined at 50 μM Cd(II) were used and a V_{max} of 472 ± 94 μmol min^{-1} mg^{-1} protein, a K_{E1} of (4.55 ± 1.88) × 10^9 M^{-1} and a K_{E2} of (0.924 ± 0.333) × 10^9 M^{-1} with an R^2 of 0.984 when activity data determined at 500 μM were used. Simulated activity was in fair agreement with determined activity (*solid lines* in (**a**) for 50 μM total Cd(II) and (**b**) for 500 μM total Cd(II)) (Ogawa et al. 2011)

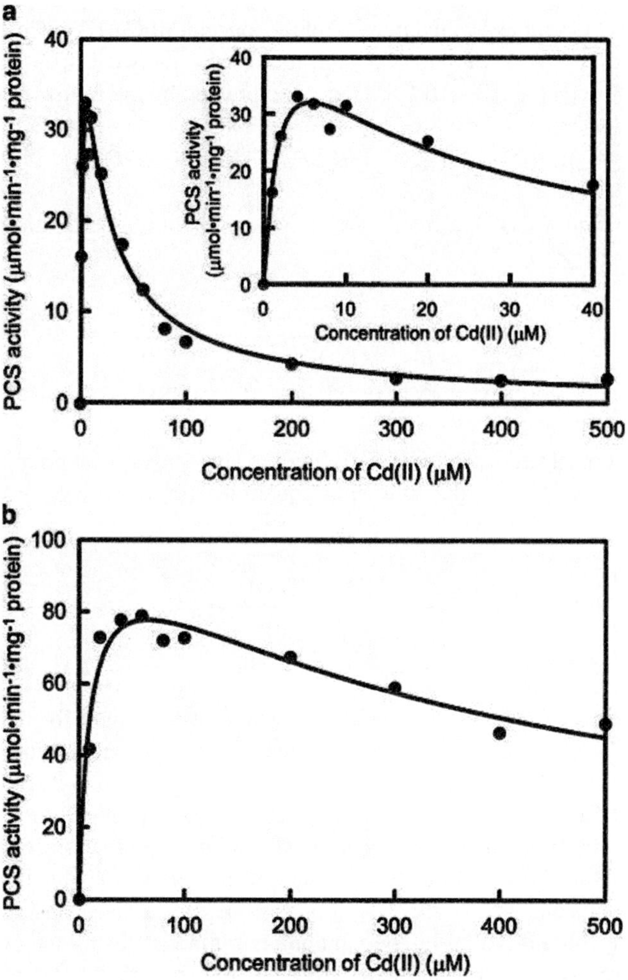

Fig. 16.6 Effects of total Cd(II) concentration on phytochelatin synthase (PCS) activity of AtPCS1 in the presence of 5 mM (**a**) and 20 mM (**b**) glutathione (GSH). The insert in (**a**) is an expansion of the figure at Cd(II) concentrations from 0 to 40 μM. The observation that the activity decreased at higher total GSH concentrations suggested that the enzyme possesses a second Cd(II)-binding site to which Cd(II) binds to inactivate the enzyme. The two-site model (16.18) was applied to a nonlinear least-square analysis, which yielded a V_{max} of 244 ± 17 μmol min^{-1} mg^{-1} protein, a K_{E1} of (23.4 ± 9.8) × 10^9 M^{-1}, and a K_{E2} of (0.557 ± 0.081) × 10^9 M^{-1} with an R^2 of 0.986 when activity data determined at 5 mM Cd(II) were used, and a V_{max} of 194 ± 14 μmol min^{-1} mg^{-1} protein, a K_{E1} of (25.3 ± 8.6) × 10^9 M^{-1}, and a K_{E2} of (0.403 ± 0.085) × 10^9 M^{-1} with an R^2 of 0.968 when activity data determined at 20 mM were used. Simulated activity was in fair agreement with determined activity ((**a**) for 5 mM total GSH and (**b**) for 20 mM total GSH) (Ogawa et al. 2011)

would inhibit PC synthesis. Given the binding constant of Cd(II) to Cd(II)-AtPCS1 at the second site, K_{E2}, the enzyme concentration with two Cd(II) binding sites occupied by Cd(II) (Cd(II)$_2$-AtPCS1) can be expressed by the following equation:

$$[\mathrm{Cd(II)}_2\text{-rAtPCS1}] = K_{E1}K_{E2}[\mathrm{Cd(II)}]^2[\mathrm{rAtPCS1}]. \tag{16.16}$$

Using (16.13) and (16.14), a fraction of the enzyme with a Cd(II) on the first site alone is shown by:

$$f_2 = \frac{1}{1 + \frac{\beta'_2}{K_{E1}} \cdot \frac{[\mathrm{GSH}]_t^2}{[\mathrm{Cd(II)}]_t} + \frac{K_{E2}}{\beta'_2} \cdot \frac{[\mathrm{Cd(II)}]_t}{[\mathrm{GSH}]_t^2}}. \tag{16.17}$$

Assuming that rAtPCS1 with two Cd(II) binding sites fully occupied by Cd(II) ions completely loses activity, the following equation from a two-site model can be expressed:

$$v = \frac{f_2 V_{max}[\mathrm{GSH}]_t}{K_{mA} + (1 + \frac{K_{mB}}{[\mathrm{Cd(II)}]_t})[\mathrm{GSH}]_t}. \tag{16.18}$$

The binding of Cd(II) to the second site would also account for the sigmoidal profile of the activity assayed in the presence of higher concentrations of Cd(II). Free Cd(II) levels may be elevated with GSH concentrations of 0–20 mM at 50 and 500 μM total Cd(II) (Fig. 16.5), in which some of the enzyme is presumed to be inactivated by bringing the Cd(II) ion onto the second site. Equation 16.18 was used to determine the Cd(II) affinity constants for the enzyme, which yielded the dissociation constants for the first site ($1/K_{E1}$) to be 81–220 pM and for the second site ($1/K_{E2}$) to be 1.0–1.1 nM. The parameters optimized using the activity data shown in Fig. 16.5 were consistent with the experimentally determined PCS activity in the presence of 5 and 20 mM GSH.

The binding constants K_{E1} and K_{E2} were also optimized to fit PCS activity determined as a function of total Cd(II) concentration according to (16.7), which gave dissociation constants for the first site ($1/K_{E1}$) of 40–43 pM and for the second site ($1/K_{E2}$) of 1.8–2.5 nM. The PCS activity estimated from the V_{max}, K_{E1}, and K_{E2} values thus obtained agreed with the experimentally obtained PCS activity at 5 and 20 mM GSH, as shown by the solid lines in Fig. 16.6a, b. Furthermore, the parameters estimated from PCS activity at 5 and 20 mM GSH were consistent with each other. This also supports the suggestion that AtPCS1 possesses two Cd (II) binding sites per molecule.

16.4 Conclusion

Producing PC peptides and complexing metal ions are crucial mechanisms for ameliorating metal toxicity in plant species. It was unequivocally demonstrated that PCS possesses a Cd(II)-binding site with a dissociation constant of a few tens to two hundred pM. Therefore, PCS activity was regulated in such a way that the enzyme was activated by Cd(II) ions by binding to the site and was deactivated by removing the ion from the activated enzyme through complexing the product PCs with Cd(II). This defense mechanism would respond promptly to a challenge of toxic metal ions, as the enzyme is constitutively expressed in cells. Although the physiological relevance of metal binding to PCS has been elucidated, the mode of metal binding to the enzyme must be resolved. A range of metal cations and metalloid oxoanions can function in the reaction initiator: metals and metalloids belonging to groups 11–15 in the fourth, fifth, and sixth periods of the periodic table are assigned to activate PCS-mediated PC synthesis (Oven et al. 2002). It is of interest that these ions with a specific chemical nature are capable of activating PCSs. Further studies will shed light on the detailed mechanisms.

References

Bisswanger H (2008) Enzyme kinetics, second, revised and updated edition. WILEY-VCH Verlag GmbH and Co., Weinheim, Germany

Clemens S, Kim EJ, Neumann D, Schroeder JI (1999) Tolerance to toxic metals by a gene family of phytochelatin synthases from plants and yeast. EMBO J 18:3325–3333

Grill E, Löffler S, Winnacker E-L, Zenk MH (1989) Phytochelatins, the heavy-metal-binding peptides of plants, are synthesized from glutathione by a specific γ-glutamylcysteine dipeptidyl transpeptidase (phytochelatin synthase). Proc Natl Acad Sci USA 86:6838–6842

Grill E, Winnacker E-L, Zenk MH (1985) Phytochelatins: the principal heavy-metal complexing peptides of higher plants. Science 230:674–676

Ha S-B, Smith AP, Howden R, Dietrich WM, Bugg S, O'Connell MJ, Goldsbrough PB, Cobbett CS (1999) Phytochelatin synthase genes from Arabidopsis and the yeast *Schizosaccharomyces pombe*. Plant Cell 11:1153–1163

Kondo N, Imai K, Isobe M, Goto T (1984) Cadystin A and B, major unit peptides comprising cadmium binding peptides induced in a fission yeast – separation, revision of structures and synthesis. Tetrahed Lett 25:3869–3872

Loeffler S, Hochberger A, Grill E, Winnacker E-L, Zenk MH (1989) Termination of phytochelatin synthase reaction through sequestration of heavy metals by the reaction product. FEBS Lett 258:42–46

Ogawa S, Yoshimura E (2010) Determination of thermodynamic parameters of cadmium(II) association to glutathione using the fluorescent reagent FluoZin-1. Anal Biochem 402:200–202

Ogawa S, Yoshidomi T, Yoshimura E (2011) Cadmium(II)-stimulated activation of *Arabidopsis thaliana* phytochelatin synthase1. J Inorg Biochem 105:111–117

Oven M, Page JF, Zenk MH, Kutchan TM (2002) Molecular characterization of the homo-phytochelatin synthase of soybean *Glycine max*. J Biol Chem 277:4747–4754

Rea PA, Vatamaniuk OK, Rigden DJ (2004) Weeds, worms, and more. Papain's long-lost cousin, phytochelatin synthase. Plant Physiol 136:2463–2474

Romanyuk ND, Rigden DJ, Vatamaniuk OK, Lang A, Cahoon RE, Jez JM, Rea PA (2006) Mutagenic definition of a papain-like catalytic triad, sufficiency of N-terminal domain for single-site core catalytic enzyme acylation, and C-terminal domain for augmentative metal activation of a eukaryotic phytochelatin synthase. Plant Physiol 141:858–869

Vatamaniuk OK, Mari S, Lang A, Chalasani S, Demkiv LO, Rea PA (2004) Phytochelatin synthase, a depeptidyltransferase that undergoes multisite acylation with γ-glutamylcysteine during catalysis. J Biol Chem 279:22449–22460

Vatamaniuk OK, Mari S, Lu Y-P, Rea PA (1999) AtPCS1, a phytochelatin synthase from *Arabidopsis*: isolation and *in vitro* reconstitution. Proc Natl Acad Sci USA 96:7110–7115

Vatamaniuk OK, Mari S, Lu Y-P, Rea PA (2000) Mechanism of heavy metal ion activation of phytochelatin (PC) synthase. J Biol Chem 275:31451–31459

Zenk MH (1996) Heavy metal detoxification in higher plants – a review. Gene 179:21–30

Chapter 17
Tolerance, Accumulation, and Detoxification Mechanism of Copper in *Elsholtzia splendens*

Yingxu Chen, Mingge Yu, and Dechao Duan

17.1 Introduction

Copper is an essential micronutrient for plant growth and development and plays a pivotal role in cellular biochemical and physiological processes, such as photosynthesis, electron transport, transcriptional regulation, and protein synthesis. Despite being essential, Cu is highly toxic to plants at elevated concentrations. Many investigations have been conducted on the effect of copper on the growth, mineral nutrition, and metabolism of plants. Excessive copper could reduce plant growth and photosynthetic activity (Lidon et al. 1993) and may also result in membrane damage by Cu binding to the sulfhydryl groups of membrane proteins (Kennedy and Gonsalves 1987).

As a general rule, plant species growing in elevated metal environments have evolved two basic strategies of metal tolerance: (1) exclusion, whereby plants avoid excessive uptake and transport of metal ions, and (2) accumulation and sequestration, whereby plants detoxify free metals by compartmentation of metals in vacuoles and complexation of metal ions by organic ligands (Clemens 2001; Hall 2002). Therefore, plants have developed both extracellular and intracellular detoxification mechanisms to reduce the phytotoxicity of free Cu. Generally, extracellular tolerance mechanisms include chelation by cell walls and extracellular exudates. The plasma membrane could also be involved in metal tolerance, either by reducing the uptake of

Y. Chen (✉) • M. Yu
Institute of Environmental Science and Technology, Zhejiang University, Hangzhou 310029, China
e-mail: yxchen@zju.edu.cn

D. Duan
Ministry of Agriculture, Key Laboratory of Non-point Source Pollution Control, Hangzhou 310029, China

Department of Environmental Engineering, Huajiachi Campus, Zhejiang University, 268 Kaixuan Road, Hangzhou, Zhejiang 310029, China
e-mail: duandechao111@163.com

I. Sherameti and A. Varma (eds.), *Detoxification of Heavy Metals*, Soil Biology 30,
DOI 10.1007/978-3-642-21408-0_17,

heavy metals or by stimulating the efflux pumping of metals that have entered the cytosol (Hall 2002). In addition to extracellular detoxification approaches, a variety of potential mechanisms also exist within the protoplast. For example, there are mechanisms for the repair of stress-damaged proteins, and for the metal chelation, or their compartmentation away from metabolic processes (Hall 2002).

Soil contamination with heavy metals has become an increasingly pressing environmental concern in China in the past several decades. The most serious problem with heavy metal contamination mainly occurs in soils near mining and wastewater irrigation areas. In addition to the relatively harsh traditional engineering and chemical soil remediation methods, phytoremediation is emerging as a promising technology with the advantages of being in situ, cost effective, and environmentally sustainable (Long et al. 2002; Salt et al. 1998).

A number of plant species endemic to metalliferous soil have been evolved to accumulate large amounts of heavy metals. *Elsholtzia splendens* (*E. splendens*), an annual herbaceous plant with an erect stem about 15–120 cm in height, belongs to the family Labiatae. *E. splendens* grows vastly over Cu-mining areas along the middle and lower streams of the Yangtze River and is identified as a Chinese-native copper-tolerant and accumulating plant species (Tang et al. 1999; Jiang and Yang 2004).

Based on the results of hydroponic experiments, Yang et al. (2002) suggested that *E. splendens* has great potential for phytoremediation of Cu-contaminated soils. To understand potential efficiency of phytoremediation using *E. splendens* as well as the copper tolerance and accumulating mechanisms, much research has been conducted with hydroponics, pot, or field experiments to investigate the physiological and biochemical response to Cu stress, and phytoremediation of metal pollutants in contaminated soil (Tang et al. 1999; Yang et al. 2002; Jiang and Yang 2004; Song et al. 2004; Ke et al. 2008; Shi et al. 2008; Zhang et al. 2008). In the subsequent sections, tolerance, detoxification, and accumulation mechanism of copper in *Elsholtzia splendens* is discussed in detail.

17.2 Behavior of Cu in Root–Soil Interface of *Elsholtzia splendens*

17.2.1 Chemical Speciation of Cu in Root–Soil Interface

Most studies dealing with particulate metals in sediment or soil systems are concerned with total metal concentration. However, it is now widely recognized that the quantification of the chemical forms of metals in soils is essential for estimating the mobility and bioavailability of the metals in the environment (Li et al. 1995). And the sequential extraction procedure of Tessier et al. (1979) has been thoroughly researched and rigorously tested and is now widely applied in sediment and soil studies by many investigators (Tessier et al. 1979; Kim and Fergusson 1991; Li et al. 1995).

Through sequential extraction, research had been conducted to evaluate the mobility and bioavailability of Cu in rhizosphere soil of *Elsholtzia splendens* (Wang 2006). The results indicated that great changes have taken place in the speciation of Cu in rhizosphere soil of *Elsholtzia splendens* and soil for not growing *Elsholtzia splendens*. Exchangeable Cu decreased and residual Cu increased in rhizosphere soil of *Elsholtzia splendens* in comparison with the control group. This shows that *Elsholtzia splendens* has a strong ability to promote the transformation of excess exchangeable Cu with great bioavailability to other Cu speciation like residual Cu with less bioavailability. This may be one of the foremost mechanisms by which *Elsholtzia splendens* is tolerant to high concentrations of Cu (Wang 2006).

17.2.2 Root Exudates and Rhizosphere Microorganisms on Cu Activation

Roots can change the major physical and chemical characteristics of the rhizosphere, which may in turn change the characteristics of trace metals present. As many researches have shown, the excretion products of root and biomolecules in metabolites from rhizosphere microorganisms include a variety of soluble (Leyval and Berthelin 1993) and insoluble organic matters (Morel et al. 1986). Root excretion products, such as acetic, oxalic, citric, tartaric acids, uronic acids, and polysaccharides, are able to form complexes and chelates with metal ions (Mench et al. 1988; Morel et al. 1986), thus modifying the fixation and mobility of heavy metal in soils.

17.2.2.1 Root Exudates on Cu Activation

Hydroponic cultures were conducted to investigate root exudates of *Elsholtzia splendens* on soil Cu activation by Shi (2004). The results from this study showed that the amount of root exudates on the dissolution of Cu from contaminated soil was significantly higher than the amount dissolved in deionized water ($P > 0.05$). The resulting differences in amounts of Cu dissolution suggested that root exudates of *Elsholtzia splendens* have a strong ability in Cu activation (Shi 2004).

17.2.2.2 Rhizosphere Microorganisms on Cu Activation

Microbes (bacteria and fungi) are ubiquitous in soils. Soil microbes play significant roles in the recycling of plant nutrients, maintenance of soil structure, detoxification of noxious chemicals, plant growth, and the plant interactions with the soil environment (Christie et al. 2004; Stepanauskas et al. 2005; Vivas et al. 2006). Some workers have claimed that arbuscular mycorrhizas may help plants to tolerate high levels of soil heavy metal contamination. For example, Heggo et al. (1990) demonstrated that at high soil heavy metal concentrations, arbuscular mycorrhizal infection reduced the concentrations of Zn, Cd, and Mn in plant leaves.

Previous research has shown that *E. splendens* is a potential copper accumulator plant for phytoremediation of Cu-contaminated soils (Tang et al. 1999; Lou et al. 2004). Whereas successful phytoremediation depends mainly on the bioavailability of heavy metals in the soil, the availability of heavy metals for plants is usually restricted by the complexation of metals within solid soil fractions. Soil microorganisms, with activity and a high surface area-to-volume ratio because of their small size and therefore providing a large contact area, may have the potential to act as microbial chelates associated with phytoremediation (Kärenlampi et al. 2000). These microbes may be crucial for establishing close interactions with *E. splendens*.

The role of rhizosphere bacteria isolated from *Elsholtzia splendens* in facilitating the solubility of copper in contaminated soil was studied (Chen et al. 2005). The results indicated that rhizosphere microbes played an important role in influencing the availability of water-soluble Cu in soils. Soils had greater concentrations of water-extractable Cu compared with axenic soils inoculated with different bacterial strains.

17.3 Impact of Cu on *Elsholtzia splendens*

Heavy metals, such as Cu, Cd, and Pb at high levels, can be toxic to living things. Recently, studies on the plants growing on metal-contaminated sites are mainly focused on the metal effects on the plant growth, mechanisms of adaptation involved, and the potential use in the remediation of metal-enriched soils. Plants growing on metal-contaminated environments showed more tolerance to heavy metals than normal plants. Although Cu is an essential micronutrient for normal plant metabolism, excessive Cu can induce a wide range of adverse effects including inhibition of pigment synthesis and photosynthesis, damage to membrane structures, and disturb uptake of mineral nutrients even in Cu-tolerant plants (Ouzounidou et al. 1995; Lidon et al. 1993; Monni et al. 2000; Brun et al. 2003).

This section summarizes the research about the impact of Cu on *Elsholtzia splendens*, including toxic symptoms, biochemical and physiological changes, and the ultrastructural damage of *Elsholtzia splendens* under Cu stress.

17.3.1 Symptoms of **Elsholtzia splendens** *Under Cu Stress*

17.3.1.1 Effects on Seed Germination

Hydroponic cultures were conducted by Lou et al. (2004) to investigate the effect of excessive Cu on the seed germination of *Elsholtzia splendens*. It was found that Cu in solution lower than 250 μmol L^{-1} seemed to stimulate seed germination, while at higher concentrations (750 μmol L^{-1}), germination percentage decreased with the increasing Cu concentrations in the solution.

17.3.1.2 Effects on *Elsholtzia splendens* Growth

Pot experiment was conducted to investigate the effects of Cu with different supply levels (control, 100, 200, 400, 600, 800, 1,000, and 1,200 mg kg^{-1}) on the growth and Cu accumulation in *E. splendens* (Jiang et al. 2008). The results showed that no reduction in shoot height and dry weight was noted when the plants were grown at Cu supply levels up to 1,000 mg kg^{-1} in soil. Slight stimulation on shoot growth was noted at Cu levels 100 mg kg^{-1}.

Tian (2005) studied the effect of different copper levels on root morphology and copper accumulation in *Elsholtzia splendens* through solution culture. The results showed that under low Cu level (<50 μmol L^{-1}), the growth of *Elsholtzia splendens* was improved, with slight increase in dry weight, root length, root surface area, and root volume. However, with an increase in Cu supply levels (100 ~ 500 μmol L^{-1}), an obvious decrease was found in dry weight, root length, root surface area, and root volume for *Elsholtzia splendens*. Similar observations were also found in other studies that low concentration Cu supply stimulated *Elsholtzia splendens* growth and high Cu supply levels suppressed that (Li et al. 2003a; Jiang et al. 2003; Peng et al. 2005a; Ke et al. 2008; Lou et al. 2004; Yang et al. 2002).

17.3.2 Biochemical and Physiological Changes of **Elsholtzia splendens** *Under Cu Stress*

17.3.2.1 Essential Nutrient Metabolism

High amount of heavy metals in plant tissues may interfere with other essential nutrients and thereby disturb the mineral nutrition of plants. Yang et al. (2002) investigated the effects of Cu on essential nutrient metabolism of *Elsholtzia splendens*. The results indicated that copper accumulation in *E. splendens* was accompanied by the ability to maintain the concentrations of other essential nutrients, except potassium, within the range considered sufficient for normal growth of higher plants. The results indicated that Cu accumulation resulted in K depletion in *E. splendens*, and the negative correlation between K and Cu concentrations was statistically significant ($r = -0.85$ in roots, $r = -0.91$ in stems, and $r = -0.76$ in leaves). The observation was in agreement with studies in *Mimulus guttatus* (Strange and Macnair 1991) and *Silene compacta* (Ouzounidou 1994).

Apart from the above study of Cu accumulation on uptake and distribution of some essential mineral nutrients, two *E. splendens* populations, copper-tolerant and nontolerant ones, were studied in hydroponic experiment for the nitrogen assimilation under excess Cu conditions (Li et al. 2007). The results demonstrated that nitrogen assimilation of copper-tolerant population was less affected by excessive Cu stress than nontolerant one.

Through quantitative imaging laser ablation inductively coupled plasma mass spectrometry (LA-ICP-MS), the effect of Cu stress on nutrient uptake and accumulation in leaves of *E. splendens* was investigated (Wu et al. 2009a). It was found that Cu stress induced accumulation of K, Mg, Mn, P, and S in the newly formed leaves, while B was not significantly affected. In the fully grown leaves, the concentrations of K, Mg, Mn, P, and S were not obviously changed after short-term treatment. In addition, a visible decrease of K and P was found in the oldest leaves, while other elements were not influenced by Cu stress.

17.3.2.2 Photosynthesis

Photosynthesis is the most important chemical reactions in nature, by which organisms convert light into chemical energy. Therefore, it is the most fundamental process of substance metabolism and energy metabolism in biological world. The mechanisms of heavy metal toxicity on photosynthesis have been intensively investigated. Deleterious effects of heavy metals, such as Cu and Cd, on various photosynthetic processes, such as biosynthesis of chlorophyll (Padmaja and Parsad 1990) and the enzymes activities of the Calvin cycle (Chugh and Sawhney 1999), have been reported in many plants. In this part, the effect of Cu on photosynthesis of *Elsholtzia splendens* was discussed.

It was suggested that Cu at lower concentration increased the content of chlorophyll a, while higher concentration of Cu decreased it. But the content of chlorophyll b was less influenced by Cu at either high or low concentration (Li et al. 2003a). In addition, photosynthesis parameters of *Elsholtzia* plants under different concentrations of Cu treatment were measured and analyzed by Tian (2005). The increase of Cu concentration ($\leq$50 μmol L^{-1}) significantly increased intercellular CO_2 concentration, photosynthetic rate, and transpiration rate of *Elsholtzia splendens* ($P < 0.01$). Four photosynthesis parameters reached the highest level at 50 μmol L^{-1} Cu, but significantly decreased at 500 μmol L^{-1} Cu ($P < 0.01$).

Copper accumulation and physiological responses to Cu stress in three populations of *Elsholtzia splendens* were comparatively studied with pot culture experiments by Liu and Xiong (2005). The result indicated that chlorophyll content was less affected in the populations from contaminated sites than those in the populations from noncontaminated sites (Liu and Xiong 2005).

17.3.2.3 Oxidative Stress

Excess Cu is toxic to plants and affects a wide range of biochemical and physiological processes. The toxicity of Cu can be considered an oxidative stress mediated by reactive oxygen species (ROS; Luna et al. 1994). Being a redox-active metal, Cu may act as an efficient generator of toxic oxygen species, such as superoxide anion ($O_2\cdot^-$), hydrogen peroxide (H_2O_2), and hydroxyl radical ($\cdot OH$) via Fenton-type reactions (Schützendübel and Polle 2002). These ROS could affect cell viability through lipid peroxidation, membrane damage, and inactivation of enzymes by reacting with lipids, proteins, and nucleic acids. It is found that ROS accumulation and lipid peroxidation by excess Cu have been observed in various plants (Hartley-Whitaker et al. 2001; Landberg and Greger 2002; Tewari et al. 2006).

Zhang et al. (2008) investigated the effects of excess Cu on the accumulation of hydrogen peroxide (H_2O_2) and antioxidant enzyme activities in roots of *Elsholtzia splendens*. The study demonstrated that excess Cu greatly increased the accumulation of superoxide anion ($O_2\cdot^-$) and H_2O_2 in *E. splendens* roots and also clarified the roles of different antioxidant enzymes in the defense against the harmful effects of Cu stress in *E. splendens*. It was suggested that the source of H_2O_2 in the cell walls could partially be NADPH oxidase, which could use cytosolic NADPH to produce O_2^-, which quickly dismutates to H_2O_2 by superoxide dismutases (SOD).

Plants responded to Cu-induced oxidative stresses by modulating antioxidant enzymes and substrates and different populations differed in the response. Compared with the control, the activities of SOD, catalase (CAT), glutathione peroxidase (GPX), and ascorbate peroxidase (APX) in the leaves of the plant increased pronouncedly at 500 μM Cu (Peng et al. 2006). At high level of copper treatment, the activities of peroxidase and SOD in the *E. splendens* populations from contaminated sites were significantly induced, while this was not the case for the population from noncontaminated site (Liu and Xiong 2005).

17.3.3 Impact of Cu on the Structure of Elsholtzia splendens

Structure is the foundations of function. Metal-induced alterations at the structural and ultrastructure level might reflect the extent of metal toxicity stress. In this part, structure changes of *Elsholtzia splendens* under Cu stress at the tissue and cell level were introduced based on the research of Ni (2004) and Peng et al. (2005c).

17.3.3.1 The Impact of Copper on Tissues Structure of *Elsholtzia splendens*

Micromorphology of root, stem, and leaf of *E. splendens* under Cu stress (50 μmol L^{-1}) was studied by optical microscope and scan electron microscope (SEM) (Ni 2004). The results indicated that root tip of *E. splendens* was not damaged, while root hair growth was severely hampered at 50 μmol L^{-1} Cu. In addition, the hierarchy of root tissue did not change much, with small irregular epidermal cells outside and clear vascular bundle inside, while cells arranged more closely under Cu stress. As for stem, there was no significant difference in its inner structure between control and plants under Cu stress, while the trichomes significantly increased after Cu exposure. Besides, it was worth noting that cells of palisade tissue and spongy tissue in leaves under 50 μmol L^{-1} Cu treatment were smaller, while cell gaps were larger than the control. Since larger gaps between cells allow more unobstructed gas exchange, 50 μmol L^{-1} Cu might stimulate the photosynthesis of *E. splendens*.

17.3.3.2 The Impact of Copper on Cells Ultrastructure of *Elsholtzia splendens*

Cu accumulation and intracellular distribution in *Elsholtzia splendens* were investigated by transmission electron microscope (TEM) and gradient centrifugation techniques by Peng et al. (2005c). The results showed that no significant toxicity was observed in the chloroplast and mitochondrion within leaf cells at 50 μmol L^{-1} Cu, but separation appeared at the cytoplasm and the cell wall within the root cells. Copper treatment at levels equal or greater than 250 μmol L^{-1} caused pronounced damage in the leaf chloroplast and root organelles by excessive Cu in vivo. And these results were similar to what has been found by Ni (2004).

By summing up the above observation of structure changes of *Elsholtzia splendens* under different concentrations of Cu stress at the tissue and cell level with optical microscope, scan electron microscope, and transmission electron microscope, we can draw the following conclusions: (1) Cu stress on the growth of root, stem, and leaf trichomes varied. At 50 μmol L^{-1} Cu, root hairs were significantly reduced, while stem trichomes pronouncedly increased. The density of leaf trichomes decreased, but the length increased. In conclusion, the response of different organs of *Elsholtzia splendens* to Cu differed, indicating that there might be diverse mechanisms of resistance to Cu stress. (2) The organelles were affected by Cu in a different order. The reduction or disappearance of Golgi apparatus preceded that of endoplasmic reticulum, while nucleus was generally the last victim organelle. Since nucleus controls protein synthesis and the cell growth, development, and heredity, damage of it will lead to cell disorder and ultimately to the cell death. Thus, the intact nucleus under Cu stress may be one of the foremost mechanisms by which *Elsholtzia splendens* is copper-resistant.

17.4 Uptake and Accumulation of Cu by *Elsholtzia splendens*

17.4.1 Copper Concentration and Accumulation of Elsholtzia splendens

Elsholtzia splendens grows vastly over copper-mined areas and was first recognized for its value in the exploration of copper ores in the 1950s (Hsieh and Hsu 1954). It has a local nickname "copper flower" because of its growth confined in highly Cu-contaminated soils. Through old mined area survey, pot experiment, and hydroponic culture, the ability of Cu accumulation by *Elsholtzia splendens* and its influencing factors was discussed in this section.

17.4.1.1 Cu Concentration of *Elsholtzia splendens* in Old Mined Area

Old mined area survey was conducted to study Cu contents of *Elsholtzia splendens* in natural conditions (Jiang and Yang 2004). The investigation area is located in Zhuji city, Zhejiang Province of China, which is 29° latitude north and is 120° longitude east. The results showed that shoot Cu concentration increased gradually with NH_4OAc-extractable soil Cu levels, but leveled off at soil available Cu levels around 300 mg kg^{-1}, whereas the root Cu concentration increased sharply with increasing soil available Cu. At the total and NH_4OAc-extractable soil Cu levels of 3454 and 660 mg kg^{-1}, copper concentration in the roots was as high as 600 mg kg^{-1}, about 4 times greater than that in the shoots. These results indicate that the extractability of Cu from the highly polluted soil is much greater by the roots than by the shoots of *E. splendens.*

17.4.1.2 Cu Concentration of *Elsholtzia splendens* in Pot Experiments

Pot experiment was conducted by Jiang et al. (2008) with the incubated soil samples and by Peng and Yang (2007) with Cu severely polluted soil because of the Cu refining activities to evaluate Cu concentration and accumulation of *E. splendens.* The study of Jiang et al. (2008) revealed that Cu concentration in roots and shoots of *E. splendens* increased sharply with increasing external Cu supply levels. The maximum Cu concentration in roots and shoots reached 1,751 and 9.45 mg kg^{-1} (DW), respectively, when the plants were grown at 1,200 mg Cu kg^{-1} for 70 days.

Compared with the study of Jiang et al. (2008), Peng and Yang (2007) evaluated the ability of Cu accumulation by *E. splendens* using polluted soil in natural conditions. The total Cu concentration of soil was 2,800 mg kg^{-1} and NH_4OAc-extractable Cu was 201 mg kg^{-1}. After 70 days of growth, it was found that Cu accumulation in *E. splendens* root was much more than that in shoot, with 827 mg kg^{-1} (DW) in root and 66.6 mg kg^{-1} (DW) in leaf, respectively.

Table 17.1 The maximum Cu concentration in root, stem, leaf, and shoot of *Elsholtzia splendens* recorded in the literature studying Cu accumulation ability by solution culture

Cu concentration in solution (mg L^{-1})	Treatment time (days)	Maximum Cu concentration in different tissues of *Elsholtzia splendens* (mg kg^{-1} dry weight) Root	Stem	Leaf	Shoot	References
6.4	25	2,294	124	147	–	Weng et al. (2005)
6.4	28	4,000	150	360	–	Peng and Yang (2007)
12.8	8	7,626	–	–	288	Qian et al. (2005)
32	6	20,200	–	–	160	Lou et al. (2004)
32	8	8,500	1,000	250	–	Peng et al. (2005c)
40	16	2,000	–	–	1,034	Jiang et al. (2003)
64	24	12,752	–	–	3,347	Yang et al. (2002)

17.4.1.3 Cu Concentration of *Elsholtzia splendens* in Hydroponic Conditions

Table 17.1 summarizes the maximum Cu concentration in different tissues of *Elsholtzia splendens* recorded in the literature studying Cu accumulation ability by solution culture. All the results showed that Cu concentration in root was much higher than that in shoot or leaf of *E. splendens*, suggesting that *E. splendens* did not hyperaccumulate Cu but behaved as a typical Cu excluder. In addition, the accumulation ability of *E. splendens* shoots or leaves varied from each experiment. Cu concentration in culture solution, treatment time, and population difference may contribute to the discrepancy of accumulation ability.

17.4.2 Factors Influencing Cu Uptake and Accumulation by Elsholtzia splendens

17.4.2.1 Physical and Chemical Properties of Soil

Most heavy metals in soil system are either precipitated by soil organic and inorganic components or absorbed on the surface of soil particulates. Thus, the proportion of soluble form is relatively small. Kumar et al. (1995) found that heavy metal content in hyperaccumulators was significantly correlated with that in soil solution. For that reason, bioavailable heavy metal in soil is an important factor influencing their uptake and accumulation by plant, which is relevant to pH, redox potential, organic matter contents, and many other physical and chemical properties of soil.

The relationships between soil solution properties and root Cu concentrations of *Elsholtzia splendens* were investigated using multiple regressions by Song et al. (2004). The results showed that root Cu concentration increased with soil solution pH when free Cu^{2+} activity was held constant, suggesting a higher phytoavailability of free Cu^{2+} at a higher pH. The positive effect of increasing pH may be explained by the increased binding capacity of a biotic surface (e.g., root cell walls and membranes) at a higher pH. Similar observations that Cu binding to the cell walls of maize roots increased with pH and that Cu toxicity to plants often increases with increasing pH in hydroponic culture experiments were also found (Plette et al. 1996).

Another interesting conclusion was that soil solution DOC appeared to play two contrasting roles on the phytoavailability of Cu: (1) reducing Cu availability by complexing Cu; and (2) increasing Cu availability by providing a strong buffer for free Cu^{2+} when free Cu^{2+} activity was in the same level.

17.4.2.2 Microorganisms

Microorganisms play an important role on the bioavailability of heavy metal in soil. On the one hand, microorganisms could decrease the solubility of heavy metals in soil solution by adsorption, absorption, complexation, and precipitation. On the other hand, they could increase heavy metal solubility through promotion of uptake and accumulation into plants by catalyzing redox or alkylation/dealkylation reaction or/and secreting proton or organic acids.

Chen et al. (2005) studied the role of rhizosphere bacteria in facilitating the solubility of copper in contaminated soil and Cu accumulation in *Elsholtzia splendens*. Soils had greater concentrations of water-extractable Cu compared with axenic soils inoculated with different bacterial strains. Further evidence for bacterial facilitation of increased solubility of Cu in the soil was obtained using the antibiotic ampicillin (0.1 mg g^{-1}). There were 36% decreases in Cu concentration in the presence of bacterial strain MS 12 and ampicillin together compared with bacterial inoculation alone. What is more, there were 2.2-fold and 2.5-fold increases in Cu accumulation in the shoots and roots of plants inoculated with strain MS2 compared to axenic controls. Furthermore, when ampicillin and the bacterial strains were added together to the nutrient solution, the Cu concentrations in roots and shoots of ampicillin-treated plants were lower than those in inoculated plants.

In addition to rhizosphere bacteria, arbuscular mycorrhizal fungi also have great significance for Cu accumulation by *Elsholtzia splendens*. Wang et al. (2005a, b, 2007) suggested that arbuscular mycorrhizal fungi inoculation increased plant biomass and enhanced shoot Cu concentrations, indicating that the arbuscular mycorrhizal fungal consortium can benefit phytoextraction of Cu and therefore play a role in phytoremediation of Cu-contaminated soils.

Rape plants inoculated with endophytic bacteria isolated from *Elsholtzia splendens* with respect to heavy metal resistance were found to increase dry weights and above-ground tissue Cu content (Sun et al. 2010). Thus, Cu-resistant endophytic bacteria are crucial for establishing close interactions with the plant species and for accelerating the efficiency of phytoremediation of Cu-contaminated soils.

17.4.2.3 Chelating Agents

Effects of chelating agents resembling ethylenediamine tetraacetic acid (EDTA), diethylenetriaminepentaacetic acid (DTPA), (S, S′)-ethylenediamine-N,N′-disuccinic acid (EDDS), and citric acid (CA) on toxicity of copper to *Elsholtzia splendens* and Cu bioaccumulation in *Elsholtzia splendens* were extensively investigated. Both positive and negative effects of chelating agents on Cu contents in *Elsholtzia splendens*, especially the above-ground parts, were observed (Chen et al. 2006; Hu et al. 2007; Jiang et al. 2004; Li et al. 2003b; Wu et al. 2007; Yang et al. 2005).

EDTA has been known to enhance the phytoextraction of metals from contaminated soil due to increased H_2O-soluble metal concentrations in the soil (Blaylock et al. 1997). Yang et al. (2005) found that EDTA application at 2.5–5.0 mmol kg^{-1} increased phytoextraction of Cu by fourfold and eightfold from both Cu-mined area soil and a paddy soil polluted by Cu relining. It was suggested that EDTA had very low or no influence on the toxicity of copper to *Elsholtzia splendens*, while EDTA increased copper transfer from the subterranean part to the aerial part of the plant (Li et al. 2003b). Compared with EDTA, however, DTPA decreased copper-induced toxicity and decreased copper accumulation in *Elsholtzia splendens* (Li et al. 2003b).

Through pot experiments, Wu et al. (2007) found that the addition of EDDS increased the amounts of Cu soluble in the soil and Cu concentrations in *Elsholtzia splendens*, especially in the leaves. Study of Hu et al. (2007) showed that EDDS enhanced metal solubility in the soil, but plant metal uptake by *Elsholtzia splendens* did not increase accordingly.

In addition, exudation of organic acids by plant roots, such as citric acid, is of great significance due to their metal chelating/complexing properties for the mobilization of heavy metals (Mench et al. 1988). It was suggested that application of citric acid (3.6 g C kg^{-1}) significantly increased the extractable Cu concentration in planted and implanted soils and facilitated Cu uptake by *Elsholtzia splendens* (Chen et al. 2006). However, Jiang et al. (2004) found that citric acid had no marked effect on both soil-extractable Cu and shoot Cu concentration or accumulation compared with the control, whereas the results from both hydroponic and pot experiments showed that CA addition decreased Cu uptake, translocation from root to shoot, and shoot Cu accumulation in *E. splendens* (Yang et al. 2005).

The observation of both positive and negative effects of chelating agents on Cu contents in *Elsholtzia splendens* is probably due to different soil properties and concentration of chelating agents used in their studies. Therefore, it might be interesting to further investigate how chelating agents affect Cu uptake by plant roots to shoots.

17.4.2.4 Other Factors

The results of Peng et al. (2005b) indicated that application of organic manure at a proper rate could enhance Cu bioavailability and increase effectiveness of Cu phytoextraction from the contaminated soil by *E. splendens*. Wang et al. (2007) found that chitosan application did not affect plant growth and Cu accumulation by *E. splendens* but increased shoot Zn, Pb, and Cd concentrations, which led to higher phytoextraction efficiencies and partitioning to shoots of Zn, Pb, and Cd.

17.5 Distribution and Location of Cu in *Elsholtzia splendens*

17.5.1 Cu Microdistribution in Different Tissues of Elsholtzia splendens

17.5.1.1 Cu Microdistribution in Cross sections of Leaf, Petiole, and Stem Using SRXRF

The accurate elemental distribution at the microscopic/submicroscopic level is of growing importance in many areas of scientific research (Adams 2003). Compared with methods resembling EDXS (electron-dispersive X-ray spectroscopy) and PIXE (proton-induced X-ray analysis), synchrotron radiation X-ray fluorescence (SRXRF) microprobe is more sensitive and less injurious to cells and also has excellent characteristics such as very high brightness, collimation, polarization, and lower bremsstrahlung, which is the electromagnetic radiation produced by an accelerated electrically charged subatomic particle (Shi et al. 2004). Thus, SRXRF microprobe is being extensively applied in many fields such as materials science, life science, and environmental science (Kang et al. 2002).

Through SRXRF microprobe, Cu, and other elements, distribution in leaf, petiole, and stem cross sections of *E. splendens* was investigated (Shi et al. 2004). The results showed that Si, P, S, Cl, K, Ca, Mn, Fe, Ni, Cu, and Zn in leaf and stem of *E. splendens* could be detected by SRXRF, and there was a significant linear correlation between the relative concentrations of elements in leaf measured by SRXRF and the concentrations measured by ICP-AES. All these results demonstrated that SRXRF microprobe was a very sensitive technique for trace element analysis and an ideal tool for localizing the sites of metal accumulation in plants.

In the leaf cross section, Cu was mainly distributed in palisade tissue, with a decrease from palisade tissue to spongy tissue. Cu contents in mesophyll were higher than that in the epidermis, and it was higher in the upper epidermis than in the lower epidermis. Similar results were also found in leaf cross section of hyperaccumulator *Arabidopsis halleri* (Küpper et al. 2000). In addition, the Cu levels in trichomes were not higher than that in other tissues. However, through micro-PIXE, Krämer et al. (1997) found that nickel was preferentially accumulated in the epidermal trichomes of the leaf in nickel hyperaccumulator *Alyssum lesbiacum*. By summing up the above observations, we may safely draw the conclusion that the compartmentation of heavy metals in peripheral, epidermal structures of leaf is not the general feature of all plants.

As for the petiole, the relative Cu concentration was higher in the vascular tissue than in other parts, and Cu levels in the adaxial cortex were higher than that in the abaxial cortex.

In the stem, the relative concentration of Cu was the highest in the vascular tissue, then in the cortex and epidermis, and the lowest concentration was in the pith. Considering the similar results of petiole, one reasonable explanation is that Cu is transported mainly in vascular tissue and there is a tendency of Cu to diffuse to the cortex and the epidermis.

17.5.1.2 Cu Distribution in Leaf Using LA-ICP-MS

Laser ablation inductively coupled mass spectrometry (LA-ICP-MS) is increasingly applied in the field of trace and isotope analysis of solid samples due to the advantage of direct solid sampling by focused laser irradiation on the sample surface and its ability to provide direct information on elemental distribution (Becker 2007). With LA-ICP-MS, quantitative imaging of Cu and its absorption dynamics in the leaves of *Elsholtzia splendens* was studied (Wu et al. 2009a, b).

The results indicated that Cu was located uniformly in the mesophyll with a slightly higher concentration in the main vein of *E. splendens* without Cu stress, and the maximum Cu content in the main vein can reach 20 $\mu g\ g^{-1}$ measured by LA-ICP-MS. By supplying enriched ^{65}Cu isotope tracer to the nutrient solution, the accumulation of ^{65}Cu in the leaves of *E. splendens* after ^{65}Cu treatment can be observed by quantitative imaging LA-ICP-MS. It was found that Cu accumulation in the newly formed leaves increased more quickly compared with the fully grown and the oldest leaves. In the newly formed leaves, the highest Cu concentration was found in the veins near the petiole and at the bottom edge around the petiole of the leaves. In the fully grown leaves, however, Cu was mainly located in the veins as well as at the tip of the leaves. By summarizing above results, we may draw the conclusion that *E. splendens* could transport excess Cu to the edge and the tip of the leaves and therefore detoxify them.

17.5.2 Subcellular Localization of Cu in Elsholtzia splendens

One of the most important principles of heavy metal tolerance and detoxification in plant cells is its selective distribution, which could avoid damaging relatively significant organelles. The cell walls of plants have the capacity to bind metal ions in negatively charged sites (Macfie and Welbourn 2000). Therefore, plant cell walls could be a critical defensive strategy of plants to heavy metal stress (Branquinho et al. 1997). In addition to cell wall compartmentation, transport of metal ions into the vacuole is also an important mechanism for heavy metal tolerance (Hall 2002).

Through transmission electron microscope (TEM) and energy-dispersive analysis of X-rays (EDX), the likely location of copper within the cells of *Elsholtzia splendens* was investigated (Ni et al. 2005). It was found that the majority of copper in *Elsholtzia splendens* was localized primarily in the vacuolar and cell wall. The study of Lou et al (2004) revealed that most of the increased Cu in the roots was found on the cell walls during the short-time culture (24 h), while copper proportion on the cell walls decreased slightly with the increasing treatment time. In addition, copper intracellular distribution in *Elsholtzia splendens* was investigated by transmission electron microscope (TEM) and gradient centrifugation techniques (Peng et al. 2005c). The highest Cu concentration was found in the plant cell wall in both the leaf and root cell of *E. splendens* at 500 μmol L^{-1} Cu, while in the leaf cell, chloroplast was the other important location site. With longer Cu exposure time, Cu location in the cell wall increased considerably and that in the chloroplast decreased accordingly. The Cu localization in the cell walls and chloroplasts could mainly account for the high detoxification of Cu in *E. splendens*.

17.6 Speciation and Biotransformation of Copper in *Elsholtzia splendens*

17.6.1 Research Methods of Heavy Metal Speciation

For a very long time, humans have been fascinated by the magical ability of plants to acquire and detoxify high concentration of essential elements (such as copper, zinc, and iron) or nonessential elements (such as arsenic, cadmium, chromium, and lead). One of the foremost explanations is that plants could control both the oxidation state and coordination environment of specific elements to maximize their detoxification, transport, or both. For detoxification of metals and metalloids, plants directly coordinate the element, using the most chemically appropriate ligand to form stable nontoxic complexes (Salt et al. 2002). Therefore, researches about the speciation and coordination environment of heavy metals are essential to clarify the mechanism of plants upon heavy metals stress.

The field of elemental speciation has been widened in the last few years due to the developments in analytical methods that can determine the molecular form of elements in real samples. ICP-MS can be used for molecule identification when combined online or offline with a molecule-specific separation technique. Liquid chromatography (LC) and gas chromatography (GC) as well as electrophoresis methods, such as capillary electrophoresis (CE) and recently gel electrophoresis (GE), have been combined extensively with ICP-MS for the determination of different elemental species (Feldmann 2005). Although the hyphenated techniques such as HPLC-ES-MS/ICP-MS system are increasingly used in speciation studies, they have not made the breakthrough in the aspect of chemical form alteration of the elements because of complicated sample pretreatments. In contrast, the in situ X-ray absorption spectroscopy (XAS) technique allows determination of element species directly in plant material with minimal sample preparation (Webb et al. 2003).

XAS is an element-specific method particularly suited to analyzing the in vivo ligand environment of metals in plants. With this technique, fractionation or extraction of the plant tissue is unnecessary, and thus the risk of artificial changes in metal complexation is minimized. Some research groups have applied this technique to investigate Ni, As, Cd, Se, and Zn ligands in some plants, especially hyperaccumulators (Gardea-Torresdey et al. 2005; Salt et al. 2002; Tian et al. 2010). In this section, the speciation and biotransformation of copper in *Elsholtzia splendens* investigated by X-ray absorption spectroscopy were introduced according to the research work of Shi et al (2008).

17.6.2 *Speciation and Biotransformation of Copper in* Elsholtzia splendens

The uniform *E. splendens* seedlings were treated with 300 μM $CuSO_4$ for either 10 days or 60 days. Samples were taken from several plant organs (roots, stems, and leaves) for XAS detection. The reference materials used in the investigation included aqueous Cu, Cu-oxalate, CuO, Cu-glutathione, Cu-histidine, Cu-cell wall of leaf, Cu-cell wall of stem, and Cu-cell wall of root. Linear combination arithmetic was applied to determine the probable Cu speciation in *E. splendens*, as analyzed by the LSFitXAFS program. The results showed that species resembling Cu-His (33–52%) and Cu-cell wall (20–44%) were the major Cu species in roots, stems, and leaves, while Cu-oxalate (9–18%) and Cu-glutathione (7–16%) species were in minor proportions. The exact proportion of different Cu species and their roles in Cu tolerance and detoxification of *E. splendens* is elaborately discussed in the following paragraphs.

17.6.2.1 Cu-Histidine

Large amounts of Cu were bound to His-like species in roots, stems, and leaves. Histidine accounts for about 33% in roots, 44% in stems, and 37% in leaves, respectively, after 10 days of Cu exposure. Similar results were also found by Wu (2009) that free histidine and cysteine might be the potential ligands chelating with Cu in *E. splendens*. The histidine concentration in stems after 500 μM Cu treatment was 11 times more than it in the stems without treatment, indicating that free histidine probably participated in the transport of Cu from the roots to the shoots.

Besides, increase of Cu-His-like species was observed in roots and leaves when Cu treatment time increased from 10 to 60 days. This could mainly be ascribed to the loss of normal intracellular compartmentation in dying cells, which allowed interaction of metals normally stored in the vacuole (Küpper et al. 1999, 2004) with cytoplasmic components. Thus, heavy metals will bind not only to free His (Krämer et al. 1996) but also to His residues of proteins and chlorophyll (Küpper et al. 2004).

Krämer et al. (1996) found that Ni bound to free His was used for Ni transport and detoxification in the Ni-hyperaccumulator *Alyssum lesbiacum*. However, this mechanism was not observed in another Ni-hyperaccumulator, *Thlaspi goesingense* (Persans et al. 1999). Increasing the Cu concentration in rooting media induces selective synthesis of histidine in xylem sap of tomato and chicory (Liao et al. 2000). However, although strong, the evidence for the function of histidine in Cu detoxification in plants needs to be further investigated.

17.6.2.2 Cu-Cell Wall

Another significant Cu species is Cu-cell wall, accounting for 39% in roots, 30% in stems, and 44% in leaves, respectively. The results confirmed that binding of Cu to the cell wall played an important role in Cu detoxification in *E. splendens*. And this was in accordance with the results described in studies of Cu ultracellular distribution. The cell wall is the first barrier against metal entering the cell. Even though it has no selective permeability, it can adsorb a certain amount of metal to decrease its activity and to inhibit its ability to permeate into the membrane, thus resisting metal toxicity. Using the methods of TEM-EDX (Ni et al. 2005), it was also found that a large amount of Cu accumulated in *E. splendens* was compartmentalized in the cell wall, especially in roots.

EXAFS spectra analysis revealed that most Cu was bound with O ligands in the cell wall of roots, stems, and leaves. These results indicated that oxygen-containing groups such as –OH and/or –COOH in cellulose, lignin, and/or pectin might be the main sites of Cu binding in the cell wall. Merdy et al. (2002) using EXAFS and XANES investigated the geometry and structure of Cu(II) bound to the cell wall residue of wheat straw; it was found that Cu(II) is surrounded by four oxygen atoms, with an average Cu–O equatorial distance equal to 1.94 Å. In the study of Shi et al. (2008), the distance of Cu–O in Cu-cell wall of root and stem complexes is also 1.94 Å.

17.6.2.3 Cu-Oxalate

Chemical species resembling Cu-oxalate held a certain proportion of Cu in *E. splendens*, accounting for 18% in roots, 15% in stems, and 9% in leaves, respectively. It was indicated that oxygen-containing complexes in the cytosol or the vacuole participated in chelating Cu when the plant was under Cu stress. Through the method of TEM-EDX, Ni et al. (2005) found that a large amount of

Cu in protoplasts of *E. splendens* was compartmented in the vacuole. However, the proportion of Cu-oxalate-like species seemed not as high as that of Cu-His-like species. Perhaps Cu compartmented in the vacuole was not only bound to organic acids, but also chelated by other ligands resembling histidine. Generally, most heavy metals are not bound to strong ligands in hyperaccumulating plants (Küpper et al. 2004). This makes sense from an energetically point of view. It is more economical for the plant to pump the metals into the epidermal vacuoles (Küpper et al. 1999) and store them, weakly bound by organic acids that also act as counterions, rather than investing energy in synthesizing the large quantities of strong ligands that would be required to bind the metal.

17.6.2.4 Cu-Glutathione

XANES fitting showed that a minor proportion of Cu (about 10%) bound to glutathione in *E. splendens*. Metallothioneins (MTs) and phytochelatins (PCs) are two metal-binding peptides playing a significant role in heavy metal detoxification, and both of them are bound with metal through S-bonds. Much evidence supports the hypothesis that MTs are involved in copper tolerance and homeostasis in plants. Zhou and Goldsbrough (1994) found that MT2 mRNA is strongly induced in *Arabidopsis* seedlings by Cu, but only slightly by Cd and Zn; when genes for MT1 and MT2 from *Arabidopsis* were expressed in an MT-deficient yeast mutant, both genes complemented the mutation and provided a high level of resistance to Cu. Through the method of XAS, it was found that PC was possibly involved in the uptake and biotransformation of Cu in *Larrea tridentata* (creosote bush), which might play an important role in copper defense and transport mechanism in the plant (Polette et al. 2000). However, in contrast, when Cu-sensitive and Cu-tolerant ecotypes of *Silene vulgaris* were exposed to concentrations of Cu giving either no or 50% inhibition of growth for each ecotype, they showed equal PC synthesis in the root tips (Schat and Kalff 1992); it was concluded that differential Cu tolerance in *S. vulgaris* does not rely on differential PC production. Similarly, Shi et al. (2008) found that the proportion of Cu bound to glutathione-like ligands was less than the proportion bound to the cell wall or His. Thus, a role for PCs in Cu detoxification in *E. splendens* is not supported. However, 300 μM Cu treatments in this study aiming at improving the sensitivity of XAS determination might result in phytotoxicity of *E. splendens*, compared to the previous studies that showed *E. splendens* can tolerate Cu up to 100 μM in hydroponic experiments (Yang et al. 2002). Thus, the XAS results may not be representative of plants not exposed to such a high concentration of Cu. Therefore, whether S ligands (e.g., MTs) play a significant role in Cu detoxification in *E. splendens* exposed to lower levels of Cu should be further studied.

In conclusion, the Cu Kedge X-ray absorption near edge structure (XANES) revealed that most copper in roots, stems, and leaves exists as divalent Cu. Cu speciation changed depending on the treatment time, but there was no unidirectional trend in roots, stems, and leaves. The percentages of potential Cu ligands in all samples were estimated by fitting the XANES spectra with linear combinations. Most Cu in roots, stems, and leaves was bound with cell wall and histidine (His)-like ligands, while a minor proportion of the Cu was bound to oxalate and glutathione-like ligands. The fitting results of Cu K-edge extended X-ray absorption fine structure (EAXFS) showed that nitrogen/oxygen (N/O) ligands were dominant in roots, stems, and leaves of the plant, while S ligands were rare. All these results suggest that Cu bound by N/O ligands plays a key role in Cu detoxification of *E. splendens*, and a role for classical metal-detoxifying S ligands, such as metallothioneins and phytochelatins, in Cu detoxification of *E. splendens* is not supported according to the study of Shi et al. (2008). Due to the phytotoxicity of 300 μM Cu to *E. splendens*, the question of whether S ligands play a significant role in Cu detoxification in *E. splendens* exposed to lower levels of Cu should be further studied.

17.7 Molecular Mechanisms of Cu Tolerance, Accumulation, and Detoxification in *Elsholtzia splendens*

17.7.1 Genomics

Genomics was established by Fred Sanger when he first sequenced the complete genomes of a virus and a mitochondrion. Since then, genomics has been developed greatly. In this section, research works of *E. splendens* in this specialized field are presented and discussed.

cDNA library construction has been widely used in functional gene screening, target genes clone, organisms functional genomics research, and many other aspects since the first cDNA library was successfully constructed (Hofstetter et al. 1976). In the study of Li (2009), root tips of *E. splendens* treated with 100 μmol L^{-1} Cu for 3 h, 6 h, and 24 h were used for construction of cDNA library. The nonnormalized full-length cDNA library was constructed based on SMART technique. The titer of unamplified Cu-excessive library was 3.44E + 07 pfu mL^{-1}. Individual clones of the library were 1.81E + 07. The titer of amplified Cu-excessive library was 4.86E + 12 pfu mL^{-1}. The percentage of recombinant clones was 99%. The length of 79% of the inserted cDNA was more than 500 bp. All the indicators had been up to standard of high-quality cDNA library, which laid solid foundations for separating and cloning copper detoxification genes and could help uncover the tolerance, accumulation, and detoxification mechanism of copper in *Elsholtzia splendens*.

In addition to the construction of full-length cDNA libraries of *E. splendens* under copper stress by Li (2009), Xia (2007) isolated a metallothionein gene named EhMT1 (GenBank accession no. DQ059081) from *Elsholtzia splendens* and conducted its functional analysis. The results showed that EhMT1 gene contains an open reading frame of 225bP encoding a putative peptide of 74 amino acid residues containing two cysteine-rich domains, which were arranged in CXCXXXCXCXXXCXXC and CXCXXXCXCXXXCXC (X are other amino acids other than cysteine) motifs at amino-terminus and carboxy-terminus, indicating that it was a type1 MT. Sequence alignment of the EhMT1 cDNA with genomic DNA showed that gene contained two introns and three extrons. The sequence comparison between EhMT1 and other MTS showed that EhMT1 shared 75% sequence similarity with *Mimulus guttatus*. EhMT1 gene expression was higher in roots than that in leaves of *Elsholtzia splendens* but no signal in stems. Semireverse transcriptional PCR indicated that the expression of EhMT1 was upregulated by Cu, and induced by heat shock and oxidative stress.

Besides, the prokaryotic expression vector was constructed and transformed into *E. coli*. It was found that the recombinant cells were tolerant to various stresses, which suggested that EhMT1 enhanced livability and adaptability of *E. coli* cells. To further study the cellular function of EhMT1, transgenic tobacco plants overexpressing EhMT1 gene were generated. The results showed that transgenic tobacco plants were more tolerant to copper at higher concentration than wild type, and the average Cu level in transgenic tobacco plants was 26% higher than that of wild type. All these results demonstrated that EhMT1 was closely related with Cu-tolerant ability of *Elsholtzia splendens*, especially for root, and might be involved in accumulation and detoxification of Cu in *Elsholtzia splendens*.

17.7.2 Transcriptomics

Gene should be firstly transcribed into mRNA, which was then translated into protein, and then control the biological metabolism. The transcriptome is the set of all RNA molecules, including mRNA, rRNA, tRNA, and other noncoding RNA produced in one or a population of cells. The study of transcriptomics, also referred to as expression profiling, examines the expression level of mRNAs in a given cell population, often using high-throughput techniques based on DNA microarray technology. Transcriptomics, as a holistic approach, is leading the plant functional genomics into a new era of rapid development.

In the transcriptomic study of Li (2009), cDNA-AFLP analysis was used to investigate the gene expression profile in *E. splendens* root tips (about 3 ~ 5 cm) at four time points (0 h, 3 h, 6 h, and 24 h) during the Cu-excess and Cu-deficiency stresses. Sixty-six transcript-derived fragments (TDFs) showing differential expression pattern were obtained. 62 unique genes were found by Blast analysis, which were composed of 43 function-known genes, 6 function-unknown genes, and 13 novel genes. Five TDFs were selected randomly and the expression levels were assessed using real-time quantitative reverse transcription PCR (qRT-PCR) and the results were consistent with that of cDNA-AFLP. Among the 62 unique genes, 12 genes were induced by Cu-excess stress, one gene was induced by Cu-deficiency stress, and other 49 genes were induced by both Cu-excess and Cu-deficiency stress. Of the 42 function-known genes, two genes are involved in redox homeostasis, six in signal transduction, three in protein metabolism, one in energy metabolism, nine in stress resistance, three in metallic tolerance and detoxification, four in cell wall metabolism, two in secondary metabolism, and include one eukaryotic translation initiation factor, two transcription factors, nine ribosomal genes, and one predominantly leaf-expressed protein gene. These Cu-induced genes indicated that Cu toxicity affected different physiological and biochemical pathways of the plant.

17.7.3 Proteomics

Proteins are vital parts of living organisms, as they are the main components of the physiological metabolic pathways of cells. The word "proteome" is a blend of "protein" and "genome" and was used for the first time in 1995 to describe the entire complement of proteins of a genome (Wasinger et al. 1995). So what is proteomics? In essence, it is the study of protein properties (expression level, posttranslational modification, interactions, etc.) on a large scale to obtain a global, integrated view of disease processes, cellular processes, and networks at the protein level (Blackstock and Weir 1999).

Many scientists have been engaged in research about morphological and physiological studies of *Elsholtzia splendens* on copper tolerance, while little is known in *E. splendens* concerning responses under Cu stress at a proteomic level. Since O'Farrell (1975) demonstrated that it was possible to separate proteins based on their isoelectric points and molecular weights by electrophoresis on polyacrylamide gels, two-dimensional polyacrylamidegel electrophoresis (2D-PAGE) has remained unchallenged as the most efficient way of separating complex protein mixtures. By this way, Li et al. (2009) investigated the Cu accumulation and proteome changes in *E. splendens* tissues upon Cu stress. SDS-PAGE analysis showed that the proteins changed more intensively in roots than did in leaves upon Cu stress. Two-dimensional gel electrophoresis (2-DE) and image analyses found that 45 protein spots were significantly changed in roots, but only six protein spots in leaves. MALDI-TOF MS and LTQ-ESI-MS/MS were also used to identify the differently expressed protein spots. The identified root proteins were involved in various cellular processes such as signal transduction, regulation of transcription and translation, energy metabolism, regulation of redox homeostasis, and cell defense. The leaf proteins were mainly degraded fragments of RuBisCo and antioxidative protein. Besides, the roles of these proteins in Cu tolerance/accumulation were also discussed by Li. The resulting differences in protein expression pattern suggested that redirection of root cellular metabolism and redox homeostasis might be important survival mechanisms of *E. splendens* upon Cu stress.

17.8 Conclusion

Copper is an essential micronutrient for plant growth and development and plays a pivotal role in cellular biochemical and physiological processes. Despite being essential, Cu is highly toxic to plants at elevated concentrations. *Elsholtzia splendens* has been proven to be a Cu-tolerant plant by many investigations and been noted as a geobotanical indicator of Cu mines.

Root exudates, rhizosphere bacteria, and arbuscular mycorrhizal fungi facilitated the solubility of copper in contaminated soil, which suggested that Cu behavior in root–soil interface was significantly influenced by *Elsholtzia splendens*. In addition, physical and chemical properties of soil and exogenous chelating agents might as well influence Cu uptake and accumulation in *E. splendens*.

Cu absorbed by *E. splendens* might in turn interfere with biochemical and physiological process resembling essential nutrient metabolism and photosynthesis in *Elsholtzia splendens*. But *Elsholtzia splendens* as a Cu-tolerant plant has evolved a series of defense strategies against Cu stress.

1. Cu was transported mainly in vascular tissue and *E. splendens* could detoxify excess Cu through diffusing them to the edge and the tip of the leaves. At the cellular level, the organelles were affected by Cu in a different order. The reduction or disappearance of Golgi apparatus preceded that of endoplasmic reticulum, while nucleus was generally the last victim organelle. Thus, the intact nucleus under Cu stress may be one of the foremost mechanisms by which *Elsholtzia splendens* is copper-resistant.
2. Plants could control both the oxidation state and coordination environment of specific elements to maximize their detoxification, transport, or both. The speciation and biotransformation of copper in *Elsholtzia splendens* investigated by X-ray absorption spectroscopy showed that species resembling Cu-His (33–52%) and Cu-cell wall (20–44%) were the major Cu species in roots, stems, and leaves, while Cu-oxalate (9–18%) and Cu-glutathione (7–16%) species were in minor proportions. Free histidine as well as histidine residues of proteins and chlorophyll probably participated in the transport of Cu from the roots to the shoots. Even though cell wall has no selective permeability, it can adsorb a certain amount of metal to decrease its activity and to inhibit its ability to permeate into the membrane, thus resisting metal toxicity. It was found that a large amount of Cu accumulated in *E. splendens* was compartmentalized in the cell wall, especially in roots. Through transmission electron microscope and energy-dispersive analysis of X-rays (TEM-EDX), it was found that a large amount of Cu in protoplasts of *E. splendens* was compartmented in the vacuole. However, the proportion of Cu-oxalate-like species seemed not as high as that of Cu-His-like species. Perhaps Cu compartmented in the vacuole was not only bound to organic acids, but also chelated by other ligands resembling histidine. In addition, the proportion of Cu bound to glutathione-like ligands was also less than the proportion bound to the cell wall or His. Thus, whether phytochelatins

(PCs) play a significant role in Cu detoxification in *E. splendens* should be further studied.

3. Molecular mechanisms of Cu tolerance, accumulation, and detoxification were also investigated. The nonnormalized full-length cDNA library of root tips of *E. splendens* treated with 100 μmol L^{-1} Cu for 3 h, 6 h, and 24 h was constructed based on SMART technique. And cDNA-AFLP analysis was used to investigate the gene expression profile in *E. splendens* root tips at four time points (0 h, 3 h, 6 h, and 24 h) during the Cu-excess stresses. Sixty-six transcript-derived fragments (TDFs) showing differential expression pattern were obtained. Besides, a metallothionein gene named EhMT1 (GenBank accession no. DQ059081) was isolated from *E. splendens* and its function was validated by generating the recombinant *E. coli* cells and transgenic tobacco plants over-expressing this gene.

With technique of two-dimensional gel electrophoresis (2-DE) combined with MALDI-TOF MS and LTQ-ESI-MS/MS, 51 differently expressed protein spots were identified. The results indicated that 45 differently expressed root proteins were involved in various cellular processes such as signal transduction, regulation of transcription and translation, energy metabolism, regulation of redox homeostasis, and cell defense. Six differently expressed leaf proteins were mainly degraded fragments of RuBisCo and antioxidative protein. The resulting differences in protein expression pattern suggested that redirection of root cellular metabolism and redox homeostasis might be important survival mechanisms of *E. splendens* upon Cu stress.

All these works help us to better understand the mechanisms of Cu tolerance, accumulation, and detoxification in *E. splendens* and lay a solid foundation for establishing engineering plants hyperaccumulating Cu in the future, which will be useful for bioconcentration and removal of excessive Cu from the environment.

References

Adams F (2003) Synchrotron radiation micro-X-ray fluorescence analysis: a tool to increase accuracy in microscopic analysis. Nucl Instrum Methods Phys Res B 199:375–381

Becker JS (2007) Inorganic mass spectrometry: principles and applications. Wiley, Chichester, UK

Blackstock WP, Weir MP (1999) Proteomics: quantitative and physical mapping of cellular proteins. Trends Biotechnol 17:121–127

Blaylock MJ, Salt DE, Dushenkov S, Zakharova O, Gussman C, Kapulnik Y, Ensley BD, Raskin I (1997) Enhanced accumulation of Pb in Indian mustard by soil-applied chelating agents. Environ Sci Technol 31:860–865

Branquinho C, Brown DH, Catarino F (1997) The cellular location of Cu in lichens and its effects on membrane integrity and chlorophyll fluorescence. Environ Exp Bot 38:165–179

Brun LA, Le Corff J, Maillet J (2003) Effects of elevated soil copper on phenology, growth and reproduction of five ruderal plant species. Environ Pollut 122:361–368

Chen YX, Wang YP, Lin Q, Luo YM (2005) Effect of copper-tolerant rhizosphere bacteria on mobility of copper in soil and copper accumulation by *Elsholtzia splendens*. Environ Int 31:861–866

Chen YX, Wang YP, Wu WX, Lin Q, Xue SG (2006) Impacts of chelate-assisted phytoremediation on microbial community composition in the rhizosphere of a copper accumulator and non-accumulator. Sci Total Environ 356:247–255

Christie P, Li X, Chen B (2004) Arbuscular mycorrhiza can depress translocation of zinc to shoots of host plants in soils moderately polluted with zinc. Plant Soil 261:209–217

Chugh LK, Sawhney SK (1999) Photosynthetic activities of *Pisum sativum* seedlings grown in presence of cadmium. Plant Physiol Biochem 37:297–303

Clemens S (2001) Molecular mechanisms of plant metal tolerance and homeostasis. Planta 212:475–486

Feldmann J (2005) What can the different current-detection methods offer for element speciation? TrAC: Trends Anal Chem 24:228–242

Gardea-Torresdey JL, Peralta-Videa JR, de la Rosa G, Parsons JG (2005) Phytoremediation of heavy metals and study of the metal coordination by X-ray absorption spectroscopy. Coord Chem Rev 249:1797–1810

Hall JL (2002) Cellular mechanisms for heavy metal detoxification and tolerance. J Exp Bot 53:1–11

Hartley-Whitaker J, Ainsworth G, Meharg AA (2001) Copper- and arsenate-induced oxidative stress in *Holcus lanatus* L. clones with differential sensitivity. Plant Cell Environ 24:713–722

Heggo A, Angle JS, Chaney RL (1990) Effects of vesicular-arbuscular mycorrhizal fungi on heavy metal uptake by soybeans. Soil Biol Biochem 22:865–869

Hofstetter H, Schamböck A, Van Den Berg J, Weissmann C (1976) Specific excision of the inserted DNA segment from hybrid plasmids constructed by the poly(dA) poly(dT) method. Biochim Biophys Acta 454:587–591

Hsieh I, Hsu H (1954) On *Elsholtzia splendens*: an indicator plant for copper (in Chinese). Bull Geol Soc China 32:360–367

Hu NJ, Luo YM, Wu LH, Song J (2007) A field lysimeter study of heavy metal movement down the profile of soils with multiple metal pollution during chelate-enhanced phytoremediation. Int J Phytoremediation 9:257–268

Jiang LY, Yang XE (2004) Chelators effect on soil Cu extractability and uptake by *Elsholtzia splendens*. J Zhejiang Univ Sci B 5:450–456

Jiang LY, Yang XE, He ZL (2004) Growth response and phytoextraction of copper at different levels in soils by *Elsholtzia splendens*. Chemosphere 55:1179–1187

Jiang LY, Yang XE, Chen JM (2008) Copper tolerance and accumulation of *Elsholtzia splendens Nakai* in a pot environment. J Plant Nutr 31:1382–1392

Kang SX, Sun X, Ju X, Huang YY, Yao K, Wu ZQ, Xian DC (2002) Measurement and calculation of escape peak intensities in synchrotron radiation X-ray fluorescence analysis. Nucl Instrum Methods Phys Res B 192:365–369

Kärenlampi S, Schat H, Vangronsveld J, Verkleij JAC, van der Lelie D, Mergeay M, Tervahauta AI (2000) Genetic engineering in the improvement of plants for phytoremediation of metal polluted soils. Environ Pollut 107:225–231

Ke WS, Xiong ZT, Chen SJ, Wang ZH (2008) Differences of Cu accumulation and cu- induced ATPase activity in roots of two populations of *Elsholtzia haichowensis* Sun. Environ Toxicol 23:193–199

Kennedy CD, Gonsalves FAN (1987) The action of divalent zinc, cadmium, mercury, copper and lead on the trans-root potential and H^+, efflux of excised roots. J Exp Bot 38:800–817

Kim ND, Fergusson JE (1991) Effectiveness of a commonly used sequential extraction technique in determining the speciation of cadmium in soils. Sci Total Environ 105:191–209

Krämer U, Cotter-Howells JD, Charnock JM, Baker AJM, Smith JAC (1996) Free histidine as a metal chelator in plants that accumulate nickel. Nature 379:635–638

Krämer U, Grime GW, Smith JAC, Hawes CR, Baker AJM (1997) Micro-PIXE as a technique for studying nickel localization in leaves of the hyperaccumulator plant *Alyssum lesbiacum*. Nucl Instrum Methods Phys Res B 130:346–350

Kumar PBAN, Dushenkov V, Motto H, Raskin I (1995) Phytoextraction: the use of plants to remove heavy metals from soils. Environ Sci Technol 29:1232–1238

Küpper H, Jie Zhao F, McGrath SP (1999) Cellular compartmentation of zinc in leaves of the hyperaccumulator *Thlaspi caerulescens*. Plant Physiol 119:305–312

Küpper H, Lombi E, Zhao FJ, McGrath SP (2000) Cellular compartmentation of cadmium and zinc in relation to other elements in the hyperaccumulator *Arabidopsis halleri*. Planta 212:75–84

Küpper H, Mijovilovich A, Meyer-Klaucke W, Kroneck PMH (2004) Tissue- and age- dependent differences in the complexation of cadmium and zinc in the cadmium/zinc hyperaccumulator *Thlaspi caerulescens* (Ganges Ecotype) revealed by X-ray absorption spectroscopy. Plant Physiol 134:748–757

Landberg T, Greger M (2002) Differences in oxidative stress in heavy metal resistant and sensitive clones of *Salix viminalis*. J Plant Physiol 159:69–75

Leyval C, Berthelin J (1993) Rhizodeposition and net release of soluble organic compounds by pine and beech seedlings inoculated with rhizobacteria and ectomycorrhizal fungi. Biol Fertil Soils 15:259–267

Li F (2009) Study of transcriptomics and proteomics on uptake and tolerance mechanism of copper in *Elsholtzia splendens* and its cDNA library construction. Environment Engineering. Doctoral Thesis, Zhejiang University, Hang Zhou

Li X, Coles BJ, Ramsey MH, Thornton I (1995) Sequential extraction of soils for multielement analysis by ICP-AES. Chem Geol 124:109–123

Li F, Xiong Z, Hu H (2003a) Copper toxicity and accumulation in *Elsholtzia splendens* (in Chinese). Environ Sci 24:30–34

Li F, Xiong Z, Hu H (2003b) Effects of chelating agents on toxicity of copper to *Elsholtzia splendens* (in Chinese). Environ Sci 24:96–100

Li MJ, Xiong ZT, Dai LP, Huang Y (2007) Effects of copper on nitrogen assimilation in copper-tolerant and non-tolerant populations of *Elsholtzia haichowensis* S. Water Air Soil Pollut 184:323–333

Liao M, Hedley M, Woolley D, Brooks R, Nichols M (2000) Copper uptake and translocation in chicory (Cichorium intybus L. cv Grasslands Puna) and tomato (Lycopersicon esculentum Mill. cv Rondy) plants grown in NFT system. II. The role of nicotianamine and histidine in xylem sap copper transport. Plant Soil 223:245–254

Lidon CF, Ramalho CJ, Henriques SF (1993) Copper inhibition of rice photosynthesis. Elsevier, Munich, Allemagne

Liu J, Xiong ZT (2005) Differences in accumulation and physiological response to copper stress in three populations of *Elsholtzia haichowensis S*. Water Air Soil Pollut 168:5–16

Long X, Yang XE, Ni W (2002) Current situation and prospect on the remediation of soils contaminated by heavy metals. Chin J Appl Ecol 13:757–762

Lou LQ, Shen ZG, Li XD (2004) The copper tolerance mechanisms of *Elsholtzia haichowensis*, a plant from copper-enriched soils. Environ Exp Bot 51:111–120

Luna CM, González CA, Trippi VS (1994) Oxidative damage caused by an excess of copper in oat leaves. Plant Cell Physiol 35:11–15

Macfie SM, Welbourn PM (2000) The cell wall as a barrier to uptake of metal ions in the unicellular green alga *Chlamydomonas reinhardtii* (Chlorophyceae). Arch Environ Contam Toxicol 39:413–419

Mench M, Morel JL, Guckert A, Guillet B (1988) Metal binding with root exudates of low molecular weight. J Soil Sci 39:521–527

Merdy P, Guillon E, Dumonceau J, Aplincurt M (2002) Spectroscopic study of copper(II) – wheat straw cell wall residue surface complexes. Environ Sci Technol 36:1728–1733

Monni S, Salemaa M, Millar N (2000) The tolerance of *Empetrum nigrum* to copper and nickel. Environ Pollut 109:221–229

Morel JL, Mench M, Guckert A (1986) Measurement of Pb^{2+}, Cu^{2+} and Cd^{2+} binding with mucilage exudates from maize (*Zea mays* L.) roots. Biol Fertil Soils 2:29–34

Ni CY (2004) Response and rhizosphere mechanism of accumulator and non-accumulator under copper Stress. Environment Engineering. Doctoral thesis, Hang Zhou, Zhejiang University

Ni CY, Chen YX, Lin Q, Tian GM (2005) Subcellular localization of copper in tolerant and non-tolerant plant (in Chinese). J Environ Sci 17:452–456

O'Farrell PH (1975) High resolution two-dimensional electrophoresis of proteins. J Biol Chem 250:4007–4021

Ouzounidou G (1994) Root growth and pigment composition in relationship to element uptake in *Silene compacta* plants treated with copper. J Plant Nutr 17:933–943

Ouzounidou G, Ciamporová M, Moustakas M, Karataglis S (1995) Responses of maize (*Zea mays* L.) plants to copper stress-I. Growth, mineral content and ultrastructure of roots. Environ Exp Bot 35:167–176

Padmaja KPD, Parsad ARK (1990) Inhibition of chlorophyll synthesis in *Phaseolus vulgaris* L. seedlings by cadmium acetate. Photosynthetica 24:399–404

Peng HY, Yang XE (2007) Characteristics of copper and lead uptake and accumulation by two species of *Elsholtzia*. Bull Environ Contam Toxicol 78:142–147

Peng HY, Tian SK, Yang XE (2005a) Changes of root morphology and Pb uptake by two species of *Elsholtzia* under Pb toxicity. J Zhejiang Univ Sci B 6:546–552

Peng HY, Yang XE, Jiang LY, He ZL (2005b) Copper phytoavailability and uptake by *Elsholtzia splendens* from contaminated soil as affected by soil amendments. J Environ Sci Health A Tox Hazard Subst Environ Eng 40:839–856

Peng HY, Yang XE, Tian SK (2005c) Accumulation and ultrastructural distribution of copper in *Elsholtzia splendens*. J Zhejiang Univ Sci B 6:311–318

Persans MW, Yan X, Patnoe J-MML, Kramer U, Salt DE (1999) Molecular dissection of the role of histidine in nickel hyperaccumulation in *Thlaspi goesingense* (Halacsy). Plant Physiol 121:1117–1126

Plette ACC, Benedetti MF, van Riemsdijk WH (1996) Competitive binding of protons, calcium, cadmium, and zinc to isolated cell walls of a gram-positive soil bacterium. Environ Sci Technol 30:1902–1910

Polette LA, Gardea-Torresdey JL, Chianelli RR, George GN, Pickering IJ, Arenas J (2000) XAS and microscopy studies of the uptake and bio-transformation of copper in *Larrea tridentata* (creosote bush). Microchem J 65:227–236

Qian M (2005) The mechanisms of Cu tolerance and the potential in the phytoremediation of *Elsholtzia Haichowensis*. Plant Nutrition. Master Thesis, Nanjing Agricultural University, Nan Jing

Salt DE, Smith RD, Raskin I (1998) Phytoremediation. Annu Rev Plant Physiol Plant Mol Biol 49:643–6683

Salt DE, Prince RC, Pickering IJ (2002) Chemical speciation of accumulated metals in plants: evidence from X-ray absorption spectroscopy. Microchem J 71:255–259

Schat H, Kalff MMA (1992) Are phytochelatins involved in differential metal tolerance or do they merely reflect metal-imposed strain? Plant Physiol 99:1475–1480

Schützendübel A, Polle A (2002) Plant responses to abiotic stresses: heavy metal-induced oxidative stress and protection by mycorrhization. J Exp Bot 53:1351–1365

Shi JY (2004) Study on uptake mechanism and molecular speciation of copper in *Elsholtzia splendens* and *Commelina communis*. Environmental Engineering. Doctoral thesis, Zhejiang University, Hang Zhou

Shi JY, Chen YX, Huang YY, He W (2004) SRXRF microprobe as a technique for studying elements distribution in *Elsholtzia splendens*. Micron 35:557–564

Shi JY, Wu B, Yuan XF, Cao YY, Chen XC, Chen YX, Hu TD (2008) An X-ray absorption spectroscopy investigation of speciation and biotransformation of copper in *Elsholtzia splendens*. Plant Soil 302:163–174

Song J, Zhao FJ, Luo YM, McGrath SP, Zhang H (2004) Copper uptake by *Elsholtzia splendens* and *Silene vulgaris* and assessment of copper phytoavailability in contaminated soils. Environ Pollut 128:307–315

Stepanauskas R, Glenn TC, Jagoe CH, Tuckfield RC, Lindell AH, McArthur JV (2005) Elevated microbial tolerance to metals and antibiotics in metal-contaminated industrial environments. Environ Sci Technol 39:3671–3678

Strange J, Macnair MR (1991) Evidence for a role for the cell membrane in copper tolerance of *Mimulus guttatus* Fischer ex DC. New Phytol 119:383–388

Sun LN, Zhang YF, He LY, Chen ZJ, Wang QY, Qian M, Sheng XF (2010) Genetic diversity and characterization of heavy metal-resistant-endophytic bacteria from two copper-tolerant plant species on copper mine wasteland. Bioresour Technol 101:501–509

Tang S, Wilke BM, Huang C (1999) The uptake of copper by plants dominantly growing on copper mining spoils along the Yangtze River, the People's Republic of China. Plant Soil 209:225–232

Tessier A, Campbell PGC, Bisson M (1979) Sequential extraction procedure for the speciation of particulate trace metals. Anal Chem 51:844–851

Tewari R, Kumar P, Sharma P (2006) Antioxidant responses to enhanced generation of superoxide anion radical and hydrogen peroxide in the copper-stressed mulberry plants. Planta 223:1145–1153

Tian SK (2005) Characteristics of Cu/Pb tolerance and accumulation by *Elsholtzia* plants and their effectiveness of heavy metal removal from polluted water. Plant Nutrition. Master thesis, Sichuan Agricultural University, Si Chuan

Tian SK, Lu LL, Yang XE, Webb SM, Du YH, Brown PH (2010) Spatial imaging and speciation of lead in the accumulator plant *Sedum alfredii* by microscopically focused synchrotron X-ray investigation. Environ Sci Technol 44:5920–5926

Vivas A, Biró B, Ruíz-Lozano JM, Barea JM, Azcón R (2006) Two bacterial strains isolated from a Zn-polluted soil enhance plant growth and mycorrhizal efficiency under Zn- toxicity. Chemosphere 62:1523–1533

Wang YP (2006) Molecular microbial ecology of heavy metals polluted soil and its effect on phytoremediation. Environmental Engineering. Doctor thesis, Zhejiang University, Hang Zhou

Wang FY, Lin XG, Yin R (2005a) Heavy metal uptake by arbuscular mycorrhizas of *Elsholtzia splendens*; and the potential for phytoremediation of contaminated soil. Plant Soil 269:225–232

Wang FY, Lin XG, Yin R (2005b) Influence of arbuscular mycorrhizal fungi on growth and Cu uptake of *Elsholtzia splendens* (in Chinese). Environ Sci 26:174–180

Wang FY, Lin XG, Yin R (2007) Role of microbial inoculation and chitosan in phytoextraction of Cu, Zn, Pb and Cd by *Elsholtzia splendens* – a field case. Environ Pollut 147:248–255

Wasinger VC, Cordwell SJ, Cerpa-Poljak A, Yan JX, Gooley AA, Wilkins MR, Duncan MW, Harris R, Williams KL, Humphery-Smith I (1995) Progress with gene-product mapping of the Mollicutes: *Mycoplasma genitalium*. Eletrophoresis 16:1090–1094

Webb SM, Gaillard JF, Ma LQ, Tu C (2003) XAS speciation of arsenic in a hyper-accumulating fern. Environ Sci Technol 37:754–760

Weng G, Wu L, Wang Z, Luo Y, Christie P (2005) Copper uptake by four *Elsholtzia* ecotypes supplied with varying levels of copper in solution culture. Environment International 31(6):880–884

Wu B (2009) Study of mechanisms for Cu tolerance and accumulation in *Elsholtzia splendens*. Environmental Engineering. Doctor thesis, Zhejiang University, Hang Zhou

Wu LH, Sun XF, Luo YM, Xing XR, Christie P (2007) Influence of S, S -EDDS on phytoextraction of copper and zinc by *Elsholtzia splendens* from metal-contaminated soil. Int J Phytoremediation 9:227–241

Wu B, Chen YX, Becker JS (2009a) Study of essential element accumulation in the leaves of a Cu-tolerant plant *Elsholtzia splendens* after Cu treatment by imaging laser ablation inductively coupled plasma mass spectrometry (LA-ICP-MS). Anal Chim Acta 633:165–172

Wu B, Zoriy M, Chen YX, Becker JS (2009b) Imaging of nutrient elements in the leaves of *Elsholtzia splendens* by laser ablation inductively coupled plasma mass spectrometry (LA-ICP-MS). Talanta 78:132–137

Xia Y (2007) The copper tolerance, cloning and functional analysis of metallothione in gene in *Elsholtzia haichowensis*. College of Life Science. Doctor thesis, Nanjing Agricultural University, Nan Jing

Yang MJ, Yang XE, Romheld V (2002) Growth and nutrient composition of *Elsholtzia splendens Nakai* under copper toxicity (in Chinese). J Plant Nutr 25:1359–1375

Yang XE, Peng HY, Jiang LY (2005) Phytoextraction of copper from contaminated soil by *Elsholtzia splendens* as affected by EDTA, citric acid, and compost. Int J Phytoremediation 7:69–83

Zhang HX, Xia Y, Wang GP, Shen ZG (2008) Excess copper induces accumulation of hydrogen peroxide and increases lipid peroxidation and total activity of copper-zinc superoxide dismutase in roots of *Elsholtzia haichowensis*. Planta 227:465–475

Zhou J, Goldsbrough PB (1994) Functional homologs of fungal metallothionein genes from *Arabidopsis*. Plant Cell 6:875–884

Chapter 18
Role of Aquatic Macrophytes in Biogeochemical Cycling of Heavy Metals, Relevance to Soil-Sediment Continuum Detoxification and Ecosystem Health

Przemysław Malec, Beata Mysliwa-Kurdziel, M.N.V. Prasad, Andrzej Waloszek, and Kazimierz Strzałka

18.1 Introduction

Soil (*per se*) is the abiotic component of life-supporting system and is considered to be the soul of life. A clean soil is an essential prerequisite for protecting the health of all biota, including humans, which is a prime concern. In recent years, dependency on heavy metals has increased manifolds particularly in a variety of industries such as plastic, textiles, microelectronics, battery industry, wood preservatives, smelting, river dredging, mining (spoils and tailings), and metallurgy. Also, heavy metals are components of, e.g., coal combustion products, urban refuses, sewage sludges, automobile exhausts, and some of fertilizers (Kabata-Pendias 2001). Under these circumstances, the potential of wetland plants in environmental protection has been widely recognized all over the world. In particular, the biodiversity of wetlands and the naturally operating principles of biogeochemical cycles have unequivocally demonstrated their significance in cleansing heavy metal contaminated water (Fig. 18.1). Water plants offer efficient and environmentally friendly solutions for clearing contaminated water, sediments, brownfields, and wastewater. In the last 5 years, the publications on various aspects of wetland science have sharply increased [data gleaned from http://www.sciencedirect.com until 2010 (Fig. 18.2)].

P. Malec • B. Mysliwa-Kurdziel • A. Waloszek • K. Strzałka (✉)
Faculty of Biochemistry, Department of Plant Physiology and Biochemistry, Biophysics and Biotechnology, Jagiellonian University, ul. Gronostajowa 7, 30–387 Kraków, Poland
e-mail: kazimierzstrzalka@gmail.com

M.N.V. Prasad
Department of Plant Sciences, University of Hyderabad, Prof. C.R. Rao Road, Gachhibowli, Central University P.O., Hyderabad 500 046, India

I. Sherameti and A. Varma (eds.), *Detoxification of Heavy Metals*, Soil Biology 30,
DOI 10.1007/978-3-642-21408-0_18,

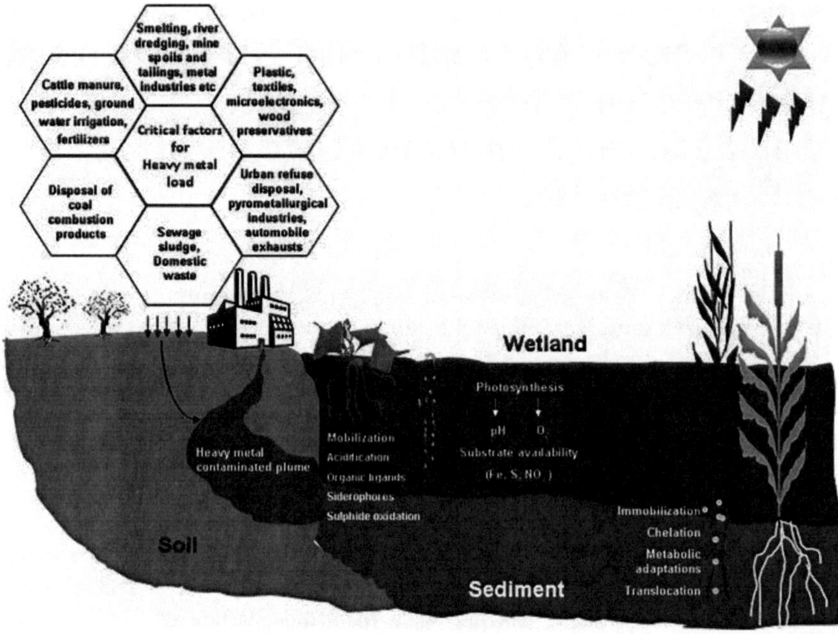

Fig. 18.1 Biogeochemistry of trace metals and their dynamics in wetland including the main detoxification strategies

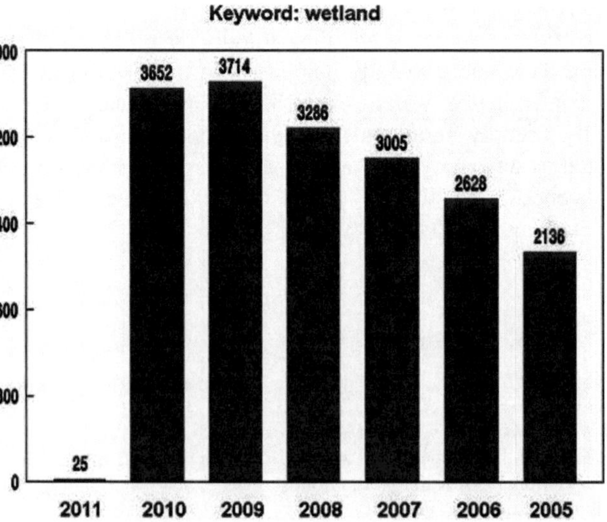

Fig. 18.2 There has been a steep rise in publications on "wetlands" during the last 5 years for their significance to ecosystem services (data source: http://www.sciencedirect.com; database status recorded on January 31st, 2011)

Many members of the angiospermous families, namely, *Cyperaceae*, *Potamogetonaceae*, *Ranunculaceae*, *Typhaceae*, *Haloragaceae*, *Hydrocharitaceae*, *Najadaceae*, *Juncaceae*, *Pontederiaceae*, *Zosterophyllaceae*, *Lemnaceae*, and *Typhaceae*, have aquatic and semiaquatic habitats. Members of these families exhibit great diversity, inhabit freshwater bodies, and complete their life cycle there. Aquatic plant lifestyles may vary greatly; for example, they may be (a) free floating, such as duckweed, (b) totally submersed, such as naiad, (c) bottom rooted and floating, such as waterlily, (d) emergent and rooted, such as quillwort, (e) totally emergent, such as cattails, and (f) streambank and wet area plants, such as alders (Jackson 1998).

Natural wetlands, especially floodplains, serve as a unique ecosystem with a wide variety of functions. One of the most important functions is their ability to act as a buffering zone to extremes associated with the discharge of rivers. The soils of floodplains also play a crucial role in the biogeocycling of toxic heavy metals and nutrients including the treatment of contaminants transported by flood and storm water. These functions lead to the formation of contaminated hot spots at selected locations in floodplain areas as a result of the deposition of heavy metal contaminated sediment. Wetland soils and sediments often function as pollutant sinks (Du Laing et al. 2009; Rinklebe et al. 2007) and these require an adequate management strategy. Understanding the main biogeochemical processes is essential in this context.

Constructed wetlands are artificial ecosystems dedicated to detoxification and the removal of contamination dissolved in the water phase. The following components that are involved in pollutant removal constitute a typical constructed wetland: substrates (soil/sediment), water, plants, and associated microbial populations. Constructed wetlands could be either aerobic (surface flow) or anaerobic (subsurface). In aerobic wetlands, plants are placed in shallow, slightly permeable soil (as e.g., clay or mine spoil). Anaerobic wetlands consist of vegetation planted in a deep, permeable mixture of substrate such as soil, peatmoss, compost, sawdust, straw/manure, hay bales, and gravel (Kosolapov et al. 2004; Gutknecht et al. 2006; Knox et al. 2006).

The determination of the chemical species of compounds present in aquatic ecosystems, both in solution and in bottom sediments, and the bioavailability of the various species and their forms of occurrence play a crucial role in assessing threats to the environment. The toxicity of elements, as well as the presence or absence of synergism and antagonism with respect to other elements, would play an important role in planning remediation strategies (Namieśnik and Rabajczyk 2010).

18.2 Soil-Sediment Continuum

Soils and sediments are connected or interlinked by a hydrological phase and these are functionally similar and share a number of common features. In fact, there is no clear borderline between soils and sediments, because both are interlinked by

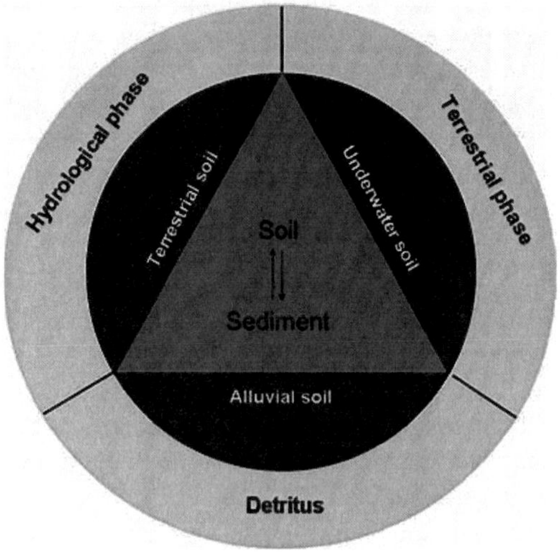

Fig. 18.3 Soils and sediments are connected and interlinked by hydrological phase and these are functionally similar and share a number of common features

hydrological and terrestrial phases, which can be explained by (a) underwater soils, (b) terrestrial soils, (c) and alluvial soils (Fig. 18.3).

Several processes take place near the sediment–water interface (Lerman 1978). There is considerable debate on the distinction between soils and sediments (Blum 2005). Sediment pollution is caused by natural and anthropogenic influence. According to Gross (1978), human activities have significantly altered sediment characteristics. Industrial effluents and urban wastes are sources of pollutants in waters and sediments. Heavy metals in floodplain soils and sediment remediation including detoxification mechanisms involving plants are gaining considerable global attention (Rinklebe et al. 2007). The metal detoxification in plants is complex and usually involves a combination of several mechanisms, both limiting the metal circulation within plant organisms and preventing damages caused by a metal-induced oxidative stress (see below).

18.3 Role of Macrophytes in Trace Metal Dynamics in Wetland Sediments

The influence of aquatic plants and their metabolism may alter the distribution of trace metals between the solid and aqueous phases. There is little knowledge on how the combined defects of wetland plants influence the biogeochemistry of wetland sediments and thereby the trace metal dynamics (Namieśnik and Rabajczyk 2010). To survive in water-saturated sediments, aquatic macrophytes have developed specialized adaptations. To support root respiration under anoxic conditions, they can effectively transfer oxygen from the surface to the roots (Armstrong 1979; Dacey 1980). A remarkable fraction of this oxygen may penetrate the rhizosphere, where it can take part in reoxidation of reduced sediment components, such as Fe^{2+} and Mn^{2+} (Mendelssohn et al. 1995). This process, called *R*adial *O*xygen *L*oss (ROL), may lead to the increase in metal mobility in wetland ecosystems. Aquatic macrophytes, translocating the oxygen into the rhizosphere, increase the redox potential and thus also decrease the pH and increase the release of metals (see: Yang and Ye 2009 and references therein). It was shown that several species of aquatic macrophytes, including *Carex rostrata, Phragmites australis, Typha angustifolia,* and *T. latifolia*, can tolerate a very low pH and have been found growing under field conditions in pH as low as 2–4.4 (Nixdorf et al. 2002). Recently, the presence of low-pH-tolerant aquatic macrophytes has been shown to alter the mobility of Cu in constructed wetlands dedicated for treating acid mine drainage (Nyquist and Greger 2009).

Physical transport processes and biogeochemical reactions driven by aquatic plants may result in extensive sulfur cycling (see e.g., Choi et al. 2006). The oxidation of sulfides in sediments results in the production of oxidized sulfur (i.e., SO_4^{2-} and S^0), which may stimulate the release of metals to the water column (Simpson et al. 1998). Also, in freshwater wetlands, the degradation of organic sulfur from ant residue may lead to an increase in sulfate concentrations in the water column (Lefroy et al. 1994; Wind and Conrad 1995). Elevated sulfate concentrations in sediments were studied by Urban et al. (1994). In the presence of organic carbon produced by plants and in the absence of other electron acceptors, organic carbon is degraded, resulting in the formation of sulfides (Urban et al. 1994). In the presence of metals, sulfides can form metal sulfide precipitates in the sediments and may control the metal concentration in interstitial water (Boulegue et al. 1982; Emerson et al. 1983; Huerta-Diaz et al. 1998). For example, the reaction of sulfide with iron compounds has been shown to lead to the precipitation of iron monosulfide (FeS) and pyrite (FeS_2) in sediments (Berner 1984; Howarth and Jørgensen 1984). It was found that the acid-volatile sulfide (AVS) constituting the most labile fraction of sediment sulfide plays a key role in the determination of metal bioavailability in sediments (Di Toro et al. 1990).

Additionally, plants affect the biogeochemical dynamics of wetland sediments via evapotranspiration-induced advection, which increases the loading of dissolved constituents into the rhizosphere. The release of organic carbon into the sediments by aquatic plants stimulates many biotic and abiotic reactions. The chemical degradation of organic carbon results in the increase in reductive properties of sediments (Du Laing et al. 2009).

In particular, organic mercurial compounds from contaminated treatment wetlands are taken up by aquatic plants, passed to the leaves, and volatilized into the atmosphere at comparatively low concentrations (Du Laing et al. 2009).

18.4 Mechanisms of Tolerance to Metals in Aquatic Macrophytes

As sessile organisms, plants must deal with environmental limitations to survive. Plants growing in metal-polluted habitats have developed complex mechanisms to tolerate elevated concentrations of metals and to control cell homeostasis in potentially harmful environment. In the case of wetland plants, metal tolerance is particularly important as they are usually exposed to both overlying water and/or sediment and they take up nutrients and other minerals from both environments (Hinman and Klaine 1992). In general, aquatic macrophytes share the following common features: (a) the root system is reduced or completely degenerated, (b) the cuticle is very thin; therefore, shoots are able to absorb metals directly from water, and (c) the vascular system is reduced and the transpiration stream is limited or lacking (Basiouny et al. 1977). Aquatic plants play an important role in the uptake, storage, and recycling of metals. The uptake of elements depends on their chemical forms available in the environment as well as on the morphology and physiology of macrophytes. Free-floating species (e.g., *Eichhornia*, *Lemna*, *Pistia*, and *Spirodela*) absorb metal ions through both the roots and leaves. In contrast, rootless species (e.g., *Ceratophyllum demersum*) absorb basically through leaves (Chandra and Kulshreshtha 2004). Therefore, the location and capacity of particular tolerance mechanisms may be different than those identified in terrestrial plants. However, in contrast to most studied terrestrial plants (reviewed by Hall 2002), the biochemical and molecular backgrounds of the tolerance of aquatic macrophytes to metals are far less known (Table 18.1).

18.4.1 Metal Immobilization

Immobilization is one of the most common mechanisms of plant adaptation to elevated concentrations of heavy metals. The avoidance of metal stress may be achieved by the binding of metal ions in the apoplast which prevents their internalization. Metals may be deposited in cell walls as metalloorganic complexes formed with ligands present in this compartment (e.g., polysaccharides and organic acids). For example, the binding of Pb in cell walls has been demonstrated in the free-floating macrophyte *Lemna minor*. In this case, elevated concentrations of (l,3)-P-glucan (callose) have been found to be accumulated in the cell walls (predominantly in root tips) of plants treated with $Pb(NO_3)_2$ (Samardakiewicz et al. 1996). Subsequent X-ray microanalysis demonstrated that the accumulation of callose correlated with the lead deposition in cell walls. In contrast, lead has not been found in the ground cytoplasm, pointing to cell wall deposition as an effective mechanism of metal avoidance (Samardakiewicz and Woźny 2000). The immobilization of metals in the cell walls of the photosynthetic tissues of aquatic plants

Table 18.1 Representative recent articles on biochemical mechanisms of heavy metal tolerance and detoxification in aquatic macrophytes

Physiological process	Plant species	Metal	Mechanism identified	References
Immobilization	*Lemna minor*	Pb	Cell wall deposition	Samardakiewicz et al. (1996)
	Spirodela intermedia, Lemna minor, Pistia stratiotes, Potamogeton lucens, Salvinia herzogii, Eichhornia crassipes	Cd, Ni, Cu, Zn, Pb, Cr	Sorption by (non-living) biomass, ion exchange	Schneider and Rubio (1999) and Miretzky et al. (2006)
	Ludwigia stolonifera, Salvinia herzogii, Pistia stratiotes	Cd, Ni, Cd, Cr	Sorption by living biomass; ion exchange	Elifantz and Tel-Or (2002) and Suñe et al. (2007)
	Hydrilla verticillata	Cu	Cell wall deposition	Xue et al. (2010)
	Fontinalis dalecarlica	Cd	Cell wall deposition	Bleuel et al. (2005)
Chelation	*Pistia stratiotes*	Cd	GSH, phytochelatins	Sanità di Toppi et al. (2007)
	Fontinalis antipyretica, Fontinalis dalecarlica	Cd	GSH	Bleuel et al. (2005)
	Elodea canadensis	Ni	Non-protein thiols	Maleva et al. (2009)
		Pb		Dogan et al. (2009)
		Cu		Malec et al. (2009a)
	Lemna trisulca	Cd	Non-protein thiols	Malec et al. (2010)
	Egeria densa	Cd	Cd-binding, metallothionein-like protein	Malec et al. (2009b)
	Azolla filiculoides	Cd, Cu, Ni, Zn	Expression of type 2 metallothionein	Schor-Fumbarov et al. (2005)
Metabolic adaptation	*Eichhornia crassipes*	Cd	AsA, APX, DHAR, MDHAR, SOD, GRD	Sanità di Toppi et al. (2007)
	Elodea canadensis	Ni	CAT, GRD	Maleva et al. (2009)
	Potamogeton crispus	Ag	AsA, POD	Xu et al. (2010)
		Cd	PAO, DAO	Yang et al. (2010)
	Potamogeton pusillus	Cu	GPX, GRD, POD	Monferrán et al. (2009)
	Fontinalis antipyretica	Cd, Cu, Pb, Zn	SOD, CAT, GRD, APX	Dazy et al. (2009)
	Alternanthera philoxeroides	Cd	POD, SOD, CAT	Ding et al. (2007)
	Lemna minor, Spirodela polyrrhiza	Cu	SOD, CAT	Kanoun-Boulé et al. (2009)
	Lemna minor	Cd, Zn, Al	SOD, CAT, APX	Tkalec et al. (2008) and Radić et al. (2010)
	Azolla filiculoides	Pb	V-H + −ATPase	Oren-Benaroya et al. (2004)
Translocation	*Potamogeton pectinatus, Potamogeton crispus*	As, Cd, Cu, Pb, Zn	Acropetal translocation	Peter et al. (1979) and Wolterbeek and van der Meer (2002)
	Elodea canadensis	Cd	Acropetal translocation	Fritioff and Greger (2007)
	Hydrilla verticillata	Cu	Acropetal and basipetal translocation	Xue et al. (2010)
	Baumea juncea, Baumea articulata Schoenoplectus, Juncus subsecundus	Cd	Acropetal translocation	Zhang et al. (2010)

has also been observed. In particular, a significant biosorption of Cd has been detected in aquatic moss *Fontinalis dalecarlica* (Bleuel et al. 2005). The submerged macrophyte *Hydrilla verticillata* has been found to accumulate Cu of up to 30 830 mg kg^{-1} dry weight. In this organism, most of the Cu has been deposited in the cell walls of shoots (Xue et al. 2010). Interestingly, a metal-binding capacity has also been demonstrated in the dead, dried biomass of aquatic plants. The sorption of Cd^{2+}, Ni^{2+},Cu^{2+}, Zn^{2+}, and Pb^{2+} by the biomass of three macrophytes (*Spirodela intermedia*, *Lemna minor*, and *Pistia stratiotes*) has been investigated. In all these plant species, effective metal sorption was observed. The *Lemna minor* biomass presented the highest mean removal percentage and *Pistia stratiotes* the lowest for all metals tested. It has been shown that this sorption is based on an ion exchange between monovalent metals (K^+ and Na^+) as counterions present in the macrophyte biomass and heavy metal ions and protons taken up from water. No significant differences were observed in the amounts of metal exchange when multimetal or individual metal solutions were used (Miretzky et al. 2006). Other authors have demonstrated that the dried biomass of *Potamogeton lucens*, *Salvinia herzogii*, and *Eichhornia crassipes* were excellent biosorbents for Cr^{2+}, Cd^{2+}, Ni^{2+}, Cu^{2+}, Zn^{2+}, and Pb^{2+}. The sorption mechanism of these biomaterials was found to proceed mainly by ion exchange reactions between metal ions and cationic weak exchanger groups present on the plant surface (Schneider and Rubio 1999). The exchange of bound heavy metal ions against the discharge of light metal ions such as Ca^{2+}, Mg^{2+}, K^+, and Na^+ has been proposed as playing a role in Cd and Ni biosorption by the living biomass of roots, floating roots, and the leaves of *Ludwigia stolonifera* (Elifantz and Tel-Or 2002) and also for Cd and Cr uptake by *Salvinia herzogii* and *Pistia stratiotes* (Suñe et al. 2007).

18.4.2 Chelation

In plants, the synthesis of thiol-containing, cation-chelating compounds, such as glutathione (GSH), phytochelatins, and metallothioneins, is a frequent response to heavy metals transported into the symplast (Cobbett and Goldsbrough 2002). Other compounds such as amino acids, organic acids, and phenols have also been identified as heavy metal chelators (Rauser 1999). Chelating agents are found to limit the circulation of toxicants within the plant organism, forming stable complexes with metal ions (Prasad 1995). For example, in *Pistia stratiotes* and *Eichhornia crassipes* exposed to Cd, glutathione levels remarkably increased. In a species more responsive to Cd exposure – *P. stratiotes* – this increase in glutathione concentration correlated with abundant phytochelatin synthesis both in roots and in leaves (Sanità di Toppi et al. 2007). Also, Cd^{2+} exposure in the mosses *Fontinalis antipyretica* and *F. dalecarlica* increased the glutathione pool. Analytical electron microscopy provided evidence that GSH may be an agent for intracellular Cd^{2+} detoxification in these bryophytes (Bruns et al. 2001). A significant increase in nonprotein thiols has been observed in *Lemna trisulca* fronds treated with low doses of Cd (Malec et al. 2010). In photosynthetic tissues of *Elodea canadensis*, the induction of nonprotein –SH groups was accompanied with an accumulation of Pb (Dogan et al. 2009) and Ni (Maleva et al. 2009) and the synthesis of low-molecular-weight peptides was observed in the presence of elevated Cu concentrations in the medium (Malec et al. 2009a). In the shoots of a closely related plant, *Egeria densa*, cadmium induced both an increase in the nonprotein thiol concentration and the synthesis of Cd-binding polypeptide, containing aromatic amino acid residues and sharing biochemical properties with metallothioneins (Malec et al. 2009b). A cDNA encoding a type 2 metallothionein, isolated from the wetland fern *Azolla filiculoides*, was termed AzMT2. The expression of AzMT2 was enhanced by the addition of Cd^{2+}, Ni^{2+}, Cu^{2+}, and Zn^{2+} to the growth medium. The increase in the transcript level of AzMT2 was proportional to the metal content in the plant. The more moderate response of AzMT2 to Zn and Cu ions, which are essential micronutrients, in comparison to toxic Cd, suggests a role for AzMT2 in metal homeostasis (Schor-Fumbarov et al. 2005).

18.4.3 Metabolic Adaptations

Heavy metal uptake is known as a factor inducing oxidative stress in plants (Mittler 2002). In response to ROS production induced by metals, aquatic plants increase the activity of enzymes involved in the scavenging of ROS, as e.g., superoxide dismutase (SOD), peroxidase (POD), and catalase (CAT). Similarly, metal stress was shown to stimulate the activity of the enzymes involved in biosynthesis and/or modification of cellular reducing agents such as ascorbate (AsA) (ascorbate peroxidase – APX, dehydroascorbate reductase – DHAR, monodehydroascorbate reductase – MDHAR), glutathione (glutathione peroxidase – GPX, glutathione reductase – GRD), and polyamines (PAs) (polyamine oxidase – PAO, diamine oxidase – DAO). In the aquatic fern *Azolla filiculoides*, the accumulation of lead in vacuoles has been found to be accompanied with the enhanced activity of tonoplast V-H^{+}-ATPase (Oren-Benaroya et al. 2004). Also, the ability to accumulate proline, being a result of enhanced 1-pyrroline-5-carboxylate-pyrroline-5-carboxylate synthetase (P5CS), as observed in salt-tolerant *Najas* sp., has been suggested to play a role in the general resistance to metal stress in aquatic macrophytes (Rout and Shaw 1998). There is an increasing number of studies confirming the role of antioxidants and the enzymes involved in their metabolism as well as other enzymes in the tolerance of aquatic plants to stress caused by heavy metals. Selected examples of the recent relevant literature are given in Table 18.1.

18.4.4 Translocation

In terrestrial plants, metal ions taken up by the root system could be effectively translocated to other organs via the xylem (Pich and Scholz 1996). Metal translocation may be responsible for a different distribution and final deposition of toxicants within the plant organism (Fritioff and Greger 2007). Accordingly, the root systems of aquatic plants play an important role in the uptake of elements (Jackson 1998), but mobility may be dependent on both the particular metal and the macrophyte species. Acropetal translocation of several elements (As, Cd, Cu, Pb, and Zn) has been observed in two *Potamogeton* species: *P. pectinatus* and *P. crispus* (Peter et al. 1979; Wolterbeek and van der Meer 2002). An apoplastic route has been proposed for Cd translocation in *Elodea canadensis* (Fritioff and Greger 2007). On the other hand, no translocation of Cu from root to other parts was found in *Potamogeton natans* (Fritioff and Greger 2006). Similarly, in *Phragmites australis* treated with Cr, Hg, Mn, or Zn, metal concentrations in plant organs decreased in the order root > rhizome > leaf > stem. All four organs showed significant differences in the concentration of these elements, which suggests low metal mobility (Bonanno and Lo Giudice 2010). Recently, it has been shown that both xylem-based acropetal and phloem-based basipetal Cu translocations may function in *Hydrilla verticillata*. However, the copper bioaccumulation factors of *Hydrilla* shoots were significantly higher than those of roots, suggesting a dominant role for acropetal translocation in this plant (Xue et al. 2010). Also, a remarkable translocation of Cd has been recently observed in four emergent wetland plants (*Baumea juncea*, *Baumea articulata*, *Schoenoplectus validus*, and *Juncus subsecundus*) (Zhang et al. 2010). However, detailed mechanisms of the transport of metals in the vascular systems of aquatic plants remain unknown.

18.5 Effect of Metals on Photosynthesis in Aquatic Macrophytes

Photosynthesis, which is the basis of all food chains up to human life, is the site of direct or indirect action of heavy metals (for a review, see Myśliwa-Kurdziel et al. 2002, 2004; Mallick and Rai 2002; DalCorso et al. 2008). As to the direct effects of heavy metals on photosynthesis, damage to the reaction centers of photosystems (PS I and PS II), the water splitting complex, the LHCII antenna complex, cyt b_6f, and other components of the photosynthetic transport chain have been observed. Heavy metals can also decrease photosynthetic efficiency through the inhibition of chlorophyll biosynthesis (for a review, see Myśliwa-Kurdziel and Strzałka 2002b), pigment degradation, metal-induced changes in chloroplast structure as well as water uptake imbalance, stomatal closure, and many others. A decline in the photosynthetic efficiency of plants exposed to different heavy metals consequently results in lower biomass production, plant growth inhibition, and can also be the reason for plant death.

The photosynthetic activity of macrophytes treated by heavy metals has long attracted attention, first because of a great need for sensitive environmental bioindicators, and next, because of the potential use of water plants in phytoremediation (for a review see Ayeni et al. 2010). In both cases, the avoidance of damage to the photosynthetic apparatus by heavy metals as well as easy, fast, and unequivocal detection of injury is the main interest of research conducted all over the world. The best candidates for phytoremediation are plants, which are able to survive and perform photosynthesis together with efficient metal accumulation. Although research on the impact of heavy metals on photosynthesis in metal-polluted plants is conducted by many groups, a comparison of the results obtained in different laboratories is often difficult due to wide variations in experimental conditions, which may involve the observation of whole plants, leaves, or isolated organelles, as well as due to different parameters measured and different techniques applied. Metal concentration and the way it is sourced result in further differences between experimental systems. Selected examples of plant species as well as the methods used in investigations of the influence of heavy metals on photosynthetic efficiency are given in Table 18.2.

There is a growing interest in using the technique of chlorophyll fluorescence measurement in the detection and investigation of the effects of heavy metals on plant physiology (Krause and Weis 1991; Joshi and Mohanty 2004). Among others, it is widely used to study macrophytes. The main advantage of this method is that it is rapid, simple, and noninvasive. Most importantly, recently developed portable instruments enable in situ measurements, and thus the monitoring of metal-induced plant injury becomes possible in real time. Fluorescence originates from chlorophyll in photosynthetic tissues exposed to light and it is widely accepted that the fluorescence emission in vivo amounts to approximately 0.6–3% of the absorbed light (Krause and Weis 1991). Chlorophyll fluorescence is a process competitive to photosynthetic reactions, and therefore, its analysis gives an insight into the physiological state of the photosynthetic apparatus. If dark-adapted leaf is illuminated, one can observe the Kautsky effect, i.e., a rapid increase in fluorescence emission from the minimum (the F_0 level), to the maximum (the F_m value), which is followed by a decrease of fluorescence accompanying the onset of the electron transport activity. The fluorescence F_v/F_m ratio ($F_v = F_m – F_0$), which gives information about the efficiency of the photochemical conversion of absorbed light energy in PS II reaction centers, is the parameter mostly used in studies of aquatic macrophytes. However, in some cases the F_v/F_0 ratio is shown to be more effective than the F_v/F_m ratio in monitoring the development of stress (see table). Appenroth et al. (2001) have investigated the toxic effects of Cr on the photosynthetic activity of *Spirodella polyrhiza* and shown that transient chlorophyll in vivo measurements, namely the O-J-I-P test, enables an indication of a decrease in the number of active reaction centers and damage to the oxygen evolving complex as the main targets of Cr. Several water plant species, known as hyperaccumulators, have the ability to accumulate nonessential metals and, up to a certain metal concentration, apparently

Table 18.2 Representative recent publications about the influence of heavy metal on photosynthesis in water plants

Plant species	Metals	What has been measured	Main observations/conclusions	References
Salvinia natans	Cr	Fluorescence ratio F_v/F_m	"Salvinia possess efficient antioxidant machinery that curtails oxidative stress caused by Cr-rich waste water and protects photosynthetic machinery from damage"	Dhir et al. (2009)
		The activity of ribulose-1,5-biphosphatecarboxylase-oxygenase (Rubisco)		
		Transthylakoidal pH gradient		
		Activities of antioxidant enzymes		
Salvinia minima	Cu	CO_2 assimilation and photosynthetic pigments	Salvinia is a good remediator to remove high Cu concentration	Al-Hamdani and Blair 2004
	1–3 mg/L	Cu uptake		
Callitriche cophocarpa	Cr (VI)	Chlorophyll fluorescence: F_v/F_m	Callitriche is a good candidate for phytoextraction of Cr	Augustynowicz et al. (2010)
		Photosynthetic pigments		
Eichhornia crassipes	Cr (III); Cr(VI) 1 and 10 mM	Chlorophyll fluorescence: F_v/F_m; F_v/F_0	Cr^{3+} was less toxic than Cr^{6+} and, in some cases, even increased photosynthesis and chlorophyll content	Paiva et al. (2009)
		Net photosynthetic rate (Pn), stomatal conductance (gs) and substomatal CO_2 concentration (Ci) photosynthetic pigments	F_v/F_0 ratio was more effective than F_v/F_m ratio in monitoring the development of Cr^{6+}-induced stress	
Eichhornia crassipes	Different metals-studies of polluted river (Brasis) – seasonal changes	Chlorophyll *a* fluorescence parameters: F_0, F_m, F_v/F_m, non-photochemical dissipation (qN and NPQ)	Stress conditions in water hyacinth along the Paraíba do Sul River (PSR) in southeastern Brazil in 2005–2006 was monitored	Lage-Pinto et al. (2008)
		Transmission electron microscopy (TEM)	Even if F_v/F_m ranged between 0.77 and 0.81, which indicated that high maximum quantum yield was maintained and the plants performed normal photosynthesis, some ultrastructural changes in chloroplasts were observed	
			Membrane integrity was maintained, which suggest an adaptation mechanism to the environmental conditions	
Wolffia globosa	Cd, Cr	Biomass production	*W. globosa* is a good Cd accumulator, but moderate Cr accumulator	Boonyapookana et al. (2002)
		Chlorophyll content	*W. globosa* was indicated as potential bioremediator of contaminated aquatic environment	

Spirodela polyrhiza	Cr	Oxygen evolution Chlorophyll fluorescence O-J-I-P test	Indicated the main sites of action of metal	Appenroth et al. (2001)
Spirodela polyrhiza *Lemna minor*	Ni	Electron microscopy for investigation of chloroplast structure Growth rate	Ni treatment resulted in changes of chloroplasts structure to chloro-amyloplasts and amylo-chloroplasts, but not to gerontoplasts The contents of the chlorophylls *a* and *b* decreased strongly; carotenoids level remained approximately constant The observed accumulation of starch without stimulation of the photosynthetic activity indicated less efficient export of carbohydrates out of the plastids *Spirodela* appeared more sensitive to nickel than *Lemna*; however, both species were sensitive, which make them not particularly suitable for phytoremediation but suitable for ecotoxicological testing instead	Appenroth et al. (2010)
Jussiaea rapens	Cd 0–60 mg/L	Chlorophyll fluorescence parameters: F_v/F_m, qP, ETR (electron transport rate)	Cd induced the decrease of the photosynthetic activity mainly due to the damage of Photosystem II Chlorophyll *a*, *b* and total chlorophyll content decreased, but carotenoid content increased	Li et al. (2008)

do not show the symptoms of toxic damage. For example, *Salvinia* sp., a fast growing free-floating aquatic plant, known for its high productivity of biomass, has been shown to accumulate different heavy metals, namely, Cr, Cu, Cd, As, and Pb, and the extent of metal uptake was higher for its higher concentration in water (Nichols et al. 2000; Al-Hamdani and Blair 2004; Hoffmann et al. 2004; Dhir et al. 2009). In the course of the accumulation, the decrease in the content of photosynthetic pigments, both chlorophylls and carotenoids, a lowered CO_2 assimilation, and an inhibition of plant growth were observed (e.g., Nichols et al. 2000; Al-Hamdani and Blair 2004). However, Dhir et al. (2009) showed an increase in the photosynthetic activity of PS I and PS II as well as transthylakoidal pH gradient and a decrease in the activity of ribulose-1,5-biphosphatecarboxylase-oxygenase (Rubisco) when measured for thylakoids isolated from Cr-treated *Salvinia natans*. They also showed only a slight decrease in chlorophyll content and no effect on the in vivo chlorophyll fluorescence parameter (F_v/F_m). From an analysis of cellular antioxidants, they concluded that *Salvinia* has an efficient antioxidant machinery that curtails oxidative stress caused by Cr in wastewater and in this way protects the photosynthetic apparatus from damage (Dhir et al. 2009).

There are many reports on the potential use of *Eichhornia crassipes* (water hyacinth), a floating macrophyte hyperaccumulator species native to South America, for the monitoring of heavy metals in aquatic environments (e.g., Lage-Pinto et al. 2008) as well as for the phytoremediation process (Lu et al. 2004; Paiva et al. 2009). Paiva et al. (2009) have shown that, with a decrease in leaf gas exchange, chlorophyll fluorescence parameters and photosynthetic pigment content in water hyacinth were more sensitive to Cr(VI) than Cr(III). In some cases, Cr(III) even increased photosynthesis and chlorophyll content. The sensitivity and effectiveness of the applied methods and examined parameters in the monitoring of Cr(VI)-induced stress development are also discussed in that work. The indirect influence of metals on photosynthetic activity via alterations of thylakoid stacking was observed in some experiments (Paiva et al. 2009; Lage-Pinto et al. 2008).

Callitriche cophocarpa is a new plant that has recently been recommended as a promising species for Cr(VI) bioremediation (Augustynowicz et al. 2010). The fluorescence F_v/F_m ratio was not affected in plants treated with 50 and 100 μM Cr (VI), although in the latter a fast recovery of the F_v/F_m ratio after an initial decrease was observed. Photosynthetic activity was completely inhibited after 8 days of treatment with 700 μM Cr(VI).

In many experiments, the inhibition of photosynthetic activity was accompanied by a decrease in chlorophyll and carotenoid content, and chlorophyll *a* seemed to be more sensitive than chlorophyll *b* (Vajpayee et al. 2000; Paiva et al. 2009; Augustynowicz et al. 2010). It has also been reported that photosynthesis in aquatic plants was also affected when Mg^{2+} in chlorophyll molecules was substituted by heavy metals, such as mercury, copper, cadmium, nickel, zinc, and lead (Küpper et al. 1996, 1998). Such substitution was observed for chlorophyll in the LHCII complex under low light intensity (Kupper et al. 2002).

It should be emphasized that photosynthetic activity or the lack of it can itself change the bioavailability of heavy metals in a water environment. There are well-known diurnal changes in CO_2 and carbonate concentrations in water, correlating with changes in pH in the course of photosynthesis. When photosynthesis operates, CO_2 and bicarbonate are consumed and water alkalinity increases. The opposite process is observed at night, when all organisms respirate. Most heavy metals (e.g., Cu, Zn, Pb, and Cd) are more soluble in low pH, so the toxicity of these metals decreases during periods of light. Additionally, extremely active photosynthesis in calcium-rich water leads to marl formation and insoluble forms of heavy metal coprecipitation, but also to phosphate coprecipitation (Otsuki and Wetzel 1972), which can profoundly change nutritive and toxicological conditions for the long period. On the other hand, the main external product of photosynthesis is oxygen, so photosynthesis makes the environment more oxidative, which can change the redox state of redox-active metals such as Fe or Mn and, in consequence, their toxicity. Wurts and Durborow (1992) provide a compact review of the interdependences of pH, CO_2, alkalinity, hardness, and gas exchange in aquatic environment. In the longer timescale, heavy metal bioavailability in water can change with periods of winter and summer stagnation and spring and autumn circulation (for review, see e.g., Lithner et al. 2000).

18.6 Polymetallic Contamination

Sources of monometallic environmental contamination are rare and have a predominantly local character, where there are mainly plants using specific metal salts for technological processes like tanneries (Cr), some wood impregnation plants (Cu, As, and Cr), and the car battery industry (Pb). Most pollution, however, is emitted from polymetallic sources, among which the mining and smelting industries have the greatest impact. So heavy metal effects on water plants should be investigated not only for particular elements but also for combinations of some metals and for groups of elements, which are geochemically and/or technologically connected. The most important groups are the "calamine" or Zn–Pb-industry group, where the main toxic elements are Zn, Pb, and Cd, often with significant additions of Tl, Ag, and other (Tremel et al. 1997; Wierzbicka et al. 2004; Aravind and Prasad 2004), and the group connected with copper ores, which also contains Zn, Pb, Cr, Mn, As, Sb, and traces of other metals (e.g., Chojnacka et al. 2005). The bioavailability of particular elements for plants in water ecosystems is very complicated and depends on many abiotic and biotic factors, often interrelated (see e.g., Weis and Weis 2004; Butler 2006). Mathematical and statistical models, which can be used for theoretical analyses of this phenomenon, were described by Samanta (2010) and Ince et al. (1999), respectively. The main metals, for example, whose excess can toxically act on plants in water contaminated by Pb–Zn industry, are zinc and cadmium, but not lead, which in natural conditions is usually cumulated mainly in sediments, because of its low solubility. Particular elements can act additively, synergically, or antagonistically, and the reaction of plants can depend on concentrations, concentration ratios, plant species, and environmental conditions.

In aquatic plants, one of the most comprehensive investigations of a two-metal effect is the protection of *Ceratophyllum demersum* (Coontail) by zinc ions during acute cadmium stress, as described in a set of papers by Aravind, Prasad, and coworkers. Part of it is summarized in a review paper of Aravind and Prasad (2005a). They show the ability of Zn^{2+} to alleviate the toxic effect of Cd^{2+} on the chloroplast membrane structure, pigment biosynthesis, activity of photosystems, photosynthetic electron transport and net photosynthesis rate (Aravind and Prasad 2004), energy transfer in photosystem II (Malec et al. 2008), as well as on the carbonic anhydrase conformation (Aravind and Prasad 2005b). Plants treated with both Cd and Zn ions showed significantly lower Cd and higher Zn uptake than those treated with the same Cd concentration only. Cd-induced oxidative stress symptoms, such as lipid peroxidation, lipoxygenase activity, and electrical conductivity, were efficiently reduced by Zn addition. On the other hand, the activity of antioxidant enzymes such as superoxide dismutase, catalase, ascorbate peroxidase, and guaiacol peroxidase showed a very high increase in Cd + Zn-treated plants as compared to Cd- or Zn-only treated ones. Interestingly, plants treated with only Zn concentrations showed neither high Zn uptake nor increase of antioxidant enzyme activity (Aravind and Prasad 2003).

The content of antioxidants, such as thiols, ascorbate, and glutathione, was decreased by Cd treatment, but Zn supplementation restored its level by enhancing the activity of enzymes of the ascorbate–glutathione cycle. This phenomenon has not been observed under only Zn treatment (Aravind and Prasad 2005c). In plants subjected to Cd stress and supplemented with zinc, a higher antioxidative status resulted in a significantly lower level of reactive oxygen species, such as superoxide anion-radical, hydroxyl radical, and H_2O_2. Also, in these plants, protein and DNA damages were remarkably lower compared to Cd-only treated material (Aravind et al. 2009). The heavy metal stress-induced formation of glutathione was significantly increased by Zn supplementation unlike the organic acids-mediated chelation mechanism, which seemed to be insensitive to Zn (Aravind and Prasad 2005d).

In the same *Ceratophyllum demersum*, Bunluesin et al. (2007) have observed a decrease in Cd accumulation in solutions with a high Zn/Cd ratio, whereas Zn accumulation decreased in the presence of Cd. A full comparison of these results with these obtained by Aravind and coworkers (Aravind and Prasad 2003, 2004, 2005a, b, c, d; Aravind et al. 2009) is impossible, however, because of different conditions of plant growth and treatment procedures. Similar results concerning Zn and Cd uptake by *Eichhornia crassipes* (water hyacinth) were obtained by Hasan et al. (2007). These authors observed that the accumulation of both Zn and Cd with an admixture was lower than with solutions of Cd or Zn only.

The set of Aravind and coworker publications (Aravind and Prasad 2003, 2004, 2005a, b, c, d; Aravind et al. 2009) is a unique analysis of different aspects of two heavy metal interactions in water plant. Summarizing these data, it seems that the decrease in cadmium toxicity by zinc is caused by an interplay of the following mechanisms: (a) cadmium's toxicity can be induced by Cd substitution of Zn in Zn-containing active sites. A Zn excess counteracts this effect; (b) a high concentration of both metals seems to act synergistically to the activation of some plant defense systems against heavy metal action, e.g., the antioxidative system. In the presence of both Cd and Zn, these systems could be more efficient than in plants under Cd stress only because of lower metabolic disturbance; (c) in some cases (as in the system investigated by Aravind and coworkers), the intensive accumulation of Zn results in the increase of Zn/Cd ratio and, in consequence, in reduction of the toxic action of cadmium.

Recently, a short study on the oxidative stress caused by the acute excess of Cu and/or Zn in *Spirodela polyrhiza* was published (Upadhyay and Panda 2010). These authors observed that the equimolar addition of Zn to Cu at high concentration (100 μM) caused a significant decrease in Cu-induced oxidative stress symptoms, namely in the concentrations of MDA, H_2O_2, and superoxide as well as a decline in the lipoxygenase activity. Activities of antioxidative enzymes (CAT, POD, and AsA POD) were strongly elevated in these plants. Interestingly, this effect correlated with an increase in both photosynthetic pigment and biomass production. Contrary to this, Zn alone caused a slight increase in oxidative stress symptoms. The induction of antioxidant system was observable only at high Zn concentrations (50–100 μM). With medium and low equimolar concentrations (10–50 μM) of both metals, a very complicated interplay was observed. These results apparently show that Cu and Zn in equimolar concentrations act synergistically in lower and antagonistically in higher concentrations. The pattern of action of the same pair of ions on *Lemna minor* (Dirilgen and Inel 1994) was different – at low concentrations (0.1–0.5 ppm) they act antagonistically to plant growth rate and dry to fresh weight ratio, whereas at higher concentrations (up to 2 ppm) they operated independently.

As far as we know, the results presented above are the only detailed investigations of effects of two metal interactions in aquatic plants. Other information on this topic is very fragmentary. For example, Charles et al. (2006) analyzed the toxicity of copper and uranium to *Lemna aequinoctialis*, by characterizing concentrations inhibiting growth for 50% and concentrations inducing minimum detectable toxic effects for Cu, U, and the combination of both. They concluded that these metals applied in combination are less toxic than individually.

The uptake and organ distribution of the four heavy metals Zn, Cu, Cd, and Pb in individual and mixed solutions in *Potamogeton natans* were analyzed by Fritioff and Greger (2006). The uptake of each metal into roots, shoots, and leaves was the same in both individual and mixed metal solutions, with the exception of Cu and Cd. Cu applied in mixed solutions had double the concentration in leaves. The accumulation of Cd in roots decreased more than half when Cd was applied in mixed solutions.

Recently, Yan et al. (2010) have published a study on the competitive effect of Cu^{2+} and Zn^{2+} on the biosorption of Pb^{2+} by *Myriophyllum spicatum.* Along with a detailed analysis of Pb biosorption equilibria in different conditions and an FT-IR analysis of Pb^{2+} binding in the biological matrix, they stated that the amount of Pb^{2+} sorbed is significantly decreased in the presence of copper and zinc ions, the former having the stronger effect.

Finally, it is necessary to point out that some aquatic plants in the presence of high Fe^{2+} concentration in substrate can form an iron oxyhydroxide root plaque as an effect of oxidation of ferrous to ferric ions by oxygen that is radially excreted from roots (ROL – see above). This plaque was described as the place where Mn and Zn ions could be immobilized (St-Cyr and Campbell 1996), or a buffer for Ca, Cu, Mn, Zn, and P (Jiang et al. 2009 and references therein). The role of this plaque in root-sediment (soil) ion exchange seems to be different for particular ions, plant species, and environmental conditions.

18.7 Conclusion

The data presented above indicate that aquatic macrophytes are organisms that may affect the cycling of heavy metals in wetland ecosystems.

Water plants possess a capacity to absorb and bioconcentrate heavy metals both from soil/sediment and/or from the surrounding water column via different mechanisms. These include cell wall binding, ion exchange-based sorption, active transport, chelation, translocation, and deposition in different organs. These processes lead eventually to the immobilization of metals in the plant biomass. Moreover, the presence of metabolic adaptations, including scavengers of ROS, antioxidants and their modifying enzymes, and specific ion transporters, allows many species of aquatic plant to survive and grow in elevated concentrations of metals. Additionally, the nonliving biomass of some wetland plants could still bind metals from a water environment. Therefore, wetland sediments are usually considered as sinks for metals. In their anoxic zone, they may contain high concentrations of reduced metals (Weis and Weis 2004).

On the other hand, aquatic macrophytes may influence many physical and chemical processes in their environment. In particular, photosynthetic activity may alter the bioavailability of heavy metals in wetland ecosystems. The *R*adial *O*xygen *L*oss (ROL) resulting in partial oxygenation of the rhizosphere may lead to profound changes in metal mobility in rhizosphere soils (Yang and Ye 2009).

Finally, competitive or synergistic effects of the elements available to aquatic macrophytes in a particular ecosystem may greatly influence the circulation and deposition of heavy metals in the environment.

Metals are important components of the biosphere. Many metals are micronutrients (e.g., Cu, Fe, Mn, Ni, and Zn) as they are involved in metabolic processes as the cofactors of enzymes in plants. At elevated concentrations, derived from e.g., anthropogenic sources, metals became toxicants causing hazardous pollution to ecosystems on a global scale (Mohan and Hosetti 1999). The use of aquatic macrophytes in the phytroremediation of metal-polluted areas is currently attracting considerable interest (Prasad 2003, 2004). Therefore, a detailed knowledge of the molecular mechanisms of the uptake, tolerance, and transport of heavy metals through aquatic plants is essential both for an understanding of the functioning of wetland ecosystems and for the development of specific phytoremediation technologies.

Acknowledgments This paper was prepared in the frames of European Regional Development Fund: the Polish Innovation Economy Operational Program (contract No. POIG.02.01.00-12-167/08, project Małopolska Centre of Biotechnology) and was financially supported by the Statutory Funds of the Faculty of Biochemistry, Biophysics and Biotechnology, Jagiellonian University.

References

Al-Hamdani SH, Blair SL (2004) Influence of copper on selected physiological responses in *Salvinia minima* and its potential use in copper remediation. Am Fern J 94:47–56

Appenroth KJ, Krech K, Keresztes Á, Fischer W, Koloczek H (2010) Effects of nickel on the chloroplasts of the duckweeds *Spirodela polyrhiza* and *Lemna minor* and their possible use in biomonitoring and phytoremediation. Chemosphere 78:216–223

Appenroth KJ, Stöckel J, Srivastava A, Strasser RJ (2001) Multiple effects of chromate on the photosynthetic apparatus of *Spirodela polyrhiza* as probed by OJIP chlorophyll *a* fluorescence measurements. Environ Pollut 115:49–64

Aravind P, Prasad MNV (2003) Zinc alleviates cadmium-induced oxidative stress in *Ceratophyllum demersum* L.: a free floating freshwater macrophyte. Plant Physiol Biochem 41:391–397

Aravind P, Prasad MNV (2004) Zinc protects chloroplasts and associated photochemical functions in cadmium exposed *Ceratophyllum demersum* L., a freshwater macrophyte. Plant Sci 166:1321–1327

Aravind P, Prasad MNV (2005a) Cadmium–zinc interactions in a hydroponic system using *Ceratophyllum demersum* L.: adaptive ecophysiology, biochemistry and molecular toxicology. Braz J Plant Physiol 17:3–20

Aravind P, Prasad MNV (2005b) Zinc mediated protection to the conformation of carbonic anhydrase in cadmium exposed *Ceratophyllum demersum* L. Plant Sci 169:245–254

Aravind P, Prasad MNV (2005c) Modulation of cadmium-induced oxidative stress in *Ceratophyllum demersum* by zinc involves ascorbate–glutathione cycle and glutathione metabolism. Plant Physiol Biochem 43:107–116

Aravind P, Prasad MNV (2005d) Cadmium-induced toxicity reversal by zinc in *Ceratophyllum demersum* L. (a free floating aquatic macrophyte) together with exogenous supplements of amino- and organic acids. Chemosphere 61:1720–1733

Aravind P, Prasad MNV, Malec P, Waloszek A, Strzałka K (2009) Zinc protects *Ceratophyllum demersum* L. (free-floating hydrophyte) against reactive oxygen species induced by cadmium. J Trace Elem Med Biol 23:50–60

Armstrong W (1979) Aeration in higher plants. Adv Bot Res 7:226–332

Augustynowicz J, Grosicki M, Hanus-Fajerska E, Lekka M, Waloszek A, Kołoczek H (2010) Chromium(VI) bioremediation by aquatic macrophyte *Callitriche cophocarpa* Sendtn. Chemosphere 79:1077–1083

Ayeni OO, Ndakidemi PA, Snyman RG, Odendaal JP (2010) Chemical, biological and physiological indicators of metal pollution in wetlands. Sci Res Essays 5:1938–1949

Basiouny FM, Haller WT, Garrard LA (1977) Evidence for root Fe nutrition in *Hydrilla verticillata* Royle. Plant Soil 48:621–627

Berner RA (1984) Sedimentary pyrite formation: an update. Geochim Cosmochim Acta 48:605–615

Bleuel C, Wesenberg D, Sutter K, Miersch J, Braha B, Bärlocher F (2005) The use of the aquatic moss *Fontinalis antipyretica* L. ex Hedw. as a bioindicator for heavy metals: 3. Cd^{2+} accumulation capacities and biochemical stress response of two *Fontinalis* species. Sci Total Environ 345:13–21

Blum WEH (2005) Do we need a journal of soils and sediments (Part 2)? J Soils Sediments 5:195–196

Bonanno G, Lo Giudice R (2010) Heavy metal bioaccumulation by the organs of *Phragmites australis* (common reed) and their potential use as contamination indicators. Ecol Indicat 10:639–645

Boonyapookana B, Upatham E, Kruatrachue M, Pokethitiyook P, Singhakaew S (2002) Phytoaccumulation and phytotoxicity of cadmium and chromium in duckweed *Wolffia globosa*. Int J Phytoremediation 4:87–100

Boulegue J, Lord CJ, Church TM (1982) Sulfur speciation and associated trace metals (Fe, Cu) in the pore waters of Great Marsh, Delaware. Geochim Cosmochim Acta 46:453–464

Bruns I, Sutter K, Menge S, Neumann D, Krauss GJ (2001) Cadmium lets increase the glutathione pool in bryophytes. J Plant Physiol 158:79–89

Bunluesin S, Pokethitiyook P, Lanza GR, Tyson JF, Kruatrachue M, Xing B, Upatham S (2007) Influences of cadmium and zinc interaction and humic acid on metal accumulation in *Ceratophyllum demersum*. Water Air Soil Pollut 180:225–235

Butler TW (2006) Geochemical and biological controls on trace metal transport in an acid mine impacted watershed. Environ Geochem Health 28:231–241

Chandra P, Kulshreshtha K (2004) Chromium accumulation and toxicity in aquatic vascular plants. Bot Rev 70:313–327

Charles AL, Markich SJ, Ralph P (2006) Toxicity of uranium and copper individually, and in combination, to a tropical freshwater macrophyte (*Lemna aequinoctialis*). Chemosphere 62:1224–1233

Choi JH, Park SS, Jaffe PR (2006) The effect of emergent macrophytes on the dynamics of sulfur species and trace metals in wetland sediments. Environ Pollut 140:286–293

Chojnacka K, Chojnacki A, Górecka H, Górecki H (2005) Bioavailability of heavy metals from polluted soils to plants. Sci Total Environ 337:175–182

Cobbett C, Goldsbrough P (2002) Phytochelatins and metallothioneins: roles in heavy metal detoxification and homeostasis. Annu Rev Plant Biol 53:159–182

Dacey JW (1980) Internal winds in the water-lilies: adaptation for life in anaerobic sediments. Science 210:1017–1019

DalCorso G, Farinati S, Maistri S, Furini A (2008) How plants cope with cadmium: staking all on metabolism and gene expression. J Integr Plant Biol 50:1268–1280

Dazy M, Masfaraud JF, Férard JF (2009) Induction of oxidative stress biomarkers associated with heavy metal stress in *Fontinalis antipyretica* Hedw. Chemosphere 75:297–302

Dhir B, Sharmila P, PardhaSaradhi P, Abdul Nasim S (2009) Physiological and antioxidant responses of *Salvinia natans* exposed to chromium-rich wastewater. Ecotoxicol Environ Saf 72:1790–1797

Ding B, Shi G, Xu Y, Hu J, Xu Q (2007) Physiological responses of *Alternanthera philoxeroides* (Mart.) Griseb leaves to cadmium stress. Environ Pollut 147:800–803

Dirilgen N, Inel Y (1994) Effects of zinc and copper on growth and metal accumulation in duckweed, *Lemna minor*. Bull Environ Contam Toxicol 53:442–449

Di Toro DM, Mahony JD, Hansen DJ, Scott KJ, Hicks MB, Mayr SM, Redmond MS (1990) Toxicity of cadmium in sediments: the role of acid volatile sulfide. Environ Toxicol Chem 9:1487–1502

Dogan M, Saygideger SD, Colak U (2009) Effect of lead toxicity on aquatic macrophyte *Elodea canadensis* Michx. Bull Environ Contam Toxicol 83:249–254

Du Laing G, Rinklebe J, Vandecasteele B, Tack FMG (2009) Trace metal behaviour in estuarine and riverine floodplain soils and sediments: a review. Sci Total Environ 407:3972–3985

Elifantz H, Tel-Or E (2002) Heavy metal biosorption by plant biomass of the macrophyte. Water Air Soil Pollut 141:207–218

Emerson S, Jacobs L, Tebo B (1983) The behavior of trace metals in marine anoxic waters: solubilities at the oxygen-hydrogen sulphide interface. In: Wong CS, Boyle E, Bruland KW, Burton JD, Goldberg ED (eds) Trace metals in sea water. Plenum Press, New York, pp 579–608

Fritioff A, Greger M (2006) Uptake and distribution of Zn, Cu, Cd, and Pb in an aquatic plant *Potamogeton natans*. Chemosphere 63:220–227

Fritioff A, Greger M (2007) Fate of cadmium in *Elodea canadensis*. Chemosphere 67:365–375

Gross MG (1978) Effects of waste disposal operations in estuaries and the coastal ocean. Annu Rev Earth Planet Sci 6:127–143

Gutknecht JLM, Goodman RM, Balser TC (2006) Linking soil process and microbial ecology in freshwater wetland ecosystems. Plant Soil 289:17–34

Hall JL (2002) Cellular mechanisms for heavy metal detoxification and tolerance. J Exp Bot 53:1–11

Hasan SH, Talat M, Rai S (2007) Sorption of cadmium and zinc from aqueous solutions by water hyacinth (*Eichchornia crassipes*). Bioresour Technol 98:918–928

Hinman ML, Klaine SJ (1992) Uptake and translocation of selected organic pesticides by the rooted aquatic plant *Hydrilla verticillata* Royle. Environ Sci Technol 26:609–613

Hoffmann T, Kutter C, Santamaria JM (2004) Capacity of *Salvinia minima* Baker to tolerate and accumulate As and Pb. Eng Life Sci 4:61–65

Howarth RW, Jørgensen BB (1984) Formation of 35 S-labelled elemental sulfur and pyrite in coastal marine sediments (Limfjorden and Kysing Fjord, Denmark) during short-term $^{35S}SO_4$ reduction measurements. Geochim Cosmochim Acta 48:1807–1818

Huerta-Diaz MA, Tessier A, Carignan R (1998) Geochemistry of trace metals associated with reduced sulfur in freshwater sediments. Appl Geochem 13:213–233

Ince NH, Dirilgen N, Apikyan IG, Tezcanli G, Ustun B (1999) Assessment of toxic interactions of heavy metals in binary mixtures: a statistical approach. Arch Environ Contam Toxicol 36:365–372

Jackson LJ (1998) Paradigms of metal accumulation in rooted aquatic vascular plants. Sci Total Environ 219:223–231

Jiang FY, Chen X, Luo AC (2009) Iron plaque formation on wetland plants and its influence on phosphorus, calcium and metal uptake. Aquat Ecol 43:879–890

Joshi MK, Mohanty P (2004) Chlorophyll a fluorescence as a probe of heavy metal ion toxicity in plants. In: Papageorgiou GC, Govindjee (eds) Chlorophyll *a* fluorescence, a signature of photosynthesis. Springer, New York

Kabata-Pendias A (2001) Trace elements in the soil and plants. CRC Press, Boca Raton

Kanoun-Boulé M, Vicente JF, Nabais C, Prasad MNV, Freitas H (2009) Ecophysiological tolerance of duckweeds exposed to copper. Aquat Toxicol 91:1–9

Knox AS, Paller MH, Nelson EA, Specht WL, Halverson NV, Gladden JB (2006) Metal distribution and stability in constructed wetland sediment. J Environ Qual 35:1948–1959

Kosolapov DB, Kuschk P, Vainshtein MB, Vatsourina AV, Wiebner A, Kasterner M, Miler RA (2004) Microbial processes of heavy metal removal from carbon deficient effluents in constructed wetlands. Eng Life Sci 4:403–411

Krause GH, Weis E (1991) Chlorophyll fluorescence and photosynthesis: the basics. Annu Rev Plant Physiol Plant Mol Biol 42:313–349

Küpper H, Küpper F, Spiller M (1996) Environmental relevance of heavy metal substituted chlorophylls using the example of water plants. J Exp Bot 47:259–266

Küpper M, Küpper F, Spiller M (1998) In situ detection of heavy metal substituted chlorophylls in water plants. Photosynth Res 58:123–133

Kupper H, Setlik I, Spiller M, Kupper FC, Prasil O (2002) Heavy metal-induced inhibition of photosynthesis: targets of in vivo heavy metal chlorophyll formation. J Phycol 38:429–441

Lage-Pinto F, Oliveira JG, Da Cunha M, Souza CMM, Rezende CE, Azevedo RA, Vitória AP (2008) Chlorophyll *a* fluorescence and ultrastructural changes in chloroplast of water hyacinth as indicators of environmental stress. Environ Exp Bot 64:307–313

Lefroy RDB, Chaitep W, Blair GJ (1994) Release of sulfur from rice residues under flooded and non-flooded soil conditions. Aust J Agric Res 45:657–667

Lerman A (1978) Chemical exchange across sediment–water interface. Annu Rev Earth Planet Sci 6:281–306

Li M, Zhang LJ, Tao L, Li W (2008) Ecophysiological responses of *Jussiaea rapens* to cadmium exposure. Aquat Bot 88:347–352

Lithner G, Borg H, Ek J, Fröberg E, Holm H, Johansson AM, Kärrhage P, Rosén G, Söderström M (2000) The turnover of metals in a eutrophic and an oligotrophic lake in Sweden. AMBIO: J Human Environ 29:217–229

Lu X, Kruatrachue M, Pokethitiyook P, Homyok K (2004) Removal of cadmium and zinc by water hyacinth *Eichhornia crassipes*. Sci Asia 30:93–103

Malec P, Maleva M, Prasad MNV, Strzałka K (2009a) Copper toxicity in leaves of *Elodea canadensis* Michx. Bull Environ Contam Toxicol 82:627–632

Malec P, Maleva MG, Prasad MNV, Strzałka K (2009b) Identification and characterization of Cd-induced peptides in *Egeria densa* (water weed): putative role in Cd detoxification. Aquat Toxicol 95:213–221

Malec P, Maleva MG, Prasad MNV, Strzałka K (2010) Responses of *Lemna trisulca* L. (Duck-weed) exposed to low doses of cadmium: thiols, metal binding complexes, and photosynthetic pigments as sensitive biomarkers of ecotoxicity. Protoplasma 240:69–74

Malec P, Waloszek A, Prasad MNV, Strzałka K (2008) Zinc reversal of cadmium-induced energy transfer changes in photosystem II of *Ceratophyllum demersum* L. as observed by whole-leaf 77 K fluorescence. Plant Stress 2:121–126

Maleva MG, Nekrasova GF, Malec P, Prasad MNV, Strzałka K (2009) Ecophysiological tolerance of *Elodea canadensis* to nickel exposure. Chemosphere 77:392–398

Mallick N, Rai LC (2002) Physiological responses of non-vascular plants to heavy metals. In: Prasad MNV, Strzalka K (eds) Physiology and biochemistry of metal toxicity and tolerance in plants. Kluwer, The Netherlands, pp 111–147

Mendelssohn IA, Kleiss BA, Wakeley JS (1995) Factors controlling the formation of oxidized root channels a review. Wetlands 15:37–46

Miretzky P, Saralegui A, Fernández-Cirelli A (2006) Simultaneous heavy metal removal mechanism by dead macrophytes. Chemosphere 62:247–254

Mittler R (2002) Oxidative stress, antioxidants and stress tolerance. Trends Plant Sci 7:405–410

Mohan BS, Hosetti BB (1999) Aquatic plants for toxicity assessment. Environ Res 81:259–274

Monferrán MV, Agudo JS, Pignata ML, Wunderlin D (2009) Copper-induced response of physiological parameters and antioxidant enzymes in the aquatic macrophyte *Potamogeton pusillus*. Environ Pollut 157:2570–2576

Myśliwa-Kurdziel B, Prasad MNV, Strzałka K (2002) Consequences of heavy metal exposure to the processes related to the light phase of photosynthesis. In: Prasad MNV, Strzałka K (eds) Physiology and biochemistry of metal toxicity and tolerance in plants. Kluwer, The Netherlands, pp 229–255

Myśliwa-Kurdziel B, Strzałka K (2002) Influence of metals on the biosynthesis of photosynthetic pigments. In: Prasad MNV, Strzalka K (eds) Physiology and biochemistry of metal toxicity and tolerance in plants. Kluwer, The Netherlands, pp 201–228

Myśliwa-Kurdziel B, Prasad MNV, Strzałka K (2004) Photosynthesis in heavy metal stressed plants. In: Prasad MNV (ed) Heavy metal stress in plants from biomolecules to ecosystems, 2nd edn. Springer, Berlin, pp 462–470, ISBN: 3-540-40131-8

Namieśnik J, Rabajczyk A (2010) The speciation and physico-chemical forms of metals in surface waters and sediments. Chem Spec Bioavailab 22(1):1–21

Nichols PB, Couch JD, Al-Hamdani SH (2000) Selected physiological responses of *Salvinia minima* to different chromium concentrations. Aquat Bot 68:313–319

Nixdorf D, Lessman D, Steinberg CEW (2002) The importance of chemical buffering for pelagic and benthic colonization in acidic waters. Water Air Soil Pollut 3:27–46

Nyquist J, Greger M (2009) A field study of constructed wetlands for preventing and treating acid mine drainage. Ecol Eng 35:630–642

Oren-Benaroya R, Tzin V, Tel-Or E, Zamski E (2004) Lead accumulation in the aquatic fern *Azolla filiculoides*. Plant Physiol Biochem 42:639–645

Otsuki A, Wetzel RG (1972) Coprecipitation of phosphate with carbonates in a marl lake. Limnol Oceanogr 17:763–766

Paiva LB, Gonçalves de Oliveira J, Azevedo RA, Ribeiro DR, da Silva MG, Vitória AP (2009) Ecophysiological responses of water hyacinth exposed to Cr^{3+} and Cr^{6+}. Environ Exp Bot 65:403–409

Peter R, Welsh H, Denny P (1979) Translocation of lead and copper in two submerged aquatic angiosperm species. J Exp Bot 30:339–345

Pich A, Scholz G (1996) Translocation of copper and other micronutrients in tomato plants (*Lycopersicon esculentum* Mill.): nicotianamine-stimulated copper transport in the xylem. J Exp Bot 47:41–47

Prasad MNV (1995) Cadmium toxicity and tolerance in vascular plants. Environ Exp Bot 35:525–545

Prasad MNV (2003) Phytoremediation of metal-polluted ecosystems: hype for commercialization. Russ J Plant Physiol 50:686–700

Prasad MNV (2004) Phytoremediation of metals and radionuclides in the environment: the case for natural hyperaccumulators, metal transporters soil-amending chelators and transgenic plants. In: Prasad MNV (ed) Heavy metal stress in plants from biomolecules to ecosystems, 2nd edn. Springer, Berlin, pp 345–391, ISBN: 3-540-40131-8

Radić S, Babić M, Skobić D, Roje V, Pevalek-Kozlina B (2010) Ecotoxicological effects of aluminum and zinc on growth and antioxidants in *Lemna minor* L. Ecotoxicol Environ Saf 73:336–342

Rauser WE (1999) Structure and function of metal chelators produced by plants: the case for organic acids, aminoacids, phytin, and metallothioneins. Cell Biochem Biophys 31:19–48

Rinklebe J, Franke Ch, Neue HU (2007) Aggregation of floodplain soils based on classification principles to predict concentrations of nutrients and pollutants. Geoderma 141:210–223

Rout N, Shaw B (1998) Salinity tolerance in aquatic macrophytes: probable role of proline, the enzymes involved in its synthesis and C4 type of metabolism. Plant Sci 136:121–130

Samanta GP (2010) A two-species competitive system under the influence of toxic substances. Appl Math Comput 216:291–299

Samardakiewicz S, Strawiński P, Woźny A (1996) The influence of lead on callose formation in roots of *Lemna minor* L. Biol Plantarum 38:463–467

Samardakiewicz S, Woźny A (2000) The distribution of lead in duckweed (*Lemna minor* L.) root tip. Plant Soil 226:107–111

Sanità di Toppi L, Vurro E, Rossi L, Marabottini R, Musetti R, Careri M (2007) Different compensatory mechanisms in two metal-accumulating aquatic macrophytes exposed to acute cadmium stress in outdoor artificial lakes. Chemosphere 68:769–780

Schneider IAH, Rubio J (1999) Sorption of heavy metal ions by the nonliving biomass of freshwater macrophytes. Environ Sci Technol 33:2213–2217

Schor-Fumbarov T, Goldsbrough PB, Adam Z, Tel-Or E (2005) Characterization and expression of a metallothionein gene in the aquatic fern *Azolla filiculoides* under heavy metal stress. Planta 223:69–76

Simpson SL, Apte SC, Batley GE (1998) Effect of short-term resuspension events on trace metal speciation in polluted anoxic sediments. Environ Sci Technol 32:620–625

St-Cyr L, Campbell PGC (1996) Metals (Fe, Mn, Zn) in the root plaque of submerged aquatic plants collected in situ: relations with metal concentrations in the adjacent sediments and in the root tissue. Biogeochemistry 33:45–76

Suñe N, Sánchez G, Caffaratti S, Maine M (2007) Cadmium and chromium removal kinetics from solution by two aquatic macrophytes. Environ Pollut 145:467–473

Tkalec M, Prebeg T, Roje V, Pevalek-Kozlina B, Ljubešić N (2008) Cadmium-induced responses in duckweed *Lemna minor* L. Acta Physiol Plantarum 30:881–890

Tremel A, Masson P, Garraud H, Donard OFX, Baize D, Mench M (1997) Thallium in French agrosystems II. Concentration of thallium in field-grown rape and some other plant species. Environ Pollut 97:161–168

Upadhyay RK, Panda SK (2010) Zinc reduces copper toxicity induced oxidative stress by promoting antioxidant defense in freshly grown aquatic duckweed *Spirodela polyrhiza* L. J Hazard Mater 175:1081–1084

Urban NR, Brezonik PL, Baker LA, Sherman LA (1994) Sulfate reduction and diffusion in sediments of Little Rock Lake, Wisconsin. Limnol Oceanogr 39:797–815

Vajpayee P, Tripathi RD, Rai UN, Ali MB, Singh SN (2000) Chromium (VI) accumulation reduces chlorophyll biosynthesis, nitrate reductase activity and protein content in *Nymphaea alba* L. Chemosphere 41:1075–1082

Weis JS, Weis P (2004) Metal uptake, transport and release by wetland plants: implications for phytoremediation and restoration. Environ Int 30:685–700

Wierzbicka M, Szarek-Łukaszewska G, Grodzińska K (2004) Highly toxic thallium in plants from the vicinity of Olkusz (Poland). Ecotoxicol Environ Saf 59:84–88

Wind T, Conrad R (1995) Sulfur compounds, potential turnover of sulfate and thiosulfate, and numbers of sulfate-reducing bacteria in planted and unplanted paddy soil. FEMS Microbiol Ecol 18:257–266

Wolterbeek HT, van der Meer AJGM (2002) Transport rate of arsenic, cadmium, copper and zinc in *Potamogeton pectinatus* L.: radiotracer experiments with As-76 Cd-109, Cd- 115, Cu-64 and Zn-65, Zn-69. Sci Total Environ 287:13–30

Wurts WA, Durborow RM (1992) Interactions of pH, carbon dioxide, alkalinity and hardness in fish ponds; SRAC Publication No, 464. Southern Regional Aquaculture Center, Princeton

Xu QS, Hu JZ, Xie KB, Yang HY, Du KH, Shi GX (2010) Accumulation and acute toxicity of silver in *Potamogeton crispus* L. J Hazard Mater 173:186–193

Xue P, Li G, Liu W, Yan C (2010) Copper uptake and translocation in a submerged aquatic plant *Hydrilla verticillata* (L.f.) Royle. Chemosphere 81:1098–1103

Yan Ch, Li G, Xue P, Wei Q, Li Q (2010) Competitive effect of Cu(II) and Zn(II) on the biosorption of lead(II) by *Myriophyllum spicatum*. J Hazard Mater 179(1–3):721–728

Yang H, Shi G, Wang H, Xu Q (2010) Involvement of polyamines in adaptation of *Potamogeton crispus* L. to cadmium stress. Aquat Toxicol 100:282–288

Yang J, Ye Z (2009) Metal accumulation and tolerance in wetland plants. Front Biol China 4:282–288

Zhang Z, Rengel Z, Meney K (2010) Cadmium accumulation and translocation in four emergent wetland species. Water Air Soil Pollut 212:239–249

Chapter 19
Role of Plant Growth Promoting Bacteria and Fungi in Heavy Metal Detoxification

Sema Camci Cetin, Ayten Karaca, Ridvan Kizilkaya, and Oguz Can Turgay

19.1 Introduction

Metals are natural parts of soils and many micronutrients including metals are required for plant growth (Sheng et al. 2008). Heavy metal ions such as Cu, Zn, Fe, Mn, Ni, and Co are essential elements, whereas Cd, Hg, Ag, and Pb are nonessential and extremely toxic elements (Williams et al. 2000; Gadd 1990). Due to industrialization and technical developments, the environment has been polluted with heavy metal by humankind (Madhaiyan et al. 2007). Unfortunately, ecosystems are still being polluted with heavy metals (Khan et al. 2000). Heavy metals are defined as elements that have an atomic numbers >20 (Jing et al. 2007) or metallic elements have specific mass higher than 5 g cm^{-3} (Gadd and Griffiths 1978). Cd, Cr, Cu, Hg, Pb, and Ni are the most common heavy metals (Jing et al. 2007).

Nowadays, metal pollution is considered as the most severe environmental problem (Jing et al. 2007; Gamalero et al. 2009; Abou-Shanab et al. 2006). Ecosystems have been contaminated with heavy metals due to different human and natural activities (Khan et al. 2000). The sources of heavy metal in the environment are: mining and smelting of metals, burning of fossil fuels, fertilizers, sewage sludge, pesticides, municipal wastes, mining, smelting, pigments, spent batteries, and vehicle exhaust (Vivas et al. 2003; Leyval et al. 1997; Denton 2007). Heavy metals are used as plant nutrients in adequate levels; however, excessive

S.C. Cetin
Faculty of Foresty, Cankiri Karatekin University, Cankiri, Turkey

A. Karaca (✉) • O.C. Turgay
Faculty of Agriculture, Department of Soil Science and Plant Nutrition, Ankara University, Diskapi, Ankara, Turkey
e-mail: akaraca@agri.ankara.edu.tr

R. Kizilkaya
Faculty of Agriculture, Department of Soil Science and Plant Nutrition, Ondokuz Mayis University, Samsun, Turkey

I. Sherameti and A. Varma (eds.), *Detoxification of Heavy Metals*, Soil Biology 30, DOI 10.1007/978-3-642-21408-0_19, © Springer-Verlag Berlin Heidelberg 2011 level of heavy metals is toxic to plants. When heavy metals are found in elevated levels in the environment, they are excessively absorbed by plant roots and transport to shoot and translocated to the shoots, where they influenced metabolism and decreased growth (Jing et al. 2007). Excessive metal concentrations in polluted soils cause a decline in soil microbial activity as well as soil fertility and yield losses (McGrath et al. 1995).

Soil-heavy metals cannot be degraded biologically but only transformed to organic complexes (Rajkumar and Freitas 2008b; Jing et al. 2007; Khan et al. 2009). Heavy metal polluted soil can be treated by chemical, physical, and biological techniques (Khan et al. 2000). In-situ techniques have some advantages such as their lower cost and reduced impact on the ecosystem compared to ex-situ techniques (Khan et al. 2000). These techniques are grouped into two different classes: ex-situ techniques (treatment on- or off-site) and in-situ techniques (treatment in-site) (Khan et al. 2000; Dary et al. 2010). Remediation methods including excavation and landfill, thermal treatment, acid leaching, and electroreclamation are not appropriate for effective remediation of heavy metals because of their high cost, low efficiency etc. (Jing et al. 2007).

Using organisms for treatment of soil pollution is termed bioremediation. An alternative method is called phytoremediation defined as "use metal accumulating plants to remove, transfer and stabilize these contaminants from soil, sediments and water" and it is a natural, clean, and economic alternative for heavy metal treatment (Khan 2005).

Metal accumulator plants are able to tolerate and concentrate high level of heavy metals. The most commonly known heavy metal accumulators are members of *Brassicaceae* family (Jing et al. 2007). *Brassica juncea* called as Indian mustard is the best-known metal-accumulator plant.

Soil can be termed as a complex interactive network (Upadhyay and Srivastava 2010; Jeffries et al. 2003) and contains a vast number of microorganisms. The rhizosphere is the site where most microorganisms thrive and is called as the largest ecosystem on earth. Bacteria, fungi, protozoa, and algae live in the rhizosphere and bacteria are the most abundant organisms among microbial communities (Ahmad et al. 2008). There is an interaction between plant and microorganisms because of rhizosphere effect. Plant root exudates many different attractive compounds into the rhizosphere for holding microbes there. Special bacteria called as plant growth promoting rhizobacteria (PGPR) that are defined as "bacteria inhabiting the rhizosphere and beneficial to plants" (Kloepper et al. 1980). PGPR is about 2–5% of rhizospheric bacteria of soil (Antoun and Prevost 2005).

Successful phytoremediation processes are required using plants, which interact with plant roots and bacteria, and studying adequate concentrations of metals in soil (Rajkumar et al. 2005). Metal-resistant microbes can help detoxification of metals in soil. Another group of microorganisms called mycorrhizae could thrive in heavy metal-contaminated soil and roots of metal accumulator plants. When mycorrhizae colonize to plant roots, it can decrease translocation of heavy metals to shoots by binding of heavy metal to the cell wall of fungal hyphae.

19.2 Heavy Metals as a Soil Pollution Agent

Heavy metals are defined as elements with metallic characteristics and an atomic number >20 (Jing et al. 2007). Heavy metals can be classified into two categories: essential (Cu, Zn, Fe, Mn, Ni, and Co) and nonessential (Cd, Hg, Ag, and Pb) elements (Williams et al. 2000; Gadd 1990). Metals are natural parts of soils (Jing et al. 2007). Heavy metal ions are main component of a variety of enzymes, transcription factors, and other proteins (Williams et al. 2000). However, excessive concentrations of both essential and nonessential heavy metals in soil can cause toxicity symptoms and inhibition of plant growth (Hall 2002). Heavy metal pollution both limits plant establishment and causes declines in numbers of soil microorganisms and their activity (Shetty et al. 1994). Also, accumulation of metals in plant organs in excessive level can limit physiological processes such as photosynthesis and synthesis of chlorophyll pigments (Wani et al. 2007b).

Influences of heavy metals on the microbial community are examined by three different ways: (1) reduction of total microbial biomass, (2) decreasing numbers of specific populations, and (3) shifting microbial community structure (Zhuang et al. 2007). Metal toxicity depends on availability, or shortly bioavailability, which refers to ability of metals from soil to living organism (Leyval et al. 1997). Factors affecting bioavailability of metals in soils are physicochemical (pH, Eh, organic matter, clay content etc.) and biological (biosorption, bioaccumulation, and solubilization) (Leyval et al. 1997). Accumulation of heavy metals by microorganisms is given in Table 19.1.

Soil-bound metals are mobilized into soil solution by three ways: (1) forming metal-chelating molecules (phytosiderophores) and releasing them into the rhizosphere to chelate and solubilize soil-bound metal, (2) decreasing soil-bound metal

Table 19.1 Accumulation of heavy metals by microorganisms (Gadd 1990)

Organism	Element	Uptake (% dry weight)
	Pb	34–40
Citrobacter sp.	Cd	13.5
Thiobacillus ferooxidans	Ag	25
Bacillus cereus	Cd	3.9–8.9
Escherichia coli	Cd	0.16–0.98
	Co	25
	Cu	34
Zoogloea sp.	Ni	13
Chlorella vulgaris	Au	10
Scenedesmus obliquus	Cd	0.3
Phoma sp.	Hg	2
	Cu	1.6
	Cd	3.0
Rhizopus arrhizus	Pb	10.4
	Cd	0.24–3.12
Saccharomyces cerevisiae	Zn	0.45

ions via plasma membrane bound metal reductases, and (3) acidifying soil environment for solubilization of metals (Raskin et al. 1994).

19.3 Phytoremediation

Heavy metal uptake by plants is a less-expensive method (Khan et al. 2000). Bioremediation is defined as "the use of organisms for the treatment of soil pollution" (Leyval et al. 2002). Phytoremediation can be defined as "the use of plants to extract, sequester and/or detoxify pollutants through physical, chemical and biological processes" (Jing et al. 2007; Glick 2003). Phytoremediation is a comparatively new approach, an environmentally friendly technique, and an emerging technology for removing pollutants from the environment (Glick 2003; Sheng et al. 2008; Garbisu and Alkorta 2001). The plants used in phytoremediation are able to tolerate and accumulate high levels of heavy metals (Nie et al. 2002). Main principles of phytoremediation are (1) extraction of pollutants from soil and translocation to shoots, (2) sequestering of pollutant via root system to prevent spreading and leaching into soil or groundwater, and (3) transformation into less toxic chemicals (Kuiper et al. 2004).

Basic phytoremediation processes are: phytoextraction, phytodegradation, rhizofiltration, phytostabilization, and phytovolatilization. Phytoremediation subgroups are defined as follows(Khan 2005): (1) phytoextraction refers to removal and concentration of metals into roots and shoot of plants, (2) phytodegradation covers degradation of contaminants by plants and their relative microbes, (3) rhizofiltration is use of plant roots for absorption of metals, (4) phytostabilization is use of plants for immobilization and reduction in mobility and bioavailability of pollutants, (5) phytovolatilization is use of plants for volatilization of pollutants from the soil into the atmosphere.

Soil contamination sources are agricultural chemicals, sewage sludge application, waste disposal, waste incineration, vehicle exhausts, anthropogenic sources etc. (Khan 2005). Using sewage sludge for agricultural purpose has been a common practice in waste disposal (Flieβbach et al. 1994). Using metals and chemicals in process industries has increasingly resulted in generation of huge amounts of effluent that contain high concentration of toxic heavy metals (Ahluwalia and Goyal 2007).

Plants that can tolerate and accumulate excessive levels of metals are defined as hyperaccumulators (Zhuang et al. 2007; Nie et al. 2002). It has been identified that approximately 400 terrestrial species are hyperaccumulators (Zaidi et al. 2006; McGrath et al. 2001). The Brassiceceae family is the most common known heavy metal accumulator (Belimov et al. 2005). Indian mustard (*Brassica juncea*) has an ability to grow in heavily contaminated soil and accumulate metals in its aboveground parts (Rajkumar et al. 2006) Hyperaccumulators are given in Table 19.2.

Plants have two main strategies for growing on metalliferous soil: (i) prevent metal from aerial part but contain high amount of metals in their root, or (ii)

Table 19.2 Some metal hyperaccumulator and their metal accumulation capacities (Lasat 2002)

Plant species	Metal	Leaf content (mg kg^{-1})
Thlaspi caerulescens	Zn, Cd	39,600–1,800
Ipomea alpine	Cu	12,300
Haumaniastrum robertii	Co	10,200
Astragalus racemosus	Se	14,900

accumulate metals in their aboveground parts (Raskin et al. 1994; Kuffner et al. 2008). Effective phytoaccumulation is dependent on two essential factors (1) having the capability of taking up and accumulating high levels of metal and (2) having the ability of producing as much biomass as possible (Burd et al. 2000; Li et al. 2007).

For adequate phytoextraction, the essential need is a phytoaccumulator that is fast growing and produces a large amount of biomass (Kamnev and Lelie 2000; Zhuang et al. 2007). Inoculation of plants with microorganisms may decrease the toxicity of heavy metals to plants in polluted soils (Madhaiyan et al. 2007). Wu et al. (2006) indicated that inoculation with PGPR may simulate plant growth and thus raise phytoremediation efficiency. Inoculation with rhizobacteria did not vastly affect the metal concentrations in plant tissues; however, the bacteria provided a much larger aboveground biomass harvest resulting in a much higher metal removal. Kumar et al. (2008) suggested that *Brassica juncea* with *Enterobacter* sp. *NBRIK28* and *Enterobacter* sp. *NBRIK28 SD1* could be used for effective phytoextraction of heavy metals from fly ash polluted sites. Nie et al. (2002) investigated the ability to proliferate in the presence of arsenate by transgenic canola with *Enterobacter cloacae UW4* ACC deaminase.

19.4 Heavy Metal Detoxification and Tolerance in Higher Plants

Potential cellular mechanisms for metal detoxification and tolerance in higher plants are summarized as (1) "restriction of metal movement to roots by mycorrhizas, (2) binding to cell wall and root exudates, (3) reduced influx across plasma membrane, (4) active efflux into apoplast, (5) chelation in cytosol by various ligands, (6) repair and protection of plasma membrane under stress conditions, and (7) transport of and accumulation of metals in vacuole" (Hall 2002).

Hyperaccumulator or accumulator plants and their related PGPR are main topic in metal detoxification. PGPR and arbuscular mycorrhizal fungi (AMF) develop plant growth and development in heavy metal polluted soil via helping root growth and branching. PGPR and AMF are able to lessen the toxicity of heavy metals either by declining the bioavailability of toxic heavy metals or raising bioavailability of nontoxic heavy metals. PGPR and AMF can alter chemical properties in the rhizosphere and stimulate metal accumulation (Denton 2007).

Menngoni et al. (2001) isolated nickel-resistant bacteria from *Alyssum bertolonii* (*Pseudomonas* and *Streptomyces*). Pseudomonas strains were located in the rhizosphere, whereas streptomyces strains were mainly found in the soil. Bacteria isolated from Alyssum murale (Ni accumulator plant) rhizosphere were *Sphingomonas macrogoltabidus, Microbacterium liquefaciens,* and *Microbacterium arabinogalactanolyticum* (Abou-Shanab et al. 2003). *Microbacterium oxydans AY509223* significantly increased nickel uptake by *Alyssum murale* from low, moderate, and high Ni soils (Abou-Shanab et al. 2006). Li et al. (2007) found that inoculation of *Sedum alfredii* with metal-tolerant bacterium, *Burkholderia cepacia* significantly stimulated plant growth, metal uptake in shoot, and provided better translocation of metals from root to shoot. Abou-Shanab et al. (2006) demonstrated that *Microbacterium oxydans AY509223* significantly enhanced Ni uptake of *Alyssum murale* by 36.1%, 39.3 %, and 27.7 % in low, medium, and high levels of Ni, respectively. Wu et al. (2006) investigated effects of inoculation of PGPR on metal uptake by *Brassica juncea.* They concluded that inoculation with PGPR might stimulate plant growth and thus raise phytoremediation efficiency. Wu et al. (2006) showed that presence of PGPR was very effective in protecting plants from metal inhibitory effects.

Application of Cd-solubilizing PGPR could enhance available Cd in rhizosphere soil and stimulate plant growth and Cd uptake (Sheng and Xia 2006). Rajkumar and Freitas (2008a) found that inoculation of metal-resistant bacteria (*Pseudomonas* sp. and *Pseudomonas jessenii*) to plants caused stimulation of metal accumulation in plant tissue and promoted shoot and root biomass. *Bradyrhizobium* sp.(vigna) RM8 isolated from greengram nodules reduced the uptake of Ni and Zn by plant organs (Wani et al. 2007b).

19.4.1 Fungi

In fact, it has been shown that the presence of ectomycorrhizal or vesicular–arbuscular fungi on the roots of plants decreased the uptake of metals by the plants and thereby increased plant biomass. Many studies about mycorrhizas have been reported in plants growing on heavy metal polluted sites. Different authors have reported most of the plants growing in heavy metal polluted sites associated with *Glomus* and *Gigaspora mycorrhizal fungal texa* (Khan et al. 2000). Weissenhorn et al. (1995) isolated *Glomus mosseae* from heavy metal polluted sites. Diaz et al. (1996) isolated *Glomus macrocarpum* from uncontaminated site, whereas *Glomus mosseae* was isolated from contaminated site.

Mycorrhizal plant metal uptake depends on several factors such as soil physicochemical properties (fertility level, pH), host plants, fungi involved, and the concentration of metal in soil (Diaz et al. 1996). Heggo et al. (1990) found that the influence of VAM fungi on heavy metal uptake is dependent upon initial soil metal concentration. Weissenhorn et al. (1994) found that heavy metals completely eliminate Arbuscular mycorrhizal (AM) colonization of plant roots in pot experiment. Gildon and Tinker (1983) found that the degree of infection with

Glomus mosseae strongly declined by adding zinc, copper, nickel, and cadmium in soil.

Fungus has an important role in detoxification that constitutes a biological barrier against transfer to the shoot. AM arbuscular mycorrhizas provide plants with essential nutrients from the soil via uptake by extraradical hyphae (Joner and Leyval 2001). Heavy metals can be transported by hyphae (Joner and Leyval 2001).

AM fungi improve plant tolerance originated heavy metal pollution (Tonin et al. 2001). Mycorrhizal fungi from uncontaminated soil provided greater biomass compared to mycorrhizal fungi from contaminated soil. Shetty et al. (1994) found that *Andropogon gerardii* only grew with the interaction of mycorrhiza in polluted soil. Mycorrhizal species such as *Suillus bovines* and *Thelephora terrestris* confined *Pinus sylvestris* against Cu toxicity (Hall 2002). Weissenhorn et al. (1995) found that the indigenous AM fungi population survived under high metal concentration in polluted soil.

Vivas et al. (2005) investigated interactive effect of *Brevibacillus brevis* and *Glomus mosseae* on plant growth. The stimulated plant Cd tolerance after coinoculation with *Brevibacillus brevis* and *Glomus mosseae* resulted in increased P and K and decreased Cd, Cr, Mn, Cu, Mo, Fe, and Ni in plant tissue.

AM fungi produce insoluble glycoprotein called glomalin that binds toxic elements such as heavy metals (Hildebrandt et al. 2007; Gonzalez-Chavez et al. 2004; Khan 2006). Glomalin can be extracted with heavy metals from soil (Gohre and Paszkowski 2006). Glomalin has a special role for filtering heavy metals at soil–hyphae interface (Khan 2006). Glomalin reduces heavy metal bioavailability via sorption and sequestration (Gonzalez-Chavez et al. 2004).

Weissenhorn et al. (1993) reported that AM infection may decline metal accumulation in plants growing in contaminated soils and thus protect the host plant against metal toxicity. Vivas et al. (2003) showed that indigenous bacteria (*Brevibacillus* sp.) coinoculated with AM fungi caused enhancing plant development under heavy metal contamination. Vivas et al. (2006) showed a protective effect of the interaction AM fungi + bacteria (*Glomus mosseae* + *Brevibacillus*) against uptake of potentially toxic Zn by Trifolium plants in moderately polluted soil. Some selected studies about mycorrhizas are given in Table 19.3.

19.4.2 Plant Growth Promoting Rhizobacteria

The rhizosphere, which is a layer of soil affected by roots and has a greater amount of microorganisms compared to surrounding bulk soil (Lugtenberg and Kamilova 2009). PGPR is defined as bacteria living the rhizosphere and beneficial to plants (Kloepper et al. 1980). PGPR improve plant health and stimulate plant growth and reduce diseases (Solano et al. 2008; Yang et al. 2009).

Inoculation of plants with microorganisms may diminish toxicity of heavy metals to plants in polluted soil (Madhaiyan et al. 2007). Bacteria are the most abundant among the other soil microbial communities (Barriuso et al. 2008) and

Table 19.3 Selected studies about mycorrhizas

Heavy metal	Mycorrhiza	Plant	Study conditions	References
Zn, Pb, Cd	*VAM*	*Andropogon gerardii*	Mine spoil	Shetty et al. (1994)
		Festuca arundinacea		
Pb	*VAM + Brevibacillus*	*Trofolium prqatense*	Pb-spiked soil	Vivas et al. (2003)
Zn, Cd, Cu, Mn, Fe	*VAM*	*Glycine max*	Zn-smelter	Heggo et al. (1990)
Zn and Pb	*Glomus mosseae*	*Lygeum sportum*	Polluted soil	Diaz et al. (1996)
	Glomus macrocorpum	*Anthyllis cytisoides*		
Zn	*AMF + Brevibacillus*	*Trifolium repens*	Zn-polluted soil	Vivas et al. (2006)
Zn, Cd	*Glomus mosseae*	*Clover*	Cd and Zn polluted soil	Tonin et al. (2001)
Cd, Ni, Cr	*Glomus mosseae*	*Cannabis sativa*	Pot experiment	Citterio et al. (2005)
Zn, Cd, Cu, Pb	*Glomus mosseae*	*Clover, maize*	Pot experiment	Joner and Leyval (2001)
Cd, Cu, Zn, Pb	*Glomus mosseae*	*Zea mays*	Pot experiment	Weissenhorn et al. (1995)

may uptake and accumulate a vast amount of metal ions (Ahluwalia and Goyal 2007). PGPR is grouped into four distinct groups which are diazototrophic PGPR, *Bacillus*, *Pseudomonas,* and *Rhizobia.* Nitrogen-fixing bacteria like *Azospirillum* sp., *Azoarcus* sp., *Burkholderia* sp., *Gluconacetobacter diazotorophic*, *Herbaspirillum* sp., *Azotobacter* sp., and *Paenibacillus polymyxa* are capable of promoting plant growth (Barriuso et al. 2008). Bacillus bacteria as defined PGPR are predominantly gram positive, whereas Pseudomonas bacteria are gram negative. Well-known PGPR are *Agrobacterium genomovars*, *Azospirillum lipoferum*, *Alcaligenes*, *Bacillus*, *Burkholderia*, *Enterobacter*, *Methylobacterium fujisawaense*, *Pseudomonas*, *Ralstonia solanacearum*, *Rhizobium*, *Rhodococcus*, *Sinorhizobium meliloti,* and *Variovorax paradoxus* (Saleem et al. 2007).

Rhizobia show high resistance to heavy metals in symbiotic plants. *Rhizobium leguminosarum bv. Trifolii* showed up to twofold times greater resistance to Cd^{2+}, Cu^{2+}, Ni^{2+} and Zn^{2+} compared to nonmucoid colonies (Purchase et al. 1997). Wani et al. (2008) found that inoculation of *Rhizobium species RP5* affected pea plants in two different ways: (1) protecting pea plants against toxic effects of Ni and Zn and (2) decreasing the uptake of Ni and Zn by plant organs.

Khan et al. (2009) reported that different symbiotic nitrogen fixers such as *Bradyrhizobium* sp.*RM8*, *Rhizobium* sp. *RP5*, *Rhizobium* sp. *RL9,* and *Mesorhizobium* sp.*RC3* were tolerant to Ni and Zn, Ni and Zn, Zn, and Cr (VI), respectively. Nitrogen-fixing bacteria *Burkholderia* sp.*J62* was found to be heavy metal resistant (Jiang et al. 2008). There is a separation for metal resistance among

PGPR. Kuffner et al. (2008) found that *Streptomyces AR16* showed extremely high Zn resistance, whereas *Pseudomonas PR04* demonstrated low tolerance to Zn.

PGPR can be classified into two main groups based on their relationship with the plants: symbiotic bacteria and free-living bacteria (Zhuang et al. 2007; Khan et al. 2009). Free-living soil bacteria are PGPR that include a number of different bacteria such as *Azotobacter*, *Azospirillum*, *Pseudomonads*, *Acetobacter*, *Burkholderia,* and *Bacilli* (Glick et al. 1998). *Allorhizobium*, *Azorhizobium*, *Bradyrhizobium*, *Mesorhizobium*, *Rhizobium*, and *Sinorrhizobium* are examples of symbiotic bacteria (Vessey 2003). During rhizoremediation, the plant releases root exudates that stimulate the survival and action of bacteria and finally result in more pollutant degradation (Kuiper et al. 2004). Plants eliminate and finally select bacteria beneficial and useful for their growth and health from rhizosphere (Barriuso et al. 2008). For successful plant growth-promoting inoculation bacteria must be able to rapidly settle root system during the growing season (Abou-Shanab et al. 2006). Some select studies about PGPR are given in Table 19.4.

PGPR may induce plant growth and development both directly and indirectly (Saleem et al. 2007). Indirect mechanisms happen outside the plant, whereas direct mechanisms occur inside the plant (Solano et al. 2008). PGPR can act in three different ways: (1) synthesizing special compounds for the plants, (2) facilitating uptake of some nutrients from soil, and (3) lessening or preventing the plants from pathogens (Zhuang et al. 2007; Glick et al. 1998; Khan et al. 2009; Safranova et al. 2006; Kloepper et al. 1980; Wei et al. 1996; Rapuach and Kloepper 1998; Sinha and Mukherjee 2008; Carillo-Gastaneda et al. 2002; Sharma et al. 2003). Indirect mechanisms include free nitrogen fixation, production of siderophores, phosphate solubilization, hydrolysis of molecules released by pathogens, synthesis of enzymes able to hydrolyze fungal cell walls, synthesis of cyanhydric acid, and improvement of symbiotic relationship with rhizobia and mycorrhizae (Solano et al. 2008). Direct mechanisms are given in Table 19.5.

Direct plant growth promotions are biofertilization (Vessey 2003), stimulation of root growth, rhizoremediation, and plant stress control (Lugtenberg and Kamilova 2009). PGPR can fix atmospheric nitrogen for providing plants, they can synthesize siderophores for solubilization and sequester iron from the soil and provide the plants, they can synthesize phytohormones like indolacetic acid (IAA). They may facilitate solubilization of minerals and may synthesize some enzymes (Glick et al. 1998; Rajkumar et al. 2005). PGPR can be used against plant pathogens such as *Bacillus pimulus*, *Bacillus subtilis,* and *Curtobacterium flaccumfaciens* are used as biological control against multiple cucumber pathogens (Rapuach and Kloepper 1998).

Rapuach and Kloepper (1998) investigated the effect of *PGPR stains INR7* (*Bacillus pumilus*), *GBO3* (*Bacillus subtilis*), and *ME1* (*Curtobacterium flaccumfaciens*) on multiple cucumber pathogens. The mixture of PGPR strains could stimulate disease protection and develop stability of biological control. So-Yeon and Cho (2009) found that a PGPR, *Serratia* sp. *SY5*, which could be applied as a promising microbial inoculant for the direct stimulation of plant biomass production and to indirectly enhance heavy-metal uptake by plants in the phytoremediation of heavy metal contaminated soils.

Table 19.4 Some selected studies about PGPR

PGPR	Plant	Heavy metal	References
Pseudomonas sp.		Cr^{6+}	Rajkumar et al. (2005)
Bacillus subtilis	*Brassica juncea*	Ni	Zaidi et al. (2006)
Pseudomonas sp. *M6*			
Pseudomonas jessenii M15	*Ricinus communis*	Ni, Cu, Zn	Rajkumar and Freitas (2008b)
Kluyvera ascorbata SUD165	*Brassica juncea*	Ni, Pb, Zn	Burd et al. (2000)
Enterobacter NBRI K28			
Enterobacter NBRI K28 SD1	*Brassica juncea*	Ni, Zn, Cr	Kumar et al. (2008)
Azotobacter chrooccum	*Triticum aestivum*	Cd, Cu, Ni, Zn, Pb, Cr	Athar and Ahmad (2002)
Pseudomonas sp. *NBRI4014*	*Glycine max*	Cd, Ni, Cr	Gupta et al. (2002)
Azospirillum brasilense Sp245			
Azospirillum brasilense Sp7		Co, Cu, Zn	Kamnev et al. (2005)
Pseudomonas sp. *29 C*			
Bacillus megaterium 4 C	*Brassica juncea*	Ni	Rajkumar and Freitas (2008a)
Bradyrhizobium sp *(vigna) RM8*	*Greengram*	Ni, Zn	Wani et al. (2007a)
Pseudomonas sp. *GRP3*	*Vigna radiata*	Fe	Sharma et al. (2003)
Pseudomonas asplenii	*Phragomites australis*	Cu	Reed et al. (2005)
Bacillus PSB1, PSB7, PSB10		Zn, Cr, Pb	Wani et al. (2007b)
Pseudomonas brassicacerum Am3			
Pseudomonas marginalis Dp1			
Rhodococcus sp. *Fp2*	*Pisum sativum*	Cd	Safranova et al. (2006)
Burkholderia sp. *J62*		Pb, Cd	Jiang et al. (2008)
Pseudomonas putida			
Pseudomonas sp.			
Alcaligenes xylosoxidans	*Brassica juncea*		
Alcaligenes sp.	*Brassica napus*		
Variovorax paradoxus			
Bacillus pimulus			
Rhodococcus sp.		Cd	Belimov et al. (2001)
Bradyrhizobium sp. *750*	*Lupinus luteus*	Cu, Cd, Pb	Dary et al. (2010)
Burkholderia cepacia	*Sedum alfredii*	Cd, Zn	Li et al. (2007)
Kluyvera ascorbata SUD165	*Tomato, canola, Indian mustard*	Ni, Pb, Zn	Burd et al. (2000)
Microbacterium oxydans AY509223			
Rhizobium galegae AY509213			
Microbacterium oxydans AY509219			
Clavibacter xyli AY509236			
Acidovorax avenae AY512827			
Microbacterium arabinogalactanolyticum AY5099225			
Microbacterium oxydans AY509222			
Microbacterium arabinogalactanolyticum AY509226	*Alyssum murale*	Ni	Abou-Shanab et al. (2006)
Achromobacter xylosoxidans Ax10	*Brassica juncea*	Cu	Ma et al. (2009b)
Bacillus sp. *I119*	*Rape, maize, sudangrass, tomato*	Cd	Sheng et al. (2008)
Pseudomonas tolaasii ACC23			
Pseudomonas fluorescens ACC9			
Alcaligenes sp. *ZN4*			
Mycobacterium sp. *ACC14*	*Brassica napus*	Cd	Dell'Amico et al. (2008)
Rhizobium sp. *RP5*	*Pisum sativum*	Ni, Zn	Wani et al. (2008)

Table 19.5 Direct PGPR mechanisms (Solano et al. 2008)

Mechanism	Effect
Plant growth regulator production	Biomass (aerial part and root), flowering
Ethylene synthesis inhibition	Root length
Induction of systemic resistance	Health
Root permeability raise	Biomass and nutrient absorption
Organic matter mineralization	Biomass and nutrient content
Mycorrhizal fungus association	Biomass and phosphorus content
Insect pest control	Health

The potential of bacterial inoculation is partly dependent on the original content of the heavy metals in soil. Wu et al. (2006) indicated that bacteria could be inhibited by the presence of extremely high metal concentrations, leading to a smaller population size and a decline in the activities of soil enzymes, e.g., N_2-fixing capacity and acid phosphatase activity. This may indicate that the potential of bacterial inoculation is greatest only in slightly or moderately metal-polluted sites.

19.4.2.1 Isolated Microorganisms

Many PGPR have been isolated from plants. *Bradyrhizobium* sp. *(vigna) RM8*, Ni, and Zn-tolerant PGPR were isolated from nodules of greengram (Wani et al. 2007b). Endophytic bacteria (*Staphylococcus*, *Microbacterium*, *Pseudomonas*, *Curtobacterium*, *Bacillus*, and *Arthrobacter*) were isolated and characterized from nickel hyperaccumulator plant *Alyssum bertolonii* (Barzanti et al. 2007). Endophytic bacteria (*Elsholtzia splendens* and *Commelina communis*) were isolated from tolerant plants (Sun et al. 2010). *Bacillus* sp., *strain EB1* was isolated from sites that were polluted with heavy metal in the southeast region of Turkey (Yilmaz 2003). Metal-tolerant PGPR, *NBRI K28 Enterobacter* sp., was isolated from fly ash contaminated soils (Kumar et al. 2008).

Characterization of bacterial communities from heavy metal contaminated soils has been widely studied. These kinds of studies showed that excessive levels of heavy metal could influence both the qualitative and quantitative structure of microbial communities with declining metabolic activity and biomass as well as suppressed diversity (Menngoni et al. 2001). Ganesan (2008) selected *Pseudomonas aeruginosa strain MKRh3* for rhizoremediation of cadmium soil. Similarly, Tripathi et al. (2005) isolated and characterized siderophore-producing lead and cadmium-resistant *Pseudomonas putida KNP9*. Trivedi et al. (2007) demonstrated chromate-reducing bacteria isolated from Himalayan Region named *Rhodococcus erythropolis MtCC 7905*.

19.4.2.2 ACC Deaminase Activity (ACC:1-Aminocyclopropane-1-Carboxylic Acid)

Dell'Amico et al. (2008) demonstrated that PGPR having ACC deaminase activity was critical for the growth of *Brassica napus*. Saleem et al. (2007)

reported that inoculation with PGPR and having ACC deaminase activity could be useful in maintaining plant growth and development under stressed conditions by decreasing stress-induced ethylene production. Inoculation with PGPR having ACC deaminase and final physiological changes in plants is given in Table 19.6.

Belimov et al. (2001) isolated 15 bacterial strains containing ACC deaminase from the rhizosphere of pea (*Pisum sativum*) and Indian mustard (*Brassica juncea*).These strains are *Pseudomonas brassicacearum*, *Pseudomonas marginalis*, *Pseudomonas oryzihabitans*, *Pseudomonas putida*, *Pseudomonas* sp, *Alcaligenes xylosoxidans*, *Alcaligenes* sp., *Variovorax paradoxus*, *Bacillus pumilus*, and *Rhodococcus* sp. They concluded that the beneficial effect of this kind of PGPR on plant growth significantly varied depending on features of the bacterial strain, plant genotype, and on the growth conditions, which may influence the ethylene status of the plant. The results suggest that PGPR containing ACC deaminase offer promise as bacterial

Table 19.6 Inoculation with PGPR having ACC deaminase and final physiological changes in plants (Saleem et al. 2007)

Plant species	PGPR	Comments	References
Brassica compestris	*Methylobacterium fujisawaense*	Bacterium induced root elongation	Madhaiyan et al. (2007)
Brassica compestris	*Bacillus ciculans DUC1* *Bacillus firmus DUC2* *Bacillus globisporus DUC3*	Bacterial inoculation stimulated root and shoot elongation	Ghosh et al. (2003)
Brassica napus	*Alcaligenes* sp. *Bacillus pumilus* *Pseudomonas* sp. *Variovorax paradoxus*	Promoted growth compared uninoculated treatment	Belimov et al. (2001)
Diathus caryophyllus L.	*Azospirillum brasilense*	Longest roots	Li et al. (2005)
Pisum sativum L.	*Rhizobium leguminosarum bv.vicae 128C53K*	Stimulated nodulation	Ma et al. (2003)
Vigna radiate L.	*Pseudomonas* sp. *Bradyrhizobium* sp.	Enhanced nodulation	Shaharoona et al. (2006)
Vigna radiate L.	*Pseudomonas putida*	Ethylene production was inhibited in inoculated cuttings	Mayak et al. (1999)
Zea mays L.	*Enterobacter sakazokii 8MR5*	Inoculation elevated agronomic properties of maize	Bababia et al. (2003)
Zea mays L.	*Pseudomonas* sp. *4MKS8* *Klebsiella oxytoca 10MKR7* *Pseudomonas* sp.	Bacterium provided root elongation	Shaharoona et al. (2006)

inocula for the improvement of plant growth, particularly under unfavorable environmental conditions.

ACC is the immediate precursor of ethylene in plants and ACC deaminase metabolizes ACC to α-ketobutyrate and ammonia (Grichko et al. 2000). *Agrobacterium rhizogenes* containing ACC deaminase was helpful to plants for increasing tolerance to nickel via inhibiting ethylene production (Stearns et al. 2005). *Phychrobacter* sp. *SRA1*, *SRA2*, *Bacillus cereus SRA10*, *Bacillus* sp. *SRP4*, and *Bacillus wehenstephanensis SRP12* containing ACC deaminase, nickel-resistant bacteria, were isolated from *Alyssum serpyllifolium* and *Phleum phleoides* (Ma et al. 2009a). Madhaiyan et al. (2007) found that the bacterial strains *CBMB20* and *CBMB40* tolerate the presence of high concentrations of $NiCl_2/CdCl_2$ in the growth media and bind considerable amounts of Ni and Cd, in part due to the PGPR containing the enzyme ACC deaminase, which can act to alter the level of ethylene in plants.

The ability of *V. paradoxus* to use ACC as both N and C sources gives these bacteria an additional competitive advantage in rhizosphere colonization. Belimov et al. (2005) suggested that taking into account that Cd induces plant stress ethylene biosynthesis and probably contributes to accumulation of ACC in roots, it is not surprising that a number of *V. paradoxus* strains were isolated among bacterial communities dominating the root zone of *B. juncea* grown in contaminated soils in their study.

19.4.2.3 Ethylene

Ethylene is a plant hormone and has vast importance in plant growth and development. Production of ethylene significantly increases under stress conditions (Belimov et al. 2001). Increased ethylene levels in plants shows exposure to various types of stress such as chilling, heat, wounding, drought, flooding, pathogen infection, and nutritional stress (Stearns et al. 2005).

Actually, ethylene is required by many plants for seed germination. However, excessive ethylene causes suppression of root elongation. PGPR with ACC deaminase promotes root elongation in a variety of plants and using of ACC deaminase that contains PGPR decreases the level of stress ethylene (Glick 2003). Heavy metals can stimulate ethylene production by plants and an excess of ethylene can inhibit plant development (Burd et al. 1998).

In the presence of excessive metals, most plants either synthesize stress ethylene or become severely depleted in iron (Glick 2003). Madhaiyan et al. (2007) showed that *Methylobacterium oryzae strain CBMB20* and *Burkholderia* sp. *Strain CBMB40* from rice decreased Ni and Cd toxicity and promoted plant growth by reducing stress ethylene. Some PGPR produce ACC deaminase that inhibits synthesis of ethylene (So-Yeon et al. 2010). Rodecap and Tigey (1981), and Fuhrer (1982) indicated that among heavy metals, cadmium (Cd) is the strongest inductor of ethylene biosynthesis in plants.

19.5 Protecting Metal Inhibitory Effect

Burd et al. (2000) mentioned that *Kluyvera ascorbata SUD 165/26* protected plants against the inhibitory effects of high levels of nickel, lead, and zinc. Sinha and Mukherjee (2008) indicated that using Cd-resistant bacterial strain provided reduction of Cd uptake in plants. Ma et al. (2009a) noted that inoculation of metal-resistant serpentine strains (*Phychrobacter* sp. *SRA1*, *SRA2*; *Bacillus cereus SRA10*; *Bacillus* sp. *SRP4*; *Bacillus weihenstephanensis SPR12*) showed to be very effective in protecting plants from Ni toxicity. Another study were studied by the same researchers investigated plant growth promoting bacterium *Achromobacter xylosoxidans strain Ax10* for improvement of copper phytoextraction by *Brassica juncea*. *A. xylosoxidans Ax10* protected plant from Cu toxicity and stimulated Cu accumulation in plant tissue (Ma et al. 2009b). *Kluyvera ascorbata SUD165/26* decreased heavy metal toxicity (Ni, Pb, and Zn) in plants (Burd et al. 2000). Ma et al. (2009b) reported that inoculation of metal-resistant serpentine strains (*Psychrobacter* sp. *SRA2*, *SRA1*, *Bacillus cereus SRA10*) effectively protected plants (*Brassica juncea* and *Brassica oxyrrhina*) from growth inhibition due to Ni. Rajkumar et al. (2006) found *Pseudomonas* sp. *PsA4* and *Bacillus* sp. *Ba32* protect plants (*Brassica juncea*) against negative effect of chromium.

Zaidi et al. (2006) indicated that inoculation of *Bacillus subtilis strain SJ-101* not only confined plant from Ni toxicity but also stimulated Ni accumulation in plant tissue with enhancement of plant growth. Burd et al. (2000) found *Kluyvera ascorbata SUD165* protected plants (tomato, canola, and Indian mustard) against negative effects of high concentrations of Ni, Pb, and Zn. Reed et al. (2005) found that inoculation of *Phragmites australis* seeds with *Pseudomonas asplenii AC* and *P. asplenii AC-1* develop seed germination and plant growth and partially defend plants from inhibition by copper. Dell'Amico et al. (2008) indicated that cadmium-resistant rhizobacteria (*Pseudomonas tolaasii ACC23*, *Pseudomonas fluorescens ACC9*, *Alcaligenes* sp. *ZN4*, and *Mycobacterium* sp. *ACC14*) protected plants (*Brassica napus*) against the inhibitory effects of cadmium possibly due to production of IAA, siderophores, and ACC deaminase activity. Wani et al. (2008) found similar results that inoculation *Rhizobium species RP5* resulted in protection to the pea plants against toxic effects of Ni and Zn. Also they noted the reduced uptake of nickel and zinc by plant organs.

19.6 Conclusions

Nature has the ability to refresh itself by various techniques. Environmental pollution, especially heavy metal pollution, can be dealt with by the activities of microorganisms through heavy metal detoxification. The most common detoxification microorganisms are bacteria PGPR, and fungi AMF. Results from the literature suggest that heavy metals may be removed from contaminated soils using PGPR and AMF by assisting root growth and branching. Interactions with detoxification microorganisms and hyperaccumulator plants may be the future solution for detoxification of metals in soils.

Acknowledgment The authors would like to thank Jennifer Tvergyak, The Soil Microbial Ecology Group, School of Environment and Natural Resources, Ohio State University, US for her comments on the manuscript.

References

Abou-Shanab RAI, Angle JS, Delorme TA, Chaney RL, Berkum P, Moavad H, Ghanem K, Ghozlan HA (2003) Rhizobacterial effects on nickel extraction from soil and uptake by *Alyssum murale*. New Phytol 158:219–224

Abou-Shanab RAI, Angle JS, Chaney RL (2006) Bacterial inoculants affecting nickel uptake by *Alyssum murale* from low, moderate and high Ni soils. Soil Biol Biochem 38:2882–2889

Ahluwalia SS, Goyal D (2007) Microbial and plant derived biomass for removal of heavy metals from wastewater. Bioresour Technol 98:2243–2257

Ahmad I, Pinchtel J, Hayat S (2008) Plant–bacteria interactions. Weinheim, Wiley-VCH

Antoun H, Prevost D (2005) Ecology of plant growth promoting rhizobacteria in PGPR. In: Siddiqui ZA (ed) Biocontrol and biofertilization. Springer, The Netherlands, pp 1–39

Athar R, Ahmad M (2002) Heavy metal toxicity effect on plant growth and metal uptake by wheat, and on free living Azotobacter. Water Air Soil Pollut 138:165–180

Bababia OO, Osir EO, Sanni AI, Odhaimbo GD, Bulimo WD (2003) Amplification of 1-aminocyclopropane-1-carboxylic (ACC) deaminase from plant growth promoting rhizobacteria in striga-infested soils. Afr J Biotechnol 2:157–160

Barriuso J, Solano BR, Lucas JA, Lobo AP, Garcia-Villanaco A, Monero FJG (2008) Ecology, genetic diversity and screening strategies of plant growth promoting rhizobacteria. In: Ahmad I, Pichtel J, Hayat S (eds) Plant–bacteria interactions. Weinheim, Wiley-VCH, pp 1–13

Barzanti R, Ozino F, Bazzicalupo M, Gabbrielli R, Galerdi F, Gonelli C, Mengoni A (2007) Isolation and characterization of endophytic bacteria from the nickel hyperaccumulator plant *Alyssum bertolonii*. Microb Ecol 53:306–316

Belimov AA, Safranova VI, Sergeyeva TA, Egorova TN, Matueyeva VA, Tsyganov VE, Borisov AY, Tikhanovich IA, Kluge C, Preisfeld A, Dietz K, Stepanok VV (2001) Characterization of plant growth promoting rhizobacteria isolated from polluted soils and containing 1-aminocyclopropane-1-caboxylate deaminase. Can J Microbiol 47:642–652

Belimov AA, Hontzeas N, Safranova VI, Demchinskoya SV, Piluzza G, Bullitta S, Glick BR (2005) Cadmium tolerant plant growth-promoting bacteria associated with the roots of Indian mustard (Brassica juncea L.Czern). Soil Biol Biochem 37:241–250

Burd GI, Dixon DG, Glick BR (1998) A plant growth-promoting bacterium that decreases nickel toxicity in seedlings. Appl Environ Microbiol 64(10):3663–3668

Burd GI, Dixon DG, Glick BR (2000) Plant growth-promoting bacteria that decrease heavy metal toxicity in plants. Can J Microbiol 46:237–245

Carillo-Gastaneda G, Munos JJ, Peralta-Videa JR, Gomez E, Tiemann KJ, Duarte-Gardea M, Gardea-Torresdey JL (2002) Alfalfa growth promotion by bacteria grown under iron limiting conditions. Adv Environ Res 6:391–399

Citterio S, Prato N, Fumagalli P, Aina R, Massa N, Santagostino A, Sgorbati S, Berto G (2005) The arbuscular mycorrhizal fungus *Glomus mosseae* induces growth and metal accumulation changes in *Cannabis sativa* L. Chemosphere 59:21–29

Dary M, Chamber-Perez MA, Palomeres AJ, Pajuela E (2010) "In situ" phytostabilisation of heavy metal polluted soils using *Lupinus luteus* inoculated with metal resistant plant-growth promoting rhizobacteria. J Hazard Mater 177:323–330

Dell'Amico E, Cavalca L, Andreoni V (2008) Improvement of *Brassica napus* growth under cadmium stress by cadmium-resistant rhizobacteria. Soil Biol Biochem 40:74–84

Denton B (2007) Advances in phytoremediation of heavy metals using plant growth promoting bacteria and fungi. Basic Biotechnol J 3:1–5

Diaz G, Azcon-Aguilar C, Honrubia M (1996) Influence of arbucular mycorrhizae on heavy metal (Zn and Pb) uptake and growth of *Lygeum spartum* and *Anthyllis cytisoides*. Plant Soil 180:241–249

Flieβbach A, Martens R, Reber HH (1994) Soil microbial biomass and microbial activity in soils treated with heavy metal contaminated sewage sludge. Soil Biol Biochem 26(9):1201–1205

Fuhrer J (1982) Ethylene biosynthesis and cadmium toxicity in leaf tissue of beans (*Phaseolus vulgaris* L.). Plant Physiol 70:162–167

Gadd GM (1990) Heavy metal accumulation by bacteria and other microorganisms. Experientia 46:834–840

Gadd GM, Griffiths AJ (1978) Microorganisms and heavy metal toxicity. Microb Ecol 4:303–317

Gamalero E, Lingua G, Berta G, Glick BR (2009) Beneficial role of plant growth promoting bacteria and arbuscular mycorrhizal fungi on plant responses to heavy metal stress. Can J Microbiol 55:501–514

Ganesan V (2008) Rhizoremediation of cadmium soil using a cadmium-resistant plant growth-promoting Rhiopseudomonad. Curr Microbiol 56:403–407

Garbisu C, Alkorta I (2001) Phytoextraction: a cost-effective plant-based technology for the removal of metals from the environment. Bioresour Technol 77(3):229–236

Ghosh S, Penterman JN, Little RD, Chavez R, Glick BR (2003) Three newly isolated plant growth-promoting bacilli facilitate the seedling growth of canola, *Brassica compestris*. Plant Physiol Biochem 41:277–281

Gildon A, Tinker PB (1983) Interactions of vesicular-arbuscular mycorrhizal infection and heavy metals in plants: I. The effects of heavy metals on the development of vesicular-arbusculat mycorrhizas. New Phytol 95:247–261

Glick BR (2003) Phytoremediation: synergistic use of plants and bacteria to clean up the environment. Biotechnol Adv 21:383–393

Glick BR (2010) Using soil bacteria to facilitate phytoremediation. Biotechnol Adv 28:367–374

Glick BR, Penrose DM, Li J (1998) A model for the lowering of plant ethylene concentrations by plant growth-promoting bacteria. J Theor Biol 190:63–68

Gohre V, Paszkowski U (2006) Contribution of arbuscular symbiosis to heavy metal phytoremediation. Planta 223:1115–112

Gonzalez-Chavez MC, Carillo-Gonzalez R, Wright SF, Nichols KA (2004) The role of glomalin, a protein produced by arbuscular mycorrhizal fungi, in sequestering potentially toxic elements. Environ Pollut 130:317–323

Grichko UP, Filby B, Glick BR (2000) Increased ability of transgenic plants expressing the bacterial enzyme ACC deaminase to accumulate Cd, Co, Cu, Ni, Pb and Zn. J Biotechnol 81:45–53

Gupta A, Meyer JM, Goel R (2002) Development of heavy metal-resistant mutants of phosphate solubilizing *Pseudomonas* sp. NBRI4014 and their characterization. Curr Microbiol 45:323–327

Hall JL (2002) Celular mechanisms for heavy metal detoxification and tolerance. J Exp Bot 53(366):1–11

Heggo A, Angle JS, Chaney RL (1990) Effects of vesicular-arbuscular mycorrhizal fungi on heavy metal uptake by soybeans. Soil Biol Biochem 22(6):865–869

Hildebrandt U, Regvar M, Bothe H (2007) Arbuscular mycorrhiza and heavy metal tolerance. Phytochemistry 68:139–146

Jeffries P, Gianinazzi S, Perotto S, Turnau K, Barea J (2003) The contribution of arbuscular mycorrhizal fungi in sustainable maintenance of plant health and soil fertility. Biol Fertil Soils 37:1–16

Jiang C, Sheng X, Qian M, Wang Q (2008) Isolation and characterization of a heavy metal-resistant *Burkholderia* sp. from heavy metal-contaminated paddy field soil and its potential in promoting plant growth and heavy metal accumulation in metal-polluted soil. Chemosphere 72:157–164

Jing Y, He Z, Yang X (2007) Role of soil rhizobacteria in phytoremediation of heavy metal contaminated soils. J Zhejiang Univ Sci B 8(3):192–207

Joner EJ, Leyval C (2001) Time-course of heavy metal uptake in maize and clover as affected by root density and different mycorrhizal inoculation regimes. Biol Fertil Soils 33:351–357

Kamnev AA, Lelie D (2000) Chemical and biological parameters as tools to evaluate and improve heavy metal phytoremediation. Biosci Rep 20(4):239–258

Kamnev AA, Tugarova AV, Antonyuk LP, Tarantilis PA, Polissiou MG, Gardiner PHE (2005) Effects of heavy metals on plant-associated rhizobacteria: comparison of endophytic and non-endophytic strains *Azospirillum brasilense*. J Trace Elem Med Biol 19:91–95

Khan AG (2005) Role of soil microbes in the rhizospheres of plant growing on trace metal contaminated soils in phytoremediation. J Trace Elem Med Biol 18:355–364

Khan AG (2006) Review: mycorrhizoremediation-an enhanced form of phytoremediation. J Zhejiang Univ Sci B 7(7):503–514

Khan AG, Kuek C, Chaundhry TM, Khoo CS, Hayes WJ (2000) Role of plants, mycorrhizae and phytochelators in heavy metal contaminated land remediation. Chemosphere 41:197–207

Khan MS, Zaidi A, Wani PA, Oves M (2009) Role of plant growth promoting rhizobacteria in the remediation of metal contaminated soils. Environ Chem Lett 7:1–19

Kloepper JW, Schroth MN, Miller TD (1980) Effect of rhizosphere colonization by plant growth-promoting rhizobacteria on potato plant development and yield. Phytopathology 70(11):1078–1082

Kuffner M, Puschenreiter M, Wieshammer G, Gorfer M, Sessitsch A (2008) Rhizosphere bacteria affect growth and metal uptake of heavy metal accumulating willows. Plant Soil 304:35–44

Kuiper I, Lagendijk EL, Bloomberg GV, Lugtenberg BJJ (2004) Rhizoremediation: a beneficial plant–microbe interaction. Mol Plant Microbe Interact 17(1):6–15

Kumar KV, Singh N, Behl HM, Srivastava S (2008) Influence of plant growth promoting bacteria and its mutant on heavy metal toxicity in *Brassica juncea* grown in fly ash amended soil. Chemosphere 72:678–683

Lasat MM (2002) Phytoextraction of toxic metals: a review of biological mechanisms. J Environ Qual 31:109–120

Leyval C, Turnau K, Haselwandter K (1997) Effect of heavy metal pollution on mycorrhizal colonization and function: physiological, ecological and applied aspects. Mycorrhiza 7:139–153

Leyval C, Joner EJ, del Val C, Haselwandter K (2002) Potential of arbuscular mycorrhizal fungi for bioremediation. In: Gianinazzi S, Schhuepp H, Barea JM, Haselwanter K (eds) Mycorrhiza technology in agriculture. Birkhauser Verlag Basel, Switzerland, pp 175–186

Li Q, Saleh-Lakka S, Glick BR (2005) The effect of native and ACC deaminase-containing *Azospirillum brasilense* Cd843 on the rooting of cornation cuttings. Can J Microbiol 51:511–514

Li WC, Ye ZH, Wong MH (2007) Effects of bacteria on enhanced metal uptake of the Cd/Zn hyperaccumulating plant, *Sedum alfredii*. J Exp Bot 58(15/16):4173–4182

Lugtenberg B, Kamilova F (2009) Plant-growth-promoting bacteria. Annu Rev Microbiol 63:541–556

Ma W, Guinel FC, Glick BR (2003) *Rhizobium leguminosarum* biovar viciae 1-aminocyclopropane-1-carboxylate deaminase promotes nodulation of pea plants. Appl Environ Microbiol 69:4396–4402

Ma Y, Rajkumar M, Freitas H (2009a) Improvement of plant growth and nickel uptake by nickel resistant-plant-growth promoting bacteria. J Hazard Mater 166:1154–1161

Ma Y, Rajkumar M, Freitas H (2009b) Inoculation of plant growth promoting bacterium *Achromobacter xylosoxidans* strain Ax10 for the improvement of copper phytoextraction by *Brassica juncea*. J Environ Manage 90:831–837

Madhaiyan M, Poonguzhali S, Sa T (2007) Metal tolerating methylotrophic bacteria reduces nickel and cadmium toxicity and promotes plant growth of tomato (*Lycopersicon esculentum* L.). Chemosphere 69:220–228

Mayak S, Tivosh T, Glick BR (1999) Effect of wild type and mutant plant growth-promoting rhizobacteria on the rooting of mungbean cuttings. J Plant Growth Regul 18:49–53

McGrath SP, Chaundri AM, Giller KE (1995) Long-term effects of metals in sewage sludge on soils, microorganisms and plants. J Ind Microbiol 14(2):94–104

McGrath S, Zhao FJ, Lombi E (2001) Plant and rhizosphere processes involved in phytoremediation of metal-contaminated soils. Plant Soil 232:207–214

Menngoni A, Barzanti R, Gonnelli C, Gabrielli R, Bazzicalupo M (2001) Characterization of nickel-resistant bacteria isolated from serpentine soil. Environ Microbiol 3(11):691–698

Nie L, Shah S, Rashid A, Burd GI, Dixon DG, Glick BR (2002) Phytoremediation of arsenate contaminated soil by transgenic canola and the plant growth-promoting bacterium *Enterobacter cloacae* CAL2. Plant Physiol Biochem 40:355–361

Purchase D, Miles RJ, Young TWK (1997) Cadmium uptake and nitrogen fixing ability in heavy-metal-resistant laboratory and field strains *of Rhizobium leguminosarum* biovar trifolii. FEMS Microbiol Ecol 22:85–93

Rajkumar M, Freitas H (2008a) Effects of inoculation of plant-growth promoting bacteria on Ni uptake by Indian mustard. Bioresour Technol 99:3491–3498

Rajkumar M, Freitas H (2008b) Influence of metal resistant-plant growth-promoting bacteria on the growth of *Ricinus communis* in soil contaminated with heavy metals. Chemosphere 71:834–842

Rajkumar M, Nagendran R, Lee KJ, Lee WH (2005) Characterization of a novel Cr^{6+} reducing *Pseudomonas* sp. with plant growth-promoting potential. Curr Microbiol 50:266–271

Rajkumar M, Nagendran R, Lee KJ, Lee WH, Kim SZ (2006) Influence of plant growth promoting bacteria and Cr^{6+} on the growth of Indian mustard. Chemosphere 62:741–748

Rapuach GS, Kloepper JW (1998) Mixtures of plant growth-promoting rhizobacteria enhance biological control of multiple cucumber pathogens. Phytopathology 88(11):1158–1164

Raskin I, Nanda Kumar PBA, Dushenkov S, Salt DE (1994) Bioconcentration of heavy metals by plants. Curr Opin Biotechnol 5:285–290

Reed MLE, Warner BG, Glick BR (2005) Plant growth-promoting bacteria facilitate the growth of the common reed *Phragmites australis* in the presence of copper or polycyclic aromatic hydrocarbons. Curr Microbiol 51:425–429

Rodecap KD, Tigey DT (1981) Stress ethylene: a bioassay for rhizosphere-applied phytotoxicants. Environ Monit Assess 1:119–127

Safranova VI, Stepanok VV, Engqvist GL, Alekseyev YV, Belimov AA (2006) Root-associated bacteria containing 1-aminocyclopropane-1-carboxylate deaminase improve growth and nutrient uptake by pea genotypes cultivated in cadmium supplemented soil. Biol Fertil Soils 42:267–272

Saleem M, Arshad M, Hussain S, Bhatti AS (2007) Perspective of plant growth promoting rhizobacteria (PGPR) containing ACC deaminase in stress agriculture. J Ind Microbiol Biotechnol 34:685–648

Shaharoona B, Arshad M, Zahir ZA (2006) Performance of *Pseudomonas* spp. containing ACC-deaminase for improving growth and yield of maize (*Zea mays* L.) in the presence of nitrogenous fertilizer. Soil Biol Biochem 38:2971–2975

Sharma A, Johri BN, Sharma AK, Glick BR (2003) Plant growth-promoting bacterium *Pseudomonas* sp. strain GRP3 influences iron acquisition in mung bean (*Vigna radiata* L. Wilzeck). Soil Biol Biochem 35:887–894

Sheng X, Xia J (2006) Improvement of rape (*Brassica napus*) plant growth and cadmium uptake by cadmium resistant bacteria. Chemosphere 64:1036–1042

Sheng X, He L, Wang Q, Ye H, Jiang C (2008) Effects of inoculation of biosurfactant-producing *Bacillus* sp. J119 on plant growth and cadmium uptake in a cadmium-amended soil. J Hazard Mater 155:17–22

Shetty KG, Hetrick BAD, Figge DAH, Schwab AP (1994) Effects of mycorrhizae and other soil microbes on revegetation of heavy metal contaminated mine spoil. Environ Pollut 86:181–188

Sinha S, Mukherjee SK (2008) Cadmium-induced siderophore production by a high Cd-resistant bacterial strain relevied Cd toxicity in plants through root colonization. Curr Microbiol 56:55–60

Solano BR, Maicas JB, Monero FJG (2008) Physiological and molecular mechanisms of plant growth promoting rhizobacteria (PGPR). In: Ahmad I, Pichtel J, Hayat S (eds) Plant–bacteria interactions. Weinheim, Wiley-VCH, pp 41–52

So-Yeon K, Cho K (2009) Isolation and characterization of a plant growth-promoting rhizobacterium, *Serratia* sp. SY5. J Microbiol Biotechnol 19(11):1431–1438

So-Yeon K, Hong SH, Ryu HW, Cho K (2010) Plant growth-promoting trait of rhizobacteria isolated from soil contaminated with petroleum and heavy metals. J Microbiol Biotechnol 20(3):587–593

Stearns JC, Shah S, Greenberg BM, Dixon DG, Glick BR (2005) Tolerance of transgenic canola expressing 1-aminocyclopropane-1-carboxylic acid deaminase to growth inhibition by nickel. Plant Physiol Biochem 43:701–708

Sun L, Zhang Y, He L, Chen Z, Wang Q, Qian M, Sheng X (2010) Genetic diversity and characterization of heavy metal-resistant-endophytic bacteria from two copper-tolerant plant species on copper mine wasteland. Bioresour Technol 101:501–509

Tonin C, Vandenkoornhuyse P, Joner EI, Straczek J, Leyval C (2001) Assesment of arbucular mycorrhizal fungi diversity in the rhizosphere of viola calaminaria and effect of these fungi on heavy metal uptake by clover. Mycorrhiza 10:161–168

Tripathi M, Munot HP, Shouche Y, Meyer JM, Goel R (2005) Isolation and functional characterization of siderophore-producing lead and cadmium resistant *Pseudomonas putida* KNP9. Curr Microbiol 50:233–237

Trivedi P, Pandey A, Sa T (2007) Chromate reducing and plant growth promoting activities of psychrotrophic *Rhodococcus erythropolis* MtCC 7905. J Basic Microbiol 47:513–517

Upadhyay A, Srivastava S (2010) Evaluation of multiple plant growth promoting traits of an isolate of *Pseudomonas fluorescens* strain Psd. Indian J Exp Biol 48:601–609

Vessey JK (2003) Plant growth promoting rhizobacteria as biofertilizers. Plant Soil 255:571–586

Vivas A, Azcon R, Biro B, Barea JM, Ruiz-Lozano JM (2003) Influence of bacterial strains isolated from lead-polluted soil and their interactions with arbuscular mycorrhizae on the growth of *Trifolium pratense* L. under lead toxicity. Can J Microbiol 49:577–588

Vivas A, Barea JM, Azcon R (2005) Interactive effect of *Brevibacillus brevis* and *Glomus mosseae* both isolated from Cd contaminated soil on plant growth, physiological mycorrhizal fungal characteristics and soil enzyme activities in Cd polluted soil. Environ Pollut 134:257–166

Vivas A, Biro B, Ruiz-Lozano JM, Barea IM, Azcon R (2006) Two bacterial strains isolated from a Zn-polluted soil enhance plant growth an mycorrhizal efficiency under Zn-toxicity. Chemosphere 62:1523–1533

Wani PA, Khan MS, Zaidi A (2007a) Chromium reduction, plant growth-promoting potential, and metal solubilization by *Bacillus* sp. isolated from alluvial soil. Curr Microbiol 54:237–243

Wani PA, Khan MS, Zaidi A (2007b) Effect of metal tolerant plant growth promoting *Bradyrhizobium* sp. (vigna) on growth, symbiosis, seed yield and metal uptake by greengram plants. Chemosphere 70:36–45

Wani PA, Khan MS, Zaidi A (2008) Effect of metal-tolerant plant growth-promoting rhizobium on the performance of pea grown in metal-amended soil. Arch Environ Contam Toxicol 55:33–42

Wei G, Kloepper JW, Tuzun S (1996) Induced systemic resistance to cucumber diseases and increased plant growth. Biol Control 86(2):221–224

Weissenhorn I, Leyval C, Berthelin J (1993) Cd-tolerant arbuscular mycorrhizal (AM) fungi from heavy metal polluted soils. Plant Soil 157:247–256

Weissenhorn I, Glashoff A, Leyval C, Berthelin J (1994) Differential tolerance to Cd and Zn of arbuscular mycorrhizal (AM) fungal spores isolated from heavy metal-polluted and unpolluted soils. Plant Soil 167:189–196

Weissenhorn I, Leyval C, Belgy G, Berthelin J (1995) Arbuscular mycorrhizal contribution to heavy metal uptake by maize (*Zea mays* L.) in pot culture with contaminated soil. Mycorrhiza 5:245–251

Williams LE, Pittman JK, Hall J (2000) Emerging mechanisms for heavy metal transport in plants. Biochim Biophys Acta 1465:104–126

Wu CH, Wood TK, Mulchandani A, Chen W (2006) Engineering plant–microbe symbiosis for rhizoremediation of heavy metals. Appl Environ Microbiol 72(2):1129–1134

Yang J, Kloepper JW, Ryu C (2009) Rhizosphere bacteria help plants tolerate abiotic stress. Trends Plant Sci 14(1):1–4

Yilmaz EI (2003) Metal tolerance and biosorption capacity of *Bacillus circulans* strain EB1. Res Microbiol 154:409–415

Zaidi S, Usmani S, Singh BR, Musarrat J (2006) Significance of *Bacillus subtilis* strain SJ-101 as a bioinoculant for concurrent plant growth promotion and nickel accumulation in *Brassica juncea*. Chemosphere 64:991–997

Zhuang X, Chen J, Shim H, Bai Z (2007) New advances in plant growth-promoting rhizobacteria for bioremediation. Environ Int 33:406–413

Chapter 20
Detoxification of Heavy Metals From Soils Through Sugar Crops

D.V. Yadav, Radha Jain, and R.K. Rai

20.1 Introduction

Heavy metals are conventionally defined as elements with metallic properties (ductility, conductivity, stability as cations, ligand specificity, etc.) and an atomic number >20. The most common heavy metal contaminants are Cd, Cr, Cu, Hg, Pb, and Ni. The other naturally occurring metallic elements with high molecular weight are also considered the important pollutants. These elements occur either naturally in soils or get deposited through the use of agricultural chemicals, urban wastes, and polluted water (Yadav et al. 2010). Such elements, when present in high quantities may be toxic and may exhibit characteristic stress symptoms like disturbance in photosynthesis, respiration, and biomass production. The increased industrial activities, indiscriminate use of inorganic and organic fertilizers and pesticides, and disposal of industrial effluents enhance the possibility of pollution and toxicity of heavy metals in agro-ecosystem and environment. Environmental pollution by toxic metals occurs through industrial, military, and agricultural processes. Metal species released into the environment by various activities tend to persist indefinitely because of their nonbiodegradable nature. These heavy metals once discharged into the waste streams get accumulated throughout the food chain, thus becoming a serious threat to the environment. Heavy metals like Ni, Zn, Cu, Cd, Cr, and Hg are toxic even in the minute quantities. Chromium, a common pollutant, gets introduced into natural waters from a variety of industrial wastewaters including those from the textiles, leather tanning, electroplating, and metal finishing industries. Chromium affects human physiology by accumulating in food chain and causes several ailments. Chromium exists in two stable oxidation states, Cr (VI) and Cr (III). The Cr (VI) state is of particular concern because of its toxicity. According to the Indian standards, the permissible limit of Cr (VI) for industrial

D.V. Yadav (✉) • R. Jain • R.K. Rai
Indian Institute of Sugarcane Research, Lucknow 226002, Uttar Pradesh, India
e-mail: dvyadav_2008@rediffmail.com

I. Sherameti and A. Varma (eds.), *Detoxification of Heavy Metals*, Soil Biology 30,
DOI 10.1007/978-3-642-21408-0_20,

effluents to be discharged to surface water is 0.1 mg l^{-1}. Chromium-contaminated wastes are usually discharged to the environment as hexavalent chromium in the form of chromate (CrO_4^{2-}) and dichromate ($Cr_2O_7^{2-\bullet}$) anions which are thermodynamically stable over a wide pH range. Therefore, removal of Cr (VI) from industrial effluents is important before discharging it into the aquatic environment or on to land. Metals are natural components in soil with a number of heavy metals being required by plants as micronutrients. However, pollution of biosphere by toxic metals has increased dramatically since the beginning of the industrial revolution. As a result of human activities such as mining and smelting of metals, electroplating, gas exhaust, energy and fuel production, fertilizer, sewage and pesticide application, municipal waste generation, etc. (Kabata-Pendias and Pendias 1989), metal pollution has become one of the most severe environmental problems today. Excessive accumulation of heavy metals is toxic to most plants. Heavy metal ions, when present at an elevated level in the environment, are excessively absorbed by roots and translocated to shoot, leading to impaired metabolism and reduced growth (Bingham et al. 1986; Foy et al. 1978). Heavy metal contamination to water and soil poses a major environmental and human health problem. In addition, excessive metal concentrations in contaminated soils result in decreased soil microbial activity, soil fertility, and crop yield (McGrath et al. 1995). For example, cadmium, as a nonessential, toxic heavy metal to plants can inhibit root and shoot growth, affect nutrient uptake and homeostasis, and is frequently accumulated by agriculturally important crops (Sanità di Toppi and Gabrielli 1999). On condition that soil Cd pollution is cumulative with levels increasing over time, the soil may eventually become unusable for crop production. Similarly, contamination of soil with Cd can negatively affect biodiversity and the activity of soil microbial communities. Further, the Cd-enriched crop products if consumed may cause diseases to animals and humans. Burd et al. (1998) revealed that the canola seeds developed normally in the presence of nickel chloride up to 1 mmol l^{-1}, but its higher levels inhibited the elongation of root and shoot of plant. The present communication deals with the detoxification of heavy metals from soils through sugar crops especially sugar cane, sugar beet, and sweet sorghum.

20.2 Remediation of Heavy Metals From Soils

Heavy metals can not be destroyed biologically since there is a lack of degradation or change in the nuclear structure of the element. Such metals can only be transformed from one oxidation state or organic complex to another (Garbisu et al. 2002). As such, the remediation of heavy metal contamination in soils is more difficult.

Until now, the methods namely excavation and land fill, thermal treatment, acid leaching and electroreclamation used for their remediation are not suitable for practical applications, because of their high cost, low efficiency, large destruction of soil structure and fertility and high dependence on the contaminants of concern, soil properties, site conditions, and so on. Thus, the development of phytoremediation strategies for heavy metal contaminated soils is necessary (Chaney et al. 2000; Cheng et al. 2002; Lasat 2002). In recent years, scientists and engineers have started to generate cost-effective technologies that include use of microorganisms/biomass or live plants to clean polluted areas. Phytoremediation is an emerging technology for cleaning up contaminated sites. It is cost-effective, in-situ nonintrusive, esthetically pleasing, ecologically benign, and socially acceptable and has long-term applicability (Alkorta and Garbisu 2001; Weber et al. 2001; Su and Wong 2004; Boonyapookana et al. 2005). The technology involves an efficient use of plants to extract, remove, sequester, and/or detoxify or immobilize environmental contaminants in a growth matrix (soil, water, or sediments) through natural, biological, chemical or physical activities, and processes of the plants (Cunningham and Ow 1996; Saxena et al. 1999; Wenzel et al. 1999; Ciura et al. 2005).

It also helps prevent landscape destruction and enhance activity and diversity of soil microorganisms to maintain healthy ecosystems, which is consequently considered a more attractive alternative than traditional methods to the approaches that are currently in use for dealing with heavy metal contamination (Bogardt and Hemmingsen 1992; Cunningham and Ow 1996; Salt et al. 1995).

Phytoremediation is best applied at sites with shallow contamination of organic, nutrient or metal pollutants that are amenable to one of several applications namely phytoextraction, rhizofiltration, phytostabilization, phytovolatilization, etc. (Schnoor 1997). In phytoextraction, the plants are used to concentrate metals from soil into their roots and shoots; in rhizofiltration, plant roots are used to absorb, concentrate or precipitate metals from effluents; in phytostabilization, plants are used to reduce the mobility or bioavailability of heavy metals through their absorption and precipitation; whereas in phytovolatilization, the absorbed volatile materials such as mercury- or arsenic-containing compounds are released into atmosphere by plants.

A plant that grows fast produces rapidly a large quantity of biomass, and is able to tolerate and accumulate greater concentrations of heavy metals in shoots is an ideal plant for phytoextraction. Most of the commonly known heavy metal accumulators belong to the Brassicaceae family (Kumar et al. 1995). Although hyperaccumulator plants have exceptionally high metal accumulating capacity, most of these have a slow growth rate and often produce limited amounts of biomass when the concentration of available metal in the contaminated soil is very high. An alternative is to use species with a lower metal accumulating capacity but higher growth rates, such as Indian mustard (*Brassica juncea*); another alternative is to provide them with an associated plant growth-promoting rhizobacteria, which also is considered an important component of phytoremediation technology (Wenzel et al. 1999; Glick 2003). Obviously, the rhizosphere contains a large microbial population with high metabolic activity compared to bulk soil (Anderson et al. 1993). Microbial populations are known to affect the mobility and availability of heavy metals to plant through release of their chelating agents, acidification, phosphate solubilization, and redox changes (Abou-Shanab et al. 2003; Smith and Read 1997). Especially, some plant growth-promoting bacteria associated with plant roots also may exert some beneficial effects on plant growth and nutrition through a number of mechanisms such as N_2 fixation, production of phytohormones and siderophores, and transformation of nutrient elements when these are either applied to seeds or incorporated into the soil (Kloepper et al. 1989). The use of rhizobacteria in combination with plants is expected to provide high efficiency for phytoremediation (Abou-Shanab et al. 2003; Whiting et al. 2001). Therefore, the potential and the exact mechanism of rhizobacteria to enhance phytoremediation of soil heavy metals pollution have recently received some attention (de Souza et al. 1999; Whiting et al. 2001). For example, Burd et al. (1998) observed that both the number of Indian mustard seeds that germinated in a nickel-contaminated soil and the attainable plant size increased by 50–100% by the addition of K. In preliminary field trials (de Souza et al. 1999), it was found that ascorbata SUD165/26, an associated plant growth-promoting rhizobacteria to the soil enhanced accumulation of Se and Hg in plant tissues from constructed wetlands.

Potential for phytoremediation depends upon the interactions among soil, heavy metals, bacteria, and plants. These complex interactions are affected by a variety of factors, such as characteristics and activity of plant and rhizobacteria, climatic conditions, soil properties, etc.

20.3 Sugar Crops

Chemically, the substance in the breakfast sugar bowl comes to us unchanged from the living organism in which it was manufactured. The familiar crystals are virtually pure sucrose, an organic chemical belonging to a large family of compounds classified as sugars.

Sucrose, almost all of which is obtained from sugar cane and sugar beet, is commercially by far the most important sweetener. Its origin in photosynthesis explains how two wholly dissimilar botanical sources can furnish a practically identical product. Sugar cane – currently providing roughly three-quarters of the world's sucrose supply – is a perennial monocotyledon, propagated from cuttings, except in the breeding of new varieties, and capable of giving repeated harvests. In contrast, sugar beet is a biennial dicotyledon, harvested in the first season and replanted annually from seed. Cane grows in tropics and subtropics, while beet in temperate climates. The sucrose comes from the stalk of sugar cane but from roots in sugar beet. Sugar cane has been exploited industrially for more than two millennia, while sugar beet for just two centuries.

The origin of sugars in photosynthesis also explains the exploitation of other, more or less important sweetener sources, some going back to ancient times, for example, boiled-down grape juice, fig and date syrup, the sap of palms, the maple tree, and sweet sorghum, and the exudations from certain trees and shrubs.

20.3.1 Sugar Cane

Sugar cane is grown in about 80 countries roughly between 35° latitude of north and south of equator. Often described as a tropical crop, much cane is actually grown in subtropical areas. The ten leading cane producers are India, Brazil, China, Thailand, Australia, Mexico, Cuba, the United States, Pakistan, and South Africa.

20.3.2 Sugar Beet

The root of sugar beet, in which sucrose accumulates, consists, from the top down, of the epicotyl or crown, the hypocotyl or neck, and a swollen taproot. Roots vary greatly in size but average about 600 g. Sugar beet (*Beta vulgaris* L.) is a member of the Chenopodiaceae or goose-foot family. Four distinct types of the species are cultivated: sugar beet, garden (red) beet, leaf beet and Swiss chard, and fodder beets. Sugar beet, the second major source of the world's sugar supply, is commercially by far the most important of the four types. Sugar beet is currently grown in nearly 50 countries. Member countries of the European Union, the United States, Turkey, Poland, Ukraine, Russia, and China are the leading producers. Beet is successfully cultivated in many soils, but a deep loam, moist yet well-drained, is best. Sown in spring, the crop is lifted before the first frosts are expected. Different from cane, harvested beet can be stored for months under suitable conditions without intolerable loss.

20.3.3 Sorghum

A near relative of sugar cane, sweet sorghum or sorghum (*Sorghum vulgare*), a native of Africa, was, for a while in the second half of the eighteenth century, thought to have the potential for becoming a mainstream source of sugar in the United States. Although it could not compete against the growing availability of sugar from beet and cane since about 1880 onwards, sorghum syrup is still produced on a small scale.

20.4 Heavy Metals Content in Sugar Crops

20.4.1 Sugar Cane

Roots, stems, and leaves of sugar cane (*Saccharum* spp.) were collected in 25 points of an area under direct influence of the municipal landfill site (MLS) and medical waste treatment system (MWTS) of Ribeirao Preto, São Paulo, Brazil (Segura-Muñoz et al. 2006). The roots contained Cd, 0.22 ± 0.12; Cr, 64.3 ± 48.7; Cu, 140.6 ± 27.7; Hg, 0.04 ± 0.02; Mn, 561.6 ± 283.3; Pb, 7.9 ± 2.1; and Zn, 177.4 ± 64.9 mg kg^{-1} dry weight. Metal levels in stems were 80–90% of those found in roots, while the concentrations detected in leaves were significantly lower than those in roots. The present results suggest that MLS and MWTS activities might have been increasing metal concentrations in edible tissues of sugar cane grown in the area under their influence. Moreover, the traditional agricultural practices in the production of sugar cane could also be another determinant factor to reach the current metal levels. The results indicate that sugar cane is a crop that is able to grow in areas where metals in soils are accumulated.

In eastern Australia, 12 sugar cane (*Saccharum officinarum* L.) varieties and their different parts were screened for plant-mobile heavy metals to assess whether genetic differences were of greater significance than the soil/environment for uptake and within plant distribution. Soil pH (1:5 soil/water) ranged from 4.5 to 6.4 and all breeding-trial sites contained relatively low levels of extractable Cd, Hg, and Pb and variable levels of Cu and Zn. Internal concentrations of Cd and Zn were more influenced by soil/environment than by variety, while the distribution of metals in plant parts was quite consistent. About 77% of the Cd and 56% of the Zn were contained in the stem, which relocates to the mill following harvest. There was a little Hg in all plant parts (concentrations < 0.05 mg kg^{-1} dry weight). From a predictive viewpoint, correlations between extractable heavy metals such as Cd in soils and corresponding plant concentrations were inconsistent, with the narrow range in soil concentrations seen as a contributing factor. On the present evidence, the uptake of heavy metals by sugar cane can be adequately managed by manipulating soil properties rather than by varietal selection. It is also clear that for each 100 tons of fresh, mature cane, about 0.2 g of Cd and 110 g of Zn will relocate to the soil surface with the trash. Corresponding quantities moved to the mill are 0.54 g of Cd and 143 g of Zn, with amounts expected to be more for cane grown in strongly acidic soils with above average levels of heavy metals (Rayment et al. 2002). The concentration of Ni and chromium in roots and leaves of sugar cane increased with their increased supply (Yadav et al. 2010). Zn concentration increased in different plant parts with an increase in Zn supply; its concentration was the highest in roots (277 μg g^{-1} dry weight) and the lowest in leaves (60.9 μg g^{-1} dry weight) which was slightly higher than control plants (Jain et al. 2010). Similar to sugar cane, highest zinc content in roots was reported in Arabidopsis (Verret et al. 2004). In the same way, roots of higher plants were considered as a barrier against heavy metal translocation to the top parts (Wallace and Romney 1977), reflecting a potential tolerance mechanism operating in the root cell.

20.4.2 Sugar Beet

A monitoring study was carried out with the aim to assess the level of toxic metals, i.e., lead (Pb), cadmium (Cd), arsenic (As), and mercury (Hg) in different vegetables grown in Sindh province of Pakistan during 2007–2008. Average concentration of Cd, Pb, As, and Hg in sugar beet was found to be 0.047, 0.001, 0.038, and 0.005 μg g^{-1}, respectively. The concentration of heavy metals found in the sugar beet samples was within the permissible limits and safe to consume (Abbas et al. 2010).

20.4.3 Sorghum

The metal concentration in different tissues of sorghum plants increased as the concentration of metals in the soil-vermicompost medium increased (Jadia and Fulekar 2008). The concentration of different heavy metals in root and shoot tissues were in the range of (ppm): Cd, 0.915–7.108 in roots, 0.927–3.260 in shoots; Cu, 1.669–12.66 in roots, 0.539–3.450 in shoots; Ni, 1.230–6.274 in roots, 0.603–1.982 in shoots; Pb, 0.603–6.078 in roots , 0.105–0.812 in shoots; Zn, 3.100–12.61 in roots; 1.820–5.864 in shoots. Zn and Cu accumulated in the largest proportions in root tissue. The heavy metals were taken up by the sorghum plants in the following order: Zn > Cu > Cd > Ni > Pb (Nandakumar et al. 1995).

20.5 Phytoremediation of Heavy Metal Pollution in Soils

20.5.1 Sugar Cane

Heavy metal pollution is a worldwide problem. Phytoremediation is an effective and low-cost interesting technology. Sugar cane (*Saccharum officinarum* L.) may be a promising candidate for phytoremediation on metal-contaminated soils due to its greater biomass, faster growth and moderate take-up and accumulation of heavy metals such as Cu, Cd, Se, Pb, and Mn. As a follow-up processing of sugar cane, bagasse may adsorb heavy metal ions in aqueous solutions and sugar cane juice may be used to produce fuel ethanol efficiently. Some new directions for further research such as biosurfactant-assisted and plant–microorganism association's phytoremediation are also prospected (Yan and Xia 2010).

Another, a batch technique has been employed for the sorption of Pb(II), Hg(II), Cd(II), and Co(II) from aqueous solutions on sulfurized steam-activated carbon (SSAC). The SSAC is prepared from sugar cane bagasse pith by single-step steam pyrolysis in the presence of SO_2 and H_2S at 400°C. The adsorption of metal ions on SSAC depends on time, concentration, pH, and temperature. In this technique, the adsorption of Pb is 99.2% , Hg(II) 97.2%, Cd(II) 93.1%, and Co(II) 81.9% in the pH range of 4.0–8.0 from their initial concentration of 100 mg l^{-1}. The selectivity order of the adsorbent is Pb(II) > Hg(II) > Cd(II) > Co(II). The maximum adsorption capacity per gram of SSAC (evaluated from the fits of the Langmuir isotherm to batch adsorption data for a contact time of 4 h at 30°C) is 200 mg Pb(II), 188.68 mg Hg(II), 153.85 mg Cd(II), and 128.70 mg Co(II). The competitive adsorption capacity of the SSAC for all metal ions is lower than noncompetitive conditions.

Heavy metal adsorption from synthetic waste waters was also studied to demonstrate its efficiency in removing metals from waste waters containing other cations and anions. The metal ions, which are bound to the SSAC, may be stripped by acidic solutions (0.2 M HCl) so that the SSAC can be recycled. Surface modification of activated carbon using steam pyrolysis in the presence of SO_2 and H_2S greatly enhances metal removal and results in a product with possible commercial potential for waste water treatment (Krishnan and Anirudhan 2002)

Different metal-complexing ligands carrying synthetic and natural adsorbents have been reported in the literature for removal of heavy metals. A new approach to obtain relatively higher adsorption capacity utilizing modified sugar cane bagasse, a by-product of cane sugar industry, as a natural metal adsorbent has been developed (Krishnani et al. 2004). The adsorption process parameters viz. pH, stirring speed, and adsorbent dose are optimized by using Taguchi method (Garg and Sud 2005). There is almost complete adsorption (98%) of Cr (VI) with sugar cane bagasse treated with citric acid at pH 2.0, stirring speed 50 rpm, and adsorbent dose 2,000 mg. Thus, modified sugar cane bagasse is an effective adsorbent for the removal of Cr (VI) from the aqueous solutions. The equilibrium of a solution between liquid and solid phases is described by Freundlich and Langmuir isotherms (Garg and Sud 2005).

Sugar cane plantlets are able to tolerate up to 100 μM of copper in nutrient solution for 33 days, with no significant reduction in fresh weight, while accumulating 45 mg Cu kg^{-1} shoot dry weight. The higher levels of copper in solution (250 and 500 μM) are lethal. Sugar cane exhibits tolerance to 500 μM Cd without symptoms of toxicity, accumulating 451 mg Cd kg^{-1} shoot dry weight after 33 days, indicating its potential as Cd phytoremediator. The DNA gel blot analyses indicate eight fragments using a metallothionein (MT) Type I probe, while ten for the MT Type II and eight for MT Type III. The number of genes for each type of MT in sugar cane may be similar to the ones identified in rice considering the inter-specific origin of sugar cane cultivars. The MT Type I gene appears to present the highest level of constitutive expression, mainly in roots, followed by MT Type II, corroborating the expression pattern described based on large-scale expressed sequence tags sequencing. The MT Type II and III genes are more expressed in shoots, where MT type I is also importantly expressed. Increasing Cu concentration has little or no effect in modulating MT genes expression, while an apparent minor modulation of some of the MT genes may be detected in Cd treatments. However, the level of response is too small to explain the tolerance and/or accumulation of Cd in sugar cane tissues. Thus, the tolerance and accumulation of cadmium in sugar cane may possibly be derived from other mechanisms, although MT may be involved in oxidative responses to high levels of Cd. As a result, sugar cane can be considered a potential candidate to be tested in Cd phytoremediation (Sereno et al. 2007).

A pot experiment was conducted in order to investigate the growth of sugar cane (*Saccharum officinarum* L.) under cadmium (Cd) contamination and its uptake and accumulation (Xia et al. 2009). The results showed that with increase of Cd concentration, the diameter and burl length of stems, and the biomass of sugar cane decreased significantly. Cd concentration in soil had obviously influenced the ground-breaking time of buds of sugar cane. The values of bioconcentration factor (BCF) were all closed to 1, which showed that sugar cane had a good absorptive capacity on Cd. The Cd concentration increased from 0.55 to 63.32 mg kg^{-1} leaves, 0.37 to 486.26 mg kg^{-1} stems, and 2.26 to 2776.22 mg kg^{-1} roots with increase in Cd concentration from 0.22 to 1,000 mg kg^{-1} soil. The Cd concentration in roots was about 6- to 45-fold more than those in stems and leaves. These results suggest that sugar cane has a high ability to tolerate and accumulate Cd, so it may be a promising plant to be used for phytoremediation of Cd-contaminated soil.

This work investigated the identification and selection of fungi isolated from sugar cane bagasse and their application for phenanthrene (Phe) removal from soil. Fungi were identified by PCR amplification of ITS regions as *Aspergillus terrus*, *Aspergillus fumigatus* and *Aspergillus niger*, *Penicillium glabrum* and *Cladosporium cladosporioides*. A primary selection of fungi was accomplished in plate, considering Phe tolerance of every strain in two different media: potato dextrose agar (PDA) and mineral medium (MM). The radial extension rate (r(r)) in PDA exhibited significant differences ($p < 0.05$) at 200 and 400 ppm of Phe. A secondary selection of *A. niger, C. cladosporoides*, and *P. glabrum* sp. was achieved based on their tolerance to 200, 400, 600 and 800 ppm of Phe, in solid culture at a sugar cane bagasse/contaminated soil ratio of 95:5, in Toyamas, Czapeck and Wunder media. Under these conditions, a maximum (70%) Phe removal by *A. niger* was obtained. In addition, *C. cladosporioides* and *A. niger* were able to remove high (800 ppm) Phe concentrations (Cortés-Espinosa et al. 2006).

An inexpensive and effective adsorbent was developed from bagasse fly ash, obtained from a sugar industry, for the dynamic uptake of lead and chromium (Gupta and Ali 2004). Lead and chromium are absorbed by the developed adsorbent up to 96–98%. The removal of these two metal ions up to 95–96% was achieved by column experiments at a flow rate of 0.5 ml min^{-1}. The adsorption was found to be exothermic in nature. The adsorbent was successfully tried for the removal of lead and chromium from wastewater in the laboratory. The developed system for the removal of two ions is very useful, economic, rapid, and reproducible.

Series of pot experiments were conducted to investigate effects of Pb progressive stress on the physiological characteristics of potted sugar cane (Wenwei and Xia 2007). The results showed that when Pb concentration was 250 mg kg^{-1} soil, the maximum biomass was obtained and the diameter and the burl length of sugar cane stems increased by 9.63% and 9.02% in comparison with the control group, respectively. There was a positive correlation ($r = +0.999$) between the relative membrane permeability and the Pb concentration. The content of chlorophyll *a* ($r = -0.994$), chlorophyll *b* ($r = -0.994$), and carotenoid ($r = -0.985$) of leaves presented significant negative correlations with the Pb concentration in the soil. With an increase of Pb concentration, protein content, activity of catalase (CAT) of the leaves and soluble sugar of stems increased in the beginning and decreased afterwards. The critical Pb concentration in soil was around 250 mg kg^{-1} soil.

While investigating the phytoremediation potential of pot-growing sugar cane (*Saccharum officinarum* L.) to lead (Pb) in soil, 0, 100, 250, 500, and 1,000 mg Pb kg^{-1} soil were used. The results showed that, sugar cane, except for the treatment of Pb 1,000 mg kg^{-1} soil, did not show any macro-toxicity. A low Pb concentration stimulated the growth of sugar cane, but its high concentration suppressed the growth. The Pb concentration increased from 13.58 to 46.56 mg kg^{-1} leaves, 5.92–26.98 mg kg^{-1} stems, and 26.65 to 1336.86 mg kg^{-1} root with increase in Pb concentration from 21.23 to 1,000 mg kg^{-1} soil. Pb concentration in roots was about 5–50 times more than those in stems and leaves. The results implied that sugar cane is a promising plant for phytoremediation of Pb-contaminated soil (Xia et al. 2009; Wenwei and Xia 2007).

20.5.2 Sugar Beet

Salinity is a limiting factor to crop production. Yields of most crops are decreased when cultivated in salt-affected areas. Generally, salinity problems are handled by chemical and biological methods. Chemical methods are usually used to reclaim sodic soils. While cultivation of salt-tolerant species on salt-affected lands forms the basis of the biological reclamation, identification of a wide variety of species with higher salt tolerance is important to achieve more success from this approach. Fodder beet is highly salt-tolerant during vegetative growth (Niazi et al. 2000). The yield of sugar beet and fodder beet is improved by application of sodium chloride as fertilizer (Draycott and Durrant 1976; Magat and Goh 1988). This improvement has been related to an increase in root sugar yield with enhanced uptake of Na^{+} from soil (Hamid and Talibudeen 1976). Chloride is an essential nutrient for plants (Marschner 1995). Cell wall extensibility is significantly affected by deposition of Ca^{2+} in the cell wall, either by forming cross bridges or by inhibiting wall-loosening enzymes (Cleland 1986). The concentration of Ca^{2+} and Mg^{2+} in reclaimed soils is reported to be higher at the top (Qadir et al. 2000). Maximum Ca^{2+} uptake by the plant from these soils can affect the plant growth. Magnesium is an important component of chlorophyll (Marschner 1995) and presence of magnesium in plant can indirectly improve the growth of plant by more chlorophyll synthesis and better rate of photosynthesis in plant. Although the halophilous nature has been reported in literature (Magat and Goh 1988), there is a little information about the role of various ions in the growth of fodder beet. These factors point toward the importance of various ions for the optimum growth of the plant tissue. Fodder beet and sea beet are the chenopod mainly grown as supplementary stock feeding coastal areas in winter when there is an acute shortage of fodder for the livestock in Pakistan. The use of crops and crop residues as feedstocks for biofuels increases domestic and global supplies, creates new industries, and may result in reduced greenhouse-gas emissions. Uncertainty about the best crop and residue sources, technologies for manufacture, future public policy, and the global supply and price of oil make it difficult to predict the best approach. California growers can produce feed stocks from grain, oilseed and woody crops and, in the Imperial Valley, from sugar cane. If the technology for making ethanol or other liquid fuels from cellulose becomes cost-effective, then saline and other waste waters may be used in biofuel feedstock production of salt-tolerant crops, particularly perennial grasses. However, recent global increases in biofuel production have raised questions about their impacts on food and feed prices, climate change, and deforestation. New state laws affecting energy use and mandating greenhouse-gas reductions require that the sustainability of all biofuels be assessed. Sustainability should take into account factors at both the global and local scales, including resource-use efficiency, cropping system adaptability, and the potential of biofuels to remediate agriculture's environmental effects (Kaffka et al. 2004).

Soil contaminated with Cd, Pb, Cu, and Zn in the Zhangshi irrigation area (Sun et al. 2007) is very hard to be remediated. Phytoextraction is considered as an efficient method to remove these toxic metals from soil. Three vegetables including sugar beet (*Beta vulgaris*), mustard (*Brassica juncea* L.), and cabbage (*Brassica oleracea* L. var. capitata Linn.) were used to bioaccumulate heavy metals in soil through pot experiment for 90 days; and nutrient elements were applied to stimulate the phytoextraction of metals. Results of bioconcentration factors (BCF) and translocation factors (TF) from this study showed that these plants could phytoextract heavy metals, but the accumulation and translocation of metals differed with species of plants, categories of heavy metals, and some environmental conditions (e.g., nutrients).

20.5.3 Sorghum

Monocotyledonous plant species are usually more tolerant to metals than dicotyledonous species (Lombi et al. 2001). For phytoremediation, grasses are the most commonly evaluated plants (Ebbs and Kochian 1998; Shu et al. 2002). The large surface area of their fibrous roots and their intensive penetration of soil reduces leaching, runoff, and erosion via stabilization of soil and offers advantages for phytoremediation. Some of the plant material may be used for nonfood purposes or it can be ashed for recycling of the metals or to be disposed in landfills (Angle and Linacre 2005). Uptake of contaminants from soil by plants occurs primarily through root system in which the principle mechanisms of preventing contaminant toxicity are found. The root system provides an enormous surface area that absorbs and accumulates the water and nutrients that are essential for growth, but also absorbs other nonessential contaminants (Arthur et al. 2005). One of the mechanisms by which uptake of metal occurs in roots may include binding of the positively charged toxic metal ions to negative charges in the cell wall (Gothberg et al. 2004). Plants that accumulate metals are suitable for extracting environmentally important toxic metals, e.g., Pb, Cd, Cu, and Ni from soil. When such plants are harvested and removed from the contaminated sites, the toxic metals are removed at reduced costs and with little loss of top soil (Nandakumar et al. 1995). It has been observed that prolonged and lavish application of chemical fertilizers reduces productivity of land and crops due to dependence on periodic inputs. Factories that manufacture fertilizers also contribute to pollution of air, land, and water. Natural fertilizers, e.g., vermicompost can provide superior nutrient value to soils, and thereby improve yields and quality of crop produce. Vermicomposting of vegetable waste with earthworm (*Eisenia foetida*) develops waste into a natural fertilizer (Maharashtra Nature Park Society 2003). The vermicompost has a high nutrient value, increases fertility of soil, and maintains soil health (Suthar et al. 2005). Application of normal compost in contaminated areas improves soil fertility and physical properties (Zheljazkov and Warman 2004). Due to increased growth of plants with more biomass attained, more metals can potentially be taken up from the contaminated soils. The use of vermicompost developed from vegetable waste using vermiculture biotechnology enhances the conditions for phytoremediation (Elcock and Martens 1995). The heavy metals from soil were taken up efficiently by the sorghum plants with application of vermicompost developed from vegetable waste in the order: Zn > Cu > Cd > Ni > Pb (Jadia and Fulekar 2008). Several studies have demonstrated that the concentration of metals in plant tissue is a function of the metal content in the growing environment (Grifferty and Barrington 2000). Bennett et al. (2003) observed that the Indian mustard translocates heavy metals to the shoot, while grasses tend to accumulate them in the root. The uptake of 8.2 mg Pb by root of *Sorghum* sp. has been reported (Nandakumar et al. 1995), indicating that lead might be found to the outer surface of plant roots, as crystalline or amorphous deposits, and could also be deposited in the cell walls or in vesicles. This might explain the higher concentrations of Pb in roots of sorghum plants (Jadia and Fulekar 2008). An ultrastructural study using transmission electron microscopy revealed the retention of unchelated Pb mainly in cell wall of roots, particularly around intercellular spaces (Wenger et al. 2003). Grasses are known to concentrate heavy metals in the roots, with only very low translocation of heavy metals to the shoot (Spier et al. 2003; Bennett et al. 2003). The large surface area of fibrous roots of sorghum and intensive penetration of roots into soil reduces leaching via stabilization of soil. The plant is further capable of immobilizing and concentrating heavy metals in roots.

In a field trial, ethylene diamine tetraacetate, ammonium nitrate and ammonium sulfate were tested for their abilities to enhance the phytoextraction efficiency of heavy metals by three varieties of sweet sorghum (*Sorghum biocolor* L.), a high biomass energy plant from a contaminated agricultural soil. It was reported that sorghum plants always achieved the greatest removal of Pb by leaves and the greatest removal of Cd, Zn, and Cu by stems. There was no significant difference among the Keller, Rio, and Mray varieties of sweet sorghums in accumulating heavy metals. EDTA treatment was more efficient than ammonium nitrate and ammonium sulfate in promoting Pb accumulation in sweet sorghum from the contaminated agricultural soil. The application of ammonium nitrate and ammonium sulfate increased accumulation of both Zn and Cd in roots of sorghum plants. These results suggest the cropping of sorghum plants facilitated by agronomic practices may be a sustainable technique for partial decontamination of heavy metal-contaminated soils (Zhuang et al. 2009).

Field crops, particularly barley, wheat (*Triticum* spp.), sorghum (*Sorghum* spp.), cotton (*Gossypium* spp.), and sugar beet have been used extensively in bioremediation of saline–sodic sites. By utilizing more water on these crops than actually needed, salts and sodium can be leached beyond the roots and the soil can be prepared for more sensitive crops (Qadir and Oster 2004). Bauder and Brock (1992) concluded that uncropped conditions, which maintain the soil at a relatively high water content and minimize repeated drying and rewetting of the soil, and crops such as sorghum-sudan grass, which cause rapid drying of the soil and create conditions conducive to leaching salts, may be the best combination of conditions to gain maximum efficiency of amendments applied to reclaim saline or sodic soil. They further suggested that a primary halophyte species or combination of like species can help to set the stage for complete restoration by amendments.

Phytoextraction has been considered for remediation of soil and water contaminated with radionuclides. Soil amendments can increase plant uptake of radionuclides. Study showed that Johnson grass (*Sorhgum halpense*) planted in soil amended with poultry litter accumulated greater amounts of cesium and strontium than did other plant species in soil amended with poultry litter or other soil amendments (EPA 2004).

Sweet sorghum is a smart crop in providing food security, energy security, ecological sustainability, and water security. In the Philippines, for one crop of 12 months, sugar cane uses 36,000 m^3 of water; for two crops of only 8–9 months, sweet sorghum sips only 8,000 m^3 (Reddy et al. 2005). That is 78% less water; in other words, sugar cane wolfs down 4.5 times more water than sweet sorghum. Sweet sorghum is tolerant to drought, soil salinity, acidity, and toxicity. It is high yielding and tolerant to drought, water-logging, and soil salinity. That means, if your soil is dry, this crop will grow well; if your soil is wet, this crop will grow as well, if not better; it will also grow well in alkaline ("sweet") and acidic ("sour") soils. In other words, sweet sorghum is adaptable to many different sites. As a dry land crop, sweet sorghum requires far less water than costly irrigated sugar cane. Sorghum grass, with its distinctive characteristics like higher biomass, fast growth and strong fibrous root system is thus proven an ideal plant for phytoremediation, which is an economical, effective, pleasing, and environmentally compatible technology for removing heavy metals from contaminated sites.

20.6 Conclusions

In this chapter, an overview of detoxification of heavy metals from soils through sugar crops especially sugar cane, sugar beet, and sweet sorghum has been presented. The findings to-date clearly indicate the need of further research on heavy metal tolerance in sugar crops that will aid in developing a system for remediation of metal-contaminated soils.

References

Abbas M, Parveen Z, Iqbal M, Riazuddin IS, Ahmed M, Bhutto R (2010) Monitoring of toxic metals (cadmium, lead, arsenic and mercury) in vegetables of Sindh, Pakistan. Kathmandu Univ J Sci Eng Technol 6:60–65

Abou-Shanab RA, Delorme TA, Angle JS, Chaney RL, Ghanem K, Moawad H, Ghozlan HA (2003) Phenotypic characterization of microbes in the rhizosphere of *Alyssum murale*. Int J Phytoremediation 5:367–379

Alkorta I, Garbisu C (2001) Phytoremediation of organic contaminants. Bioresour Technol 79:273–276

Anderson TA, Guthrie EA, Walton BT (1993) Bioremediation in the rhizosphere: plant roots and associated microbes clean contaminated soil. Environ Sci Technol 27:2630–2636

Angle JS, Linacre NA (2005) Metalphytoextraction – a survey of potential risks. Int J Phytoremediation 7:241–254

Arthur EL, Rice PJ, Anderson TA, Baladi SM, Henderson KLD, Coats JR (2005) Phytoremediation – an overview. Crit Rev Plant Sci 24:109–122

Bauder JW, Brock TA (1992) Crop species, amendments, and water quality effects on selected soil physical properties. Soil Sci Soc Am J 56:1292–1298

Bennett LE, Burkhead JL, Hale KL, Terry N, Pilon M, Pilon-Smits EAH (2003) Analysis of transgenic Indian mustard plants for phytoremediation of metals-contaminated mine tailings. J Environ Qual 32:432–440

Bingham FT, Pereyea FJ, Jarrell WM (1986) Metal toxicity to agricultural crops. Metal Ions Biol Syst 20:119–156

Bogardt AH, Hemmingsen BB (1992) Enumeration of phenanthrene-degrading bacteria by an over layer technique and its use in the evaluation of petroleum-contaminated sites. Appl Environ Microbiol 58:2579–2582

Boonyapookana B, Parkpian P, Techapinyawat S, Delaune RD, Jugsujinda A (2005) Phytoaccumulation of lead by sunflower (*Helianthus annuus*), tobacco (*Nicotiana tabacum*) and vetiver (*Vetiveria zizanioides*). J Environ Sci Health A Tox Hazard Subst Environ Eng 40:117–137

Burd GI, Dixon DG, Glick BR (1998) A plant growth-promoting bacterium that decreases nickel toxicity in seedlings. Appl Environ Microbiol 64:3663–3668

Chaney RL, Brown SL, Li YM (2000) US-EPA "Phytoremediation: state of science." Progress in risk assessment for soil metals, and in-situ remediation and phytoextraction of metals from hazardous contaminated soils, Boston, MA, 2000 May 1–2

Cheng S, Grosse W, Karrenbrock F, Thoennessen M (2002) Efficiency of constructed wetlands in decontamination of water polluted by heavy metals. Ecol Eng 18:317–325

Ciura J, Poniedzialek M, Sekara A, Jedrszezyk E (2005) The possibility of using crops as metal phytoremediants. Pol J Environ Stud 14:17–22

Cleland RE (1986) The role of hormones in wall loosening and plant growth. Aust J Plant Physiol 13:93–103

Cortés-Espinosa DV, Fernández-Perrino FJ, Arana-Cuenca A, Esparza-García F, Loera O, Rodríguez-Vázquez R (2006) Selection and identification of fungi isolated from sugar cane bagasse and their application for phenanthrene removal from soil. J Environ Sci Health A Tox Hazard Subst Environ Eng 41(1532–4117):475–486

Cunningham SD, Ow DW (1996) Promises and prospects of phytoremediation. Plant Physiol 110:715–719

de Souza MP, Chu D, Zhao M, Zayed AM, Ruzin SE, Schichnes D, Terry N (1999) Rhizosphere bacteria enhance selenium accumulation and volatilization by Indian mustard. Plant Physiol 119:565–573

Draycott AP, Durrant MJ (1976) Response by sugar beet to potassium and sodium fertilizers, particularly in relation to soils containing little exchangeable potassium. J Agric Sci (Camb) 87:105–112

Ebbs SD, Kochian LV (1998) Phytoextraction of zinc by oat (*Avina sativa*), barley (*Hordeum vulgare*), and Indian mustard (*Brassica juncea*). Environ Sci Technol 32:802–806

Elcock G, Martens J (1995) Composting with redwiggler worms. City Farmer, Canada's Office of Urban Agriculture, Canada, p 1

EPA US (2004) Radionuclide biological remediation resource guide, EPA 905-B-04-001. August. http://www.cluin.org/download/remed/905b04001.pdf

Foy CD, Chaney RL, White MC (1978) The physiology of metal toxicity in plants. Annu Rev Plant Physiol 29:511–566

Garbisu C, Hernandez-Allica J, Barrutia O, Alkorta I, Becerril JM (2002) Phytoremediation: a technology using green plants to remove contaminants from polluted areas. Rev Environ Health 17:173–188

Garg UK, Sud D (2005) Optimization of process parameters for removal of Cr (VI) from aqueous solutions using modified bagasse. Electr J Environ Agric Food Chem 4:1150–1160

Glick BR (2003) Phytoremediation: synergistic use of plants and bacteria to clean up the environment. Biotechnol Adv 21:383–393

Gothberg A, Greger M, Holm K, Bengtsson BE (2004) Influence of nutrient levels on uptake and effects of mercury, cadmium, and lead in water spinach. J Environ Qual 33:1247–1255

Grifferty A, Barrington S (2000) Zinc uptake by young wheat plants under two transpiration regimes. J Environ Qual 29:443–446

Gupta VK, Ali I (2004) Removal of lead and chromium from wastewater using bagasse fly ash: a sugar industry waste. J Colloid Interface Sci 271:321–328

Hamid A, Talibudeen O (1976) Effects of sodium on the growth and ion uptake by barley, sugar beet and broad beans. J Agric Sci (Camb) 86:49–56

Jadia CH, Fulekar MH (2008) Vermicomposting of vegetable waste: a bio-physicochemical process based on hydro-operating bioreactor. Afr J Biotechnol 7:3723–3730

Jain R, Srivastava S, Solomon S, Shrivastava AK, Chandra A (2010) Impact of excess zinc on growth parameters, cell division, nutrient accumulation, photosynthetic pigments and oxidative stress of sugar cane. Acta Physiol Plant 32:979–986

Kabata-Pendias A, Pendias H (1989) Trace elements in the soil and plants. CRC Press, Boca Raton, FL

Kaffka SR, Oster JD, Corwin DL (2004) Forage production and soil reclamation using saline drainage water. In: Proceedings National Alfalfa 475 Symposium, San Diego, CA, 13–15 Dec 2004. UC Cooperative 476 Extension, University of California, Davis, pp 247–253

Kloepper JW, Lifshitz R, Zablotowicz RM (1989) Free-living bacterial inocula for enhancing crop productivity. Trends Biotechnol 17:39–44

Krishnan AK, Anirudhan TS (2002) Uptake of heavy metals in batch systems by sulfurized steam activated carbon prepared from sugar cane bagasse Pith. Ind Eng Chem Res 41:5085–5093

Krishnani K, Parimala V, Meng XG (2004) Detoxification of chromium(VI) in coastal water using lignocellulosic agricultural waste. Water Soil Air 30:541–545

Kumar PBA, Dushenkov V, Motto H, Raskin I (1995) Phytoextraction: the use of plants to remove heavy metals from soils. Environ Sci Technol 29:1232–1238

Lasat HA (2002) Phytoextraction of toxic metals: a review of biological mechanisms. J Environ Qual 31:109–120

Lombi E, Zhao FJ, Dunham SJ, McGrath SP (2001) Phytoremediation of heavy metal contaminated soil: natural hyperaccumulation versus chemically enhanced phytoextraction. J Environ Qual 30:1919–1926

Magat SS, Goh KM (1988) Effects of chloride fertilizers on yield and uptake of chloride, potassium and sodium by fodderbeet (*Beta vulgaris* L.) in two New Zealand soils. J Agric Sci (Camb) 111:207–216

Maharashtra Nature Park Society (2003) Methods in biostatistics for medical students and research workers, 6th edn. Jaypee Brothers, New Delhi, India

Marschner H (1995) Mineral nutrition of higher plants. Academic, London, p 889

McGrath SP, Chaudri AM, Giller KE (1995) Long-term effects of metals in sewage sludge on soils, microorganisms and plants. J Ind Microbiol 14(2):94–104

Nandakumar PBA, Dushenkov V, Motto H, Raskin I (1995) Phytoextraction: the use of plants to remove heavy metals from soils. Environ Sci Technol 29:1232–1238

Niazi BH, Rozema J, Broekman RA, Salim M (2000) Dynamics of growth and water relations of fodderbeet and seabeet in response to salinity. J Agron Crop Sci 184:101–109

Qadir M, Oster JD (2004) Crop and irrigation management strategies for saline–sodic soils and waters aimed at environmentally sustainable agriculture. Sci Total Environ 323:1–19

Qadir M, Ghafoor A, Murtaza G (2000) Amelioration strategies for saline soils: a review. Land Degrad Dev 11:501–521

Rayment GE, Jeffrey AJ, Barry GA (2002) Heavy metals in Australian sugar cane. Commun Soil Sci Plant Anal 33:3203–3212

Reddy BVS, Ramesh S, Sanjana Reddy P, Ramaiah B, Salimath PM, Kachapur R (2005) Sweet sorghum – a potential alternate raw material for bio-ethanol and bio-energy. Int Sorghum Millets Newsl 46:79–86

Salt DE, Blaylock M, Kumar NPBA, Dushenkov V, Ensley BD, Chet I, Raskin I (1995) Phytoremediation: a novel strategy for the removal of toxic metals from the environment using plants. Biotechnology 13:468–474

Sanità di Toppi L, Gabrielli R (1999) Response to cadmium in higher plants. Environ Exp Bot 41:105–130

Saxena PK, Krishnaraj S, Dan T (1999) Phytoremediation of heavy metal contaminated and polluted soils. In: Prasad MNV, Hagemeyer J (eds) Heavy metal stress in plants: from molecules to ecosystems. Springer, Berlin, pp 305–329

Schnoor JL (1997) Abstract: Design of phytoremediation at contaminated sites. Presentation 48. In: 12th Annual Conference on Hazardous Waste Research – Abstracts Book, Kansas City, MO, May 19–22, 1997

Segura-Muñoz SI, da Silva Oliveira A, Nikaido M, Trevilato TM, Bocio A, Takayanagui AM, Domingo JL (2006) Metal levels in sugar cane (*Saccharum* spp.) samples from an area under the influence of a municipal landfill and a medical waste treatment system in Brazil. Environ Int 32(1):52–57

Sereno ML, Almeida RS, Nishimura DS, Figueira A (2007) Response of sugar cane to increasing concentrations of copper and cadmium and expression of metallothionein genes. J Plant Physiol 164:1499–1515

Shu WS, Xia HP, Zhang ZQ, Lan CY, Wong MH (2002) Use of vetiver and three other grasses for revegetation of Pb/Zn mine tailings: field experiment. Int J Phytoremediation 4:47–57

Smith SE, Read DJ (1997) Mycorrhizal symbiosis. Academic, San Diego

Spier TW, Van Schaik AP, Percival CME, Pang L (2003) Heavy metals in soil, plants and groundwater following high-rate sewage sludge application to land. Water Air Soil Poll 150:349–358

Su DC, Wong JWC (2004) Selection of Mustard Oilseed Rape (*Brassica juncea L.*) for Phytoremediation of cadmium contaminated soil. Bull Environ Contam Toxicol 72:991–998

Sun L, Niu Z, Sun T (2007) Effects of amendments of N, P, Fe on phytoextraction of Cd, Pb, Cu, and Zn in soil of Zhangshi by mustard, cabbage, and sugar beet. Environ Toxicol 22:565–571

Suthar SS, Watts J, Sandhu M, Rana S, Kanwal A, Gupta D, Meena MS (2005) Vermicomposting of kitchen waste by using *Eisenia foetida* (SAVIGNY). Asian J Microbiol Biotechol Environ Sci 7:541–544

Verret F, Gravot A, Auroy P, Leonhardt N, David P, Nussaume L, Vavasseur A, Richaud P (2004) Overexpression of AtHMA4 enhances root-to-shoot translocation of zinc and cadmium and plant metal tolerance. FEBS Lett 576:306–312

Wallace A, Romney EM (1977) Roots of higher plants as a barrier to translocation of some metals to shoots of plants. In: Biological implications of metals in the environment. Proceedings of the 15th annual Hanford life science symposium, Richland, Washington, pp 370–379, 29 Sept to 10 Oct

Weber O, Scholz RW, Bvhlmann R, Grasmuck D (2001) Risk perception of heavy metal soil contamination and attitudes toward decontamination strategies. Risk Anal 21:967–977
Wenger K, Gupta SK, Furrer G, Schulin R (2003) The role of nitrilotriacetate in copper uptake by tobacco. J Environ Qual 32:1669–1676
Wenwei C, Xia H (2007) Physiological responses of sugar cane to Pb progressive stress. College of Environmental Science and Engineering, Zhejiang Gongshang University, Hangzhou 310012, Zhejiang, China
Wenzel WW, Lombi E, Adriano DC (1999) Biochemical processes in the rhizosphere: role in phytoremediation of metal-polluted soils. In: Prasad MNV, Hagemeyer J (eds) Heavy metal stress in plants: from molecules to ecosystems. Springer, Berlin, pp 273–303
Whiting SN, de Souza MP, Terry N (2001) Rhizosphere bacteria mobilize Zn for hyperaccumulation by *Thlaspi caerulescens*. Environ Sci Technol 35:3144–3150
Xia H, Yan Z, Chi X, Wenwei C (2009) Evaluation of the phytoremediation potential of *Saccharum officinarum* for Cd-contaminated soil. Energy and environment technology, ICEET '09, International conference on issue, vol 3. Guilin, Guangxi, 16–18 Oct, pp 314–318
Yadav DV, Jain R, Rai RK (2010) Impact of heavy metals on sugar cane. In: Sherameti I, Varma A (eds) Soil heavy metals, vol 19, Soil biology series. Springer, Germany, pp 339–367, pp 492
Yan Z, Xia H (2010) Bioinformatics and biomedical engineering (iCBBE). 4th International Conference, Chengdu, 18–20 June
Zheljazkov VD, Warman PR (2004) Application of high – Cu compost to dill and peppermint. J Agric Food Chem 52:2615–2622
Zhuang P, Shu W, Li Z, Liao B, Li J, Shao J (2009) Removal of metals by sorghum plants from contaminated land. J Environ Sci 21:1432–1437

Chapter 21
Detoxification of Heavy Metals Using Earthworms

Oguz Can Turgay, Ridvan Kizilkaya, Ayten Karaca, and Sema Camci Cetin

21.1 Introduction

No doubt that earthworms are yet most valuable creatures on our planet with their 600 million years history. Factors such as soil characteristics, agricultural/industrial activities, and environmental pollution have significant effects on population dynamics and the biomasses of these invertebrates. Previous researches carried on earthworms indicated that they improve soil quality and fertility through their feeding, burrowing, and casting activities. Recently, there is a growing attention on a distinctive characteristic of earthworms. Many researches have revealed that earthworms have a capability to change the availability, uptake, and accumulation of heavy metals by passing and accumulating toxic metals through their body tissues. This chapter aims to provide an overview of earthworms in relation to their contribution to heavy metal detoxification in soil.

O.C. Turgay • A. Karaca (✉)
Faculty of Agriculture, Department of Soil Science and Plant Nutrition, Ankara University, Diskapi, Ankara, Turkey
e-mail: akaraca@agri.ankara.edu.tr

R. Kizilkaya
Faculty of Agriculture, Department of Soil Science and Plant Nutrition, Ondokuz Mayis University, Samsun, Turkey

S.C. Cetin
Faculty of Forestry, Cankiri Karatekin University, Cankiri, Turkey

I. Sherameti and A. Varma (eds.), *Detoxification of Heavy Metals*, Soil Biology 30,
DOI 10.1007/978-3-642-21408-0_21, © Springer-Verlag Berlin Heidelberg 2011

21.2 Ecological Classification of Earthworms

An earthworm substantially enhances physical, chemical, and biological characteristics of soil through their feeding, casting, and burrowing activities. Different earthworm species differ in their ecological strategies which influence main physical features of soil (soil aggregation and porosity) in different degrees and therefore classify them as epigeic, endogeic, and anecic (Lee 1985; Lavelle and Spain 2001; Karaca et al. 2010a; Kizilkaya et al. 2011) (Fig. 21.1). Epigeic species such as *Lumbricus rubellus*, *Eisenia fetida*, *Dendrodrilus rubidus* live in soil humus zone. They are fed from organic materials accumulated above mineral soil layer and therefore occasionally digest mineral soil particles. The typical habitats of epigeic species are manure masses and plant debris layers in forest ecosystems. They build their burrows in organic material layer or in a depth of 0–2.5 cm mineral soil and substantially feed on organic compounds rich in microorganisms. Epigeics are small worms (usually shorter than 7.5 cm) with a reddish brown color. Anecic species such as *Lumbricus terrestris*, *Aporrectodea longa*, *Dendrobaena platyura* are reddish brown color worms with the largest and longest sizes ranging between 12.5 and 20.0 cm. They live in permanent or semi-permanent burrows reached to a depth of 2 m, feed on decaying organic material on soil surface and leave their castings at the mouth of their burrows located on the surface. Endogeic species (*Aporrectodea caliginosa*, *Allolobophora chlorotica*, *Octolasion lacteum* v.b.) live on organic compounds found in mineral soil layers, and inhabit the top 0–50 cm of the soil. They are distinguished from epigeic and anecic species by their distinct color characteristics such as lack of red-brown skin pigmentation and appearance of very pink color on the head and gray color on the body. Adult endogeic species can range from 3 to 12.5 cm (Lee 1985; Lavelle and Spain 2001; Karaca et al. 2010a; Kizilkaya et al. 2011).

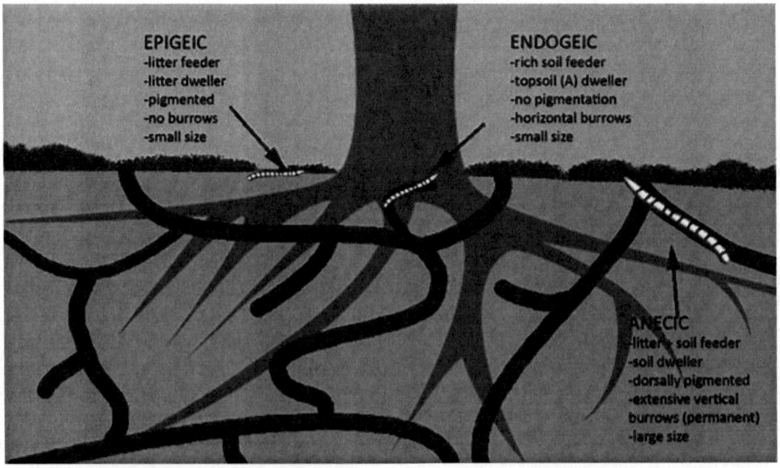

Fig. 21.1 Ecological classification of soil earthworms

21.3 Earthworm Distribution in Soil

In general, there are considered to be around 6,000 species of earthworms (Fragoso et al. 1999; Lavelle and Spain 2001). Apart from extreme environments such as desert and glacial soils, earthworms can inhabit a wide variety of soil environments including agriculture, forest, and pasture ecosystems. The size of adult earthworms ranged from few mm to 2 m while their body mass changes between 10 mg and 1 kg. Giant earthworm species usually inhabit in tropical regions of southern hemisphere such as South America and Africa, Southeastern Asia, Australia, and New Zealand. Some other earthworms living in other regions of the earth are usually comparatively smaller and have lower body mass. The Lumbricidae are dominant species in temperate areas while remaining families inhabit predominantly tropical or subtropical areas (Lavelle et al. 1999). The number of different earthworm species living in a certain soil environment can be three or five and occasionally more than ten. In general, similar earthworm species exist in similar soil and climate conditions (Edwards and Lofty 1982a, b; Edwards and Bohlen 1996). The soil can contain 10–1,000 individual earthworms or 1–200 g earthworm biomass per square meter soil, depending on time and soil environmental characteristics. Earthworm biomass and diversity in soil is closely associated with plant vegetation and climate. Even forest type, deciduous versus coniferous can be significant factor affecting earthworm abundance and species diversity in soil. The temperate region soils were found to be predominated by 12 Lumbricidae species while 7 species were identified from Africa's soils (Edwards and Lofty 1982a, b; Edwards and Bohlen 1996). Tropical agroecosystems have been reported to be more diverse than these regions with 20 earthworm species (Barois et al. 1999). Lifespan of an earthworm usually ranges between 10 and 12 years but many species are eaten by the predator such as large insects, moles, and birds and therefore survive 1–2 years (Edwards and Bohlen 1996). The main soil characteristics that control the earthworm number and biomass are soil organic matter (SOM), texture, pH, water holding capacity, and soil temperature (Lee 1985; Bernier and Ponge 1994; Lavelle and Spain 2001; Karaca et al. 2010a).

21.4 Factors Affecting Earthworm Population and Activity

The factors affecting earthworm populations and activities in soil can be discussed under two general perspective as surrounding environmental factors (i.e., climate, soil characteristics, and plant vegetation) and biological relationships (i.e., competition, hunting, and parasitism).

21.4.1 Climate

Climate has a direct influence on earthworm biology and life cycle while their habitats and feeding activities are indirectly related to climate. For example, temperature affects either earthworm metabolic activities individually or their distribution globally (Lavelle 1983; Lavelle et al. 1989, 1999). Epigeic and anecic species rapidly decompose organic matter at high temperatures and hence surface debris availability diminishes. This process occurs more rapidly in tropical region soils compared those in temperate region soils (Lavelle et al. 1999). Insufficient soil moisture at high temperatures affects earthworm populations and their activities negatively (Gerard 1967; Phillipson et al. 1976). Depending on the variability among different species, earthworm growth is optimum at the field capacity (10 kPa tension), rapidly declines over 100 kPa moisture tension and completely ceases below the permanent wilting point (1,500 kPa) (Nordström and Rundgren 1974; Nordström 1975; Baker et al. 1993). Earthworm activity changes greatly between seasons in temperate regions. It is usually higher in spring and autumn and lower during winter months when they penetrate deeper into soil. In dry summers they also move through deeper soil layers and can survive by forming stationary coils until environmental conditions become favorable. Cocoon production is generally seasonal but can also be produced any time of the year. In temperate zones, cocoon is mostly produced during spring and early summer while second cocoon production is comparatively lower and achieved in autumn. The number of the cocoons produced changes between 1 and 20 depending on earthworm species. Earthworm activity is closely associated with soil moisture and temperature. Insufficient and oversaturated moisture conditions have negative effect on earthworms and most species cannot survive below 1°C and above 30–35°C (Edwards 1983). Earthworms produce cocoons where the environmental conditions are not favorable enough to survive.

21.4.2 Soil Properties

The physical and chemical characteristics of soils (i.e., texture, depth, pH, and organic matter) affect earthworm activity significantly. These characteristics are largely influenced by climatic factors (i.e., precipitation and temperature). For example, basic cations such as Na^+, Mg^{2+}, and Ca^{2+} are leached through the soil profile, replaced with H^+ and eventually results in soil acidification in regions with heavy rainfall. Earthworm activity declines rapidly below pH 4.5 and most species ceases to exist under strongly acidic conditions i.e., pH < 3.5. The ideal pH conditions vary depending on earthworm species but most earthworms living in temperate regions favor the range between pH 5.0 and 7.4 (Satchell 1967). On the other hand, earthworm populations and activities are also low under highly alkaline soils. This is due to the fact that soils developed in arid and semi-arid regions contain inadequate organic matter in relation to low amount of rainfall. The texture is one of the physical aspects of soil influential on earthworm populations and activities. Comparing with sandy and clayey soils, loamy textured soils are more preferable for earthworms (Guild 1948). Long-term anoxic conditions may occur after heavy rain falls in heavy clay soils and sandy soils may have lower water holding capacity causing decreases in earthworm activities. Soil depth is another important factor effective on earthworm distribution in moderate and tropical regions (Phillipson et al. 1976; Fragoso and Lavelle 1992; Lavelle et al. 1999). Earthworms build their burrows in different depths of soil and the activities of the species living in lower soil layers are usually poor comparing with those growing upper soil layers (Curry and Cotton 1983).

Earthworms feed on decomposing organic compounds in soil and their populations and activities are largely depend on SOM content and quality which are closely associated with plant vegetation growing on soil surface. The litter layer of plant residues composed of grass, herbaceous plants, and deciduous trees contains a nutrient-rich organic matter with a lower C:N ratio less than 20:1 and therefore more preferable than those with high C:N ratios more than 60:1 (Edwards and Bohlen 1996; Hendrix et al. 1992). Organic matter amendments may have a favorable effect on earthworm life in soil. However, animal wastes can reversely affect earthworms and decrease their activities when applied to soil due to their high salt and ammonia contents. Liming acidic soils and chemical fertilization applications yield increases in plant biomass production and hence increase earthworm biomass.

21.5 Earthworm Castings

Earthworms usually can digest soil or organic materials at the rate of 60% of their body mass and defecate their fecal pellets which are called earthworm castings into their burrows in soil. There are basically four types of earthworm castings (Lee 1985; Lavelle 1988; Edwards and Bohlen 1996). The first type is globular and usually produced by large earthworm species i.e., anecics and endogeics. Second type is also formed by endogeics and anecics but it is usually shapeless. Third, these species also cast spherical pellets at the soil surface. The fourth type appears to be granular or pellet and is formed by smaller earthworm species such as epigeics, small endogeics, and various anecic species. The earthworm casts produced by different species differ in their effect on soil structure. The first three types of casts are larger, heavier, and more compact whereas granular type is smaller, lighter, and more crumbly (Blanchart et al. 1997, 1999). Although the characteristics of earthworm excrements largely resemble to the composition of organic material that earthworms feed on, they may differ in chemical and biological characteristics (Kizilkaya et al. 2010b). This feature is also closely related to the mechanical grinding step and the activities of the microorganisms that live in the earthworm's intestinal system. Compared with the organic material consumed by earthworms, the cast of earthworms has a narrow C:N ratio and higher available nutrient content, more stabile microbial characteristics, and higher extracellular enzyme activities (Kizilkaya and Hepsen 2004, 2007; Kizilkaya 2008; Hepsen and Kizilkaya 2010). Therefore, earthworm casts are commonly accepted as a natural fertilizer due to their high nutrient capacity and called vermicompost in agriculture (Kizilkaya et al. 2010a).

21.6 Earthworm Effects on Soil Characteristics

The influences of earthworms on soil characteristics are mainly driven by their feeding, casting, and burrowing activities (Lavelle and Spain 2001). Depending on the mineral soil and/or organic material digested by earthworms, their casts accumulated at soil surface are rich in organic compounds and may change soil properties. Organic constituents in earthworm castings are readily mineralized due to their high C, N, and water contents (Kizilkaya 2008) and this increases stabile aggregate formation in soil. Fungal hyphae and various microbial metabolites in the excrement provide more strong binding between soil particles and hence play an important role in the formation of stabile soil aggregates (Chan and Heenan 1995; Tomlin et al. 1995; Haynes and Fraser 1998). The gallery building activities of earthworms vary in relation to their species and soil type and increase macropores in soil. Depending on ecological categories of the earthworm, these galleries can range between 1 and 10 mm (Edwards and Shipitalo 2004). Epigeic species are small species living near the soil surface. They create small galleries within the first few cm of soil either vertically or horizontally. Endogeics burrow continuously and form a network of channels. These species usually leave their feces within their galleries and this may limit water percolation through lower soil layers. Anecic earthworms create deep vertical galleries that may reach to 2 m into the soil. The bulk density of soil surrounding these galleries is higher and water movement is more stabile in these galleries (Lee 1985; Tomlin et al. 1995). Earthworms mix mineral soil with organic residues accumulated on soil surface and usually leave their excrements that are rich in readily decomposable organic compounds near the soil surface. The role of earthworms in organic material decomposition and nutrient conversion processes is driven by their population density and feeding status (Lee 1985; Barois et al. 1999). Epigeic earthworm species usually feed on organic residues on the soil surface or mineral soil through O and A horizons. They help to the process of mixing organic materials with topsoil and thereby stimulate mineralization of organic compounds due to increasing access between organic compounds and microbial populations. While anecic species carry organic materials from soil surface to lower layers through their deep vertical galleries they leave their feces at soil surface. Epigeic and anecic earthworm activities contribute the formations of "mull" soil horizon and a well developed A horizon. Moreover, they lead to formation of organo-mineral complexes (vermimul) consisting of earthworm excrements in Ah horizon (Green et al. 1993). Endogeic species feed primarily on mineral soil, decaying surface organic residues and symbiotic microorganisms. In endogeic species, the mineralization of organic compounds of the casts and nutrient contents of their burrow walls are much larger than in the surroundings. This earthworm secretion-rich soil compartment and the gallery zone burrowed by epigeic species is termed as "drilosphere" (Lavelle et al. 1998). The excrements and burrow walls of both endogeic and anecic species are rich in microbial populations and have higher enzyme activities. The characteristics of organic materials consumed by earthworms determine the microbiological properties of their excrements. In general, the excrements of the earthworms feed on organic materials with narrow C:N ratios are more readily decomposed and have better microbiological properties such as higher microbial biomass and enzyme activities (Kizilkaya and Hepsen 2004, 2007; Kizilkaya 2008). Soil pH is also influenced from earthworm activities. This is mainly related to higher basic cation content (Ca^{2+}, Mg^{2+}, and K^{+}) of earthworm casts than their surroundings. Earthworm activities also increase available nutrient pool in soil (Mackay et al. 1983; Lopez-Hernandez et al. 1993; Parkin and Berry 1994; Zhang et al. 2000; Karaca et al. 2010b).

21.7 Earthworm–Heavy Metal Relationships and Accumulation and Detoxification of Heavy Metals by Earthworms

Although heavy metals exist lithologically in the ground, their concentrations in soil increase through various industrial emissions (Cemek and Kizilkaya 2006), commercial fertilizers, (Karaca et al. 2002) and sewage sludges (Kizilkaya and Bayrakl 2005). Depending on soil characteristics, heavy metals accumulated in food chain and this affects soil life and especially biological–biochemical reactions negatively (Kizilkaya and Askin 2002; Kizilkaya et al. 2004; Karaca et al. 2010b). The earthworms are used as a cursor when assessing the effects of heavy metals on various ecosystems. Heavy metals influence earthworm life by killing (Fitzpatrick et al. 1996; Neuhauser et al. 1985; Spurgeon and Hopkin 1995; Kizilkaya et al. 2009,) or inhibiting their growth (Khalil et al. 1996; Van Gestel et al. 1991), cocoon production (Ma 1988; Spurgeon and Hopkin 1996), and activity (Siekierska and Urbanska-Jasik 2002).

In ecosystem risk assessments, concentrations of mobile or available heavy metals are more significant than their total amounts (Nahmani et al. 2007). Earthworms can affect either available or total metal concentrations in soil in that they are capable of accumulating heavy metals in their tissues (Kizilkaya et al. 2009; Karaca et al. 2010a), and hence reduce their involvement in soil food chain. During their feeding activities, earthworms can change either available or total metal concentrations in soil due to their metal accumulation capability (Beyer et al. 1987; Wang et al. 1998; Paoletti 1999; Nahmani et al. 2007; Hepsen and Kizilkaya 2007) and hence reduce their involvement in soil food chain. They also leave a portion of heavy metals to soil environment by their casting activities. Earthworms can change both available and total metal concentrations in soil. During their feeding activities, earthworms partially accumulate heavy metals in their tissues (Beyer et al. 1987; Wang et al. 1998; Paoletti 1999; Nahmani et al. 2007; Hepsen and Kizilkaya 2007) and also leave a portion of heavy metals to soil environment by their casting activities (Lee 1985; Lavelle et al. 1998, 1999) (Fig. 21.2) and hence reduce their involvement in soil food chain. Heavy metals can be concentrated in earthworm tissues in high concentrations, whereas their excrements can contain lower amount of metals (Kizilkaya 2004). In this case, new generations can move through less polluted or unpolluted soils (Karaca et al. 2010a). The accumulation of heavy metals by earthworms is mainly associated with the factors such as type of mineral soil, organic matter content, and metal concentrations of their living environment (Table 21.1) and has been often applied as biological monitoring of various metal pollutions (Rozen and Mazur 1997; Wang et al. 1998) resulted from industrial activities (Wright and Stringer 1980; Bengtsson et al. 1983; Beyer et al. 1985) and heavy road traffic (Gish and Christensen 1973; Gullvag 1979).

Another factor affecting accumulation of heavy metals in earthworms is their ecological category (Nahmani et al. 2007). For example, *Lumbricus rubellus* and *Aporrectodea caliginosa* are the members of different ecological categories and their metal accumulation capacities were found to be different from each other

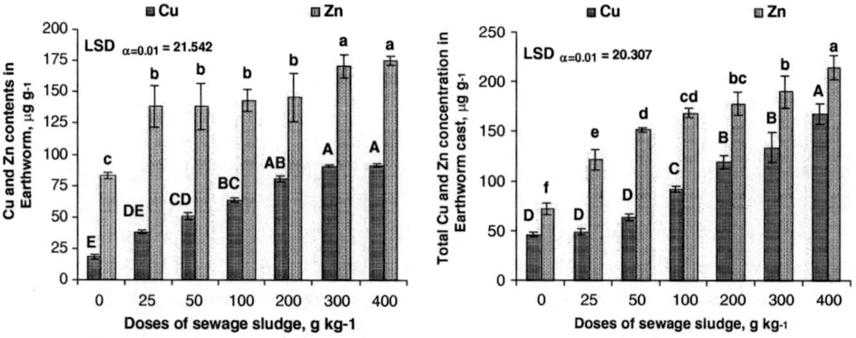

Fig. 21.2 Changes in Cu and Zn concentrations of earthworm tissue (**a**) and excrement (**b**) under increasing doses of sewage sludge treatments (Kizilkaya 2004)

Table 21.1 Heavy metal contents of soil and earthworm (Wang et al. 1998)

Soil	Cu (mg kg^{-1})		As (mg kg^{-1})		Cd (mg kg^{-1})		Pb (mg kg^{-1})		Zn (mg kg^{-1})		Hg (mg kg^{-1})	
No	S	E	S	E	S	E	S	E	S	E	S	E
1	74.68	4.58	56.58	16.32	9.18	47.3	670.5	35.2	657.8	67.0	0.95	0.57
2	67.25	3.10	57.78	4.77	3.81	15.8	325.3	8.70	367.8	53.0	0.39	0.27
3	51.25	3.00	51.78	5.86	6.26	15.4	459.0	13.5	479.0	80.0	0.75	0.36
4	25.00	3.06	28.59	1.41	1.87	1.9	204.6	20.8	159.3	35.3	0.30	0.30
5	25.63	2.00	23.56	1.29	0.52	5.8	55.0	2.46	91.1	47.0	0.27	0.05
6	40.63	2.28	24.56	1.72	0.53	2.1	61.1	0.81	220.8	41.0	0.74	0.11

S soil; *E* earthworm

(Morgan and Morgan 1999). Suthar et al. (2008) also observed significant differences between metal accumulation capacities of endogeic *Metaphire posthuma* and anecic *Lampito mauritii* collected from different soils (agricultural and orchard) and sewage sludge. The effect of ecological category on heavy metal accumulation by earthworms is related to the facts that different earthworm species inhabit and burrow at different depths through soil profile and the organic materials consumed by earthworms have different metal contents (Ireland and Richars 1977; Ash and Lee 1980; Morgan and Morgan 1999). In terms of ecology, epigeics are the species capable to accumulate highest amount of metals in their tissues while anecics are the species with least metal accumulation. However, there may be differences between metal accumulation capacities of different earthworms within the same ecological category. Among epigeics, the most capable species is *Eisenia fetida* which is applied as reference earthworm during heavy metal toxicity tests due to its advantages such as being easily culturable, short regeneration, rapid reproduction, and response to various heavy metals in laboratory conditions (International Standard Organization 1993, 1998). The level of heavy metals accumulated in earthworm body depends on the quality of organic materials used by earthworms (Fig. 21.3). It has been shown that the earthworms feeding on the organic materials with larger C:N ratio accumulated higher concentrations of metals compared to those using organic materials with narrow C:N ratio (Kizilkaya 2005). Moreover, heavy metals concentrated in the species of *Lumbricus rubellus* and *Aporrectodea tuberculata* found to decrease depending on the increase in organic matter content of metal

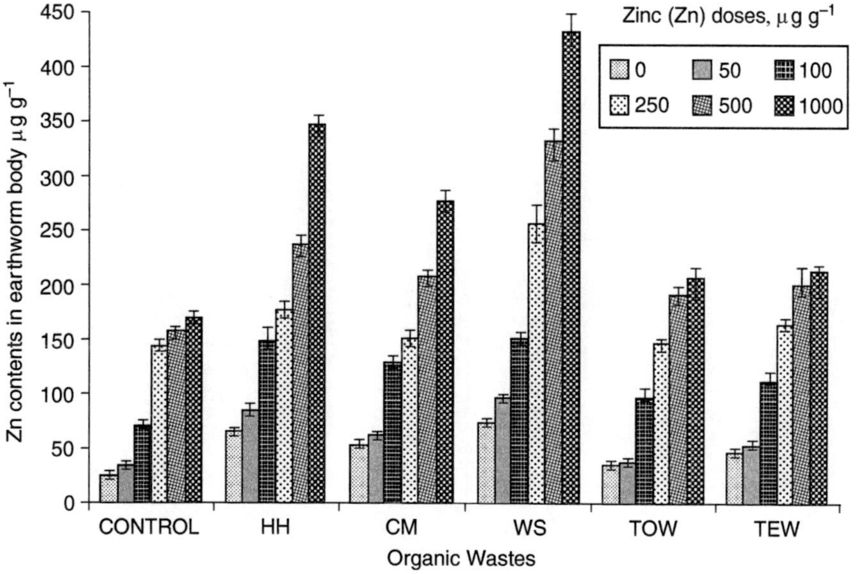

Fig. 21.3 The changes in Zn concentrations of earthworm tissues under different organic residue applications combined with increasing Zn doses (*HH* hazelnut husk, *CM* cow manure, *WS* wheat straw, *TOW* tobacco production waste, *TEW* tea production waste) (Kizilkaya 2005)

polluted soils (Ma 1982; Beyer et al. 1987). However, if the amount of metals binding to organic compounds increases in a polluted soil, metal accumulation in earthworms increases since they naturally prefer high quality organic compounds rather than mineral soil. Vermicomposting of sewage sludges having high heavy metal content and their applications in increasing doses have been shown to increase the metal concentration accumulated in earthworm tissues (Kizilkaya 2004, 2005).

The soil pH is one of the important soil chemical characteristics that can control heavy metal accumulation by earthworms (Morgan 1985; Morgan and Morgan 1988). The solubility of heavy metals is closely related to soil pH and metal uptake by living organism's increases with decreasing pH (Herms and Brümner 1984). Therefore, earthworm metal accumulation increases in lower pH conditions. For example, in acidic soils with metal pollution, *Lumbricus rubellus* (Ma 1982; Ma et al. 1983) and *Aporrectodea caliginosa* were found to accumulate higher metal concentrations (Peramaki et al. 1992). Soil moisture content also affects metal content accumulated in earthworm tissues (Marinussen and van der Zee 1997) (Fig. 21.4). Soil moisture primarily affects earthworm activity and then fate of

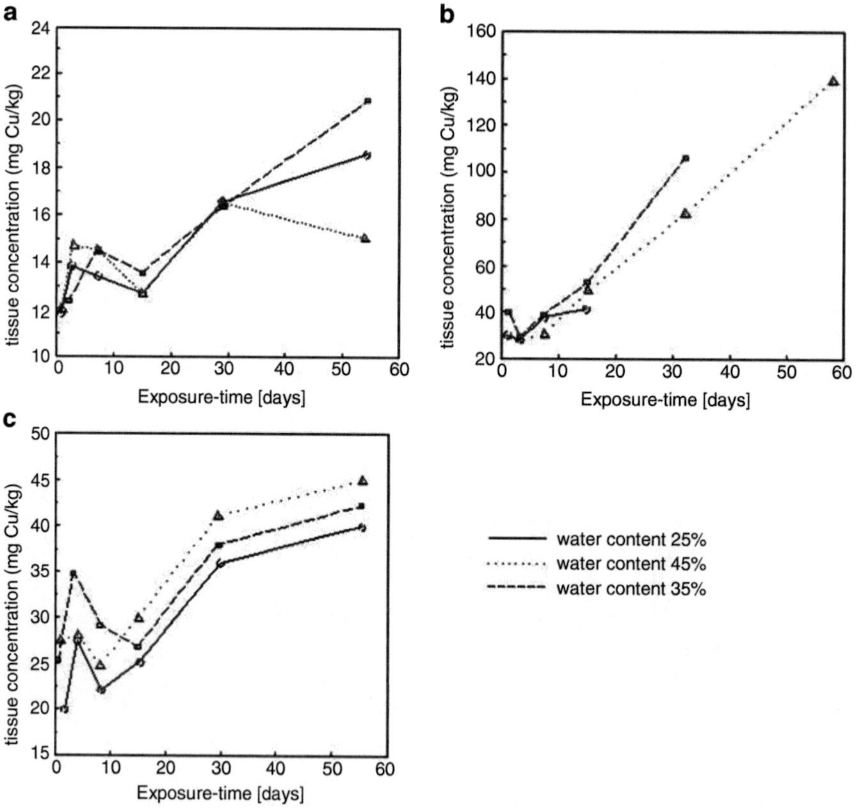

Fig. 21.4 Copper accumulation by *L. rubellus* under laboratory conditions; (**a**) reference soil (10 mg kg^{-1} Cu), (**b**) soil No 3 (132 mg kg^{-1} Cu), and (**c**) soil No 11 (80 mg kg^{-1} Cu)

heavy metals in soil. Water content near field capacity is the most appropriate soil water condition favoring an ideal earthworm activity whereas saturated and insufficient water conditions have negative influence on both earthworm activities and their metal accumulation capability. High soil salinity decreases metal accumulation by earthworms (Chang et al. 1997). Similarly, high level of ammonium ions in earthworm feeding environment affects earthworm activities (Masciandaro et al. 2002; Kizilkaya et al. 2009) and hence their metal accumulation capacity negatively.

21.8 Conclusion

Many researches revealed that earthworms can improve soil fertility by stimulating physical, chemical, and biological characteristics of the soil. They can also change soil ecology by suppressing plant pathogens and promoting the growth of soil microflora and fauna. More recently, earthworms have been shown to be not only resistant to metal toxicity but also capable of accumulating heavy metals in their body tissues and increasing metal uptake. This newly explored feature of earthworms may provide many advantages for monitoring of soil environmental quality, pollution assessment, and phytoremediation. However, it should be kept in mind that earthworm–heavy metal relationships are mostly driven by soil characteristics and their ecological category.

References

Ash CPJ, Lee DL (1980) Lead, cadmium, copper and iron in earthworms from roadside sites. Environ Pollut 22:59–67

Baker GH, Barrett VJ, Carter PJ, Williams PML, Buckerfield JC (1993) Seasonal changes in the abundance of earthworms (Annelida: Lumbricidae and Acanthodrilidae) in soils used for cereal and lucerne production in South Australia. Aust J Agric Res 44:1291–1301

Barois I, Lavelle P, Brossand M, Tondal L, Martinez M, Rossi JP, Senapati BK, Angeles A, Fragoso C, Jimienez JJ, Decaens T, Lattand C, Kamyono J, Chapius L, Brown GE, Monerno A (1999) Ecology of earthworms with large environmental tolerance and extended distribution. In: Lavelle P, Brussaard L, Hendrix P (eds) Earthworm management in tropical ecosystems. CABI, Wallingford, Oxford, UK, pp 57–86

Bengtsson G, Nordstrom S, Rundgren S (1983) Population density and tissue metal concentration of lumbricids in forest soils near a brass mill. Environ Pollut 30:87–108

Bernier N, Ponge JF (1994) Humus form dynamics during the sylvogenetic cycle in a mountain spruce forest. Soil Biol Biochem 26:183–220

Beyer WN, Pattee OH, Sileo L, Hoffman DJ, Mulhern BM (1985) Metal contamination in wildlife living near two zinc smelters. Environ Pollut 38:63–86

Beyer WN, Hensler G, Moore I (1987) Relation of pH and other soil variables to concentrations of Pb, Cu, Zn, Cd and Se in earthworms. Pedobiologia 30:167–172

Blanchart E, Lavelle P, Braudeau E, Le Bissonnais Y, Valentin C (1997) Regulation of soil structure by geophagous earthworm activities in humid savannas of Côte d.Ivoire. Soil Biol Biochem 29:431–439
Blanchart E, Albrecht A, Alegre J, Duboisset A, Gilot C, Pashanasi B, Lavelle P, Brussaard L (1999) Effects of earthworms on soil structure and physical properties. In: Lavelle P, Brussaard L, Hendrix P (eds) Earthworm management in tropical ecosystems. CABI, Wallingford, Oxford, UK, pp 149–172
Cemek B, Kizilkaya R (2006) Spatial variability and monitoring of Pb contamination of farming soils affected by industry. Environ Monit Assess 117:357–375
Chan KY, Heenan DP (1995) Occurrence of enchytraeid worms and some properties of their casts in an Australian soil under cropping. Aust J Soil Res 33:651–657
Chang LW, Meier JR, Smith MK (1997) Application of plant and earthworm bioassays to evaluate remediation of a lead-contaminated soil. Arch Environ Contam Toxicol 32:166–171
Curry JP, Cotton DCF (1983) Earthworms and land reclamation. In: Satchell JE (ed) Earthworm ecology from Darwin to vermiculture. Chapman and Hall, London, pp 215–228
Edwards CA (1983) Earthworm ecology in cultivated soils. In: Satchell JE (ed) Earthworm ecology from Darwin to vermiculture. Chapman and Hall, London, pp 123–137
Edwards CA, Bohlen PJ (1996) Biology of and ecology earthworms, 3rd edn. Chapman and Hall, London
Edwards CA, Lofty JR (1982a) The effect of direct drilling and minimal cultivation on earthworm populations. J Appl Ecol 19:723–724
Edwards CA, Lofty JR (1982b) Nitrogenous fertilizers and earthworm populations in agricultural soils. Soil Biol Biochem 14:515–521
Edwards WM, Shipitalo MJ (2004) Consequences of earthworms in agricultural soils: aggregation and porosity. In: Edwards CA (ed) Earthworm ecology. Lewis Publishers, Boca Raton, pp 147–161
Fitzpatrick LC, Muratti-Ortiz JF, Venables BJ, Goven AJ (1996) Comparative toxicity in earthworms *Eisenia fetida* and *Lumbricus terrestris* exposed to cadmium nitrate using artificial soil and filter paper protocols. Bull Environ Contam Toxicol 57:63–68
Fragoso C, Lavelle P (1992) Earthworm communities of tropical rain forests. Soil Biol Biochem 24:1397–1408
Fragoso C, Lavelle P, Blanchart E, Senapati BK, Jimenez JJ, Martinez M, Decaens T, Tondoh J (1999) Earthworm communities of tropical agroecosystems: origin, structure and influence of management practices. In: Lavelle P, Brussaard L, Hendrix P (eds) Earthworm management in tropical agroecosystems. CAB International, Wallingford
Gerard BM (1967) Factors affecting earthworms in pastures. J Anim Ecol 36:235–252
Gish CD, Christensen RE (1973) Cadmium, nickel, lead and zinc in earthworms from roadside soil. Environ Sci Technol 7:1060–1062
Green RN, Trowbridge RL, Klinka K (1993) Towards a taxonomic classification of humus forms. Forest Sci Monogr 29:1–49
Guild WJMcL (1948) Studies on the relationship between earthworms and soil fertility. III. The effect of soil type on the structure of earthworm populations. Ann Appl Biol 35:181–192
Gullvag BM (1979) Subcellular localization of polluting metals in roadside earthworms exposed to traffic exhaust gases. Cytobios 22:141–153
Haynes R, Fraser P (1998) A comparison of aggregate stability and biological activity of the earthworms *Lumbricus terrestris* and *Aporrectodea giardi* and consequences on C transfer in soil. Eur J Soil Biol 36:27–34
Hendrix PF, Mueller BR, Bruce RR, Langdale GW, Parmelee RW (1992) Abundance and distribution of earthworms in relation to landscape factors on the Georgia Piedmont, USA. Soil Biol Biochem 24:1357–1361
Hepsen FS, Kizilkaya R (2007) Determination of sewage sludge soil treatments effect on Cu bioaccumulation by earthworms and Cu contents of casts and surrounding soil.

14th International conference of students, PhD students and young scientists, Lomosonov-2007, Moscow State University, Moscow, Russia, 10–14 April 2007

Hepsen TFS, Kizilkaya R (2010) Effects of different application doses of sewage sludge on microbial biomass C and basal respiration in soil and in earthworm *L. terrestris* L. cast. In: Kizilkaya R, Gulser C, Dengiz O (eds) Proceedings of the international soil science congress on management of natural resources to sustain soil health and quality. Ondokuz Mayis University, Samsun, Turkey, pp 1111–1117, 26–28 May 2010

Herms U, Brümner G (1984) Solubility and retention of heavy metals in soils. J Plant Nutr Soil Sci 147:400–424

International Standard Organization (1993) Standard number No.11268-1. Soil quality-effects of pollutants on earthworms (*Eisenia fetida*) – part I: determination of acute toxicity using artificial soil substrate. ISO, Geneva

International Standard Organization (1998) Standard number No.11268-2. Soil quality-effects of pollutants on earthworms (*Eisenia fetida*) – part II: method for the determination of effects on reproduction. ISO, Geneva

Ireland MP, Richars KS (1977) The occurrence and localization of heavy metals and glycogen in the earthworms *Lumbricus rubellus* and *Dendrobaena rubida* from a heavy metal site. Histochemistry 51:153–166

Karaca A, Naseby D, Lynch J (2002) Effect of cadmium-contamination with sewage sludge and phosphate fertiliser amendments on soil enzyme activities, microbial structure and available cadmium. Biol Fertil Soils 35:435–440

Karaca A, Kizilkaya R, Turgay OC, Cetin SC (2010a) Effects of earthworms on the availability and removal of heavy metals in soils. In: Sherameti I, Varma A (eds) Soil heavy metals, soil biology, vol 19. Springer, Berlin, pp 369–388

Karaca A, Cetin SC, Turgay OC, Kizilkaya R (2010b) Effects of heavy metals on soil enzyme activities. In: Sherameti I, Varma A (eds) Soil heavy metals, soil biology, vol 19. Springer, Berlin, pp 237–262

Khalil MA, Abdel-Lateif HM, Bayoumi BM, van Straalen NM (1996) Analysis of separate and combined effects of heavy metals on the growth of *Aporrectodea caliginosa* (Oligochaeta; Annelida), using the toxic approach. Appl Soil Ecol 4:213–219

Kizilkaya R (2004) Cu and Zn accumulation in earthworm *Lumbricus terrestris* L. in sewage sludge amended soil and fractions of Cu and Zn in casts and surrounding soil. Ecol Eng 22:141–151

Kizilkaya R (2005) The role of different organic wastes on zinc bioaccumulation by earthworm *Lumbricus terrestris* L. (Oligochaeta) in successive Zn added soil. Ecol Eng 25:322–331

Kizilkaya R (2008) Dehydrogenase activity in *Lumbricus terrestris* casts and surrounding soil affected by addition of different organic wastes and Zn. Bioresour Technol 99:946–953

Kizilkaya R, Askin T (2002) Influence of cadmium fractions on microbiological properties in Bafra plain soils. Arch Agro Soil Sci 48:263–272

Kizilkaya R, Bayrakl B (2005) Effects of N-enriched sewage sludge on soil enzyme activities. Appl Soil Ecol 30:192–202

Kizilkaya R, Hepsen S (2004) Effect of biosolid amendment on enzyme activities in earthworm (*Lumbricus terrestris*) casts. J Plant Nutr Soil Sci 167:202–208

Kizilkaya R, Hepsen S (2007) Microbiological properties in earthworm *Lumbricus terrestris* L. cast and surrounding soil amended with various organic wastes. Commun Soil Sci Plant Anal 38:2861–2876

Kizilkaya R, Askin T, Bayrakl B, Sağlam M (2004) Microbiological characteristics of soils contaminated with heavy metals. Eur J Soil Biol 40:95–102

Kizilkaya R, Hepsen S, Akca I, Bayrakli B, Askin T, Turkmen C (2009) Determination of total and mobile Pb fractions during vermicomposting in sewage sludge. International symposium on environment, Kyrgyzstan – Turkey Manas University, Faculty of Engineering, Bishkek, Kyrgyzstan, 20–23 May 2009

Kizilkaya R, Hepsen Turkay FS, Turkmen C, Durmus M (2010a) Vermicompost effects on the wheat yield and nutrient contents in soil and plant. International conference on soil fertility and soil productivity, Berlin, Germany, 17–20 March 2010

Kizilkaya R, Hepsen Turkay FS, Bayrakli B, Askin T, Turkmen C, Akca I, Ceyhan V (2010b) Enzyme and earthworm activities during vermicomposting in sewage sludge, In: Kizilkaya R, Gulser C, Dengiz O (eds) Proceedings of the international soil science congress on management of natural resources to sustain soil health and quality. Ondokuz Mayis University, Samsun, Turkey, pp 1047–1054, 26–28 May 2010

Kizilkaya R, Karaca A, Turgay OC, Cetin SC (2011) Earthworm interactions with soil enzymes. In: Karaca A (ed) Biology of earthworms, soil biology, vol 24. Springer, Berlin, pp 141–158

Lavelle P (1983) The structure of earthworm communities. In: Satchell JE (ed) Earthworm ecology from Darwin to vermiculture. Chapman and Hall, London, pp 449–466

Lavelle P (1988) Earthworms and the soil system. Biol Fertil Soils 6:237–251

Lavelle P, Spain AV (2001) Soil ecology. Kluwer Academic Publishers, Dordrecht, The Netherlands

Lavelle P, Barois I, Martin A, Zaidi Z, Schaefer R (1989) Management of earthworm populations in agroecosystems: a possible way to maintain soil quality? In: Clarholm M, Bergström L (eds) Ecology of arable land. perspectives and challenges. Kluwer Academic Publishers, Dordrecht, pp 109–122

Lavelle P, Pashanasi B, Charpentier F, Gilot C, Rossi J, Derouard L, Andre J, Ponge J, Bernier N (1998) Influence of earthworms on soil organic matter dynamics, nutrient dynamics and microbiological ecology. In: Edwards CA (ed) Earthworm ecology. Lewis Publishers, Boca Raton, FL, pp 103–122

Lavelle P, Brussaard L, Hendrix P (1999) Earthworm management in tropical agroecosystems. CABI Publishing, Wallingford, Oxford, UK

Lee KE (1985) Earthworms: their ecology and relationships with soils and land use. Academic, Sydney, Australia

Lopez-Hernandez D, Lavelle P, Fardeau JC, Nino M (1993) Phosphorous transformations in two P-sorption contrasting tropical soils during transit through *Pontoscolex corethrurus* (Glossoscolecidae: Oligochaeta). Soil Biol Biochem 25:789–792

Ma WC (1982) The influence of soil properties and worm-related factors on the concentrations of heavy metals in earthworms. Pedobiologia 24:109–119

Ma WC (1988) Toxicity of copper to lumbricid earthworms in sandy agricultural soils amended with Cu-enriched organic waste materials. Ecol Bull 39:53–56

Ma WC, Edelman Th, van Beersum I, Jans Th (1983) Uptake of cadmium, zinc, lead and copper by earthworms near a zinc-smelting complex: influence of soil pH and organic matter. Bull Environ Contam Toxicol 30:424–427

Mackay AD, Springgett JA, Syers JK, Gregg PEH (1983) Origin of the effect of earthworms on the availability of phosphorus in a phosphate rock. Soil Biol Biochem 15:63–73

Marinussen MPJC, van der Zee SEATM (1997) Cu accumulation by *Lumbricus rubellus* as affected by total amount of Cu in soil, soil moisture and soil heterogeneity. Soil Biol Biochem 29:641–647

Masciandaro G, Ceccanti D, Garcia C (2002) In situ vermicomposting of biological sludges and impacts on soil quality. Soil Biol Biochem 32:1015–1024

Morgan JE (1985) The interactions of exogenous and endogenous factors on the uptake of heavy metals by the earthworm *Lumbricus rubellus*. In: Lekkas TD (ed) Proceedings of the International conference of heavy metals in the environment, vol 1. CEP Consultants Ltd, Edinburg, pp 736–738

Morgan JE, Morgan AJ (1988) Calcium-lead interactions involving earthworms. Part 2. The effects of accumulated lead on endogenous calcium in *Lumbricus rubellus*. Environ Pollut 55:41–54

Morgan JE, Morgan AJ (1999) The accumulation of metals (Cd, Cu, Pb, Zn and Ca) by two ecologically contrasting earthworm species (*Lumbricus rubellus* and *Aporrectodea caliginosa*): implications for ecotoxicological testing. Appl Soil Ecol 13:9–20

Nahmani J, Hodson ME, Black S (2007) A review of studies performed to assess metal uptake by earthworms. Environ Pollut 145:402–424

Neuhauser EF, Loehr RC, Milligan DL, Malecki MR (1985) Toxicity of metals to the earthworm *Eisenia fetida*. Biol Fertil Soils 1:149–152

Nordström S (1975) Seasonal activity of lumbricids in southern Sweden. Oikos 26:307–315

Nordström S, Rundgren S (1974) Environmental factors and lumbricid associations in southern Sweden. Pedobiologia 14:1–27

Paoletti MG (1999) The role of earthworms for assessment of sustainability and as bioindicators. Agric Ecosyst Environ 74:137–155

Parkin TB, Berry EC (1994) Nitrogen transformations associated with earthworm casts. Soil Biol Biochem 26:1233–1238

Peramaki P, Itamies J, Karttunen V, Lajunen LHJ, And PE (1992) Influence of pH on the accumulation of cadmium and lead in earthworms (*Aporrectodea caliginosa*) under controlled conditions. Ann Zool Fenn 29:105–111

Phillipson J, Abel R, Steel J, Woodell SRJ (1976) Earthworms and the factors governing their distribution in an English Beachwood. Pedobiologia 16:258–285

Rozen A, Mazur L (1997) Influence of different levels of traffic pollution on haemoglobin content in the earthworm *Lumbricus terrestris*. Soil Biol Biochem 29:709–711

Satchell JE (1967) Lumbricidae. In: Burges A, Raw F (eds) Soil biology. Academic, London, pp 259–322

Siekierska E, Urbanska-Jasik D (2002) Cadmium effect on the ovarian structure in earthworm *Dendrobaena veneta* (Rosa). Environ Pollut 120:289–297

Spurgeon DJ, Hopkin SP (1995) Extrapolation of the laboratory based OECD earthworm toxicity test to metal-contaminated field sites. Ecotoxicology 4:190–205

Spurgeon DJ, Hopkin SP (1996) Effects of metal-contaminated soils on the growth, sexual development, and early cocoon production of the earthworm *Eisenia fetida*, with particular reference to zinc. Ecotoxicol Environ Saf 35:86–95

Suthar S, Singh S, Dhawan S (2008) Earthworms as bioindicator of metals (Zn, Fe, Mn, Cu, Pb and Cd) in soils: is metal bioaccumulation affected by their ecological category? Ecol Eng 32:99–107

Tomlin AD, Shipitalo MJ, Edwards WM, Protz R (1995) Earthworms and their influence on soil structure and infiltration. In: Hendrix PF (ed) Earthworm ecology and biogeography in North America. Lewis Publishers, Boca Raton, FL, pp 159–183

Van Gestel M, van Dis WA, Dirven-van Breemen EM, Sparenburg PM, Baerselman R (1991) Influence of cadmium copper and pentachlorophenol on growth and sexual development of *Eisenia andrei* (Oligochaeta, Annelida). Biol Fertil Soils 12:117–121

Wang Z, Zhang Y, Guo Y, Xia W, Li Z (1998) Monitoring of soil heavy metal pollution by earthworm. J Environ Sci 10:437–444

Wright MA, Stringer A (1980) Lead, zinc and cadmium content of earthworms from pasture in the vicinity of an industrial smelting complex. Environ Pollut 23:313–321

Zhang B, Li G, Shen T, Wang J, Sun Z (2000) Changes in microbial biomass C, N and P and enzyme activities in soil incubated with the earthworms *Metaphire guilelmi* or *Eisenia fetida*. Soil Biol Biochem 32:2055–2062

Chapter 22
Heavy Metal Stabilization by Promoting Zeolite Synthesis in Soil

Roberto Terzano and Matteo Spagnuolo

22.1 Introduction

22.1.1 Zeolites

Zeolites are crystalline hydrated aluminosilicates of alkaline and alkaline-earth cations. They are characterized by their ability to hydrate and dehydrate reversibly and to exchange some of their constituent cations, both without major change of structure (Gottardi and Galli 1985). About 50 natural zeolites have been recognized during the past 200 years and at least 150 species having no natural counterparts have been synthesized in the laboratory. Both natural and synthetic zeolites are used commercially because of their unique adsorption, ion-exchange, molecular sieve, and catalytic properties.

In the same way as quartz and feldspar, zeolites are *tectosilicates*. Tectosilicates consist basically of three-dimensional frameworks of SiO_4^{4-} tetrahedra, wherein all oxygens of each tetrahedron are shared with adjacent tetrahedra. In zeolite structures, however, some of the quadrivalent Si is replaced by trivalent Al, giving rise to a deficiency of positive charge in the framework. This charge is balanced by monovalent and divalent cations, mainly Na^+, K^+, Ca^{2+}, and Mg^{2+}, elsewhere in the structure. Gottardi in 1978 proposed the following general formula for a zeolite:

R. Terzano (✉) • M. Spagnuolo
Department of Agroforestry and Environmental Biology and Chemistry, University of Bari "Aldo Moro", Via Amendola 165/A, 70126 Bari, Italy
e-mail: r.terzano@agr.uniba.it

I. Sherameti and A. Varma (eds.), *Detoxification of Heavy Metals*, Soil Biology 30,
DOI 10.1007/978-3-642-21408-0_22, © Springer-Verlag Berlin Heidelberg 2011

$$\left(M_x{}^+, M_y{}^{2+}\right)\left(Al_{(x+2y)}Si_{n-(x+2y)}O_{2n}\right)\cdot mH_2O$$

where M^+ and M^{2+} are monovalent and divalent cations, respectively.

Cations within the first set of parentheses are called *exchangeable cations*, while those within the second set are known as *structural cations* because, together with the oxygen, they make up the framework of the structure. The Si/Al atomic ratio of a zeolite is ≥1:1. The three-dimensional framework structures of quartz and feldspar are relatively dense and tightly packed (d = 2.6–2.7 kg dm^{-3}), whereas those of zeolite materials are remarkably open (d = 1.9–2.3 kg dm^{-3}). Zeolite structures may be visualized as SiO_4 and AlO_4 tetrahedra (so-called primary building units, PBU) linked together into simple geometrical forms or more complex polyhedra (secondary building units, SBU). The SBUs may range in complexity from simple rings of 4 or 6 tetrahedra (4-rings or 6-rings) up to cubo-octahedra (sodalite unit, β), as shown in Fig. 22.1.

Secondary building units may be linked together in a variety of ways, each giving rise to a crystal structure possessing a unique set of physical and chemical properties. In synthetic zeolite A, for example, sodalite units (β) are linked by double 4-rings of tetrahedra (D4R), whereas in zeolite faujasite, the same sodalite units (β) are connected by double 6-rings of tetrahedra (D6R) (Fig. 22.1). The porous nature of zeolite structures is readily apparent from the reported illustrations. The internal pores or cages and the connecting channels in a zeolite depend on the specific arrangement of SBUs in that species. Channel systems may be one-dimensional (i.e., unconnected parallel channels in a single direction), two-dimensional (i.e., connecting channels in a single plane), or three-dimensional. Diffusion rates, for adsorption and ion-exchange of molecules and ions, are functions of the spatial distribution of the channels. The system of channels and cages is different in each zeolite structure, giving rise to a variety of materials, each

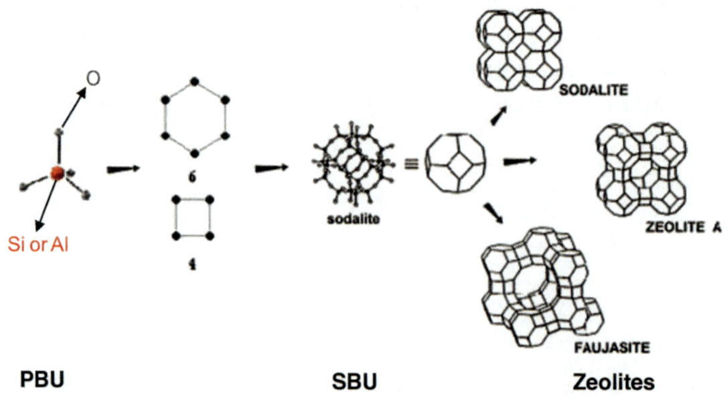

Fig. 22.1 Example of combination of tetrahedral units (PBU) to form secondary building units (SBU) and different zeolite structures. Each circle or corner in the structures represents the center of a Si- or Al-containing tetrahedron

capable of screening molecules or cations by molecular or ion sieving in slightly different ways.

Crystalline zeolites are some of the most effective cation exchangers known, commonly having cation-exchange capacities (CEC) of 200–300 cmol kg^{-1} (two to three times those of the most reactive soil minerals such as smectites or vermiculites). Synthetic zeolite A, for example, with a 1:1 Si/Al ratio, has a calculated CEC of about 540 cmol kg^{-1}. These attractive cation-exchange properties have been widely investigated in many agricultural and environmental applications including slow-release fertilizers and/or in-soil reservoir for K^+ and NH_4^+, traps for heavy metals, soil conditioners, growth media, dietary supplements in animal nutrition, water purifiers, odor controllers, and fungicide or pesticide carriers.

Zeolites can naturally occur in soil, especially in volcanic soils. Some of these zeolites have been inherited from the parent materials, while others appear to be of pedogenic origin. The most commonly reported zeolites in soils are analcime, chabazite, clinoptilolite, erionite, faujasite, heulandite, laumontite, mordenite, and phillipsite (Table 22.1). Most natural zeolites are synthesized by the hydrothermal alteration of volcanic glass.

Zeolites can also be easily synthesized. The main pathway to zeolite synthesis is that of hydrothermal synthesis (Cundy and Cox 2003), i.e., achieved in the presence of considerable amounts of water, (high) temperature, and alkaline conditions.

During the past 20 years, many patents and technical articles have proposed different hydrothermal activation methods to synthesize different zeolites from various aluminosilicate source materials including clay minerals such as kaolinite (Murat et al. 1992; Chandrasekhar and Pramada 1999), montmorillonite (Lee et al. 2002), bentonite (Ruiz et al. 1997; de la Villa et al. 2001; Ramirez et al. 2002), halloysite (Gualtieri 2001) or interstratified illite-smectite (Baccouche et al. 1998), and glasses of various origins (Wirsching and Holler 1989). Another interesting and

Table 22.1 Representative unit-cell formulae and selected physical and chemical properties of some important zeolites (modified from Ming and Mumpton 1989)

Zeolite	Representative unit-cell formula	Si/Al ratio	Density (kg dm^{-3})	Void volume (%)	Cation-exchange capacity (cmol kg^{-1})[a]
Analcime	$Na_{16}(Al_{16}Si_{32}O_{96})\cdot 16H_2O$	1.8–2.8	2.25	18	460
Chabazite	$(Na_2Ca)_6(Al_{12}Si_{24}O_{72})\cdot 40H_2O$	1.6–3.0	2.05–2.10	47	420
Clinoptilolite	$(Na_3K_3)(Al_{12}Si_{24}O_{72})\cdot 40H_2O$	4.3–5.3	2.16	34	220
Erionite	$(Na,Ca_{0.5},K)_9(Al_9Si_{27}O_{72})\cdot 27H_2O$	3.0–3.5	2.02	35	320
Faujasite	$(Ca,Na_2)_{29}(Al_{58}Si_{134}O_{384})\cdot 240H_2O$	2.25	1.91	47	340
Heulandite	$Ca_4(Al_8Si_{28}O_{72})\cdot 24H_2O$	2.5–3.7	2.2	39	290
Laumontite	$Ca_4(Al_8Si_{16}O_{48})\cdot 16H_2O$	1.8–2.3	2.3	34	420
Mordenite	$Na_8(Al_8Si_{40}O_{96})\cdot 24H_2O$	4.2–5.0	2.13	28	220
Phillipsite	$(Na,K)_5(Al_5Si_{11}O_{32})\cdot 20H_2O$	1.7–2.4	2.15	31	380
Zeolite A (LTA)	$Na_{96}(Al_{96}Si_{96}O_{384})\cdot 216H_2O$	1.0–1.2	1.99	47	540
Zeolite X (LTX)	$Na_{86}(Al_{86}Si_{106}O_{384})\cdot 264H_2O$	1.0–1.5	1.93	50	470
Zeolite P (NaP1)	$Na_6(Al_6Si_{10}O_{32})\cdot 12H_2O$	1.5–2.0	2.26	32	270

[a]Calculated value

largely used starting material is coal fly ash, the major by-product of coal burning in electric power plants consisting mainly of aluminosilicate glass. The interest for such material resides in its inexpensiveness and in the opportunity of partly solving the problem of its disposal (Hemni 1987; Mondragon et al. 1990; Shigemoto et al. 1993; Park and Choi 1995; Singer and Berkgaut 1995; Amrhein et al. 1996; Berkgaut and Singer 1996; Querol et al. 2002; Hollman et al. 1999; Poole et al. 2000; Murayama et al. 2002).

All the developed synthetic methodologies are based on the dissolution of Al–Si-bearing phases in the starting material by means of alkaline solutions (mainly NaOH and KOH solutions) and the subsequent precipitation of zeolitic material.

22.1.2 Aim of the Remediation Process

In order to remediate a soil contaminated with heavy metals (HM), it is possible to manage the contamination in three different ways: (1) by reducing metal concentration to an acceptable level; (2) by isolating the contamination or the contaminated soil to prevent any further spreading of the pollutant toward other environmental compartments or humans; or (3) by reducing the biological availability (i.e., bioavailability) of the metal.

The first approach is the most desirable since the pollutant is partially or completely removed from the soil. However, it is often not feasible owing to different problems such as long accomplishment times, high costs, the impact on soil, and the geological features of the site.

The second strategy involves the sacrifice of the contaminated soil which is disposed in special landfills with consequent problems related to the scarcity of this kind of structures, the huge volumes of materials to mobilize with relative costs, and the adverse public opinion.

The third approach aims to minimize the bioavailable HM fraction by means of processes that chemically and/or physically stabilize the contaminant in soil, that is, to transform mobile or potentially mobile metal forms into definitively immobile or not available forms. Obviously, this approach does not lower the concentration of the metals in soil, but it can considerably reduce the risk for the environment and for living organisms. This methodology can be also applied in situ with relatively moderate costs.

In general, the remediation technique most accepted from a social point of view is the bioremediation that refers to the attenuation of the pollutant by natural processes using living organisms such as microbes or plants. However, the use of microbes or plants to remediate HM-polluted soils has several limitations and drawbacks which make these techniques not always feasible.

Among the available remediation technologies, one that is at the same time cheap, versatile, applicable in situ, and rapid, is the so-called solidification/stabilization (S/S). S/S aims to minimize the solubility of toxic species and/or to immobilize them into inert and insulating structures. In this contest, HM can be transformed into less toxic forms by precipitating them as less soluble metal-hydroxides, carbonates, phosphates, silicates, or sulfides, by substitution of the HM into stable mineral structures, by sorption reactions with available surfaces, and by physical encapsulation.

Within this context, the HM stabilization process exploiting the direct zeolite synthesis in soil aims at precipitating HM as insoluble hydroxides/oxides and simultaneously entrapping the HM precipitates inside the forming zeolite structures. Differently from the "traditional" use of zeolites in soil remediation technologies as cation exchangers, the process described herein does not exploit the high CEC typical of zeolites but rather the large cavities and pores characteristic of zeolite structures. It is in these cavities that the HM precipitates can be trapped. Differently from cation-exchange processes, which are reversible, metal precipitation and sequestration inside stable minerals such as zeolites can ensure for a long-term immobilization of HM.

22.2 Promoted Zeolite Synthesis in Soil

In order to promote both metal precipitation and zeolite synthesis in soil, an alkaline pH is needed. Therefore, alkalizing agents such as NaOH, KOH, and lime must be added to soil. In addition, for zeolite synthesis, its constituting "building blocks" (PBU) are required, i.e., Si and Al containing framework tetrahedra (Fig. 22.1). These "building blocks" can be potentially provided by several Si- and Al-containing materials as reported in Sect. 22.1.1 and, in principle, by several soil constituents such as clay minerals, amorphous and crystalline silicates, aluminates, and aluminosilicates (Terzano et al. 2005a). Anyway, to get a sufficient amount of zeolites for HM stabilization, further reacting phases must be added to soil. Published literature presents data obtained by using coal fly ash and NaOH to induce zeolite synthesis in soil (Terzano et al. 2005a, b; Belviso et al. 2010a).

Coal fly ash is mainly constituted of aluminosilicate glass, mullite, and quartz with smaller amounts of residual coal and ore minerals. The glass accounts for 60–80% of fly ash (Singer and Berkgaut 1995) and makes them a readily available source of Si and Al. Anyway, to increase the amorphous phase and therefore the amount of available Si and Al for zeolite synthesis, coal fly ash can be fused at 550°C with NaOH prior to hydrothermal reaction in soil (Shigemoto et al. 1993; Chang and Shih 1998; Terzano et al. 2005b, c; Belviso et al. 2010b).

An agricultural silty-clay soil characterized by pH 7.9 and 2.4% organic carbon was treated with 10% (in weight) fused coal fly ash (Terzano et al. 2005a) and, after the addition of water (2:1, w/w), soil was incubated at 30 and 60°C at atmospheric pressure. Soil samples were then collected after 1 h, 24 h, 1 week, 1 month, 6 months, and 1 year and analyzed to detect zeolite formation by means of X-ray diffraction (XRD) and scanning electron microscopy coupled with energy-dispersive X-ray microanalysis (SEM-EDX).

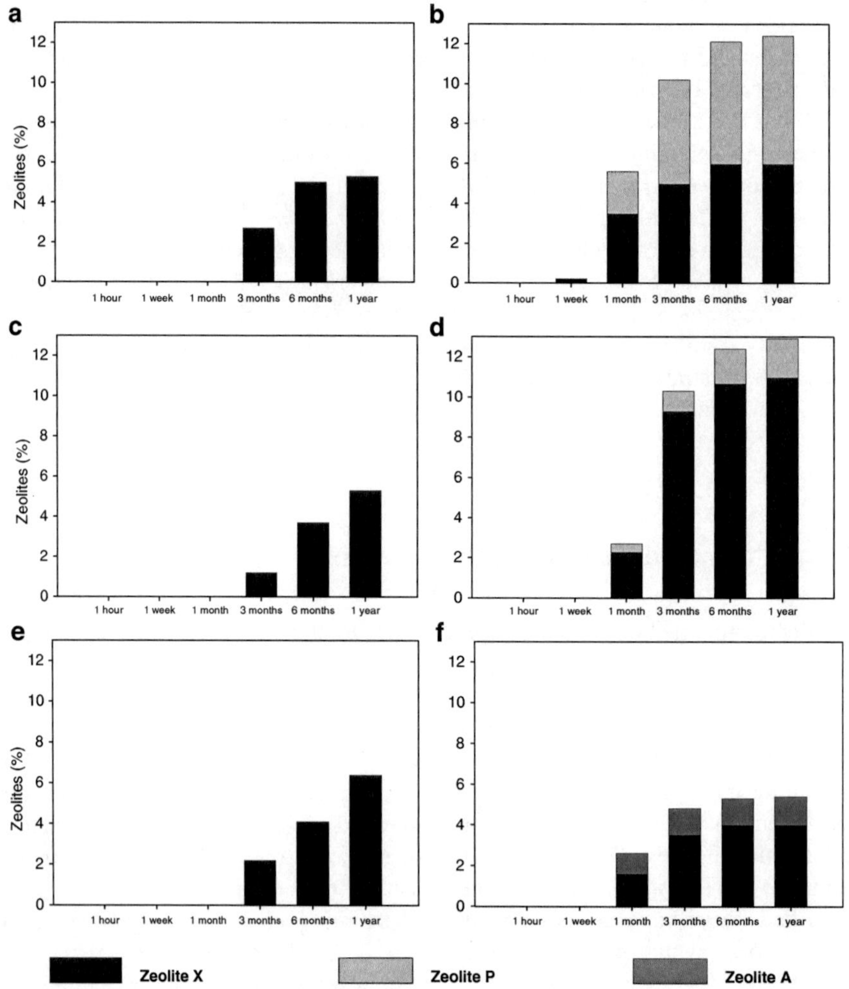

Fig. 22.2 Amount of zeolites, as a percentage of the soil total dry weight, synthesized at different incubation times by treating with fused coal fly ash an agricultural soil at 30 and 60°C. (**a**, **b**) Uncontaminated soil treated at 30 and 60°C, respectively. (**c**, **d**) Soil contaminated with 15 mg g^{-1} of Cu treated at 30 and 60°C, respectively. (**e**, **f**) Soil contaminated with 15 mg g^{-1} of Cd treated at 30 and 60°C, respectively. Modified from Terzano et al. 2005a, 2007

At 60°C, zeolites formed already after 1 week (Fig. 22.2b) and the amount of the zeolites increased gradually over time. The synthesized zeolites were identified by XRD as zeolite X and zeolite P. Zeolite X is a synthetic analogue of faujasite, while zeolite P belongs to the Gismondine series. After 6 months of incubation at 60°C, the amount of synthesized zeolites accounted for ca. 12% of the soil dry weight. Therefore, added coal fly ash (10%) was almost completely converted to zeolites, with a contribution also from the reaction of other soil minerals. After 1 year, only a slight further increase was observed.

When the same soil was treated at 30°C, zeolite formation was observed only after 3 months (Fig. 22.2a). At this temperature, only zeolite X formed and the amount synthesized after 6 months was slightly more than 5%. At this temperature, the formation of zeolite P was thermodynamically not favored.

In the view of the application of direct zeolite synthesis in soil to stabilize HM, the formation of zeolite X is more desirable since it has a larger surface area and bigger internal cavities to allocate HM precipitates.

These results demonstrate that it is possible to directly synthesize zeolites in soil by appropriate soil treatment. The type of zeolite formed in soil seems to be regulated by both incubation temperature and Si/Al molar ratio in solution, as evidenced by Terzano et al. (2005a).

To achieve zeolite synthesis in soil, a pH above 12 seems to be mandatory since at pH 11 this process was not observed (Terzano et al. 2004a). However, it has been observed that after the treatment pH naturally decreased over time, reaching values of 10–11 after 1 year (Terzano et al. 2005b). After this period, suitable soil amendments and conditioners (e.g., compost, industrial sludges, acidic wastewaters, calcium sulfate) could easily restore the soil pH to more sustainable values.

22.3 Promoted Zeolite Synthesis in HM-Polluted Soils

Up to now, this stabilization process has been applied to copper (Terzano et al. 2005b), cadmium (Terzano et al. 2004b, 2007), nickel (Belviso et al. 2010a), and lead (Belviso et al. 2008) contaminated soils.

The same agricultural soil described in Sect. 22.2 was artificially contaminated with 15 mg of Cu/g of soil dry weight and treated with fused coal fly ash at 30 and 60°C (Terzano et al. 2005b). At 60°C, zeolite synthesis was observed after 1 month and its amount increased with the incubation time reaching, after 1 year, ca. 13% of the soil dry weight (Fig. 22.2d). Most of the synthesized zeolite was zeolite X (11%). These data suggest that the presence of Cu caused a delay in the starting of zeolite synthesis at 60°C, but did not hinder the formation of the minerals. In addition, Cu seemed to favor the formation of zeolite X over that of zeolite P in soil (Sect. 22.2).

At 30°C, zeolite synthesis started after 3 months and reached its maximum after 1 year, slightly more than 5%, as in the uncontaminated soil. At this temperature, only zeolite X formed (Fig. 22.2c). In a Ni-polluted soil, Belviso et al. (2010a) observed zeolite formation at 30°C already after 1 month.

Simultaneously with the process of zeolite synthesis, the partitioning of Cu between the solid phase and the soil solution (K_d) was determined over a period of 1 year (Fig. 22.3a). K_d is the ratio of the mass of metal in the solid phase per unit mass of solid phase to the mass of metal in solution per unit volume of the liquid phase. The higher the K_d, the higher the amount of metal in the solid phase. Immediately after the treatment, the K_d value is very high, due to the high pH (ca. 13) which generates a high rate of sorption phenomena (precipitation, in

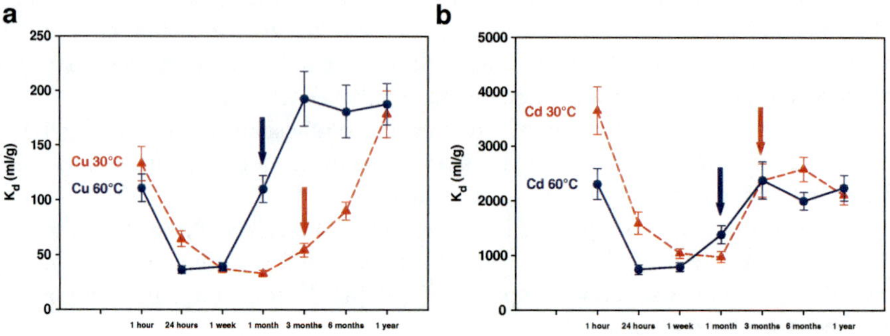

Fig. 22.3 Variation over time in the distribution coefficient (K_d) of Cu (**a**) and Cd (**b**), after the treatment of the contaminated soil with fused coal fly ash at 30 and 60°C. A higher K_d value means a higher amount of metal in the solid phase. The *arrows* indicate when the zeolite synthesis started. K_d = [mass of the metal in the solid phase by unit mass of the solid phase (μg g^{-1})]/[mass of metal in solution by unit volume of liquid phase (μg ml^{-1})]

particular). However, during the first month of incubation, for the soil treated at 30°C, and the first week, for the soil treated at 60°C, the K_d value strongly decreased, thus indicating increased Cu redissolution. This process was caused by the high alkaline conditions that probably promoted the solubilization of soil organic matter and therefore the dissolution of Cu bound to it. In fact, it is well known that soil-soluble organic matter forms strong complexes with copper (Xue and Sigg 1999). This observation is confirmed by looking at the K_d values for Cd (Fig. 22.3b) for the same soil artificially polluted with Cd (15 mg g^{-1}) and treated in the same way as for Cu. The K_d values are always ca. 20 times higher than for Cu. Cd has a lower affinity for soluble organic matter (Xue and Sigg 1999); thus, its concentration in solution is less influenced by the dissolution of organic matter in alkaline conditions.

What is significant is that the K_d values, for both the Cu- and Cd-polluted soils and for both the temperatures, start to increase again simultaneously with the beginning of zeolite synthesis (indicated by the arrows in Fig. 22.3). The high K_d values observed after 1 year, also in comparison with K_d values determined for the same soil treated only with NaOH (Terzano et al. 2005b), suggest that zeolite synthesis promoted a higher partitioning of the metals in/on the solid phase, thus drastically reducing HM solubility.

Also, Cu and Cd availability was evaluated by five sequential EDTA (5 mM, pH 7.5) extractions (Ford et al. 1999; Terzano et al. 2004b, 2005b). It was observed that Cu and Cd became increasingly resistant, over time, to extraction with EDTA, both at 30 and 60°C. The fraction of metal remaining in the solid phase after the five sequential extractions with EDTA is called residual and can be considered representative of the unavailable metal fraction (Garrabrants and Kosson 2000). After 1 year from the treatment, ca. 40% and 30% of the initial Cu became unavailable at 60°C and 30°C, respectively, while ca. 20% was quantified as unavailable for Cd, both at 30°C and 60°C (Terzano et al. 2004b).

The difference between the zeolite synthesis process in the Cu- and Cd-contaminated soils resides in the fact that at 60°C, besides zeolite X, zeolite A is formed in the presence of Cd instead of zeolite P and that after 1 year the amount of synthesized zeolites is ca. half of that synthesized with Cu (Fig. 22.2). In both cases, it was observed that at 60°C no or just a small amount of zeolite P was formed. This result is in contrast with what was observed by treating the same soil, under the same conditions, in the absence of metal contamination and with thermodynamic considerations, which, at 60°C, would favor the formation of the more stable zeolite P minerals over both zeolite X and A (Petrovic et al. 1993; Chang and Shih 1998). Therefore, it appears that the presence of high concentrations of Cu or Cd ions, while not significantly influencing the type and the amount of zeolites which can be obtained by treating the soil at 30°C, could drive the synthesis toward the formation of sodalite unit-based zeolites (X and A), at 60°C (Terzano et al. 2007).

Belviso et al. (2010a) studied the effect of the same treatment at 27–30°C on a clay-loam soil with a pH of 7.88 and 1.75% organic carbon, artificially contaminated with 10 mg g^{-1} of Ni, both at a laboratory and bench scale. They observed that concomitantly with the beginning of zeolite X synthesis (after 1 month), the amount of Ni extractable with ammonium acetate at pH 7.0 was ca. 50% smaller than immediately after the treatment. In addition, they observed that in the soil samples with the higher amount of synthesized zeolites (3, 6, 9, and 12 months), more than 60% of Ni could be extracted only with the third step of the BCR sequential extractions protocol (Quevauviller et al. 1997), i.e., treating the sample with H_2O_2 at 85°C for 1 h in acidic conditions (pH 2). XRD analyses confirmed that Ni was solubilized when the zeolite X structure collapsed after the second BCR step. Similar results were obtained by Belviso et al. (2008) also for a Pb-polluted soil.

22.4 Microscopic Observations

Scanning Electron Microscope (SEM) images of zeolite particles directly synthesized in soil in the presence of polluting metals are reported in Fig. 22.4. These zeolite minerals have been characterized for chemical composition and size by using an SEM automated single particle analysis (ASPA) method (Terzano et al. 2007). A summary of these data is reported in Table 22.2.

Zeolite X (Fig. 22.4a) synthesized at 30°C and 60°C corresponds to a mineral characterized by the general formula $NaAlSi_{1.3}O_{4.6}$, with the additional presence of small amounts of Ca and Fe. This formula corresponds almost exactly to that of zeolite X synthesized directly from coal fly ash ($NaAlSi_{1.23}O_{4.46}$; Querol et al. 2002). Differences among the synthesized zeolite X minerals can be found in the average diameter of the crystals formed in the presence of Cu or Cd ions. Zeolite X particles crystallized in the presence of Cd are generally smaller (average diameter ca. 5 μm) than zeolite X formed in the Cu-contaminated soil (average diameter ca. 10 μm).

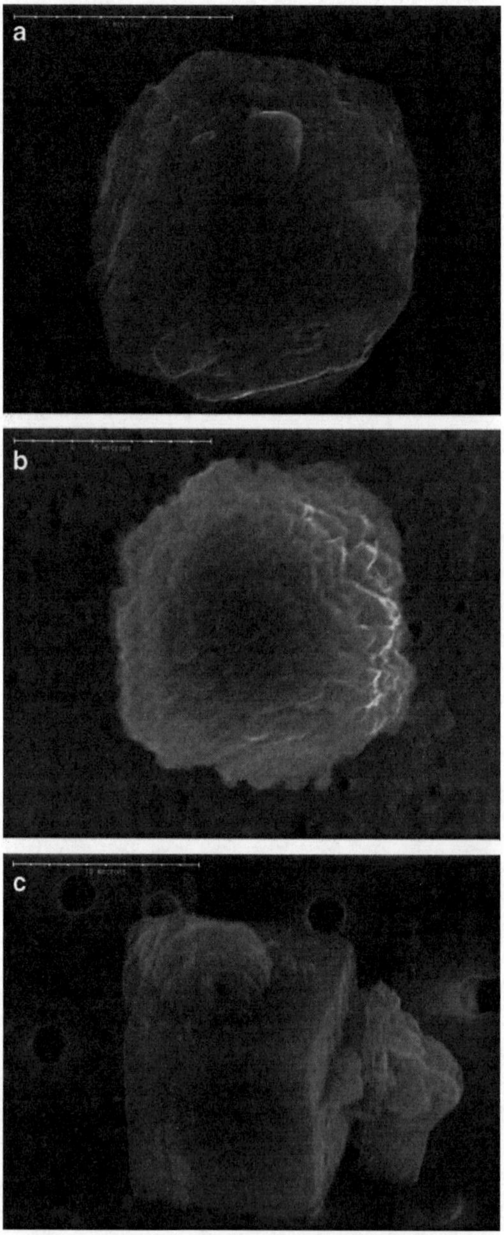

Fig. 22.4 SEM images of zeolite X (**a**), zeolite P (**b**), and zeolite A (**c**) crystals formed in the soil samples treated with fused coal fly ash (Terzano et al. 2007, with permission from Elsevier, copyright 2007)

Zeolite P (Fig. 22.4b) synthesized at 60°C in the presence of Cu is characterized by the general formula $Na_{1.2}AlSi_{1.6}O_{5.4}$, with an average crystal diameter of 7 μm. Also for this zeolite, the obtained general formula is very similar to that of zeolite NaP1 synthesized directly from coal fly ash ($NaAlSi_{1.67}O_{5.34}$; Querol et al. 2002).

Table 22.2 Average composition and diameter of zeolite X, zeolite P and zeolite A formed in soil samples contaminated with Cu or Cd and treated with fused coal fly ash at 30°C (Cu 30°C, Cd 30°C) and at 60°C (Cu 60°C, Cd 60°C)

	Cu 30°C	Cu 60°C		Cd 30°C	Cd 60°C	
	Zeolite X	Zeolite X	Zeolite P	Zeolite X	Zeolite X	Zeolite A
Na_2O (%wt)	17 ± 5	18 ± 8	19 ± 4	17 ± 4	20 ± 6	22 ± 5
Al_2O_3 (%wt)	29 ± 4	30 ± 2	27 ± 2	31 ± 2	30 ± 2	29 ± 2
SiO_2 (%wt)	47 ± 6	47 ± 3	52 ± 2	48 ± 3	46 ± 4	47 ± 4
K_2O (%wt)	0.2 ± 0.1	0.3 ± 0.1	<MDL	0.4 ± 0.2	<MDL	<MDL
CaO (%wt)	5 ± 2	3 ± 1	0.7 ± 0.4	2 ± 2	3 ± 1	0.3 ± 0.2
Fe_2O_3 (%wt)	1.0 ± 0.6	0.9 ± 0.7	0.6 ± 0.4	0.7 ± 0.5	0.5 ± 0.2	0.8 ± 0.5
Other (%wt)	0.8	0.8	0.6	0.7	0.5	0.9
Diameter (μm)	10 ± 2	9 ± 5	7 ± 2	3 ± 1	6 ± 2	4 ± 1
Si/Al	1.35 ± 0.20	1.33 ± 0.07	1.62 ± 0.05	1.29 ± 0.05	1.32 ± 0.07	1.35 ± 0.05

Modified from Terzano et al. 2007
<MDL: below the minimum detection limit

At last, zeolite A (Fig. 22.4c) synthesized at 60°C in the presence of Cd is characterized by the formula $Na_{1.2}AlSi_{1.3}O_{4.8}$, quite different from that reported in the literature ($NaAlSi_{1.1}O_{4.2}$; Querol et al. 2002) for the same type of zeolite synthesized directly from coal fly ash but almost identical to that of zeolite X, even if with a lower Ca content (Table 22.2).

Zeolite X particles synthesized in a soil polluted by Cu and treated with fused coal fly ash were analyzed by synchrotron X-ray microspectroscopy techniques to determine the spatial distribution and speciation of Cu in and/or on the formed minerals (Terzano et al. 2005b). The zeolite X particles analyzed had a diameter of ca. 10–20 μm and their nature was confirmed by μ-XRD and SEM observations. To image elemental distribution inside such small particles synchrotron μ-XRF (micro-X-ray fluorescence) tomography was used, utilizing a submicron X-ray beam. With this technique, by exploiting the highly penetrative capacity of X-rays, it was possible to directly investigate cross sections of the zeolite particles without any physical cutting of the sample or other modifications of the chemical and morphological characteristics of the particles themselves. As can be seen in Fig. 22.5, Cu was heterogeneously distributed both inside and outside the zeolite mineral. This result provides direct evidence that Cu was really trapped inside the newly formed zeolite particles. Also, by looking at this image, it is reasonable to think that Cu is not located in the zeolite framework, but is most likely buried as micro- or (nano-) clusters inside the newly formed minerals. This kind of distribution could be explained by considering that zeolite X has been formed from an amorphous aluminosilicate matrix in which Cu has been precipitated as a consequence of the addition of alkaline-fused coal fly ash to the soil.

The chemical state of Cu in and/or on the formed zeolite X was also investigated by means of micro-X-ray absorption near edge structure (μ-XANES) spectroscopy (Terzano et al. 2005b). The comparison between the linear combinations of Cu standard spectra and the measured XANES spectra for the selected particles

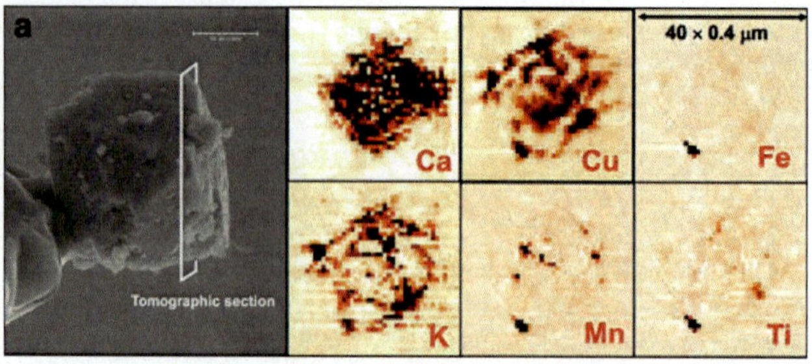

Fig. 22.5 Reconstructed elemental maps from XRF microtomography on a single zeolite X particle (**a**), directly isolated from a Cu-contaminated soil treated with fused coal fly ash at 30°C. (Terzano et al. 2005, with permission from American Chemical Society, copyright 2005)

allowed for an estimation of the concentrations of the possible copper compounds in zeolite particles. In all the analyzed zeolite X particles, Cu appeared to be mainly present as a mixture of $Cu(OH)_2$ (from 30 to 60%) and CuO (from 40 to 70%). Bonding of Cu hydr(oxides) clusters to the zeolite framework was suggested by extended X-ray absorption fine structure (EXAFS) spectroscopy investigations on zeolite X particles synthesized at low temperature in the presence of Cu (Terzano et al. 2006). These hypotheses are also supported by the data obtained from Fourier Transformed-Infra Red (FT-IR) and electron paramagnetic resonance (EPR) investigations on pure zeolite X minerals (Terzano et al. 2006).

22.5 Description of the Stabilization Process

All the data available in the literature provide sufficient information to hypothesize concurrent physicochemical phenomena responsible for HM stabilization when promoting zeolite synthesis in soil.

First of all, the precipitation of metal hydroxides, owing to the high increase in pH after the addition of alkali-activated Si–Al-containing materials, is certainly the main process responsible for the observed marked reduction in HM solubility.

However, processes such as chemisorption and especially coprecipitation can be also involved. In particular, coprecipitation can be extremely effective since HM can coprecipitate with the amorphous aluminosilicates coming from the dissolution of alkali-activated Si and Al.

It is well known on thermodynamic basis that coprecipitation is an environmental process that could reduce heavy metal solubility below that of the least soluble pure HM mineral phases likely to form under environmental conditions (Martinez and McBride 2000).

Cation-exchange reactions cannot be invoked since at the high alkaline pH of the experiments, the investigated HM (Cu, Cd, Ni, and Pb) are present in solution predominantly as hydroxy-anionic forms.

HM stabilization inside the solid phase can be partly ascribed to the formation of the zeolitic structures that originates from the alkali-activated matrix of the added Si–Al-containing material or from the partial dissolution of soil minerals. During zeolite synthesis, Al changes its coordination from octahedral to tetrahedral, the same coordination of Si. The hypothesis of the active role of zeolites in immobilizing HM inside their structure is supported by the sudden increase in the K_d values observed just as zeolite synthesis starts (Terzano et al. 2005b), the observations by Belviso et al. (2010a) that HM mobilization takes place when zeolite structures are chemically destroyed, and, mainly, by the results obtained from synchrotron X-ray-based microanalyses of the zeolitic particles, directly isolated from the treated soils. They show clear evidence that HM can be trapped inside the mesopores of the zeolitic structures as microscopic or nano-inclusions in the mineral (Wark et al. 1997), mainly as heterogeneously distributed clusters of hydroxides or oxides. As a consequence of this, HM can be immobilized inside stable crystalline alluminosilicate materials.

Another factor influencing HM stabilization in soil is the formation of the so-called geopolymers. Geopolymers can be viewed as the amorphous equivalent of certain synthetic zeolites and have more or less the same chemical composition, although the absence of the distinctive long-range zeolite structure makes them amorphous to XRD detection (Van Jaarsveld et al. 1997). It has been shown that geopolymerization reactions can lead to the immobilization of toxic metals through a combination of physical encapsulation and chemical bonding into the amorphous phase of the geopolymeric matrix (Van Jaarsveld et al. 1999).

The formation of amorphous "geopolymers," especially in the fraction of activated Si and Al that has not been transformed into more crystalline minerals (at least as can be evidenced by XRD), should be therefore mentioned in order to exhaustively describe the physicochemical processes that cooperate in HM stabilization.

A schematic overview of all the processes likely to contribute to HM stabilization in soil after the treatment to promote direct zeolite synthesis is reported in Fig. 22.6.

22.6 Conclusions

This chapter evidences the actual possibility for zeolites to be synthesized directly on site by treating polluted soils with Si–Al-containing materials in alkaline conditions, even at low temperatures (27–30°C). These Si–Al-containing materials can be also waste materials such as coal fly ash, blast furnace slag, building wastes, glass, and aluminum wastes, which can thus find useful applications and be subtracted from disposal. This process can be a promising methodology to effectively stabilize HM in polluted soils, especially in combination with other physicochemical or biological remediation processes.

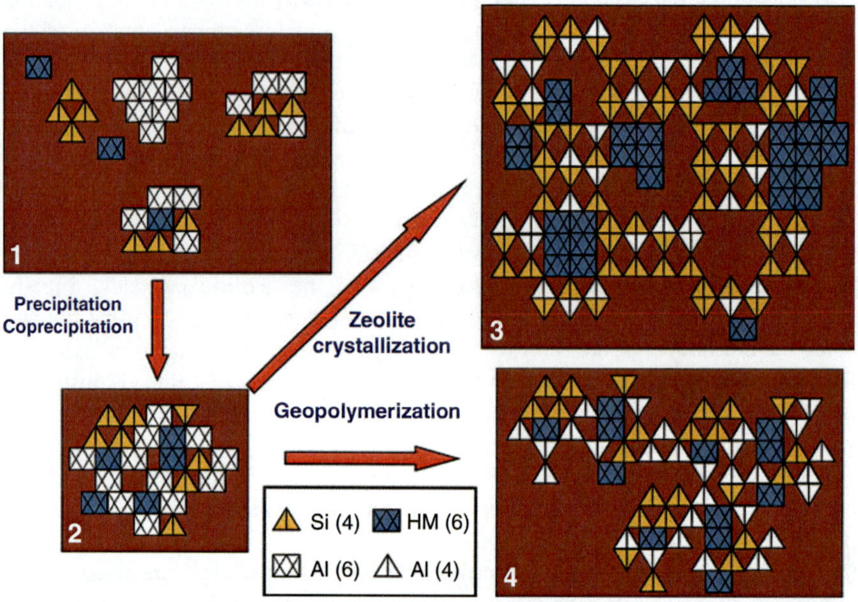

Fig. 22.6 Schematic representation of the HM stabilization processes in treated soils: 1) Si–Al-containing materials are added to HM-polluted soil together with alkalizing agents; 2) HM are precipitated or coprecipitated in an amorphous aluminosilicate phase; according to the reaction conditions, 3) crystalline zeolites can form over time and trap HM precipitates inside their internal cavities, and/or 4) noncrystalline geopolymers can form thus chemically and physically immobilizing HM. (4) and (6) in the legend and the related symbols stand for tetrahedral and octahedral coordination, respectively

Further research is needed to assess the effectiveness of the process in the field, with real contaminated soils. Also, reactive aluminosilicate materials other than fly ash and NaOH should be tested to have more efficient and safer stabilization processes.

Acknowledgments This chapter is dedicated to Prof. Pacifico Ruggiero in the year of his retirement, for actively contributing to this research with his precious scientific guide, open mind for novelties, continuous supporting, and polite moods.

References

Amrhein Ch, Haghnia GH, Kim TS, Mosher PA, Gagajena RC, Amanios T, de la Torre L (1996) Synthesis and properties of zeolites from coal fly ash. Environ Sci Technol 30:735–742

Baccouche A, Srasra E, El Maaoui M (1998) Preparation of Na-P1 and sodalite octahydrate zeolites from interstratified illite-smectite. Appl Clay Sci 13:255–273

Belviso C, Cavalcante F, Fiore S (2008) The zeolites synthesised in contaminated soil treated with coal fly ash. What is their role? 33rd International geological congress, Oslo, Norway, 6–14 Aug 2008. Last accessed 28 Nov 2010

Belviso C, Cavalcante F, Ragone P, Fiore S (2010a) Immobilization of Ni by synthesizing zeolite at low temperatures in a polluted soil. Chemosphere 78:1172–1176

Belviso C, Cavalcante F, Fiore S (2010b) Synthesis of zeolite from Italian coal fly ash: differences in crystallization temperature using seawater instead of distilled water. Waste Manage 30:839–847

Berkgaut V, Singer A (1996) High capacity cation exchanger by hydrothermal zeolitization of coal fly ash. Appl Clay Sci 10:369–378

Chandrasekhar S, Pramada PN (1999) Investigation on the synthesis of zeolite NaX from Kerala kaolin. J Porous Mater 6:283–297

Chang HC, Shih WH (1998) A general method for the conversion of fly ash into zeolites as ion exchangers for cesium. Ind Eng Chem Res 37:71–78

Cundy CS, Cox PE (2003) The hydrothermal synthesis of zeolites: history and development from the earliest days to the present time. Chem Rev 103:663–702

de la Villa RV, Cuevas J, Ramirez S, Leguey S (2001) Zeolite formation during the alkaline reaction of bentonite. Eur J Mineral 13:635–644

Ford RG, Scheinost AC, Scheckel KG, Sparks DL (1999) The link between clay mineral eathering and the stabilization of Ni surface precipitates. Environ Sci Technol 33:3140–3144

Garrabrants AC, Kosson DS (2000) Use of a chelating agent to determine the metal availability for leaching from soils and wastes. Waste Manage 20:155–165

Gottardi G (1978) Mineralogy and crystal chemistry of zeolites. In: Sand LB, Mumpton FA (eds) Natural zeolites: occurrence, properties, use. Pergamon Press Inc., Elmsford, NY, pp 31–44

Gottardi G, Galli E (1985) Natural zeolites. Springer, Heidelberg, Germany

Gualtieri AF (2001) Synthesis of sodium zeolites from a natural halloysite. Phys Chem Miner 28:719–728

Hemni T (1987) Synthesis of hydroxy-sodalite ("zeolite") from waste coal ash. Soil Sci Plant Nutr 33(3):517–521

Hollman GG, Steenbruggen G, Janssen-Jurkovicova M (1999) A two-step process for the synthesis of zeolites from coal fly ash. Fuel 78:1225–1230

Lee S-R, Han Y-S, Choy J-H (2002) 2D → 3D transformation of layered aluminosilicate upon base treatment. Solid State Ionics 151:343–346

Martinez CE, McBride MB (2000) Aging of coprecipitated Cu in alumina: changes in structural location, chemical form, and solubility. Geochim Cosmochim Acta 64(10):1729–1736

Ming DW, Mumpton MA (1989) Zeolites in soil. In: Dixon JB, Weed SB (eds) Minerals in soil environments, 2nd edn. Soil Science Society of America, Madison, WI, USA, pp 873–911

Mondragon F, Rincon F, Sierra L, Escobar C, Ramirez J, Fernandez J (1990) New perspectives for coal ash utilization: synthesis of zeolitic materials. Fuel 69:263–266

Murat M, Amokrane A, Bastide JP, Montanaro L (1992) Synthesis of zeolites from thermally activated kaolinite. Some observations on nucleation and growth. Clay Miner 27:119–130

Murayama N, Yamamoto H, Shibata J (2002) Mechanism of zeolite synthesis from coal fly ash by alkali hydrothermal reaction. Int J Miner Process 64:1–17

Park M, Choi J (1995) Synthesis of phillipsite from fly ash. Clay Sci 9(4):219–229

Petrovic I, Navrotsky A, Davis ME, Zones SI (1993) Thermochemical study of the stability of frameworks in high-silica zeolites. Chem Mater 5:1805–1813

Poole C, Prijatama H, Rice NM (2000) Synthesis of zeolite adsorbents by hydrothermal treatment of PFA wastes: a comparative study. Miner Eng 13(8–9):831–842

Querol X, Moreno N, Umana JC, Alastuey A, Hernandez E, Lopez-Soler A, Plana F (2002) Synthesis of zeolites from coal fly ash: an overview. Int J Coal Geol 50:413–423

Quevauviller Ph, Rauret G, Lopez-Sanchez J-F, Rubio R, Ure A, Muntau H (1997) Certification of trace metal extractable contents in a sediment reference material (CRM 601) following a three-step sequential extraction procedure. Sci Total Environ 205:223–234

Ramirez S, Cuevas J, Vigil R, Leguey S (2002) Hydrothermal alteration of "La Serrata" bentonite (Almeria, Spain) by alkaline solutions. Appl Clay Sci 21:257–269

Ruiz R, Blanco C, Pesquera C, Gonzalez F, Benito I, Lopez JL (1997) Zeolitization of a bentonite and its application to the removal of ammonium ion from waste water. Appl Clay Sci 12:73–83
Shigemoto N, Hayashi H, Miyaura K (1993) Selective formation of Na-X, zeolite from coal fly ash by fusion with sodium hydroxide prior to hydrothermal reaction. J Mater Sci 28:4781–4786
Singer A, Berkgaut V (1995) Cation exchange properties of hydrothermally treated coal fly ash. Environ Sci Technol 29(9):1748–1753
Terzano R, Spagnuolo M, Medici L, Ruggiero P (2004a) Stabilization of Cu and Cd in the presence of montmorillonite by means of coal fly ash. Fresen Environ Bull 13(10):995–999
Terzano R, Spagnuolo M, Ruggiero P, Medici L, Janssens K, Vekemans B, Vincze L (2004b) Stabilizzazione di metalli pesanti in seguito a processi di formazione di zeoliti in suoli trattati con coal fly ash. In: Morselli L (ed) Atti dei seminari di Ecomondo 2004, vol 2. Maggioli Editore, Italy, pp 230–238, in Italian
Terzano R, Spagnuolo M, Medici L, Tateo F, Ruggiero P (2005a) Zeolite synthesis from pre-treated coal fly ash in presence of soil as a tool for soil remediation. Appl Clay Sci 29(2): 99–110
Terzano R, Spagnuolo M, Medici L, Vekemans B, Vincze L, Janssens K, Ruggiero P (2005b) Copper stabilization by zeolite synthesis in polluted soils treated with coal fly ash. Environ Sci Technol 39:6280–6287
Terzano R, Spagnuolo M, Medici L, Tateo F, Ruggiero P (2005c) Characterisation of different coal fly ashes for their application in the synthesis of zeolite X as cation exchanger for soil remediation. Fresen Environ Bull 14(4):263–267
Terzano R, Spagnuolo M, Medici L, Tateo F, Vekemans B, Janssens K, Ruggiero P (2006) Spectroscopic investigation on the chemical forms of Cu during the synthesis of zeolite X at low temperature. Appl Geochem 21:993–1005
Terzano R, Spagnuolo M, Medici L, Dorriné W, Janssens K, Ruggiero P (2007) Microscopic single particle characterization of zeolites synthesized in a soil polluted by copper or cadmium and treated with coal fly ash. Appl Clay Sci 35:128–138
Van Jaarsveld JGS, Van Deventer JSJ, Lorenzen L (1997) The potential use of geopolymeric materials to immobilise toxic metals: part I. Theory and applications. Miner Eng 10(7): 659–669
Van Jaarsveld JGS, Van Deventer JSJ, Schwartzman A (1999) The potential use of geopolymeric materials to immobilise toxic metals: part II. Material and leaching characteristics. Miner Eng 12(1):75–91
Wark M, Kessler H, Schulz-Ekloff G (1997) Growth and reactivity of zinc and cadmium oxide nano-particles in zeolites. Microporous Mater 8:241–253
Wirsching U, Holler H (1989) Experimental studies on zeolite formation conditions. Eur J Mineral 1:489–506
Xue H, Sigg L (1999) Comparison of the complexation of Cu and Cd by humic or fulvic acids and by ligands observed in lake waters. Aquat Geochem 5:313–335

Index

I. Sherameti and A. Varma (eds.), *Detoxification of Heavy Metals*, Soil Biology 30, DOI 10.1007/978-3-642-21408-0, © Springer-Verlag Berlin Heidelberg 2011

R

S

T